State-of-the-Art Membrane Science and Technology in the Iberian Peninsula 2021-2022

State-of-the-Art Membrane Science and Technology in the Iberian Peninsula 2021-2022

Guest Editors

Isabel Coelhoso
Clara Casado-Coterillo
Diogo M.F. Santos
José Ignacio Calvo
Liliana C. Tomé
Svetlozar Velizarov

MDPI

Basel • Beijing • Wuhan • Barcelona • Belgrade • Novi Sad • Cluj • Manchester

Guest Editors
Isabel Coelhoso
Department of Chemistry
Universidade Nova de Lisboa
Caparica
Portugal

Clara Casado-Coterillo
Department of Chemical and Biomolecular Engineering
University of Cantabria
Santander
Spain

Diogo M.F. Santos
Chemical Engineering Department
Universidade de Lisboa
Lisbon
Portugal

José Ignacio Calvo
Department of Applied Physics
University of Valladolid
Valladolid
Spain

Liliana C. Tomé
Department of Chemistry
Universidade Nova de Lisboa
Caparica
Portugal

Svetlozar Velizarov
Faculty of Science and Technology
New University of Lisbon
Caparica
Portugal

Editorial Office
MDPI AG
Grosspeteranlage 5
4052 Basel, Switzerland

This is a reprint of the Special Issue, published open access by the journal *Membranes* (ISSN 2077-0375), freely accessible at: https://www.mdpi.com/journal/membranes/special_issues/iberian_2122.

For citation purposes, cite each article independently as indicated on the article page online and as indicated below:

Lastname, A.A.; Lastname, B.B. Article Title. *Journal Name* **Year**, *Volume Number*, Page Range.

ISBN 978-3-7258-8039-3 (Hbk)
ISBN 978-3-7258-8040-9 (PDF)
https://doi.org/10.3390/books978-3-7258-8040-9

Contents

About the Editors

Isabel Coelhoso

Isabel Coelhoso is Associate Professor with Habilitation in Chemical Engineering at Universidade Nova de Lisboa (NOVA FCT), Portugal, in the areas of Transport Phenomena and Separation Processes. Her research interests are related to membrane separation processes, membrane contactors for process intensification and the design of biopolymer membranes for biotechnology, food and health applications.

She has authored 120 peer-reviewed articles with ca. 4200 citations and has an h-index of 39. She has participated in 21 research projects (national, with industry, European) and was an invited expert for several projects' evaluation: EU—PRIMA and PATHFINDER, FCT—Portugal, ANI, ANEP—Spain, UQ—Australia, RGC—Hong Kong, SAR—China.

Clara Casado-Coterillo

Clara Casado-Coterillo received her PhD in Chemical Engineering at the age of 27 years from the University of Cantabria Spain, where she returned as Senior Researcher in 2012, after pursuing research on the synthesis and characterization of different membrane materials for diverse molecular separation applications in international research centres at the University of Hiroshima (Japan), the European Membrane Institute at the University of Twente (the Netherlands), the Institut Européen des Membranes (Montpellier, France), the University of Zaragoza and the Institute of Chemical Technology (Spain). She has written more than 60 peer-reviewed publications and has participated in more than 20 projects in competitive calls, nine of them as the principal investigator. She became Associate Professor in 2019. She has directed three PhD international theses. She was elected member of the council of the European Member Society in 2024.

Diogo M.F. Santos

Diogo M.F. Santos is a Principal Researcher at the CeFEMA—Center of Physics and Engineering of Advanced Materials—and an Invited Assistant Professor at Instituto Superior Técnico (Universidade de Lisboa, Portugal). His research focuses on developing materials and processes for electrochemical energy conversion and storage, especially electrodes and membranes for fuel cells and electrolyzers. D.M.F. Santos has published over 200 papers with approximately 6500 citations; his current h-index is 42. He has been listed on Stanford University's "World's Top 2% Scientists list" since 2020.

José Ignacio Calvo

José Ignacio Calvo (BSc and PhD on Physics) is Professor of Applied Physics at the University of Valladolid (Campus of Palencia) and member of the ISP (Institute of Sustainable Processes) of the same university. He has devoted his 35 years of scientific career to research in membranes, with special focus in membrane characterization. He is the author of some 79 papers in international journals (20 in *Journal of Membrane Science*) and three books, with more than 3100 citations, corresponding to an h-index of 31, according to Scopus. Many of these papers have been related with porosimetric techniques to determine the pore sizes and pore size distribution of synthetic membranes.

Liliana C. Tomé

Liliana C. Tomé is an Assistant Professor in Chemical Engineering at the University of Coimbra (Portugal). She holds a Master's degree in Materials Derived from Renewable Resources from the University of Aveiro (2008) and a PhD in Engineering and Technology Sciences–Chemical Engineering from ITQB NOVA (2014). She was awarded a Marie Skłodowska-Curie Individual Fellowship (2018–2020) and received the L'Oréal Portugal Medal of Honor for Women in Science in 2020. She has authored 76 peer-reviewed publications, with over 3,700 citations and an h-index of 37 (Scopus). Her research focuses on ionic liquids, eutectic mixtures, and polymer-based systems, aiming to develop more efficient and sustainable materials and separation processes.

Svetlozar Velizarov

Svetlozar Velizarov was born and educated in Bulgaria. He holds PhD degrees in Chemistry (1993) and (Bio)Chemical Engineering (2003). He has conducted research in Bulgaria, Germany, Denmark, and, since 1999, Portugal. He is currently a tenured researcher at the Associated Laboratory for Green Chemistry (LAQV), NOVA School of Science and Technology at Universidade NOVA de Lisboa. Since 2011, he has been teaching Membrane Engineering in the same university. So far, he has authored 121 peer-reviewed publications (Scopus), with ca. 4300 citations. His current h-index is 39. His research focuses on developing efficient and sustainable membrane-based processes addressing urgent challenges at the water–food–energy nexus.

Preface

Despite the absence of membrane companies fabricating and commercializing membranes in the Peninsula until recently, many Portuguese and Spanish researchers have followed the pioneering steps of the membrane science and technology coined in Europe in the 1970s, contributing to the study and application of membranes in existing and new processes relevant to the reality of both Portugal and Spain.

The Reprint provides a clear and up-to-date overview of Iberian research into membrane science and technology, as well as its most significant applications. Furthermore, the Reprint includes, as a final appendix, a directory of the main research groups active in the field of membranes across the Iberian Peninsula. We believe the reprint will be a valuable resource for Iberian researchers wishing to keep abreast of the latest work carried out in both countries and sustain the network of present and future collaborations.

Isabel Coelhoso, Clara Casado-Coterillo, Diogo M.F. Santos, José Ignacio Calvo, Liliana C. Tomé, and Svetlozar Velizarov

Guest Editors

membranes

MDPI

Editorial

State of the Art Membrane Science and Technology in the Iberian Peninsula 2021–2022

Clara Casado-Coterillo [1], Diogo M. F. Santos [2], Liliana C. Tomé [3], Svetlozar Velizarov [3], Isabel Coelhoso [3,*] and José Ignacio Calvo [4,*]

[1] Department of Chemical and Biomolecular Engineering, University of Cantabria, Av. Los Castros s/n, 39005 Santander, Spain
[2] Center of Physics and Engineering of Advanced Materials, Laboratory for Physics of Materials and Emerging Technologies, Chemical Engineering Department, Instituto Superior Técnico, Universidade de Lisboa, 1049-001 Lisbon, Portugal
[3] LAQV/REQUIMTE, Department of Chemistry, NOVA School of Science and Technology, Universidade Nova de Lisboa, 2829-516 Caparica, Portugal
[4] Department of Applied Physics, Institute for Sustainable Processes, Universidad de Valladolid, 34071 Palencia, Spain
* Correspondence: imrc@fct.unl.pt (I.C.); joseignacio.calvo@uva.es (J.I.C.)

This Special Issue of the journal *Membranes* arises from the need to highlight the developments in the field of membrane research and membrane processes that have been emerging in recent years by researchers and research groups based in the Iberian Peninsula. Despite the absence of membrane companies fabricating and commercializing membranes in the Peninsula until last year, many Portuguese and Spanish researchers have followed the pioneering steps of the membrane science and technology coined in Europe in the 1970s, contributing to the study and application of membranes in existing and new processes relevant to the reality of both Portugal and Spain. This Special Issue is composed of 14 research articles plus 2 reviews, each focusing on the perspective of one country (Spain or Portugal) but adding to the historical perspective of the beginning of membrane technology in each country.

Thus the review by Calvo et al. [1] compiles the initial milestones of membrane research in Spain. It comments on the singularity of the first steps in membrane materials and mechanisms, by researchers from the field of Applied Physics and Irreversible Thermodynamics, mainly focused on ion separation processes. This was followed by applied disciplines such as chemical engineering, aimed at implementing membranes to improve existing or new industrial and environmental applications. These opened up myriad fields of knowledge in materials science, polymer chemistry, ceramic and inorganic materials, to design novel membrane materials to face the challenges experienced by foreign commercial membranes in practical applications. To highlight the increased attention accorded to membrane technology in Spain, this review also contains a directory of all the research groups active nowadays in membrane research in Spain.

In line with the origins of membrane technology studies in Spain and the current state of the art, among the Spanish articles included in this Special Issue, an important number of these articles report on the applications linked to the use of natural resources and the contribution of membranes to the solution of environmental problems related to this use.

Soto-Herranz et al. [2] investigated the use of gas permeable tubular membranes to optimise the reduction in ammonium from manure generated in livestock facilities. Using various e-PTFE membrane systems, NH_3 in an acidic solution from a pig farm is extracted for its use as a fertiliser. In a second work, Soto et al. [3] continued this study by analysing the optimal conditions for NH_3 capture in suspended gas membrane systems, analysing the usefulness of different capture solutions in terms of NH_3 captured and economic cost.

In relation to the application of membrane processes to the partial dealcoholisation of wines, Calvo et al. [4] studied various alternatives for the partial dealcoholisation of white wines. They addressed the dealcoholisation of wines from Verdejo grapes by combining nanofiltration with pervaporation (PV), or single step PV, as well as a last possibility based on membrane dialysis. Of the three processes, dialysis performed best in terms of an adequate reduction in alcohol content, in the shortest time, and with the least disturbance to the organoleptic characteristics of the resulting wines.

The particularities of the production of lactic acid derivatives from the dairy industry were addressed by González Díaz et al. [5]. In their work, the pervaporation process was used for the dehydration of mixtures containing ethanol, ethyl lactate, and lactic acid, in addition to water, at high concentrations. Various membranes and the optimal conditions for the desired separation were analysed.

The growing interest in the production of biofuels has motivated numerous studies on separation techniques that allow the separation/concentration of organics produced by fermentation, improving productivity and performance. In this light, Arregoitia-Sarabia et al. [6] also studied novel pervaporation hollow fibre membranes composed of a PEBA layer and an external PDMS layer for the separation of butanol in order to evaluate their performance in the selective removal of organics from a synthetic ABE solution. In their study, they developed hollow fibres composed of a PEBA layer and an external PDMS layer, allowing an excellent selectivity to butanol (due to PEBA) together with the protective role of the PDMS layer, which allows sealing small holes or gaps in the fibres.

Regarding membrane modification, Jiménez-Robles et al. [7] modified the surface of PVDF membranes applied to the recovery of dissolved methane in aqueous streams. A three-step procedure (activation, functionalisation, and curing) allowed maximising the hydrophobicity of the starting PVDF membranes in order to increase the membrane performance in the dissolved methane recovery process.

The always interesting and active topic of membrane reactor research has also been addressed in this Special Issue through two articles from Spanish-based researchers. Particularly, Lejarazu-Larrañaga et al. [8], in an article selected as the Editor's Choice by the journal's editorial team, in a collaborative study performed in Portugal explored the application of reverse osmosis (RO) membranes in a process, patented by Portuguese researchers, known as Ion Exchange Membrane Bioreactor (IEMB). The specific application was nitrate removal from polluted water sources. For this purpose, end-of-life RO membranes were recycled, chemically modified and thus converted into anion-exchange membranes, which allowed for preparing them at potentially lower prices than those of commercial ones, also contributed to a more circular economy. Moreover, Yang and López-Grimau [9] addressed the development of a hybrid membrane bioreactor composed of a moving bed biofilm reactor combined with a membrane bioreactor (MBBR-MBR), applied to the treatment of wastewater from a Spanish textile company.

The presence in the Special Issue of several works dealing with the study of liquid membranes revealed the continuous interest in a promising and particularly active field of research in Spain. Firstly, León et al. [10] studied the kinetic and mechanistic modelling of emulsion liquid membranes. These models allowed to study the process of adsorption to the carrier-mediated transport in emulsion liquid membranes and their application to Cu(II) removal from aqueous solutions. On the other side, Alguacil and López [11] analysed the transport of iron (III) in mixed solutions obtained from the treatment of exhausted alkaline batteries by means of supported liquid membranes. In particular, the use of the carrier Cyanex 923, coupled to a liquid membrane with a flat solid support, allows the selective separation of Iron (III) from manganese (II) in solutions generated after the treatment of discharged batteries. Finally, Fontás et al. [12] used different ionic liquids as carriers for the surface characterisation of polymer inclusion membranes (PIMs). Techniques such as XPS or SEM were used to characterise the samples and estimate the possible losses of ionic liquids and the surface modifications performed. This last paper is certainly more related to membrane characterization, a field in which Spanish researchers have traditionally been very active, as can be seen in the Membrane Groups directory included in the Spanish-side review [1]. In this area, Lima-Rodríguez et al. [13] proposed a novel method for the non-destructive study of the mechanical properties of membranes. This procedure was based on the use of vibro-acoustic tests and the dynamic analysis of the behaviour of the membrane to these vibrational sound waves. This paper was also chosen as the Editor's Choice.

Regarding the Portuguese-side contributions, the review by Tomé et al. [14] addressed the pioneering work on membrane science and technology in Portugal by the research groups of Instituto Superior Técnico—Universidade de Lisboa (IST), NOVA School of Science and Technology—Universidade Nova de Lisboa (FCT NOVA) and Faculdade de Engenharia—Universidade do Porto (FEUP) aiming to provide a historical perspective on the topic. Then, it overviewed the trends and challenges in membrane processes and materials involving Portuguese researchers, contributing to a more sustainable water–energy–material–food nexus.

The work by Janeca et al. [15] presented the synthesis and characterisation of novel cellulose acetate-based monophasic hybrid skinned amine-functionalized CA-SiO_2-$(CH_2)_3NH_2$ membranes using an innovative method that combines phase inversion and sol-gel techniques. The permeation properties of these membranes were assessed using an in-house-built single hemodialysis membrane module (SHDMM) under dynamic conditions. The CA-SiO_2-$(CH_2)_3NH_2$ membranes fully permeated urea, creatinine, and uric acid, while completely retaining albumin. Long-term filtration studies of albumin solutions indicated that fouling did not occur at the membrane surface.

Rita et al. [16] explored the use of membrane filtration to treat spent caustic effluents from Sines Refinery, in Portugal. These effluents are very challenging due to their very hazardous nature in terms of toxicity (rich in mercaptans, sulfides and other aromatic compounds) as well as their extreme pH (approximately 12–14). The potential of nanofiltration (NF) polymeric and ceramic membranes was assessed by membrane life expectancy. Although the polymeric membrane performance was very good, its lifespan was reduced after 6 weeks of contact with spent caustic effluents. Contrary to expectations, the ceramic membrane tested was not chemically more resistant than the polymeric one, suggesting that a pH of 13.9 is very aggressive, even for ceramic membranes.

We can summarize this Special Issue by concluding that, in the Iberian Peninsula, active membrane-related research is now being carried out in a variety of applications, ranging from the production of clean water for potable use, agriculture, and irrigation; to the recovery of valuable compounds from natural sources and industrial effluents; and membrane reactors to energy generation and storage, as well as the integration of membranes and membrane technology in carbon capture and utilization strategies. These cover a multi-scale approach from membrane materials synthesis to process development, involving fundamental science laboratories, chemical engineering departments and technological centres, and research institutes. As it has been shown in the articles composing the Issue, membrane research in the Iberian Peninsula is strongly concerned with Sustainable Development Goals and the evolution towards a sustainable and circular economy where membrane processes could be a unique tool.

Acknowledgments: We want to express our deepest acknowledgment to all authors participating in the Special Issue. Their contributions and that of the devoted reviewers allowed to meet the high standards of quality of the Journal.

Conflicts of Interest: The authors declare no conflict of interest.

References

1. Calvo, J.I.; Casado-Coterillo, C.; Hernández, A. Past, Present and Future of Membrane Technology in Spain. *Membranes* **2021**, *11*, 808. [CrossRef] [PubMed]
2. Soto-Herranz, M.; Sánchez-Báscones, M.; García-González, M.C.; Martín-Ramos, P. Comparison of the Ammonia Trapping Performance of Different Gas-Permeable Tubular Membrane System Configurations. *Membranes* **2022**, *12*, 1104. [CrossRef] [PubMed]

3. Soto-Herranz, M.; Sánchez-Báscones, M.; Antolín-Rodríguez, J.M.; Martín-Ramos, P. Evaluation of Different Capture Solutions for Ammonia Recovery in Suspended Gas Permeable Membrane Systems. *Membranes* **2022**, *12*, 572. [CrossRef] [PubMed]
4. Calvo, J.I.; Asensio, J.; Sainz, D.; Zapatero, R.; Carracedo, D.; Fernández-Fernández, E.; Prádanos, P.; Palacio, L.; Hernández, A. Membrane Dialysis for Partial Dealcoholization of White Wines. *Membranes* **2022**, *12*, 468. [CrossRef] [PubMed]
5. González Díaz, E.; Álvarez-García, S.; Luque, S.; Álvarez, J.R. Low-Temperature Hydrophilic Pervaporation of Lactic Acid Esterification Reaction Media. *Membranes* **2022**, *12*, 96. [CrossRef] [PubMed]
6. Arregoitia-Sarabia, C.; González-Revuelta, D.; Fallanza, M.; Ortiz, A.; Gorri, D. PEBA/PDMS Composite Multilayer Hollow Fiber Membranes for the Selective Separation of Butanol by Pervaporation. *Membranes* **2022**, *12*, 1007. [CrossRef] [PubMed]
7. Jiménez-Robles, R.; Moreno-Torralbo, B.M.; Badia, J.D.; Martínez-Soria, V.; Izquierdo, M. Flat PVDF Membrane with Enhanced Hydrophobicity through Alkali Activation and Organofluorosilanisation for Dissolved Methane Recovery. *Membranes* **2022**, *12*, 426. [CrossRef] [PubMed]
8. Lejarazu-Larrañaga, A.; Ortiz, J.M.; Molina, S.; Pawlowski, S.; Galinha, C.F.; Otero, V.; García-Calvo, E.; Velizarov, S.; Crespo, J.G. Nitrate Removal by Donnan Dialysis and Anion-Exchange Membrane Bioreactor Using Upcycled End-of-Life Reverse Osmosis Membranes. *Membranes* **2022**, *12*, 101. [CrossRef] [PubMed]
9. Yang, X.; López-Grimau, V. Reduction of Cost and Environmental Impact in the Treatment of Textile Wastewater Using a Combined MBBR-MBR System. *Membranes* **2021**, *11*, 892. [CrossRef] [PubMed]
10. León, G.; Gómez, E.; Miguel, B.; Hidalgo, A.M.; Gómez, M.; Murcia, M.D.; Guzmán, M.A. Feasibility of Adsorption Kinetic Models to Study Carrier-Mediated Transport of Heavy Metal Ions in Emulsion Liquid Membranes. *Membranes* **2022**, *12*, 66. [CrossRef] [PubMed]
11. Alguacil, F.J.; Lopez, F.A. Separation Iron(III)-Manganese(II) via Supported Liquid Membrane Technology in the Treatment of Spent Alkaline Batteries. *Membranes* **2021**, *11*, 991. [CrossRef]
12. Fontàs, C.; Vera, R.; Anticó, E.; Martínez de Yuso, M.d.V.; Rodríguez-Castellón, E.; Benavente, J. New Insights on the Effects of Water on Polymer Inclusion Membranes Containing Aliquat 336 Derivatives as Carriers. *Membranes* **2022**, *12*, 192. [CrossRef]
13. Lima-Rodriguez, A.; Garcia-Manrique, J.; Dong, W.; Gonzalez-Herrera, A. A Novel Methodology to Obtain the Mechanical Properties of Membranes by Means of Dynamic Tests. *Membranes* **2022**, *12*, 288. [CrossRef] [PubMed]
14. Tomé, L.C.; Santos, D.M.F.; Velizarov, S.; Coelhoso, I.M.; Mendes, A.; Crespo, J.G.; de Pinho, M.N. Overview of Membrane Science and Technology in Portugal. *Membranes* **2022**, *12*, 197. [CrossRef] [PubMed]
15. Janeca, A.; Rodrigues, F.S.C.; Gonçalves, M.C.; Faria, M. Novel Cellulose Acetate-Based Monophasic Hybrid Membranes for Improved Blood Purification Devices: Characterization under Dynamic Conditions. *Membranes* **2021**, *11*, 825. [CrossRef] [PubMed]
16. Rita, A.I.; Nabais, A.R.; Neves, L.A.; Huertas, R.; Santos, M.; Madeira, L.M.; Sanches, S. Assessment of the Potential of Using Nanofiltration Polymeric and Ceramic Membranes to Treat Refinery Spent Caustic Effluents. *Membranes* **2022**, *12*, 98. [CrossRef] [PubMed]

membranes

MDPI

Article

Comparison of the Ammonia Trapping Performance of Different Gas-Permeable Tubular Membrane System Configurations

María Soto-Herranz [1], Mercedes Sánchez-Báscones [1,*], María Cruz García-González [1] and Pablo Martín-Ramos [1,2]

1 Department of Agroforestry Sciences, ETSIIAA, University of Valladolid, Avenida de Madrid 44, 34004 Palencia, Spain
2 Instituto Universitario de Investigación en Ciencias Ambientales de Aragón (IUCA), EPS, Universidad de Zaragoza, Carretera de Cuarte s/n, 22071 Huesca, Spain
* Correspondence: mercedes.sanchez@uva.es

Abstract: The technology of gas-permeable tubular membranes (GPMs) is promising in reducing ammonia emissions from livestock manure, capturing NH_3 in an acidic solution, and obtaining final products suitable for valorization as fertilizers, in line with the principles of the circular economy. This study aimed to evaluate the performance of several e-PTFE membrane systems with different configurations for the recovery of NH_3 released from pig slurry. Ten different configurations were tested: only a submerged membrane, only a suspended membrane in the same chamber, only a suspended membrane in an annex chamber, a submerged membrane + a suspended membrane in the same chamber, and a submerged membrane + a suspended membrane in an annex chamber, considering in each case the scenarios without and with agitation and aeration of the slurry. In all tests, sulfuric acid (1N H_2SO_4) was used as the NH_3 capture solution, which circulated at a flow rate of 2.1 $L \cdot h^{-1}$. The results showed that NH_3-N removal rates ranged from 36–39% (for systems with a single submerged or suspended membrane without agitation or aeration of the slurry) to 70–72% for submerged + suspended GPM systems with agitation and aeration. In turn, NH_3-N recovery rates were found to be between 44–54% (for systems with a single membrane suspended in an annex compartment) and 88–91% (for systems based on a single submerged membrane). However, when choosing a system for farm deployment, it is essential to consider not only the capture and recovery performance of the system, but also the investment and operating costs (ranging from 9.8 to 21.2 €/kg N recovered depending on the selected configuration). The overall assessment suggests that the simplest systems, based on a single membrane, may be the most recommendable.

Keywords: ammonia recovery; gas-permeable membrane; submerged GPM system; suspended GPM system

1. Introduction

Ammonia (NH_3) emissions from livestock farming are a major source of odor and environmental pollution [1,2]. In addition, they lead to a significant loss of N, a valuable plant nutrient. Therefore, there is great interest in the application of control technologies to reduce NH_3 emissions through N capture and recovery, which would partially offset the implementation and operation costs associated with such control technologies through the revenue obtained from the sale of the fertilizer product [3,4].

Among the technologies used to remove NH_3 generated in livestock houses are the treatment of the exhaust air from the house using scrubbing or filtration techniques (biotrickling or biofilters) and chemical amendments [5–7]. However, in these cases, either the retention of N involves high costs, or, even if the N is retained without volatilizing in the litter, the NH_3 is not recovered as a separate product, as is the case with scrubbing techniques.

This study investigates the use of an alternative technology based on expanded polytetrafluoroethylene (e-PTFE) gas-permeable passive membranes [3,8,9] that allows capturing NH_3 (g) in an acidic solution and recovering N in a concentrated and purified form, not dependent on intense air movement. This membrane material was selected as it is microporous, flexible, and hydrophobic, with a high permeability rate for low-pressure gas flow differentials between the inside and outside of the tube. In previous studies, e-PTFE membranes have led to high NH_3 recoveries from different sources, such as chicken manure, pig manure, and anaerobically digested slurry [10–12], effectively reducing the TAN concentration in the sources. Further, it has also demonstrated high NH_3 recovery efficiencies from livestock housing air [3,13,14].

The performance of this technology depends on the availability of NH_3 in the TAN source, where NH_3 and NH_4^+ are in equilibrium, which in turn depends on factors such as the pH and temperature of the TAN source [3,15]. The concentration of NH_3 in the emitting source, the surface area of the membrane, the composition of the capture solution used, and its rate of circulation also influence the efficiency of the process [9,15–20].

To date, these membranes have generally been used in the submerged configuration for the capture of N from slurry and digestate [16,21]. More recently, the use of suspended membrane configurations has also been proposed [3,13,14], and some authors have explored the possibility of combining both systems [9]. Submerged GPM configurations yield higher NH_3 capture efficiencies than suspended GPM systems, as they are in direct contact with the N-emitting source. However, agitation and aeration of the slurry to achieve these efficiencies involve a significant discharge of NH_3 (g) into the medium, and the submerged GPM system is unable to capture all the N from the source before the discharge of NH_3 occurs. To improve the NH_3 capture performance, the possibility of combining both GPM systems (submerged and suspended) has been considered.

Nevertheless, it is difficult to compare the efficiency of the different configurations due to the diversity of working conditions in the different studies (very different N concentrations in the emitting solution, different membrane characteristics and surfaces, variety of capture solutions and their pumping speeds, etc.).

The aim of this research was to carry out a comparative study in systematized conditions that would allow evaluating the response of different gas-permeable membrane systems on the NH_3 diffusion flux and the NH_3 extraction efficiency from pig slurry in a reliable way, to select the most advantageous ones with a view to their application on an industrial scale (for NH_3 capture from livestock buildings, in manure storage facilities, in composters, etc.).

2. Materials and Methods

2.1. Origin of the Slurry

Pig slurry was collected from a slurry pond of a mother sow farm located in Dueñas (Palencia, Spain) and was transported to the laboratory and kept at 4 °C until the experiments were carried out.

2.2. Experimental Design

The schematics of the different GPM membrane systems used in these experiments for NH_3 recovery are shown in Figures S1–S10.

The assembly consisted of a closed methacrylate chamber with a capacity of 30 L, inside which 2 L of pig slurry were placed together with different membrane systems, installed inside or in a 15 L annex compartment. The evaluated systems were the following:

- S1: Submerged GPM without slurry agitation or aeration.
- S2: Suspended GPM installed in the headspace of the treatment chamber without slurry agitation and aeration.
- S3: Suspended GPM installed in the headspace of a chamber attached to the treatment chamber where the slurry was without agitation or aeration.

- S4: Submerged and suspended GPMs in the same treatment chamber without agitation or aeration of the slurry.
- S5: Submerged GPM without slurry agitation or aeration and suspended GPM installed in the headspace of a chamber attached to the treatment chamber where the slurry was located.
- S6: GPM submerged with slurry agitation and aeration of 0.24 L air/(L slurry·min).
- S7: Suspended GPM installed in the headspace of the treatment chamber with slurry agitation and aeration of 0.24 L air/(L slurry·min).
- S8: Suspended GPM installed in the headspace of a chamber attached to the treatment chamber where the slurry was located with agitation and aeration of 0.24 L air/(L slurry·min).
- S9: Submerged and suspended GPMs in the same treatment chamber with slurry agitation and aeration of 0.24 L air/(L slurry·min).
- S10: Submerged GPM with slurry agitation and aeration of 0.24 L air/(L slurry·min) and suspended GPM installed in the headspace of a chamber attached to the treatment chamber where the slurry was located.

All experiments were carried out at the same time in the same laboratory, in controlled conditions. The temperature of the slurry and uptake solutions was monitored using a pH and temperature electrode, giving mean values of 23.4 ± 2.5 and 24.4 ± 2.4 °C, respectively. The pH of the slurry in the chambers was maintained during the experiments at an average value of 8.3 ± 0.2.

In all adjoining chamber systems, the air was drawn in at a flow rate of 200 $L \cdot h^{-1}$ from the treatment chamber to the adjoining compartment, with air recirculation back to the treatment chamber.

The airflow rate selected for slurry aeration was low to effectively maintain pH and avoid NH_4^+ nitrification (which lowers slurry pH). This selected aeration rate showed good results in previous studies [22,23]. In the aerated treatments, the air was supplied by an aquarium air pump from the bottom of the experimental chamber through a porous stone, and the supply was controlled by an airflow meter (Instruments Direct, Canton, GA, USA). Given that aeration was combined with agitation, uniformity throughout the volume was assured. Additionally, 10 $mg \cdot L^{-1}$ of a commercial nitrification inhibitor (allylthiourea) was added to the slurry in all experiments.

For NH_3 capture, an e-PTFE membrane (Zeus Industrial Products Inc., Orangeburg, SC, USA) with an outer diameter of 5.2 mm, a wall thickness of 0.64 mm, a polymer density of 0.95 $g \cdot cm^{-3}$, a porosity < 60%, an average pore length of 12.7 ± 5.9 μm, and an average pore width of 1.3 ± 0.9 μm was used. The pores of the e-PTFE membrane were elongated in the extrusion process during its fabrication [24]. The length of the GPM tubing was 100 cm in all experiments, giving a surface area of 163.4 cm^2 per installed membrane (i.e., 163.4 cm^2 for single membrane systems and 326.8 cm^2 for double membrane systems). The choice of this membrane length/surface was supported by previous studies [19].

Each membrane circuit had an associated 1-L 1N H_2SO_4 reservoir for NH_3 capture. Acid was pumped through the membranes using a peristaltic pump (Pumpdrive 5001, Heidolph, Schwabach, Germany) at a flow rate of 2.1 $L \cdot h^{-1}$.

2.3. Analysis Methodology

An orifice in the chambers, which was opened for a few seconds and in a punctual way, was used to carry out the slurry sampling. This also served to balance the pressure between the inside and outside of the chamber. Slurry and acidic solution samples were collected in triplicate (25 mL) three times a week.

Analyses of pH, temperature, and NH_3-N concentration were performed on all samples. pH and temperature were measured with a Crison GLP22 m electrode (Crison Instruments S.A., Barcelona, Spain), and the NH_3-N concentration was determined by distillation (Kjeltec™ 8100; Foss Iberia S.A., Barcelona, Spain) by collecting the distillate in borate buffer and subsequent titration with 0.2N HCl [25].

2.4. Data Calculation

Free ammonia (FA) was quantified according to Equation (1) [26]:

$$[NH_3]/[TNH_3] = \left(1 + \left(\frac{10^{-pH}}{10^{-(0.09018+2729.92/T)}}\right)\right)^{-1}, \quad (1)$$

where $[NH_3]$ is the concentration of free ammonia, $[TNH_3]$ is the total concentration of ammonia, and T is the temperature (K).

The amount of NH_3 removed (mg NH_3-N) was determined as the difference between the amount of NH_3 in the slurry at the beginning and the end of the experiment. The mass of NH_3 recovered (mg NH_3-N) refers to the amount of NH_3-N captured at the end of the experiment in the acidic solution. The N recovery efficiency (%) was estimated by dividing the recovered mass by the removed mass. The mass flux of NH_3-N across the membrane (J, expressed in mg $NH_3\text{-N}\cdot cm^{-2}\cdot d^{-1}$) was determined considering the mass of N captured per day and the surface area of the GPM pipe.

2.5. Statistical Analyses

Data were tested for homogeneity and homoscedasticity using the Shapiro–Wilk and Levene tests, respectively, after which a two-way analysis of variance (ANOVA) was performed, followed by a post hoc comparison of means using Tukey's test at $p < 0.05$. The statistical software R was used for these statistical analyses [27].

2.6. Economic Analysis of the Different GPM Systems

To compare the different GPM systems used to reduce NH_3 emissions, an economic analysis of each system was performed. Operating costs associated with the electrical energy consumption of the elements used in each case (peristaltic pump, agitator, aeration pump, and suction pump) and costs associated with the price of the membrane and acid were considered. The values used in these calculations were based on the experimental data from this study, along with the following assumptions:

- The annual production of pig slurry on the farm is 2795 m^3 per year (i.e., 7.7 $m^3\cdot d^{-1}$), resulting from 1300 animals producing 2.15 m^3 slurry$\cdot$year^{-1} each.
- Pig slurry contains between 931.6 and 1032.1 mg TAN$\cdot L^{-1}$ for each system evaluated.
- A TAN removal target for pig slurry of approximately 35% is proposed.
- TAN recovery efficiencies and TAN recovery rates at the end of the 7-day experimental period have been taken as a reference for each evaluated system.
- The cost of the membrane is 115 €$\cdot m^{-2}$ (1.88 €$\cdot m^{-1}$) [12] and a 10% replacement per year is considered.
- Annualized equipment costs are calculated assuming a useful life of 10 years and an interest rate of 8% [12].
- The amount of H_2SO_4 (98%) required to capture TAN ranges from 446.5 to 494.7 kg acid/kg recovered TAN for the systems evaluated in this study.

3. Results and Discussion

3.1. Nitrogen Removal from Pig Slurry

The concentration of NH_3-N present in the slurry decreased during the experimental period for all GPM systems (Table 1), with reductions ranging from 36 to 72%.

Table 1. Mass balance of NH_3-N recovered using the different configurations of the GPM system. Mass of NH_3-N in the slurry at the beginning of the experiment (Initial NH_3-N mass), mass of NH_3-N emitted (NH_3-N removed), mass of NH_3-N recovered in the acidic solution (NH_3-N recovered), NH_3 removal and NH_3 recovery efficiencies (NH_3-N removal and NH_3-N recovered, respectively), and N flux rate (N flux).

Conditions	GPM System	Initial NH_3-N Mass (mg)	NH_3-N Removed (mg)	NH_3-NRecovered (mg)	NH_3-N Removal (%)	NH_3-N Recovered (%)	N Flux ($g \cdot m^{-2} \cdot d^{-1}$)
Without agitation + aeration	S1	4125 ± 227	1599 ± 136 b	1414 ± 21 cd	39	88	12.4 ± 0.2 cd
	S2	4308 ± 47	1567 ± 96 b	1083 ± 8 e	36	69	9.5 ± 1.7 e
	S3	4340 ± 16	2680 ± 119 a	1453 ± 111 c	62	54	12.7 ± 1.0 c
	S4	4125 ± 227	2710 ± 172 a	2361 ± 61 a	66	87	10.3 ± 0.3 cde
	S5	4006 ± 116	2692 ± 241 a	2278 ± 41 a	67	85	10.0 ± 0.2 de
With agitation + aeration	S6	4125 ± 227	2545 ± 66 a	2328 ± 10 a	62	91	20.4 ± 0.0 a
	S7	4122 ± 46	2696± 142 a	1918 ± 197 b	65	71	16.8 ± 0.1 b
	S8	4006 ± 116	2901 ± 327 a	1272 ± 20 de	72	44	11.4 ± 0.2 cde
	S9	4437 ± 20	3211 ± 29 a	2545 ± 259 a	72	79	11.1 ± 0.2 cde
	S10	4006 ± 116	2824 ± 254 a	2434 ± 2 a	70	86	10.6 ± 0.0 de

Means (n = 3) followed by a common lowercase letter are not significantly different according to Tukey's test ($p < 0.05$). Pairs of homologous systems are highlighted in the same colors. The meaning of the abbreviations is presented in Section 2.2.

The most efficient systems in terms of NH_3-N removal were S8, S9, and S10, all of which are systems with slurry agitation and aeration. In general, agitation and aeration of the N emission source allow the pH of the slurry to be kept high while preventing the nitrification of NH_4^+. Aeration causes the release of OH^-, which raises the pH of the slurry, reducing its alkalinity according to Equation (2):

$$HCO_3^- + \text{air} \rightarrow OH^- + CO_2, \quad (2)$$

The increase in pH due to aeration, therefore, affects the formation of NH_3 and the efficiency of the membrane system to recover N from the slurry, acting as indicated in Equation (3):

$$NH_4^+ + OH^- \leftrightarrow NH_3 + H_2O, \quad (3)$$

The recovery of NH_3 (g) through the gas-permeable membrane causes an increase of acidity in the slurry, as H^+ does not permeate the hydrophobic membrane. Therefore, the recovery of the NH_3 by the membrane, from a net perspective, causes an acidification of the slurry and a reduction of the alkalinity. Daguerre et al. [28] studied this behavior in further detail, concluding that the release of OH^- from natural carbonates increased the manure pH and promoted gaseous ammonia formation and membrane uptake, and, at the same time, the recovery of gaseous ammonia through the membrane acidified the slurry. Similar behavior was observed by other authors, such as Oliveira et al. [29].

Another effect induced by aeration was the increase of FA content in the slurry, which increased in values ranging between 203.2 and 359.4 $mg \cdot L^{-1}$ from the beginning to the end of the experiment in the systems with slurry aeration (due to the increase in slurry pH), while in the systems without aeration it ranged between 42.7 and 315.7 $mg \cdot L^{-1}$. Three cases were observed in which the FA content decreased at the end of the trial in comparison to the beginning, these being systems S3, S6, and S7. According to García-González and Vanotti [30], since the FA content was always above 40 $mg \cdot L^{-1}$ (except for the three cases mentioned), active permeation of NH_3 through the membranes would be guaranteed, favoring an optimal NH_3 capture process.

Our results were consistent with those obtained by García-González et al. [22], who determined a 33% higher NH_4^+ removal efficiency for a submerged e-PTFE membrane treatment in aerated slurry than in non-aerated slurry during an 18-day experiment. In our case, aeration using a submerged membrane (S6) improved the NH_3-N removal efficiency by 23% compared to the same system without aeration (S1) in 7 days. In addition, the aforementioned authors indicated that, in the test with submerged membrane without slurry aeration, the FA concentration remained below 100 mg $N \cdot L^{-1}$ during the experiment, compared to FA concentrations of up to 250 mg $N \cdot L^{-1}$ in the treatment with aerated slurry. In this regard, the results of this study differed from those mentioned above: in the submerged membrane trial without slurry aeration (S1), an increase in FA concentration was observed as the experiment progressed, while in the submerged membrane trial with slurry aeration (S6) this concentration decreased over time.

Likewise, the S4 configuration was similar to the one used by Majd and Mukhtar [9], who used suspended and submerged membrane systems at the same time for the capture of NH_3 from slurry without agitation and aeration. In this study, our results showed a removal of NH_3-N from the slurry of 66% in 7 days, higher than that found by the authors mentioned above, who determined a removal percentage of NH_3-N of 48% in 15 days of experimental operation with pH control. This difference can be attributed to the fact that the pig slurry used as a source of N emission in this study presented a notably higher initial NH_3-N concentration (2170 $mg \cdot L^{-1}$) than the one used in their experiments (117 $mg \cdot L^{-1}$).

The lowest NH_3-N reductions in the slurry were observed in systems S1 and S2 (39% and 36%, respectively) (Table 1), corresponding to submerged and suspended membrane systems without slurry agitation and aeration, respectively. In both systems, a low amount of NH_3 was released into the air in the chamber and, therefore, a lower amount was recovered in the acidic solution compared to other systems. For comparison purposes, other authors such as Vanotti and Szogi [11] and García-González and Vanotti [30] obtained NH_4^+ removal percentages in treatments with submerged e-PTFE membranes without aeration or agitation of the slurry of 50 and 57% with respect to the initial content of NH_4^+ in the slurry in 9 and 20 days, respectively.

The rest of the systems (S3, S5, and S7) showed intermediate NH_3-N removal values, in the 60–70% interval.

3.2. Nitrogen Recovery from Pig Slurry

The concentration of NH_3-N in the capture solution increased in all cases during the experimental period, exceeding the threshold of 1000 mg $N \cdot L^{-1}$ (Figure S11). The highest NH_3-N concentration was obtained with systems S9, S10, S5, and S4. All of these systems had two membranes (submerged and suspended), either in the same compartment or in an annexed one, with and without continuous agitation of the slurry, respectively. Such a high concentration would be due to the sum of NH_3 recoveries in both membranes. If the NH_3-N recovery per specific membrane surface area is plotted instead (Figure 1), it becomes apparent that these double-membrane systems would not be so efficient, as will be discussed below. Comparing homologous pairs without and with aeration, the highest increase in NH_3-N recovery per membrane surface area was obtained in the S2–S7 pair (45%), followed by S1–S6 (38%).

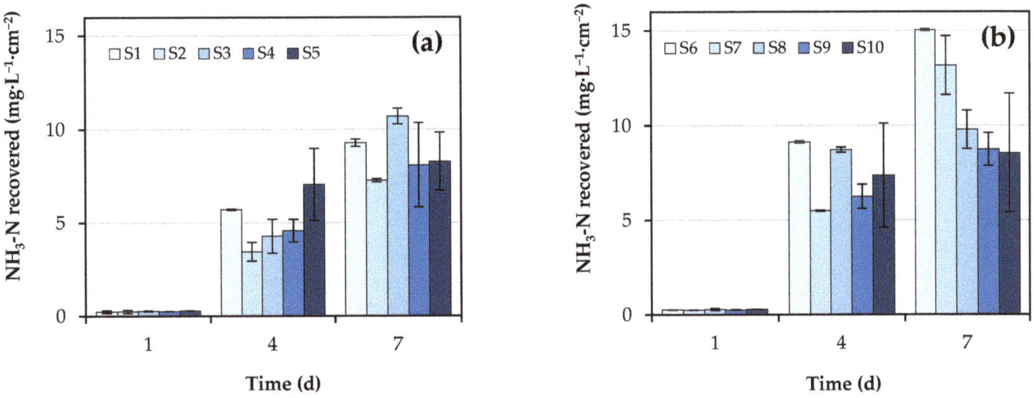

Figure 1. Evolution of NH_3-N concentration in the acid capture solution per membrane surface area in (**a**) S1–S5 systems without sludge aeration and agitation; and (**b**) S6–S10 systems with slurry agitation and aeration. The meaning of the abbreviations is presented in Section 2.2.

Regarding the mass of recovered NH_3-N and the NH_3-N recovery efficiency, it is interesting to make a comparison between homologous systems, in pairs: S1–S6, S2–S7, S3–S8, S4–S9, and S5–S10. Regarding the first parameter, significant differences were observed only in the following pairs of counterparts: S1–S6, S2–S7, and S3–S8. In contrast, no significant changes in the mass of NH_3-N recovered were observed for the S4–S9 and S5–S10 pairs.

In the three cases in which significant differences were observed, they were systems with a single membrane, and the difference between the systems was a result of the presence or absence of agitation and aeration in the slurry. As indicated above, aeration allows the slurry pH to be kept high, which favors the presence of more NH_3 (g) in the air of the treatment chamber, which explains why a significantly higher recovery was obtained than in the same type of system in which there was no aeration and agitation of the slurry.

As for the recovery efficiency, it should be noted that the differences were minimal, within measurement error (±10%): 3% for S1–S6, 2% for S2–S7, 10% for S3–S8, 8% for S4–S9, and 1% for S5–S10. This suggests that the efficiency of NH_3-N recovery depends on the system configuration (i.e., a submerged GPM, a suspended GPM, or a suspended GPM in an annex chamber, or two GPM, either in the same chamber or in an annex chamber) and not on the presence/absence of agitation and aeration of the slurry.

Comparing the results obtained with those mentioned in the literature, the recovery efficiency values for systems based on a single submerged GPM (S1–S6), of 88 and 91% (Table 1), were comparable to those reported by Vanotti et al. [31], Oliveira Filho et al. [29], and Riaño et al. [23], who also obtained high ammonia recovery (~90%) using a submerged membrane with low aeration of raw or digested swine manure. However, they were lower than those reported by García-González et al. [22], who, in a process of capture of ammonia from raw swine manure using a submerged gas-permeable membrane, were able to achieve an overall recovery of ammonia of 99% with respect to the initial NH_4^+ present in the manure, regardless of the presence or absence of aeration. Dube et al. [12] also achieved 96–98% recoveries of NH_4^+ present in digested swine manure in 5 days using a submerged tubular gas membrane, together with low-rate aeration and a nitrogen inhibitor, while recoveries obtained using submerged membrane without digested swine manure aeration were 76–95% over a period of 25 days.

In the S2–S7 systems, based on a single GPM suspended in the air chamber in which the slurry was located, the efficiencies were 69 and 71%, notably lower than that obtained by Soto-Herranz et al. [19] (96%) for a similar suspended e-PTFE system for an acid flow rate of 2.1 $L \cdot h^{-1}$ in an experimental period of 7 days, similar to the one used in this study.

The amount of NH_3-N recovered was also notably lower (1083 vs. 1602 mg N). These differences can be attributed to the fact that in that experiment a synthetic solution was used as a source of N emission instead of pig slurry, with a TAN of 6000 mg NH_3-N·L^{-1} (vs. 4308 mg NH_3-N·L^{-1} in this study). On the other hand, the values obtained in this study were significantly higher than those obtained in an NH_3 capture system that used a non-hydrophobic and non-microporous membrane and a 2:3 (v/v) ratio of emitting solution (pig slurry) to acidic solution: 46% in 24 h and 59% in 48 h [32].

On the other hand, when comparing the results of S1–S2 (submerged vs. suspended GPM without slurry agitation and aeration) and S6–S7 (submerged vs. suspended GPM with slurry agitation and aeration), significant differences in ammonia capture were observed as a function of membrane position. Such differences can be attributed to the fact that the membranes suspended in the headspace of the chamber are not in direct contact with the emission source and, therefore, the NH_3 capture is lower than in submerged GPM systems, given that the concentration in the air is lower than that present in the liquid. However, it should be noted that this result does not coincide with that reported by Rothrock et al. [3], who found that the relative position of the tubular membranes (above, directly on, or below the poultry litter surface) did not significantly affect the total mass of NH_3 recovered by the system or the mass of NH_4-N remaining in the poultry litter after volatilization. It is also worth mentioning that, in the present work, no reduction in the TAN recovery rate was detected over time in those configurations in which the membrane was submerged, suggesting that fouling—which determines the lifetime of GPMs and affects their economic viability [33]—was not relevant. However, further studies with longer run times would be needed to determine the effect of membrane configuration on the anti-fouling performance.

In S3–S8 systems, in which the membrane was suspended in an annex compartment, forced ventilation toward this compartment (aeration), as previously indicated, should favor the presence of more NH_3 (g) available to be aspirated and concentrated in the annex chamber. It should be noted that the capture process of NH_3 in the acidic solution worked better in the S3 system than in S8, which had agitation and aeration of the slurry. This unexpected result could have its origin in the fact that agitation leads to a higher NH_3 emission, part of which would be lost before being recovered by the membrane, while in the system without agitation only the phenomenon of gas diffusion through the membrane would influence NH_3 capture. The NH_3-N recovery efficiencies achieved in these systems, 54 and 44%, cannot be reliably compared with the literature: to the best of the authors' knowledge, this type of system has only been previously tested by our group, with values of 4108 g NH_3-N in 232 days in a pig farm and 794 g NH_3-N in 256 days in a poultry farm under real conditions, not under laboratory ones.

On the other hand, the results obtained for systems S4–S9 and S5–S10 are interesting. In both cases, double membranes were used: in the first pair of systems, both membranes (submerged and suspended) were located inside the same treatment chamber and, in the second pair, the suspended GPM was installed in an annex compartment. Each pair of systems was studied with the presence or absence of agitation and aeration. No significant differences in NH_3 capture were observed between homologous or non-homologous pairs. This suggests that when using a double membrane system, it would not be necessary to apply aeration and agitation to the slurry to favor the capture of NH_3-N. Furthermore, concerning the systems in which air was extracted from the treatment chamber to an external compartment in which a suspended GPM membrane was installed (S5–S10), no significant differences were observed in the mass of NH_3-N recovered in the capture solution between homologous pairs. This is especially interesting, since air extraction into an external compartment would reduce the effect of agitation and aeration on the NH_3 recovery performance of the systems. The NH_3-N recovery efficiencies achieved in these systems were 87% and 79% for S4–S9 and 85–86% for S5–S10 (Table 1). Majd and Mukhtar [9] used an experimental design similar to S4 and, although they did not expressly report recovery efficiency data, it is possible to calculate it based on the information provided in their study, obtaining a joint NH_3-N recovery efficiency of 79%. Therefore, the results obtained in this study would be consistent with those reported by these authors.

If the impact on NH_3-N recovered from the use of the double membrane system compared to the single membrane system is evaluated, significant differences can be observed between systems without aeration or agitation of the slurry with a single membrane (S1 and S2) and the homologous double membrane system (S4). Such differences are reasonable, given that in the latter system the membrane surface area in contact with NH_3, either in the slurry or in the air, is twice than in the former two systems. On the other hand, when comparing systems with aeration and slurry agitation with a single membrane (S6 and S7) vs. their double membrane counterpart (S9), differences were only observed between S7 (suspended membrane) and S9 (double membrane), not between S6 (submerged membrane) and S9 (double membrane). This result is interesting, given that it raises the possibility that, in cases in which agitation and aeration are applied to the slurry, it would not be necessary to use a double membrane to attain a high NH_3-N capture: using only a membrane submerged in the slurry would be enough to obtain good capture yields.

3.3. *N Flux Rates*

In the submerged GPM systems (S1 and S6), without and with slurry aeration and agitation, respectively, the N flux rates were 12.4 and 20.4 g $N \cdot m^{-2} \cdot d^{-1}$ (Table 1), higher than those obtained by González-García et al. [34] (8 g $N \cdot m^{-2} \cdot d^{-1}$) and García-González and Vanotti [22] (9.5 g $N \cdot m^{-2} \cdot d^{-1}$) in submerged systems with low slurry aeration.

N flux rates in suspended GPM systems (S2 and S7), without and with aeration and agitation, respectively, were 9.5 and 16.8 g $N \cdot m^{-2} \cdot d^{-1}$, both within the range of values provided by Rothrock et al. [13] for flat suspended membranes (1.25–17.78 g $N \cdot m^{-2} \cdot d^{-1}$) and by Rothrock et al. [3] for tubular membranes (1.29–16.52 g $N \cdot m^{-2} \cdot d^{-1}$) without agitation and aeration of the N emitting source. The values obtained with suspended systems arranged in an annex chamber without or with slurry aeration and agitation in the treatment chamber (S3 and S8) were 12.7 and 11.4 g $N \cdot m^{-2} \cdot d^{-1}$, also within the range of values found for suspended systems.

Finally, as for GPM systems combining two membrane systems (S4–S9 and S5–S10), the values obtained in the present study, of 10.3 and 11.1 g $N \cdot m^{-2} \cdot d^{-1}$ in S4 and S9, and 10.0 and 10.6 g $N \cdot m^{-2} \cdot d^{-1}$ in S5 and S10, were notably higher than those obtained by Majd and Mukhtar [9], who obtained 0.9 g $N \cdot m^{-2} \cdot d^{-1}$ for a GPM system similar to S4. The difference can be attributed to the fact that the concentration of TAN in the slurry used by these authors was very low compared to the initial concentration of the slurry used in these trials, as explained above.

3.4. *Economic Cost Analysis of the Different GPM Systems*

Based on the goal removal and recovery rates, between 2.5 and 2.8 kg of $TAN \cdot d^{-1}$ (i.e., 911.3 and 1009.6 kg of $TAN \cdot year^{-1}$, respectively) should be recovered. With TAN recovery rates ranging from 9.5 to 20.3 g $TAN \cdot m^{-2} \cdot d^{-1}$, membrane surfaces between 136.1 and 282.6 m^2 would be required to achieve this recovery.

The cost of the membrane to start up the different systems at the pilot scale would range from 15,690 € for S6 to 32,573 € for S2. The additional components, taking indicative prices reported for pilot-scale prototypes for the capture of NH_3 from the air and the recovery of TAN from liquid media [35,36], would represent annualized costs of 18,792 € for S1, 18,792 € for S2, 30,154 € for S3, 27,654 € for S4, 30,154 € for S5, 23,007 € for S6, 23,007 € for S7, 24,994 € for S8, 31,869 € for S9, and 34,369 € for S10 (Table 2). In addition, the supplementary 10% expenses considered for the annual replacement of the membrane would range from 1569 € for S6 to 3257 € for S2.

Table 2. Summary of investment and operating costs and revenues for each gas-permeable membrane configuration.

	S1	S2	S3	S4	S5	S6	S7	S8	S9	S10
Membrane cost (€/m)					1.88					
Membrane cost (€/m^2)					115.24					
Investment										
Membrane initial cost (€)	25,686	32,573	24,547	30,923	28,774	15,690	17,625	25,240	28,719	27,145
Total initial investment (including membranes and other equipment) (€)	44,478	51,365	54,701	58,577	58,928	38,697	40,632	50,234	60,588	61,514
Annualized costs 8% interest, 10-year life (€/year)	6629	7655	8152	8730	8782	5767	6055	7486	9029	9167
Operating Costs										
Membrane replacement 10% (€)	2569	3257	2455	3092	2877	1569	1763	2524	2872	2715
Chemicals (€)	143	139	140	143	130	143	133	130	144	130
Power (€)	1610	1610	3329	3329	3329	6537	6537	6811	8256	8256
Total annualized costs (€/year)	10,950	12,661	14,076	15,294	15,118	14,017	14,488	16,951	20,301	20,268
Revenue for the sale of the fertilizer product (€/year)	1049	1019	1027	1049	948	1049	975	948	1050	948
Net annual cost (€/year)	9901	11,642	13,049	14,245	14,170	12,967	13,513	16,003	19,251	19,320
Cost per N recovered (€/kg N)	9.8	11.9	13.2	14.1	15.5	12.9	14.4	17.6	19.1	21.2
Cost per treated slurry (€/m^3)	3.5	4.2	4.7	5.1	5.1	4.6	4.8	5.7	6.9	6.9
Net cost (€/(place·year))	7.6	9.0	10.0	11.0	10.9	10.0	10.4	12.3	14.8	14.9

The annual cost associated with the consumption of H_2SO_4 in the capture solution would range from 129.5 to 143.5 € (taking a guide price of 0.29 €·kg^{-1} [12]).

The approximate total energy consumption of the equipment used was 14.7 kWh·d^{-1} for S1 and S2; 30.4 kWh·d^{-1} for S3, S4, and S5; 59.7 kWh·d^{-1} for S6 and S7; 62.2 kWh·d^{-1} for S8; and 75.4 kWh·d^{-1} for S9 and S10. This represents annual electricity costs in the 1610 € (S1 and S2) to 8256 € (S9 and S10) range, assuming an average unit cost in Spain of 0.3 kWh·d^{-1}.

Adding up all the aforementioned expenses, the estimated annual expenses for a pilot plant using gas-permeable membranes in a 1300-head pig farm would be in the 10,950 € (S1) to 20,301 € (S9) interval.

Concerning revenues, the ammonium sulfate potentially recovered per year in each system would have a value (as fertilizer) ranging from 948 € (S5, S8, and S10) to 1050 € (S9), assuming a value of 1.0 €·kg^{-1} of N as ammonium sulfate.

Therefore, the estimated net cost of ammonia recovery per year would be in the 9901 € (S1) to 19,320 € (S10) range.

A summary of investment costs, operating costs, and revenues is shown in Table 2.

No costs have been found in the literature concerning the operation of prototypes combining systems for NH_3 capture from the air with systems for N reduction in manure, either at the laboratory or pilot scale, so this would be the first study in which such economic cost data are provided for a technology combining both systems.

The annual cost values per animal place (€·place^{-1}·year^{-1}) obtained in this study are of the order of those obtained with technologies such as biotrickling filters (13.2 €·place^{-1}·year^{-1}) or acid scrubbers (13.7 €·place^{-1}·year^{-1}) used to clean the air in pig housing [5]. Likewise, the values would also fall within the range of costs reported for acid scrubbers (up to 26 €/kg NH_3 treated) [37].

The energy consumption per kg of recovered nitrogen obtained in the present study ranged from 1.86 kWh/kg N (S1) to 10.62 kWh/kg N (S10). Other technologies employed for the recovery of nitrogen from the slurry, such as stripping technology, have an associated energy consumption between 3.1 and 8.65 kWh/kg N recovered [38,39]. Therefore, systems with configurations S1 and S2 would be more efficient, configurations S9 and S10 would be more energy-intensive, and the rest of the configurations would be comparable to stripping technologies.

3.5. Practical Implications of the Study

In an ammonia capture system using GPMs, the membrane is the most important and expensive component of the whole system. The amount of membrane required per m^3 of slurry to be treated and its arrangement are key elements in terms of achieving economically viable ammonia capture systems. For this reason, different membrane configurations have been investigated in this work to obtain high ammonia captures at the lowest possible cost. The reported results have an important practical implication, given that this study has been carried out in the framework of the LIFE Green Ammonia project, in which ammonia capture systems with a technology readiness level (TRL) 9 are going to be designed and built, with a view to the placing on the market of these systems.

4. Conclusions

Comparison of the five proposed GPM installation configurations, with and without slurry agitation and aeration, under comparable laboratory-scale working conditions, suggests that: (1) agitation and aeration, which result in the release of more NH_3 from the slurry to the environment, leads to statistically significant differences in the mass of NH_3-N recovered and N flux rate in the case of systems based on a single submerged GPM, a single suspended GPM, or a single suspended GPM installed in an annex compartment, but not in systems with double membranes, in which NH_3-N removal efficiencies of $\approx$66.5% can be achieved without incurring extra energy costs; (2) the NH_3-N recovery efficiency depends mainly on the chosen configuration, being higher those achieved by single submerged GPM systems and by double membrane systems (between 79 and 91%); (3) the higher capture efficiency of the more complex systems (with double membranes and/or with slurry agitation and aeration) does not compensate the associated investment and/or operating costs, respectively. Thus, with a view to the application of GPM technology at a farm scale, the simplest systems, based on single submerged or suspended GPMs installed in the same chamber where the slurry is located, would be the most advisable, presenting lower or comparable costs per kg of N recovered than those of biotrickling filters and acid scrubbers, and an energy consumption per kg of N recovered which is significantly lower than that of stripping systems.

Supplementary Materials: The following supporting information can be downloaded at: https://www.mdpi.com/article/10.3390/membranes12111104/s1, Figure S1. Scheme of the NH_3 capture process in the GPM S1 system, consisting of a submerged membrane without agitation or aeration of pig slurry; Figure S2. Scheme of the NH_3 capture process in the GPM S2 system, consisting of a suspended membrane without agitation or aeration of the pig slurry; Figure S3. Scheme of the NH_3 capture process in the GPM S3 system, consisting of a membrane suspended in a compartment attached to the slurry treatment chamber, from where the NH_3-laden air is sucked in; Figure S4. Scheme of the NH_3 capture process in the GPM S4 system, consisting of a submerged and a suspended membrane without agitation and aeration of pig slurry; Figure S5. Scheme of the NH_3 capture process in the GPM S5 system, consisting of a suspended membrane in a compartment attached to the slurry treatment chamber, from which the NH_3-laden air is sucked in, and a submerged membrane without agitation and aeration of pig slurry; Figure S6. Scheme of the NH_3 capture process in the GPM S6 system, consisting of a submerged membrane with agitation and aeration of pig slurry; Figure S7. Scheme of the NH_3 capture process in the GPM S7 system, consisting of a suspended membrane with agitation and aeration of pig slurry; Figure S8. Scheme of the NH_3 capture process in the GPM S8 system, consisting of a membrane suspended in a compartment attached to the slurry treatment chamber with agitation and aeration of the slurry, from which NH_3-laden air is sucked in; Figure S9. Scheme of the NH_3 capture process in the GPM S9 system, consisting of a submerged and a suspended membrane with agitation and aeration of pig slurry; Figure S10. Scheme of the NH_3 capture process in the GPM S10 system, consisting of a suspended membrane in a compartment attached to the slurry treatment chamber, from where the NH_3-laden air is sucked in, and a submerged membrane with agitation and aeration of pig slurry; Figure S11. Evolution of NH_3-N concentration in the acid capture solution in systems S1–S5 without slurry aeration and agitation, and in systems S6–S10 with slurry agitation and aeration.

Author Contributions: Conceptualization, M.S.-B. and P.M.-R.; data curation, M.S.-H. and M.C.G.-G.; formal analysis, M.S.-H., M.C.G.-G. and P.M.-R.; funding acquisition, M.S.-B.; investigation, M.S.-H., M.S.-B. and P.M.-R.; methodology, M.S.-H.; resources, M.S.-B.; supervision, M.S.-B.; validation, M.S.-B., M.C.G.-G. and P.M.-R.; visualization, M.S.-H.; writing-original draft, M.S.-H., M.S.-B., M.C.G.-G. and P.M.-R.; writing-review and editing, M.S.-H., M.C.G.-G. and P.M.-R. All authors have read and agreed to the published version of the manuscript.

Funding: This work was funded by the European Union in the framework of the LIFE Project Green Ammonia (LIFE20 ENV/ES/000858, 'Market technology based on membranes for the reduction of ammonia in livestock farms').

Institutional Review Board Statement: Not applicable.

Informed Consent Statement: Not applicable.

Data Availability Statement: The data presented in this study are available on request from the corresponding author. The data are not publicly available due to their relevance as part of an ongoing Ph.D. thesis.

Conflicts of Interest: The authors declare no conflict of interest.

References

1. Mukhtar, S.; Mutlu, A.; Capareda, S.C.; Parnell, C.B. Seasonal and spatial variations of ammonia emissions from an open-lot dairy operation. *J. Air Waste Manag. Assoc.* **2008**, *58*, 369–376. [CrossRef] [PubMed]
2. Ni, J.Q.; Cortus, E.L.; Heber, A.J. Improving ammonia emission modeling and inventories by data mining and intelligent interpretation of the national air emission monitoring study database. *Atmosphere* **2011**, *2*, 110–128. [CrossRef]
3. Rothrock, M.J.; Szögi, A.A.; Vanotti, M.B. Recovery of ammonia from poultry litter using gas-permeable membranes. *Trans. ASABE* **2010**, *53*, 1267–1275. [CrossRef]
4. Hristov, A.N.; Hanigan, M.; Cole, A.; Todd, R.; McAllister, T.A.; Ndegwa, P.M.; Rotz, A. Review: Ammonia emissions from dairy farms and beef feedlots. *Can. J. Anim. Sci.* **2011**, *91*, 1–35. [CrossRef]
5. Melse, R.W.; Ogink, N.W.M. Air scrubbing techniques for ammonia and odor reduction at livestock operations: Review of on-farm research in the Netherlands. *Trans. Am. Soc. Agric. Eng.* **2005**, *48*, 2303–2313. [CrossRef]
6. Pagans, E.L.; Font, X.; Sánchez, A. Biofiltration for ammonia removal from composting exhaust gases. *Chem. Eng. J.* **2005**, *113*, 105–110. [CrossRef]
7. Rothrock, M.J.; Cook, K.L.; Warren, J.G.; Sistani, K. The effect of alum addition on microbial communities in poultry litter. *Poult. Sci.* **2008**, *87*, 1493–1503. [CrossRef]
8. Mukhtar, S.; Samani Majd, A.M.; Borhan, M.S.; Beseda, J.F. An investigation of ammonia extraction from liquid manure using a gas-permeable membrane. In Proceedings of the 2011 ASABE Annual International Meeting, Louisville, KY, USA, 7–10 August 2011; paper number 1111178. American Society of Agricultural and Biological Engineers: St. Joseph, MI, USA, 2011. [CrossRef]
9. Samani Majd, A.M.; Mukhtar, S. Ammonia recovery enhancement using a tubular gas-permeable membrane system in laboratory and field-scale studies. *Trans. ASABE* **2013**, *56*, 1951–1958. [CrossRef]
10. Sürmeli, R.Ö.; Bayrakdar, A.; Çalli, B. Ammonia recovery from chicken manure digestate using polydimethylsiloxane membrane contactor. *J. Clean. Prod.* **2018**, *191*, 99–104. [CrossRef]
11. Vanotti, M.B.; Szogi, A.A. Use of gas-permeable membranes for the removal and recovery of ammonia from high strength livestock wastewater. *Proc. Water Environ. Fed.* **2012**, *2011*, 659–667. [CrossRef]
12. Dube, P.J.; Vanotti, M.B.; Szogi, A.A.; García-González, M.C. Enhancing recovery of ammonia from swine manure anaerobic digester effluent using gas-permeable membrane technology. *Waste Manag.* **2016**, *49*, 372–377. [CrossRef] [PubMed]
13. Rothrock, M.J.; Szögi, A.A.; Vanotti, M.B. Recovery of ammonia from poultry litter using flat gas permeable membranes. *Waste Manag.* **2013**, *33*, 1531–1538. [CrossRef]
14. Soto-Herranz, M.; Sánchez-Báscones, M.; Antolín-Rodríguez, J.M.; Martín-Ramos, P. Pilot plant for the capture of ammonia from the atmosphere of pig and poultry farms using gas-permeable membrane technology. *Membranes* **2021**, *11*, 859. [CrossRef] [PubMed]
15. Ahn, Y.T.; Hwang, Y.H.; Shin, H.S. Application of PTFE membrane for ammonia removal in a membrane contactor. *Water Sci. Technol.* **2011**, *63*, 2944–2948. [CrossRef] [PubMed]
16. Garcia-González, M.C.; Vanotti, M.B. Recovery of ammonia from swine manure using gas-permeable membranes: Effect of waste strength and pH. *Waste Manag.* **2015**, *38*, 455–461. [CrossRef]

17. Fillingham, M.; VanderZaag, A.C.; Singh, J.; Burtt, S.; Crolla, A.; Kinsley, C.; MacDonald, J.D. Characterizing the performance of gas-permeable membranes as an ammonia recovery strategy from anaerobically digested dairy manure. *Membranes* **2017**, *7*, 59. [CrossRef]
18. Nagy, J.; Kaljunen, J.; Toth, A.J. Nitrogen recovery from wastewater and human urine with hydrophobic gas separation membrane: Experiments and modelling. *Chem. Pap.* **2019**, *73*, 1903–1915. [CrossRef]
19. Soto-Herranz, M.; Sánchez-Báscones, M.; Antolín-Rodríguez, J.M.; Vanotti, M.B.; Martín-Ramos, P. Effect of acid flow rate, membrane surface area, and capture solution on the effectiveness of suspended GPM systems to recover ammonia. *Membranes* **2021**, *11*, 538. [CrossRef]
20. Soto-Herranz, M.; Sánchez-Báscones, M.; Antolín-Rodríguez, J.M.; Martín-Ramos, P. Evaluation of different capture solutions for ammonia recovery in suspended gas permeable membrane systems. *Membranes* **2022**, *12*, 572. [CrossRef]
21. González-García, I.; Riaño, B.; Molinuevo-Salces, B.; Vanotti, M.B.; García-González, M.C. Improved anaerobic digestion of swine manure by simultaneous ammonia recovery using gas-permeable membranes. *Water Res.* **2021**, *190*, 116789. [CrossRef]
22. García-González, M.C.; Vanotti, M.B.; Szogi, A.A. Recovery of ammonia from swine manure using gas-permeable membranes: Effect of aeration. *J. Environ. Manag.* **2015**, *152*, 19–26. [CrossRef] [PubMed]
23. Riaño, B.; Molinuevo-Salces, B.; Vanotti, M.B.; García-González, M.C. Application of gas-permeable membranes for-semi-continuous ammonia recovery from swine manure. *Environments* **2019**, *6*, 32. [CrossRef]
24. Ebnesajjad, S. *Expanded PTFE Applications Handbook: Technology, Manufacturing and Applications*; Andrew, W., Ed.; Elsevier Science: New York, NY, USA, 2016; ISBN 1437778569/9781437778564.
25. Rice, E.W.; Baird, R.B.; Eaton, A.D. *Standard Methods for the Examination of Water and Wastewater*, 23rd ed.; American Public Health Association: Washington DC, USA; American Water Works Association: Denver, CO, USA; Water Environment Federation: Washington DC, USA, 2017.
26. Hansen, K.H.; Angelidaki, I.; Ahring, B.K. Anaerobic digestion of swine manure: Inhibition by ammonia. *Water Res.* **1998**, *32*, 5–12. [CrossRef]
27. R Core Team, R. *A Language and Environment for Statistical Computing*; R Foundation for Statistical Computing: Vienna, Austria, 2020.
28. Daguerre-Martini, S.; Vanotti, M.B.; Rodriguez-Pastor, M.; Rosal, A.; Moral, R. Nitrogen recovery from wastewater using gas-permeable membranes: Impact of inorganic carbon content and natural organic matter. *Water Res.* **2018**, *137*, 201–210. [CrossRef]
29. Oliveira Filho, J.D.S.; Daguerre-Martini, S.; Vanotti, M.B.; Saez-Tovar, J.; Rosal, A.; Perez-Murcia, M.D.; Bustamante, M.A.; Moral, R. Recovery of ammonia in raw and co-digested swine manure using gas-permeable membrane technology. *Front. Sustain. Food Syst.* **2018**, *2*, 30. [CrossRef]
30. García-González, M.C.; Vanotti, M.B. Nitrogen recovery from liquid manure using gas-permeable membranes: Effect of wastewater strength and pH control. In Proceedings of the 2014 ASABE and CSBE/SCGAB Annual International Meeting, Montreal, QC, Canada, 13–16 July 2014; American Society of Agricultural and Biological Engineers: Montreal, QC, Canada, 2014; Volume 7004.
31. Vanotti, M.B.; Dube, P.J.; Szogi, A.A.; García-González, M.C. Recovery of ammonia and phosphate minerals from swine wastewater using gas-permeable membranes. *Water Res.* **2017**, *112*, 137–146. [CrossRef]
32. Serra-Toro, A.; Astals, S.; Madurga, S.; Mata-Álvarez, J.; Mas, F.; Dosta, J. Ammoniacal nitrogen recovery from pig slurry using a novel hydrophobic/hydrophilic selective membrane. *J. Environ. Chem. Eng.* **2022**, *10*, 108434. [CrossRef]
33. Zarebska, A.; Nieto, D.R.; Christensen, K.V.; Norddahl, B. Ammonia recovery from agricultural wastes by membrane distillation: Fouling characterization and mechanism. *Water Res.* **2014**, *56*, 1–10. [CrossRef]
34. González-García, I.; Oliveira, V.; Molinuevo-Salces, B.; García-González, M.C.; Dias-Ferreira, C.; Riaño, B. Two-phase nutrient recovery from livestock wastewaters combining novel membrane technologies. *Biomass Convers. Biorefinery* **2022**, *12*, 4563–4574. [CrossRef]
35. Soto-Herranz, M.; Sánchez-Báscones, M.; Antolín-Rodríguez, J.M.; Martín-Ramos, P. Reduction of ammonia emissions from laying hen manure in a closed composting process using gas-permeable membrane technology. *Agronomy* **2021**, *11*, 2384. [CrossRef]
36. Molinuevo-Salces, B.; Riaño, B.; Vanotti, M.B.; Hernández-González, D.; García-González, M.C. Pilot-scale demonstration of membrane-based nitrogen recovery from swine manure. *Membranes* **2020**, *10*, 270. [CrossRef] [PubMed]
37. United Nations Economic and Social Council. *ECE/EB.AIR/120: Guidance Document on Preventing and Abating Ammonia Emissions from Agricultural Sources*; United Nations Economic and Social Council, Economic Commission for Europe, Executive Body for the Convention on Long-Range Transboundary Air Pollution: Geneva, Switzerland, 2014. Available online: https://unece.org/fileadmin/DAM/env/documents/2012/EB/ECE_EB.AIR_120_ENG.pdf (accessed on 3 November 2022).
38. Zarebska, A.; Romero Nieto, D.; Christensen, K.V.; Fjerbæk Søtoft, L.; Norddahl, B. Ammonium fertilizers production from manure: A critical review. *Crit. Rev. Environ. Sci. Technol.* **2015**, *45*, 1469–1521. [CrossRef]
39. De Vrieze, J.; Colica, G.; Pintucci, C.; Sarli, J.; Pedizzi, C.; Willeghems, G.; Bral, A.; Varga, S.; Prat, D.; Peng, L.; et al. Resource recovery from pig manure via an integrated approach: A technical and economic assessment for full-scale applications. *Bioresour. Technol.* **2019**, *272*, 582–593. [CrossRef] [PubMed]

membranes

MDPI

Article

PEBA/PDMS Composite Multilayer Hollow Fiber Membranes for the Selective Separation of Butanol by Pervaporation

Carla Arregoitia-Sarabia, Daniel González-Revuelta, Marcos Fallanza, Alfredo Ortiz and Daniel Gorri *

Departamento de Ingenierías Química y Biomolecular, Universidad de Cantabria, Av. de los Castros s/n, 39005 Santander, Spain

* Correspondence: gorrie@unican.es

Abstract: The growing interest in the production of biofuels has motivated numerous studies on separation techniques that allow the separation/concentration of organics produced by fermentation, improving productivity and performance. In this work, the preparation and characterization of new butanol-selective membranes was reported. The prepared membranes had a hollow fiber configuration and consisted of two dense selective layers: a first layer of PEBA and a second (outer) layer of PDMS. The membranes were tested to evaluate their separation performance in the selective removal of organics from a synthetic ABE solution. Membranes with various thicknesses were prepared in order to evaluate the effect of the PDMS protective layer on permeant fluxes and membrane selectivity. The mass transport phenomena in the pervaporation process were characterized using a resistances-in-series model. The experimental results showed that PEBA as the material of the dense separating layer is the most favorable in terms of selectivity towards butanol with respect to the other components of the feed stream. The addition of a protective layer of PDMS allows the sealing of possible pinholes; however, its thickness should be kept as small as possible since permeation fluxes decrease with increasing thickness of PDMS and this material also has greater selectivity towards acetone compared to other feed components.

Keywords: hydrophobic pervaporation; dip-coating; selective membranes; concentration polarization; biobutanol recovery

1. Introduction

In recent years, a series of initiatives have been carried out to support research and development that allow the replacement of fossil fuels with biofuels produced from renewable resources [1–3]. Among the various biofuels that have attracted attention, biobutanol stands out as it has, compared to ethanol, a higher energy density, lower miscibility with water, and lower vapor pressure. In addition, important advances have taken place in the production of biobutanol from different feedstocks [4,5]. As commented by Iyyappan et al. [6], a process involving cost-effective substrate and efficient biobutanol recovery methods could help with the implementation of the biobutanol production industry. Sugarcane bagasse, algal biomass, crude glycerol, and lignocellulosic biomass are considered potential cost-effective substrates for the production of butanol, which could replace glucose-based substrates.

Although, to a large extent, n-butanol is currently produced by chemical synthesis, there is a growing interest in its production by a biochemical route, mainly through the process known as ABE fermentation, in which, in addition to n-butanol, microorganisms also produce acetone and ethanol [7]. The biological production of butanol is specific to several Clostridia species. Among these, *Clostridium acetobutylicum* is considered the main species for biobutanol production; although, in recent years, other options have also aroused interest, such as the use of *Escherichia coli* or genetically modified organisms [6, 8]. ABE fermentation by *C. acetobutylicum* takes place in two phases: an acidogenesis phase wherein the microbes mainly produce acetic acid and butyric acid, followed by a solventogenesis phase wherein the microbes mainly produce ABE compounds [9].

The industrial implementation of the fermentation process has faced several obstacles [10], including the inhibition of microorganisms by the butanol formed during fermentation. This phenomenon causes the ABE fermentation to stop when the concentration of the solvents produced is around 2 wt%, causing low utilization of the substrate and the costly recovery of butanol from diluted solutions. The obstacle due to the low solvent tolerance could be solved with the continuous butanol removal from the fermentation broth. Various separation techniques have been proposed and studied to remove products and increase the efficiency of the fermentation process, including gas-stripping, liquid–liquid extraction, and membrane technologies such as pervaporation and membrane distillation [11–15]. Some technical-economic studies indicate that the in situ product recovery (ISPR) from the ABE fermentation broth not only increases the productivity and the yield of ABE by eliminating the product inhibition, but it also reduces the energy consumption and the separation cost [16].

Among the separation techniques, pervaporation has attracted sustained interest from the scientific community in recent times. Pervaporation is a technique that allows the separation of liquid mixtures by permeating their components at different rates through a dense selective membrane, applying a certain vacuum on the downstream side of the membrane to establish the driving force for mass transfer. A series of advantages are attributed to pervaporation, such as high selectivity, low operating temperature, reasonable performance to cost ratio, possibility of modular design, and the absence of a separating agent that could cause product contamination [17,18]. Particularly, the low/moderate operating temperatures make pervaporation especially useful to work coupled with a bioreactor without harming the activity of microorganisms [7]. In addition, potential drawbacks associated with the use of membranes should be taken into account, such as membrane fouling, high equipment cost, and low/moderate productivity conditioned by permeation fluxes through the membrane [15,16].

For the case that interests us, which involves the selective removal of organic compounds from an aqueous solution, hydrophobic pervaporation membranes are used. Various polymeric materials for membranes have been tested for the separation of alcohols in the studies reported in the literature, among which poly(dimethylsiloxane) (PDMS) is the most used and, to a lesser extent, there are poly(ether-block-amide) (PEBA), poly(octylmethylsiloxane) (POMS), poly[1-(trimethylsilyl)-1-propyne] (PTMSP), polyurethane, and poly(vinylidene fluoride) (PVDF) [19–21]. In addition, in recent years, there has been an increasing interest in composite membranes formed with a polymer matrix with the addition of fillers such as zeolites [22,23], ZIF [24], MOF [25], and graphene oxide [26]. From all these polymer materials, PEBA has been especially selective towards butanol [27] due to its affinity for this compound that allows an appreciable solubility of butanol in the polymeric matrix, while the PEBA's hydrophobic character limits the transport of water through the membrane.

Although most of the studies on pervaporation reported in the literature are carried out with flat sheet membranes, there is a growing interest in the development of membranes with a hollow fiber configuration that allow the construction of compact membrane modules with high membrane surface areas [28]. Fibers can be made as isotropic membranes with a uniformly dense structure [29], but they are preferably formed as a microporous structure with a dense selective layer on the outside or inside surface (anisotropic membranes) [30]. The dense surface layer can be either integral with the fiber or a separate layer coated on the porous support fiber. Given that our purpose is to obtain hollow fibers with a thin dense layer of PEBA, it was found in previous studies [31] that a suitable way of doing this is by depositing a thin layer of PEBA by dip-coating on a porous support of a suitable material, such as polypropylene. However, the aim of achieving PEBA dense layer thicknesses equal to or lower than 1 μm makes the presence of pinholes on the membrane surface more likely, causing selectivity losses. One way to repair the presence of pinholes is to cover the membrane with a second thin coating layer of a relatively permeable material such as silicone rubber to seal defects. Thus, a defect-free membrane would be obtained where the silicone rubber protective layer improves selectivity at the cost of a certain decrease in permeability.

Several examples of studies on multilayer membranes, including a protective layer of PDMS, are mentioned in the review papers by Dai et al. [32] and by Kujawski and col. [33]. Yahaya [34] reported a study on the separation of phenol from aqueous streams by PV using PDMS/PEBA two-layer hollow fiber membranes. From the pervaporation experiments, it was found that a significant improvement in the phenol/water separation factor and phenol flux was achieved with two-layer (PDMS/PEBA) membranes compared to that achieved using only the PDMS membrane. These results indicated that two-layer membranes combine the unique features of PDMS (exhibiting high permeability) and PEBA (exhibiting high permselectivity) to achieve this improvement in membrane performance. Jiang and Song [35] prepared polysulfone (outer layer)/Matrimid (inner) dual-layer hollow fiber PV membranes, applying them for tert-butanol dehydration. This work is a useful reference for mass transfer modeling since it uses a resistances-in-series model adapted to the cylindrical configuration of hollow fibers.

In the present work, a study about the preparation of multilayer membranes with hollow fiber configuration is reported. The fibers have two separating dense layers: a first layer of PEBA and a second (outer) layer of PDMS. The membranes were characterized in terms of their morphology and were subsequently tested to evaluate their separation performance in the selective removal of organics from a synthetic ABE solution. Membranes with various thicknesses of PDMS were prepared in order to evaluate the effect of the PDMS protective layer on permeant fluxes and membrane selectivity.

2. Materials and Methods

2.1. Materials

For the preparation of the composite HF membranes, Celgard X-20 polypropylene (PP) commercial HF membranes (supplied by Celanese) were used as support. These fibers had an internal diameter of 400 μm and a wall thickness of 30 μm, the porosity was equal to 40%, and the pores in the membrane were approximately 0.115 μm in diameter.

For the formation of the dense layers, the coating solutions were prepared with the following polymers: (a) Pebax 2533 (Hexanedioic acid, polymer with azacyclotridecan-2-one and .alpha.-hydro-.omega.-hydroxypoly(oxy-1,4-butanediyl)) (CAS No. 77402-38-1), which was kindly supplied by Arkema, France; (b) PDMS (Sylgard 184 silicone elastomer kit) was supplied by Dow Corning Company. n-Hexane (analytical grade, Honeywell) was used as the solvent for PDMS.

The aqueous feed solution for the PV experiments was prepared by mixing n-butanol (Merck), ethanol (Merck), and acetone (Riedel-de Haën) with ultrapure water Milli Q obtained from a Merck-Millipore system (supplied by Merck KGaA, Darmstadt, Germany). The acetone/butanol/ethanol content in the feed solution was 1:2:1 wt%. All of the materials were of analytical grade and were used without further purification.

2.2. Fabrication of Multilayer Hollow Fibers

As previously indicated, the purpose of this work was the preparation of multilayer membranes with hollow fiber configuration, incorporating two separating dense layers: a first layer of PEBA and a second (outer) layer of PDMS. Since both PEBA and PDMS polymers are rubbery materials, the deposition of two thin dense layers can be adequately carried out by dip-coating. The fabrication of ultrathin skin layer hollow fiber membranes implies a certain complexity in terms of the operating variables to be taken into account, as pointed out by Chung and col. [36]. Here, the dip-coating procedure consisted of immersing the commercial fibers in the coating solution for few seconds (3–5 s) to allow a thin film formation on the outer phase of the fibers. Covering of the ends of the fibers was previously performed to prevent the dip-coating solution from entering the fiber lumen.

The first layer was placed on the outside of the support by the dip-coating of a polymeric solution containing PEBA, following the procedure described in our previous work [31]. The coating solution used was prepared with a concentration of 2 wt% of PEBA in n-butanol, keeping it under stirring at 70 °C for 24 h. Then, it was left to rest for natural degasification and cooling for 24 h at room temperature. The viscosity was measured at room temperature using a rotational viscometer (Model Alpha Series L, Fungilab S.A., Spain), obtaining a value of 7.5 cP. The procedure that we followed allowed us to obtain a dense layer thickness of about 1.6 μm, as measured by SEM analysis. After depositing the PEBA layer, the fibers were left to stand for 1–2 days at room temperature for drying.

The second dense layer of the HF membranes was obtained by dip-coating with a PDMS coating solution. In all cases, the same PP support was used, and the dense layer thickness of PEBA was 1.6 μm. The PDMS polymer solution prepared from the two-component kit was slowly diluted in n-hexane to prepare various coating solutions, depending on the thickness of the dense layer to be obtained. The viscosity of the mixture was measured at room temperature using a rotational viscometer. Finally, the dip-coating of the hollow fibers was conducted by immersing a fiber for a few seconds (3–5 s) in a specific coating solution according to the intended dense layer thickness, making sure the whole fiber was covered. The fibers were placed such that they did not have contact in between and were left for curing at room temperature for 1–2 days. Thermal accelerated aging treatment was carried out by securely fastening the fibers in an oven and heating it up to 100 °C for 1 h for the complete crosslinking of the polymer. After, they were left for cooling at room temperature for one day. A uniform film of PDMS was formed on top of the fiber.

The fibers were potted into modules for pervaporation tests, for which epoxy adhesive (DP 105, 3M Scotch-Weld) was used to seal the ends of the module. Each module consists of 15 fibers with a length of 15 cm, and the total membrane area was 28.3 cm^2 based on the internal diameter of the HF.

In addition, PDMS dense films were prepared to be used in other characterization tests such as contact angle measurements and to establish an indicative relationship between the viscosity of the PDMS polymer solution and the thickness of the film. A small glass plate was immersed in the same PDMS coating solution used for the second layer formation and followed by the same crosslinking process after which the film thickness was measured by using a digital micrometer (Mitutoyo, Germany).

2.3. Membrane Characterization

The thickness of the dense layers and cross-sectional morphologies of the composite membranes were determined by means of scanning electron microscopy (SEM, model Zeiss EVO MA15). The samples were prepared by immersing and fracturing the membranes in liquid nitrogen, followed by gold thin film deposition using a sputter coater.

The static contact angle for a PDMS film was measured by the sessile liquid drop method using a contact angle measurement system (DSA25, Krüss, Germany) in order to obtain information about hydrophobicity/hydrophilicity and the wetting behavior of the membranes prepared. For the goniometric measurements, a flat sheet membrane was prepared as described in the previous section. A 2.0 μL drop of different pure solvents (water, acetone, butanol, and ethanol) and ABE solution were deposited on the material membrane's surface at five different sites. Each value was obtained using the software provided through image recognition. The average value for the contact angle was then considered.

Additionally, the results of TGA and FTIR analyses for the membranes used in this work were included as Supplementary Materials. The thermal stability of the components used in this work (Pebax 2533, PDMS, and PP) was examined by thermogravimetric analysis (TGA), both individually and together on the hollow fiber membrane. The TGA experiments were performed using a DTG-60H Shimadzu thermobalance. Membrane samples with an initial mass between 3 and 15 mg were placed on an alumina cell. The samples were heated up to 700 °C at a heating rate of 5 °C min^{-1} in nitrogen (25 mL min^{-1}). In the specific case of PDMS, the final temperature was 1000 °C because this material has a higher resistance to degradation in a nitrogen atmosphere. On the other hand, an ATR-FTIR analysis of the polymers used for the fabrication of the selective layer of the membranes developed in this work, Pebax 2533 and PDMS (as dense homogeneous films), were carried out using a Perkin Elmer spectrum 65 Fourier Transform Infrared Spectrometer in the region of 400–3900 cm^{-1}. In addition, this analysis was carried out on one of the hollow fiber membranes composed of a polypropylene support, a dense Pebax layer, and a second (outer) dense PDMS layer.

2.4. PV Experiments

The pervaporation experiments were carried out in a laboratory-scale unit supplied by Sulzer Chemtech (Germany) that was previously used by the authors in other pervaporation studies [37,38]. The separation process takes place when the feed mixture goes through the shell side, and the permeate comes out of the lumen side when the vacuum is applied (Figure 1). Unless otherwise stated, pervaporation tests were performed at 40 °C, since this temperature is within the appropriate temperature range to carry out the ABE fermentation process as indicated in the literature [39]. The permeate side was kept below 10 mbar using a diaphragm vacuum pump (Vacuubrand PC 3004 VARIO), and samples of the feed and permeate were collected every 30 min and were weighed and analyzed by gas chromatography (GC). After condensation, the permeate undergoes a separation of phases into an n-butanol-rich phase and a water-rich phase. After measuring the mass, the permeate was diluted with water to form a single-phase solution and then ABE concentrations were measured by GC. Samples were analyzed by duplicate in a headspace gas chromatograph (GCMS-QP2010, Ultra Shimadzu) equipped with a flame ionization detector (FID). Compounds were separated into a DB-Wax 30 m × 0.25 mm × 0.25 μm column with a detector temperature of 270 °C. Helium was used as a carrier gas at a flow rate of 82 mL min^{-1}. The oven temperature was initially set at 80 °C and was subsequently increased to 150 °C at 10 °C min^{-1}. GC calibration was performed with external standards. Each PV experiment lasted for at least 4 h after the stabilization process (stable operating temperature) to be sure that a pseudo steady state was reached.

The PV performance of a membrane was evaluated in terms of permeate flux J, membrane selectivity α, separation factor β, and pervaporation separation index (PSI). The total flux J (kg m^{-2} h^{-1}) across the membrane is obtained by relating the mass of permeate collected with the time interval and the membrane area, as follows:

$$J = \frac{m}{A_m \, \Delta t} \tag{1}$$

After that, the flux for each component J_i is calculated from the total flux and the permeate composition obtained by chromatographic analysis. Membrane selectivity α, separation factor β, and the pervaporation separation index (PSI) are calculated using Equations (2)–(4), respectively:

$$\alpha_{i,j} = \frac{P_i}{P_j} \tag{2}$$

$$\beta_{i,j} = \frac{y_i/y_j}{x_i/x_j} \tag{3}$$

$$PSI = J\left(\beta_{i,j} - 1\right) \tag{4}$$

In this work, Equation (5) was adopted to describe the flux of a permeant species, where the driving force for mass transfer is a function of permeant activity and the overall resistance ($R_{i,OV}$) includes contributions of resistances from the liquid boundary layer and the membrane itself, as shown in the next section.

$$J_i = \frac{1}{R_{i,\,OV}}\left(a_{i,\,F} - a_{i,\,P}\right) \tag{5}$$

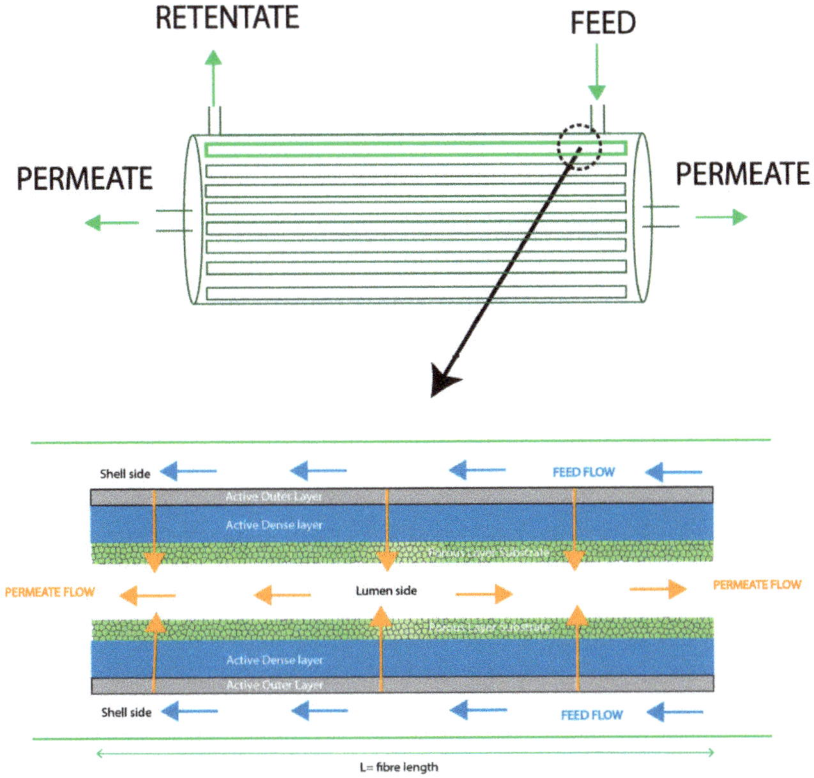

Figure 1. Pervaporation module.

3. Results and Discussion

3.1. Membrane Characterization

In order to establish a frame of reference for the subsequent preparation of hollow fibers with PDMS dense protective layers with different thicknesses, we started by preparing a series of homogeneous PDMS films. As previously detailed in Section 2.2, polymer solutions with different PDMS contents were used, whose viscosity were measured using a rotational viscometer. Subsequently, the thickness of the PDMS films obtained was measured using a digital micrometer. Table 1 shows the results obtained, where it can be seen how the thickness of the films ranged from 2 to 80 μm as the concentration of the PDMS polymer solution and its viscosity increased.

Table 1. Thickness of films obtained from polymeric solutions with different PDMS contents.

PDMS (wt%) [(a)]	Viscosity (cP)	Film Thickness (μm)
5	1 ± 0.06	2 ± 1
10	1.3 ± 0.06	3 ± 1
15	2 ± 0.06	4 ± 1
20	2.5 ± 0.06	8 ± 1
40	11 ± 0.2	20 ± 1
60	200 ± 3	50 ± 1
80	560 ± 7.5	70 ± 1.5
100	4900 ± 60	80 ± 1.5

[(a)] It corresponds to the dilution of the two-component kit in n-hexane.

Figure 2 shows how the thicknesses of the dense PDMS films obtained by dip-coating depend on the viscosity of the polymer solution at room temperature. This trend is an indicative useful guide for selecting the characteristics of the coating solution when it is intended to obtain dense layers with a thickness less than 5 μm, especially due to the practical difficulty in handling polymer solutions with high viscosity and non-Newtonian behavior [40]. Thus, PDMS polymer solutions in the same range of viscosities shown in Figure 2 were later used as coating solutions to cover the hollow fibers with a thin outer dense layer.

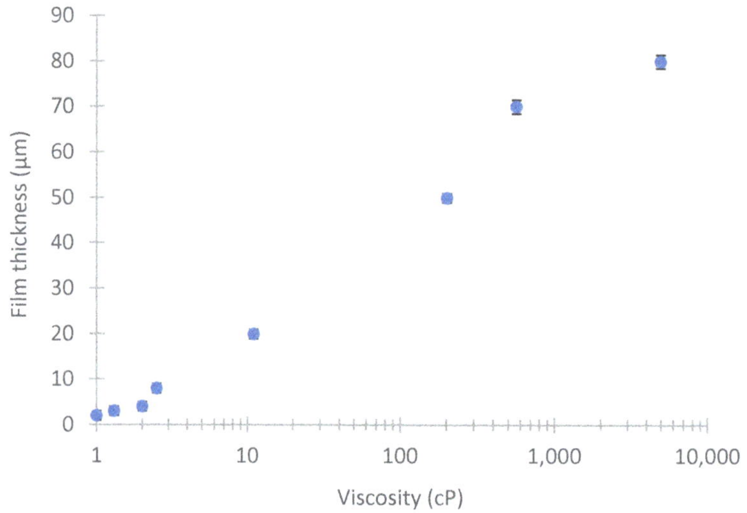

Figure 2. Influence of the viscosity of the PDMS polymer solution on the film thickness.

In order to obtain additional information about the characteristics of the polymeric materials used, we measured the contact angle between the PDMS film and different chemical compounds that make up the ABE solution. The measured contact angles for PDMS are shown in Table 2, where the measured values for Pebax 2533 films, which were already reported in a previous work [31], were also included as a reference. The water contact angle measured for PDMS was 101°, highlighting the hydrophobic character of the surface. Data reported in previous studies show some variability, which has been attributed to surface roughness or experimental difficulties, with data ranging from 95° to 120° [41]. As Knozowska et al. [42] mentioned, the contact angle for a given material depends on the degree of crosslinking, and this also corresponds to an increase in roughness. The contact angles for organics were always <90°, while, for ABE mixture, the measured value was very close to that of water. Also included in Table 2 are the surface tension values for pure compounds (from Dortmund Data Bank), showing that the contact angle is influenced by the surface tension but does not depend solely on it but on the affinity between the components of the solution and the surface of the polymeric material. Several authors have reported the relationship between the contact angle and the surface free energy (SFE). In general, wetting and lower contact angles occur when the surface and the liquid have similar surface energies (surface tensions, in the case of the liquids). In the case of water, a lower contact angle value was observed for Pebax 2533-based membranes compared to PDMS membranes, the SFE values being in the opposite order, according to data calculated by Knozowska et al. [42].

Table 2. Contact angle measured for selective polymers and ABE components.

Liquid	Measured Contact Angle [°] PDMS		Pebax 2533 [31]		Surface Tension ($mN\ m^{-1}$), 293.15 K
	Value	SD	Value	SD	
Butanol	35.7	3.8	38.4	6.2	24.6
Ethanol	42.3	4.4	55.0	8.8	22.3
Acetone	38.6	9.6	32.4	13.2	23.3
Water	101.0	3.3	76.8	8.2	72.8
ABE mixture	100.1	11.7	70.2	7.3	

SD: standard deviation.

3.2. Membrane Performance in Pervaporation of ABE Solutions

This section reports the results corresponding to pervaporation tests carried out with membrane modules built with different sets of hollow fibers. It was worked with various thicknesses of the dense layers in order to be able to determine the contribution of each material to the membrane performance.

PV experiments were performed with each membrane module flowing (1:2:1 wt%) ABE solutions at 40 °C. For each experiment, the partial permeation fluxes were related to the driving force (activity gradient) to obtain the overall resistance to mass transfer (see Equation (5)). The activity coefficients for the components in the liquid feed mix were evaluated by the NRTL method using the Aspen Plus software.

It is well known that, in separation operations with selective membranes, the resistance to mass transfer in the fluid phase of the feed adjacent to the membrane can have a notable influence on the separation performance. Usually, hydrophobic membranes are significantly more permeable to dissolved organic compounds than to water, causing a depletion of the former compounds in the liquid boundary layer. This phenomenon is known as concentration polarization. Such effects depend mainly on the hydrodynamic conditions in the liquid phase and are usually investigated by changing the feed flow rate in pervaporation experiments. Therefore, it is essential to be able to quantify the incidence of the concentration polarization phenomenon in our tests in order to later be able to analyze the intrinsic resistance of the membrane. In a previous study [31] working with PEBA thin-film composite hollow fiber membranes, it was reported that the total resistance was fitted (Wilson plot) by the reciprocal of the lineal velocity u ($m\ min^{-1}$) through the membrane module raised to an exponent of 0.9, a factor frequently adopted for parallel flow in membrane contactors [43,44]. In this work, further analysis of those permeation data for organic compounds was made in order to obtain a correlation that describes the transport parameters in terms of characteristic dimensionless numbers. Given the cylindrical configuration of the hollow fibers and taking into account that, in this study, it was adopted referring the permeation fluxes in all cases to the internal area of the support (PP), the contribution of the individual mass transfer resistances to the overall resistance for the case of fibers with a single selective dense layer of PEBA is given by the following equation:

$$R_{i,\ ov} = \frac{1}{k_{i,\ ov}} = \left(\frac{\gamma_{i,\ F}}{k_{i,\ bl}\ \rho_m}\right) \times \frac{A_{supp,\ in}}{A_{fiber,\ out}} + \frac{\delta_{PEBA}}{P_{i,\ PEBA}} \times \frac{A_{supp,\ in}}{A_{LM,\ PEBA}} + R_{i,\ supp} \tag{6}$$

where $k_{i,ov}$ is the mass transfer coefficient for component i; $k_{i,bl}$ is the mass transfer coefficient for component i at the liquid boundary layer; $\gamma_{i,F}$ is the activity coefficient for component i in the liquid phase; ρ_m is the molar density of feed liquid; $P_{i,PEBA}$ is the membrane permeability for component i through the *PEBA* layer; δ_{PEBA} is the thickness of the *PEBA* layer; $R_{i,supp}$ is the mass transfer resistance in membrane support; $A_{supp,in}$ is the membrane area based on the internal diameter of the porous support; $A_{fiber,out}$ is the membrane area based on the outer diameter; $A_{LM,PEBA}$ is the logarithmic mean area of the *PEBA* layer.

The mass transfer coefficient at the feed boundary layer ($k_{i,bl}$) depends on the circulation configuration of the ABE solution in the membrane module. In this system, the ABE solution circulates through the shell side in order to maximize mass transfer coefficients, increasing the transfer area and improving the hydraulic conditions [45]. Usually, the mass transfer coefficients in the liquid boundary layer are correlated through the Sherwood number (Sh) as a function of the dimensionless Reynolds (Re) and Schmidt (Sc) numbers. Numerous papers have dealt with reviewing and analyzing the correlations proposed for predicting shell-side mass-transfer. Among them, the papers by Lipnizki and Field [46], Shen et al. [47], and the recent work by Estay et al. [45] deserve to be highlighted. For the calculation of the Reynolds number, the definition of equivalent diameter (d_{eq}) for the shell-side flow proposed by Dahuron and Cussler [48] was adopted, as follows:

$$d_{eq} = \frac{4 \times (cross - sectional\ area\ of\ flow)}{wetted\ perimeter} = \frac{{d_{shell}}^2 - NF \times {d_{out}}^2}{d_{shell} + NF \times d_{out}} \tag{7}$$

Thus, the Reynolds number is calculated as follows:

$$S_{shell} = \frac{\pi}{4}{d_{shell}}^2 - NF \times \frac{\pi}{4}{d_{out}}^2 \tag{8}$$

$$u_{shell} = F/S_{shell} \tag{9}$$

$$Re_{shell} = \frac{u_{shell} \times \rho_{mass} \times d_{eq}}{\mu} \tag{10}$$

The mass transfer coefficient in the liquid boundary layer (shell-side) is related to the Sherwood number as follows:

$$Sh = \frac{k_{i,\ bl} \times d_{eq}}{D_i} \tag{11}$$

where the equivalent diameter is used as the characteristic length. The values of the diffusion coefficients (D_i) for organic compounds in aqueous solution was estimated using the Wilke-Chang correlation.

Thus, an estimation of parameters was carried out to achieve a correlation that links the mass transfer coefficient at the liquid phase with the properties and operating conditions. For this, a set of experimental data was used that was obtained from PV tests with various modules built with hollow fibers coated with dense layers of PEBA with various thicknesses, working with ABE solutions at 40 °C and different flow rates (0.2, 0.3, 0.5, 1.2, 2.0, and 4.5 L min^{-1}). Reynolds numbers for the experiments were in the range of 340–7800, and the length of the modules was 15 cm in all cases. The estimation procedure established that the correlation that best describes the mass transfer in the liquid phase circulating through the shell-side in flow parallel to the fibers is the following:

$$Sh = 0.025\ (1 - \varphi)\ Re^{0.9}\ Sc^{0.33} \tag{12}$$

where the packing fraction (φ) was calculated as follows:

$$\varphi = \frac{NF \times {d_{out}}^2}{{d_{shell}}^2} \tag{13}$$

The parameter estimation procedure allowed, at the same time, to determine the permeability values of PEBA and resistance in the support for each one of the permeants, whose values are shown in Table 3. In the case of water as permeant, the mass transfer resistance in the liquid phase was assumed negligible. The overall mass transfer resistance values for butanol calculated with the model for three membrane modules were plotted against the experimental data to build the model parity graph (Figure 3), proving that the fit of the model can be taken as satisfactory, with good agreement between the experimental and simulated data. The small difference in the $R_{i,supp}$ and $P_{i,PEBA}$ values reported in Table 3 with respect to those reported in previous work [31] is due to the fact that, in this work, in Equation (6) the ratio of areas in each term was included so that the mass transfer resistances were referred in all cases to the internal area of the support. These data confirm that the order in which organic compounds permeate through PEBA in terms of permeabilities (kmol m^{-1} s^{-1}) is given by: $P_{BuOH} > P_{EtOH} > P_{Acet}$. The intrinsic selectivity of PEBA membranes towards n-butanol is explained by the preferential sorption of n-butanol over acetone and ethanol, as reported by experimental studies by Liu and Feng [49] and Heitmann et al. [50].

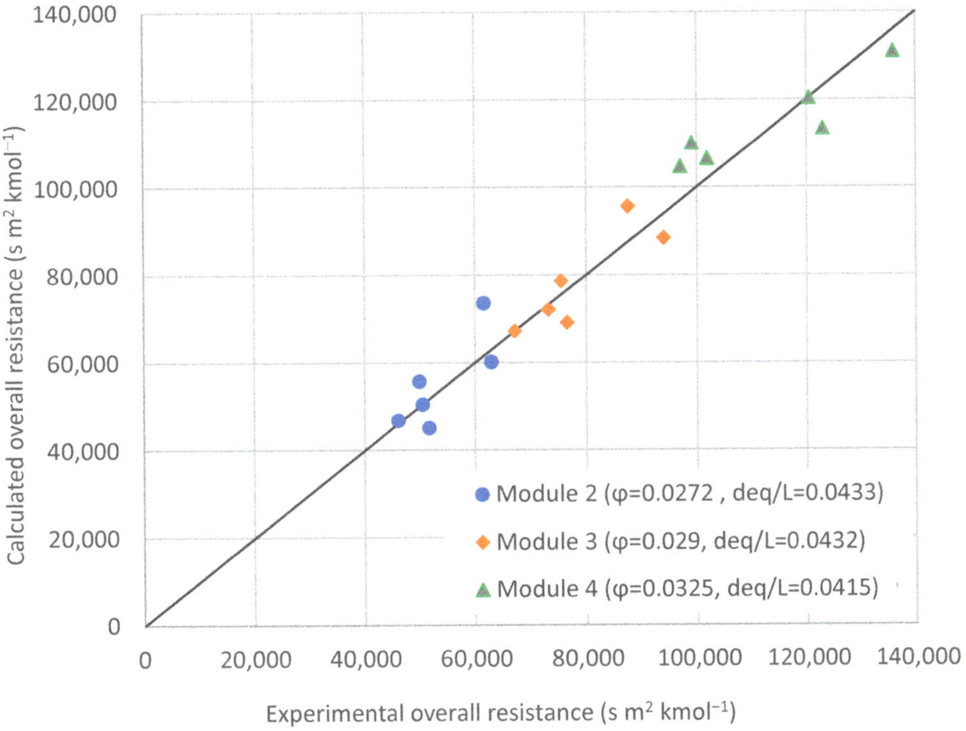

Figure 3. Parity graph of simulated overall resistances vs. experimental overall resistances for butanol permeation. Values were obtained and calculated for three modules at different flow rates.

Below are the results obtained in PV tests working with membrane modules made of hollow fibers consisting of two dense layers of selective material: a first dense layer of PEBA deposited on the PP support and a second dense (outer) layer of PDMS. Different thicknesses of PDMS for the outer layer were deposited by dip-coating in the range of 3–80 μm. Thus, the contribution of the individual mass transfer resistances to the overall resistance for the case of fibers with two selective dense layers is given by the following equation:

$$R_{i,\,ov} = \frac{1}{k_{i,\,ov}} = \left(\frac{\gamma_{i,\,F}}{k_{i,\,bl}\,\rho_m}\right) \times \frac{A_{supp,\,in}}{A_{fiber,\,out}} + \frac{\delta_{PDMS}}{P_{i,\,PDMS}} \times \frac{A_{supp,\,in}}{A_{LM,\,PDMS}} + \frac{\delta_{PEBA}}{P_{i,\,PEBA}} \times \frac{A_{supp,\,in}}{A_{LM,\,PEBA}} + R_{i,\,supp} \quad (14)$$

where $P_{i,PDMS}$ is the membrane permeability for component i through the PDMS layer; δ_{PDMS} is the thickness of the PDMS layer

Membrane modules were assembled with each type of fiber, which, in all cases, contained 15 fibers with a useful length of 15 cm. The PV experiments were performed at 40 °C, while the feed flowrate was 4.5 L min^{-1}, which was the maximum possible flow rate for the feed liquid on the shell side in order to minimize the mass transfer resistance in the liquid boundary layer. The data collected in these experiments made it possible to evaluate the PDMS permeability values for the different permeants by means of Equation (14), taking into account that the permeabilities of PEBA were previously determined. Analyzing the PDMS permeability values for the organic compounds reported in Table 3, these follow the order acetone > n-butanol > ethanol when the driving force for mass transfer is the difference in activities for the permeant species across the membrane. Although it is not easy to make a comparison with previous studies reported in the literature due to the influence of different materials and working conditions, the higher permeability through PDMS for acetone over n-butanol and ethanol is consistent with the results presented by Rozicka et al. [51] and van Wyk et al. [52]. In this sense, Rozicka et al. [51] reported the performance of three commercial PDMS-based membranes (Pervatech, Pervap 4060 and PolyAn) in the pervaporation removal of acetone, butanol, and ethanol from binary aqueous mixtures at 25 °C. Using the data reported in that study to evaluate permeabilities (recalculating them to consider the difference in activities for the permeant species across the membrane as the driving force for mass transfer), it turns out that the highest permeability value corresponds to acetone, as well as higher separation factors for acetone over n-butanol and ethanol. Results of the same order were also reported by van Wyk et al. [52] in a study on the separation of ABE model solutions with PDMS membranes (Pervatech) in the range of 30–50 °C. The behavior of PDMS membranes more favorable to acetone permeation compared to PEBA membranes can be largely attributed to the different solvent uptake of both membrane materials. As mentioned above, the experimental studies by Liu and Feng [49] and Heitmann et al. [50] showed that the solvent uptake in PEBA 2533 is considerably higher for n-butanol than for acetone or ethanol. However, swelling studies for PDMS membranes, carried out with both pure solvents [51,53] and aqueous solutions [50], have shown that the acetone uptake is at least similar to or even higher than that of butanol, while the ethanol uptake is lower than that of the other two organics. The higher affinity of PDMS for acetone compared to the other permeants was also reported by Yang et al. [54] in a study that used inverse gas chromatography for the characterization of the solubility thermodynamics and diffusion of solvent-PDMS systems.

Table 3. Mass transfer resistance of the support and permeability of the membrane selective layers.

Compound	Resistance in Support (m^2 s $kmol^{-1}$)		PEBA Permeability (kmol m^{-1} s^{-1})		PEBA Selectivity	PDMS Permeability (kmol m^{-1} s^{-1})		PDMS Selectivity
	Value	S.E.	Value	S.E.	$\alpha_{i/water}$	Value	S.E.	$\alpha_{i/water}$
n-butanol	35.2×10^3	2.6×10^3	3.0×10^{-10}	1.9×10^{-11}	0.64	1.1×10^{-10}	2.1×10^{-11}	1.67
ethanol	42.5×10^3	3.6×10^3	2.0×10^{-10}	1.1×10^{-11}	0.41	5.3×10^{-11}	1.8×10^{-11}	0.80
acetone	34.6×10^3	4.2×10^3	7.6×10^{-11}	3.3×10^{-12}	0.16	3.2×10^{-10}	3.8×10^{-11}	4.85
water	74.2×10^3	2.1×10^3	4.7×10^{-10}	3.8×10^{-11}		6.6×10^{-11}	2.0×10^{-11}	

S.E.: standard error.

In order to have a broader view of the variables that influence the separation performance of a PV module with dual-layer hollow fibers, a sensitivity analysis performed with simulation tools is presented below. Since the fiber length is usually around 1.0–1.5 m in membrane modules for industrial applications, we performed a simulation study to evaluate the influence of the PDMS dense layer thickness and the feed flow rate on the separation performance for a hypothetical membrane module 1 m in length. In all cases, fibers that included the porous support of PP and a selective dense layer of PEBA with a thickness of 1.6 μm were considered, on which a dense protective layer of PDMS was deposited, with variable thicknesses in the range of 0 to 20 μm. The equations that describe the mass transfer in the HF membranes (Equations (5)–(14)) together with the material and energy balances were included in a distributed parameter model that was implemented in Aspen Custom Modeler, making use of Aspen Plus subroutines for the estimation of thermodynamic properties (densities, vapor pressures, activity coefficients, enthalpies) and transport properties (viscosity, diffusion coefficients), and then a series of simulations were run. Setting the separation/concentration of n-butanol as a priority, the most favorable results in terms of PSI and butanol content (wt%) in the permeate stream correspond to the hollow fibers with the thinnest thickness of the PDMS dense layer and the highest feed flow rate (4.5 L min^{-1}), as shown in Figure 4. It is important to highlight the relevant effect that the fluid dynamic conditions in the liquid phase of the feed can have on the separation performance when selective membranes are used for the removal of organics from dilute solutions. Thus, considering the case of a dual-layer HF membrane with a PDMS layer thickness of 1 µm and by increasing the flow rate from 0.2 to 4.5 L min^{-1} (which corresponds to a Reynolds number interval of 340–7690), the contribution of the boundary layer resistance to the overall mass transfer resistance decreased from 39% to 3.7% for butanol, while the butanol content in the permeate increased from 21.2 to 28.7 wt%. This phenomenon is probably due to the transition from a laminar to a turbulent flow for the feed liquid on the shell-side.

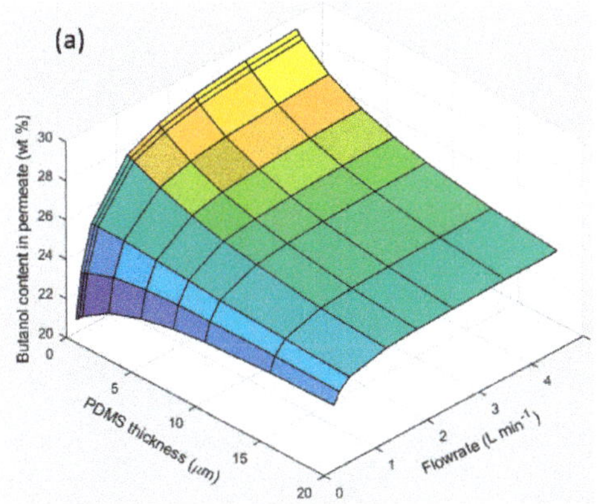

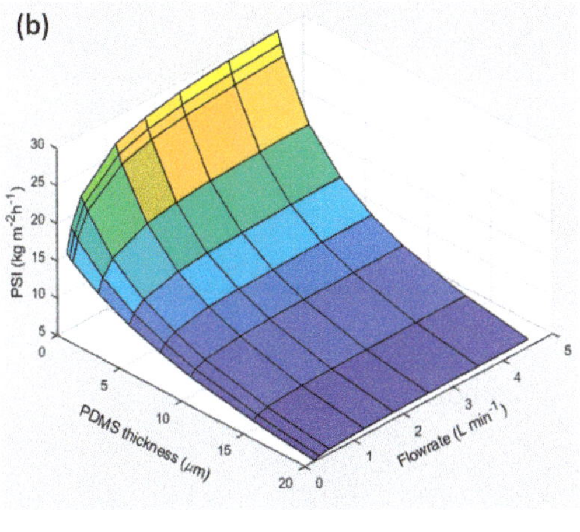

Figure 4. Simulation showing the influence of the feed flow rate and thickness of the PDMS dense layer on (**a**) butanol content in permeate (wt%) and (**b**) PSI, for a membrane module 1 m in length (operating conditions: 40 °C; downstream pressure: 10 mbar; feed mixture: (1:2:1 wt%) ABE solutions).

From the simulations, it was observed that, as the PDMS layer becomes thicker, it becomes more difficult to remove the butanol; thus, the boundary layer resistance becomes insignificant, regardless of its magnitude. We can see that, in thicker PDMS layers, there is less variation in the content of n-butanol in the permeate with a change in the feed flowrate due to the lower contribution of the boundary layer resistance to the overall resistance.

4. Conclusions

The experimental results allowed for finding a relationship between the viscosity of the PDMS coating solution and the thickness of the dense layer that can be achieved by the dip-coating procedure. The mass transport phenomena in the pervaporation process were characterized using a resistances-in-series model. Knowing that, under working conditions, fluid dynamic conditions can significantly influence the separation achieved, a correlation for the mass transfer coefficients in the liquid boundary layer as a function of the dimensionless Reynolds (Re) and Schmidt (Sc) numbers was developed. The parameter estimation procedure allowed to determine that the correlation that best describes the mass transfer in the liquid phase circulating through the shell side in the flow parallel to the fibers is $Sh = 0.025\ (1 - \varphi)\ Re^{0.9}\ Sc^{0.33}$ for HF modules with low packing fraction and Reynolds numbers in the range 340–7800. The PV results showed that PEBA, as the material of the dense separating layer, is the most favorable in terms of selectivity towards butanol with respect to the other organics. The addition of a protective layer of PDMS allows the sealing of possible pinholes; however, its thickness should be kept as thin as possible since permeation fluxes decrease with the increasing thickness of PDMS, and this material also shows greater selectivity towards acetone compared to other organics.

Supplementary Materials: The following supporting information can be downloaded at: https://www.mdpi.com/article/10.3390/membranes12101007/s1, Figure S1: Influence of the PDMS content in the polymer solution on the film thickness; Figure S2: TGA of polypropylene, Pebax and PDMS polymers, and a hollow fiber membrane with two dense layers (Pebax + PDMS) on a PP support; Figure S3: Differential thermogravimetric (DTG) curves of the studied materials; Figure S4: ATR-FTIR analysis for Pebax, for PDMS and for the hollow fiber composite membrane.

Author Contributions: C.A.-S.: conceptualization, investigation, writing—original draft. D.G.-R.: investigation. M.F.: methodology, supervision. A.O.: investigation, funding acquisition. D.G.: writing—review and editing, supervision, funding acquisition. All authors have read and agreed to the published version of the manuscript.

Funding: This research was funded by Agencia Estatal de Investigación (PID2019-104369RB-I00/AEI/10.13039/501100011033). This work was also partially funded by the European Union through the project "HYLANTIC"-EAPA_204/2016, which is cofinanced by the European Regional Development Fund in the framework of the INTERREG Atlantic program, and by the Project ENERGY PUSH SOE3/P3/E0865, which is cofinanced by the European Regional Development Fund (ERPF) in the framework of the INTERREG SUDOE Programme. The Spanish Ministry of Science and Innovation is also gratefully acknowledged for the FPI research scholarship BES-2017-081708 (C.A.-S.).

Institutional Review Board Statement: Not applicable.

Informed Consent Statement: Not applicable.

Data Availability Statement: Data is contained within the article or Supplementary Materials.

Conflicts of Interest: The authors declare no conflict of interest.

Abbreviations

The following abbreviations are used in this manuscript:

HF	hollow fiber
PDMS	poly(dimethylsiloxane)
PEBA	poly(ether-block-amide)
PP	polypropylene
PSI	Pervaporation Separation Index (kg m^{-2} h^{-1})
PV	pervaporation
Symbols and Units	
a_i	activity of component *i* (mole fraction, dimensionless)
A	area (m^2)
d_{eq}	equivalent diameter (m)
D_i	diffusion coefficient for component *i* (m^2 s^{-1})

F	liquid flow rate (m^3 s^{-1})
J	total transmembrane flux (kg m^{-2} s^{-1})
J_i	transmembrane flux of each component (kg m^{-2} s^{-1})
$k_{i,bl}$	mass transfer coefficient for component (i) at the liquid boundary layer (m s^{-1})
NF	number of fibers in a module
P_i	permeability for component (i) (kmol s^{-1} m^{-1})
Re	Reynolds number
$R_{i,\ ov}$	overall mass transfer resistance (s m^2 $kmol^{-1}$)
$R_{i,supp}$	mass transfer resistance in membrane support (s m^2 $kmol^{-1}$)
S_{shell}	shell-side cross-sectional area of flow (m^{-2})
Sc	Schmidt number
Sh	Sherwood number
u	velocity (m s^{-1})
x_i	mole fraction of component (i) in the feed stream
y_i	mole fraction of component (i) in the permeate stream
$\alpha_{i/j}$	selectivity (-)
$\beta_{i/j}$	separation factor (-)
γ_i	activity coefficient (-)
δ	thickness (m)
ρ_m	molar density of feed liquid (kmol m^{-3})
φ	packing fraction

References

1. Dürre, P. Biobutanol: An attractive biofuel. *Biotechnol. J.* **2007**, *2*, 1525–1534. [CrossRef]
2. Kim, J.H.; Cook, M.; Peeva, L.; Yeo, J.; Bolton, L.W.; Lee, Y.M.; Livingston, A.G. Low energy intensity production of fuel-grade bio-butanol enabled by membrane-based extraction. *Energy Environ. Sci.* **2020**, *13*, 4862–4871. [CrossRef]
3. European Commission Biofuels. Available online: https://energy.ec.europa.eu/topics/renewable-energy/bioenergy/biofuels_en (accessed on 15 July 2022).
4. Bankar, S.B.; Nimbalkar, P.R.; Khedkar, M.A.; Chavan, P. V Biobutanol: Research breakthrough for its commercial interest. In *Liquid Biofuel Production*; Singh, L.K., Chaudhary, G., Eds.; Scrivener Publishing LLC: Beverly, MA, USA, 2019; pp. 237–283; ISBN 9781119459866.
5. Veza, I.; Muhamad Said, M.F.; Latiff, Z.A. Recent advances in butanol production by acetone-butanol-ethanol (ABE) fermentation. *Biomass Bioenergy* **2021**, *144*, 105919. [CrossRef]
6. Iyyappan, J.; Bharathiraja, B.; Vaishnavi, A.; Prathiba, S. Overview of current developments in biobutanol production methods and future perspectives. In *Biofuels and Biodiesel*; Basu, C., Ed.; Springer: New York, NY, USA, 2021; pp. 3–21; ISBN 978-1-0716-1323-8.
7. Shin, C.; Baer, Z.C.; Chen, X.C.; Ozcam, A.E.; Clark, D.S.; Balsara, N.P. Block copolymer pervaporation membrane for in situ product removal during acetone–butanol–ethanol fermentation. *J. Memb. Sci.* **2015**, *484*, 57–63. [CrossRef]
8. Wang, Y.; Janssen, H.; Blaschek, H.P. Fermentative biobutanol production: An old topic with remarkable recent advances. In *Bioprocessing of Renewable Resources to Commodity Bioproducts*; Bisaria, V.S., Kondo, A., Eds.; John Wiley & Sons: Hoboken, NJ, USA, 2014; pp. 227–260.
9. Hestekin, J.; Lopez, A.; Clausen, E.; Potts, T. Biobutanol: An important biofuel and bio-product. In *Applications of Microbial Engineering*; Gupta, V.K., Schmoll, M., Mazutti, M.A., Mäki, M., Tuohy, M.G., Eds.; CRC Press: Boca Raton, FL, USA, 2013; pp. 450–487; ISBN 978-1-4665-8578-2.
10. Green, E.M. Fermentative production of butanol- the industrial perspective. *Energy Biotechnol. Environ. Biotechnol.* **2011**, *22*, 337–343. [CrossRef]
11. Oudshoorn, A.; van der Wielen, L.A.M.; Straathof, A.J.J. Assessment of options for selective 1-butanol recovery from aqueous solution. *Ind. Eng. Chem. Res.* **2009**, *48*, 7325–7336. [CrossRef]
12. Outram, V.; Lalander, C.-A.; Lee, J.G.M.; Davies, E.T.; Harvey, A.P. Applied in situ product recovery in ABE fermentation. *Biotechnol. Prog.* **2017**, *33*, 563–579. [CrossRef]
13. Cabezas, R.; Suazo, K.; Merlet, G.; Quijada-Maldonado, E.; Torres, A.; Romero, J. Performance of butanol separation from ABE mixtures by pervaporation using silicone-coated ionic liquid gel membranes. *RSC Adv.* **2019**, *9*, 8546–8556. [CrossRef]
14. Cabezas, R.; Duran, S.; Zurob, E.; Plaza, A.; Merlet, G.; Araya-Lopez, C.; Romero, J.; Quijada-Maldonado, E. Development of silicone-coated hydrophobic deep eutectic solvent-based membranes for pervaporation of biobutanol. *J. Memb. Sci.* **2021**, *637*, 119617. [CrossRef]
15. Cai, D.; Wen, J.; Zhuang, Y.; Huang, T.; Si, Z.; Qin, P.; Chen, H. Review of alternative technologies for acetone-butanol-ethanol separation: Principles, state-of-the-art, and development trends. *Sep. Purif. Technol.* **2022**, *298*, 121244. [CrossRef]

16. Knozowska, K.; Kujawska, A.; Li, G.; Kujawa, J.; Bryjak, M.; Kujawski, W.; Lipnizki, F.; Ahrné, L.; Petrinić, I.; Kujawski, J.K. Membrane assisted processing of acetone, butanol, and ethanol (ABE) aqueous streams. *Chem. Eng. Process. Process Intensif.* **2021**, *166*, 108462. [CrossRef]
17. Abdehagh, N.; Tezel, F.H.; Thibault, J. Separation techniques in butanol production: Challenges and developments. *Biomass Bioenergy* **2014**, *60*, 222–246. [CrossRef]
18. Jaimes, J.H.B.; Torres Alvarez, M.E.; de Moraes, E.B.; Maciel, M.R.W.; Filho, R.M. Separation and semi-empiric modeling of ethanol–water solutions by pervaporation using PDMS membrane. *Polymers* **2021**, *13*, 93. [CrossRef]
19. Ong, Y.T.; Tan, S.H. Pervaporation separation of a ternary azeotrope containing ethyl acetate, ethanol and water using a buckypaper supported ionic liquid membrane. *Chem. Eng. Res. Des.* **2016**, *109*, 116–126. [CrossRef]
20. Serna-Vázquez, J.; Zamidi Ahmad, M.; Castro-Muñoz, R. Simultaneous production and extraction of bio-chemicals produced from fermentations via pervaporation. *Sep. Purif. Technol.* **2021**, *279*, 119653. [CrossRef]
21. Si, Z.; Wu, H.; Qin, P.; Van der Bruggen, B. Polydimethylsiloxane based membranes for biofuels pervaporation. *Sep. Purif. Technol.* **2022**, *298*, 121612. [CrossRef]
22. Cheng, C.; Liu, F.; Yang, H.K.; Xiao, K.; Xue, C.; Yang, S.-T. High-performance n-butanol recovery from aqueous solution by pervaporation with a PDMS mixed matrix membrane filled with zeolite. *Ind. Eng. Chem. Res.* **2020**, *59*, 7777–7786. [CrossRef]
23. Castro-Muñoz, R.; Boczkaj, G. Pervaporation zeolite-based composite membranes for solvent separations. *Molecules* **2021**, *26*, 1242. [CrossRef] [PubMed]
24. Davey, C.J.; Leak, D.; Patterson, D.A. Hybrid and mixed matrix membranes for separations from fermentations. *Membranes* **2016**, *6*, 17. [CrossRef]
25. Çalhan, A.; Deniz, S.; Romero, J.; Hasanoğlu, A. Development of metal organic framework filled PDMS/PI composite membranes for biobutanol recovery. *Korean J. Chem. Eng.* **2019**, *36*, 1489–1498. [CrossRef]
26. Tang, W.; Lou, H.; Li, Y.; Kong, X.; Wu, Y.; Gu, X. Ionic liquid modified graphene oxide-PEBA mixed matrix membrane for pervaporation of butanol aqueous solutions. *J. Memb. Sci.* **2019**, *581*, 93–104. [CrossRef]
27. Arregoitia-Sarabia, C.; González-Revuelta, D.; Fallanza, M.; Gorri, D.; Ortiz, I. Polymer inclusion membranes containing ionic liquids for the recovery of n-butanol from ABE solutions by pervaporation. *Sep. Purif. Technol.* **2020**, *248*, 117101. [CrossRef]
28. Lau, H.S.; Lau, S.K.; Soh, L.S.; Hong, S.U.; Gok, X.Y.; Yi, S.; Yong, W.F. State-of-the-art organic- and inorganic-based hollow fiber membranes in liquid and gas applications: Looking back and beyond. *Membranes* **2022**, *12*, 539. [CrossRef] [PubMed]
29. Urtiaga, A.; Gorri, D.; Ortiz, I. Mass-transfer modeling in the pervaporation of VOCs from diluted solutions. *AIChE J.* **2002**, *48*, 572–581. [CrossRef]
30. Baker, R.W. *Membrane Technology and Applications*, 3rd ed.; John Wiley & Sons: Chichester, UK, 2012.
31. Arregoitia-Sarabia, C.; González-Revuelta, D.; Fallanza, M.; Ortiz, A.; Gorri, D. Polyether-block-amide thin-film composite hollow fiber membranes for the recovery of butanol from ABE process by pervaporation. *Sep. Purif. Technol.* **2021**, *279*, 119758. [CrossRef]
32. Dai, Z.; Ansaloni, L.; Deng, L. Recent advances in multi-layer composite polymeric membranes for CO2 separation: A review. *Green Energy Environ.* **2016**, *1*, 102–128. [CrossRef]
33. Li, G.; Kujawski, W.; Válek, R.; Koter, S. A review—The development of hollow fibre membranes for gas separation processes. *Int. J. Greenh. Gas Control* **2021**, *104*, 103195. [CrossRef]
34. Yahaya, G.O. Separation of phenol from aqueous streams by a composite hollow fiber based pervaporation process using polydimethyl siloxane (PDMS)/polyether-block-amide (PEBA) as two-layer membranes. *Sep. Sci. Technol.* **2009**, *44*, 2894–2914. [CrossRef]
35. Jiang, L.Y.; Song, Z.W. Interfacial resistance of dual-layer asymmetric hollow fiber pervaporation membranes formed by co-extrusion. *J. Polym. Res.* **2011**, *18*, 2505–2514. [CrossRef]
36. Chung, T.-S.; Feng, Y. *Hollow Fiber Membranes: Fabrication and Applications*; Elsevier: Amsterdam, The Netherlands, 2021; ISBN 978-0-12-821876-1.
37. Gorri, D.; Urtiaga, A.; Ortiz, I. Pervaporative recovery of acetic acid from an acetylation industrial effluent using commercial membranes. *Ind. Eng. Chem. Res.* **2005**, *44*, 977–985. [CrossRef]
38. Urtiaga, A.M.; Gorri, E.D.; Gómez, P.; Casado, C.; Ibáñez, R.; Ortiz, I. Pervaporation technology for the dehydration of solvents and raw materials in the process industry. *Dry. Technol.* **2007**, *25*, 1819–1828. [CrossRef]
39. Mayank, R.; Ranjan, A.; Moholkar, V.S. Mathematical models of ABE fermentation: Review and analysis. *Crit. Rev. Biotechnol.* **2013**, *33*, 419–447. [CrossRef] [PubMed]
40. Li, G.; Kujawski, W.; Knozowska, K.; Kujawa, J. The effects of PEI hollow fiber substrate characteristics on PDMS/PEI hollow fiber membranes for CO2/N2 separation. *Membranes* **2021**, *11*, 56. [CrossRef]
41. Owen, M.J.; Dvornic, P.R. General introduction to silicone surfaces. In *Silicone Surface Science*; Owen, M.J., Dvornic, P.R., Eds.; Springer: Dordrecht, The Netherlands, 2012; pp. 1–21; ISBN 978-94-007-3876-8.
42. Knozowska, K.; Li, G.; Kujawski, W.; Kujawa, J. Novel heterogeneous membranes for enhanced separation in organic-organic pervaporation. *J. Memb. Sci.* **2020**, *599*, 117814. [CrossRef]
43. Franco, J.; DeMontigny, D.; Kentish, S.; Perera, J.; Stevens, G. A study of the mass transfer of CO2 through different membrane materials in the membrane gas absorption process. *Sep. Sci. Technol.* **2008**, *43*, 225–244. [CrossRef]
44. Fallanza, M.; Ortiz, A.; Gorri, D.; Ortiz, I. Using membrane reactive absorption modeling to predict optimum process conditions in the separation of propane-propylene mixtures. *Ind. Eng. Chem. Res.* **2013**, *52*, 8843–8855. [CrossRef]

45. Estay, H.; Troncoso, E.; Ruby-Figueroa, R.; Romero, J. Performance evaluation of mass transfer correlations in the GFMA process: A review with perspectives to the design. *J. Memb. Sci.* **2018**, *554*, 140–155. [CrossRef]
46. Lipnizki, F.; Field, R.W. Mass transfer performance for hollow fibre modules with shell-side axial feed flow: Using an engineering approach to develop a framework. *J. Memb. Sci.* **2001**, *193*, 195–208. [CrossRef]
47. Shen, S.; Kentish, S.E.; Stevens, G.W. Shell-side mass-transfer performance in hollow-fiber membrane contactors. *Solvent Extr. Ion Exch.* **2010**, *28*, 817–844. [CrossRef]
48. Dahuron, L.; Cussler, E.L. Protein extractions with hollow fibers. *AIChE J.* **1988**, *34*, 130–136. [CrossRef]
49. Liu, F.; Liu, L.; Feng, X. Separation of acetone–butanol–ethanol (ABE) from dilute aqueous solutions by pervaporation. *Sep. Purif. Technol.* **2005**, *42*, 273–282. [CrossRef]
50. Heitmann, S.; Krüger, V.; Welz, D.; Lutze, P. Experimental investigation of pervaporation membranes for biobutanol separation. *J. Membr. Sep. Technol.* **2013**, *2*, 245–262.
51. Rozicka, A.; Niemistö, J.; Keiski, R.L.; Kujawski, W. Apparent and intrinsic properties of commercial PDMS based membranes in pervaporative removal of acetone, butanol and ethanol from binary aqueous mixtures. *J. Memb. Sci.* **2014**, *453*, 108–118. [CrossRef]
52. van Wyk, S.; van der Ham, A.G.J.; Kersten, S.R.A. Pervaporative separation and intensification of downstream recovery of acetone-butanol-ethanol (ABE). *Chem. Eng. Process. Process Intensif.* **2018**, *130*, 148–159. [CrossRef]
53. Mai, N.; Kim, S.; Ha, S.; Shin, H.; Koo, Y.-M. Selective recovery of acetone-butanol-ethanol from aqueous mixture by pervaporation using immobilized ionic liquid polydimethylsiloxane membrane. *Korean J. Chem. Eng.* **2013**, *30*, 1804–1809. [CrossRef]
54. Xia, Y.; Chen, J.; Wu, Z.; Wang, T.; Li, J. Measurement of solubility thermodynamic and diffusion kinetic characteristic of solvents in PDMS by inverse gas chromatography. *Eur. Polym. J.* **2015**, *73*, 259–267. [CrossRef]

membranes

MDPI

Article

Evaluation of Different Capture Solutions for Ammonia Recovery in Suspended Gas Permeable Membrane Systems

María Soto-Herranz [1,*], Mercedes Sánchez-Báscones [1], Juan Manuel Antolín-Rodríguez [1] and Pablo Martín-Ramos [2]

1 Departamento de Ciencias Agroforestales, ETSIIAA, Universidad de Valladolid, Avenida de Madrid 44, 34004 Palencia, Spain; mercedes.sanchez@uva.es (M.S.-B.); juanmanuel.antolin@uva.es (J.M.A.-R.)

2 Instituto Universitario de Investigación en Ciencias Ambientales de Aragón (IUCA), EPS, Universidad de Zaragoza, Carretera de Cuarte s/n, 22071 Huesca, Spain; pmr@unizar.es

* Correspondence: maria.soto.herranz@alumnos.uva.es

Abstract: Gas permeable membranes (GPM) are a promising technology for the capture and recovery of ammonia (NH_3). The work presented herein assessed the impact of the capture solution and temperature on NH_3 recovery for suspended GPM systems, evaluating at a laboratory scale the performance of eight different trapping solutions (water and sulfuric, phosphoric, nitric, carbonic, carbonic, acetic, citric, and maleic acids) at 25 and 2 °C. At 25 °C, the highest NH_3 capture efficiency was achieved using strong acids (87% and 77% for sulfuric and nitric acid, respectively), followed by citric and phosphoric acid (65%) and water (62%). However, a remarkable improvement was observed for phosphoric acid (+15%), citric acid (+16%), maleic acid (+22%), and water (+12%) when the capture solution was at 2 °C. The economic analysis showed that water would be the cheapest option at any working temperature, with costs of 2.13 and 2.52 €/g N (vs. 3.33 and 3.43 €/g N for sulfuric acid) in the winter and summer scenarios, respectively. As for phosphoric and citric acid, they could be promising NH_3 trapping solutions in the winter months, with associated costs of 3.20 and 3.96 €/g N, respectively. Based on capture performance and economic and environmental considerations, the reported findings support that water, phosphoric acid, and citric acid can be viable alternatives to the strong acids commonly used as NH_3 adsorbents in these systems.

Keywords: ammonia adsorbents; ammonia recovery; citric acid; organic acids; phosphoric acid; water

1. Introduction

The application of manure as a fertilizer in areas with a high concentration of livestock can lead to a surplus of nutrients in the soil [1] and damage ecosystem quality, causing eutrophication of surface and groundwater bodies, soil acidification, and global warming [2]. In the European Union (EU), the livestock sector is estimated to contribute to 78% of biodiversity loss, 73% of N and P water pollution, 80% of soil acidification and air pollution (NH_3 and NO_x emissions), and 81% of global warming caused by the whole agricultural sector [3]. In response to this situation, the EU has promoted policies to improve the utilization of manure nutrients in agriculture and reduce their environmental impact through the Nitrates Directive (Directive 91/676/EEC EC) and the National Emission reduction Commitments Directive (Directive (EU) 2016/2284). In the framework of meeting the objectives set in these policies, and due to the high price of fertilizers, there is a growing interest in developing technologies to reduce emissions of NH_3 and to recover total ammonia nitrogen (TAN) from animal manure as a concentrated fertilizer product [4]. This would allow the production of nitrogen fertilizers that can compete with conventional synthetic fertilizers produced by the Haber-Bosch process, contributing to the achievement of circular economy goals [5].

A wide variety of technologies are currently available for TAN recovery, such as air scrubbers [6], reverse osmosis [7], adsorbent materials [8], evaporation techniques [9], struvite precipitation [10], nitrification-denitrification processes [11], the anaerobic ammonium oxidation (anammox) process [12], microbial fuel cells [13], or treatment with microalgae and photosynthetic bacteria [14,15]. However, they have important limitations that hinder their widespread application: air scrubbers and zeolite adsorption techniques require manure pre-treatment [16], reverse osmosis requires high working pressures, struvite precipitation requires additives [17], and biological treatments are only effective at low to medium TAN concentrations [18], given that bacterial activity can be inhibited in the presence of high organic matter and nitrogen contents [19].

In recent years, gas-permeable membrane (GPM) technology has emerged as an alternative to the above-mentioned approaches. This technology, based on the passage of NH_3 through a hydrophobic microporous membrane and its subsequent capture and concentration in an extraction solution on the other side of the membrane, offers advantages such as high transfer surface area, low working pressures, no need for manure clarification pre-treatment or additives [16], low energy consumption, and the possibility of being used in combination with other treatment technologies [20]. Its main drawback is the cost of installation and maintenance of the membranes [18].

Laboratory and pilot-scale studies have demonstrated the suitability of GPM technology based on expanded polytetrafluoroethylene (e-PTFE) membranes for N recovery from liquid effluents [16,21], wastewater from food industries [22], or sewage sludge [23]. Promising results have also been reported for the capture of NH_3 volatilized from animal manures [24–26].

Concerning the optimization of the capture process, issues such as the effects of pH, temperature, and airflow on the recovery efficiency of TAN from manure have been discussed in the literature [27,28]. However, the impact on process performance associated with the use of extraction solutions alternative to sulfuric acid (normally used to recover NH_3 in the form of $(NH_4)_2SO_4$, despite its safety and environmental risks) has seldom been analyzed in such previous research. Some authors have explored the possibility of using other mineral acids (nitric acid, phosphoric acid, acid mixtures, etc.) [29–32] for NH_3 removal from tertiary effluents, wastewater, or human urine, while other researchers have used organic acids, carbonic acid, or water [33–35] for NH_3 recovery from anaerobic sewage sludge digestate, landfill leachate, or wastewater. Nonetheless, the afore mentioned studies explored those alternative capture solutions in air stripping systems, liquid-liquid membrane contactors, hollow fiber membrane contactors, scrubbing systems, flow-electrode capacitive deionization systems, or capacitive membrane extraction systems (CapAmm), not on air-suspended GPM systems.

In view of this research gap, the present study aimed to investigate—on a laboratory scale—the NH_3 capture efficiency of different extraction solutions in a suspended GPM system. Specifically, a comparison in terms of performance was made between all the stripping solutions proposed in the above-referred studies (viz. sulfuric, phosphoric, nitric, carbonic, acetic, citric, and maleic acids, and deionized water). As a secondary objective, the influence of the temperature of the capture solution on NH_3 recovery was analyzed, an important aspect for the application of this technology in real livestock facilities conditions that had not been addressed in previous works. Thirdly, the operating costs, together with the associated chemical expenditure, were estimated for each trapping solution. The reported technical and preliminary economic evaluation may be useful not only to researchers working on more sustainable membrane-based systems but also to livestock farm owners and agricultural operators interested in producing certified organic fertilizers (which is not possible if mineral acids are used).

2. Materials and Methods

2.1. Reagents

Synthetic ammonia solutions were prepared with 18.2 MΩ·cm Milli-Q water (Merck Millipore, Burlington, MA, USA). Other reagents used were: sulfuric acid (CAS 7664-93-9, 98%), phosphoric acid (CAS 7664-38-2, 85%), nitric acid (CAS 7697-37-2, 65%), acetic acid

(CAS 64-19-7, 99.7%), citric acid (CAS 77-92-9, 99.5%), maleic acid (CAS 110-16-7, 99%), and N-allylthiourea (CAS 109-57-9, 98%). All chemicals were analytical grade reagents and were supplied by Panreac Química S.L.U. (Barcelona, Spain).

2.2. Experimental Conditions

The experimental setup was described in detail in previous work by our group [36]. The diagrams of the setups used for the NH_3 capture assays performed at 25 °C, at 2° C, and using carbonic acid are presented in Figures S1–S3, respectively.

It should be clarified that the selected temperatures for the capture solutions (25 and 2 °C) correspond to the average temperatures, during the period 1999–2019, of the warmest month (August) and the coldest month (January) in the town of Santa María la Real de Nieva (Segovia, Spain), where it is planned to carry out pilot-scale tests in the near future. Given that capture solution tanks are installed outside the livestock buildings, these would be the approximate temperatures at which the capture solutions would be in summer and winter.

To minimize variability resulting from the use of manures with inconstant NH_3 concentrations, a 6000 mg NH_3-N·L^{-1} synthetic solution was used in all the tests, consisting of 24.6 g $NH_4Cl \cdot L^{-1}$ + 43.2 g $NaHCO_3 \cdot L^{-1}$ + 10 mg N-allylthiourea·L^{-1} (as a nitrification inhibitor). The pH of the synthetic solutions was kept above 8 in all experiments (similar to those of real emitting sources such as pig slurry, chicken manure, laying hen manure, etc. [16, 20,37] and the temperature was kept at 25 °C (which is the usual setpoint temperature in many Spanish farms [38]), to replicate real farm conditions. It is worth noting that the pH affects the TAN (NH_4^+/NH_3) equilibrium, in such a way that values above 8 promote the conversion of NH_4^+ to NH_3, resulting in a higher presence of free ammonia and favoring mass transfer through the membrane [39].

Any capture solution can be used in the NH_3 extraction process as long as its pH is below the value of the acidic dissociation constant of the ammonium/ammonia pair (pKa = 9.24), where pK is the logarithm of the reciprocal of the acidic dissociation constant (Ka). Therefore, for all NH_3 capture solutions tested, it was necessary to keep the pH below this pKa value [32,40]. However, in this study, pH corrections with concentrated acid were not necessary in any case (except in the case of carbonated water traps, where CO_2 was dosed at a pressure of 0.1 bar depending on the pH present in the medium (pH < 6.36)).

The 1N capture solutions were continuously circulated inside the membrane using a Pumpdrive 5001 peristaltic pump (Heidolph, Schwabach, Germany) at a flow rate of 2.1 L·h^{-1}.

The absorption surface (163.4 cm^2) was the same in all experiments, and the characteristics of the suspended e-PTFE membrane at the beginning of the experiment (supplied by Zeus Industrial Products Inc., Orangeburg, SC, USA) are summarized in Table S1. Given that the membranes were not reused, a morphological analysis of the membranes was not conducted after the 7-day experiments.

For each capture solution-temperature combination, three runs were carried out over 7 days.

2.3. Analysis Methodology

During the experiments, pH, electrical conductivity, temperature, and NH_3-N concentration of the capture solutions and the synthetic N-emitting solution were monitored. pH, electrical conductivity, and temperature were measured with a GLP22 electrode (Crison Instruments S.A., Barcelona, Spain). NH_3-N concentration was determined by steam distillation, subsequent collection of distillates in borate buffer, and titration with 0.2 mol·L^{-1} HCl. A Kjeltec 8100 apparatus (Foss Iberia S.A., Barcelona, Spain) was used for distillation.

2.4. Data Calculations

The amount of NH_3-N emitted (expressed in mg NH_3-N) was determined by the difference between the amount of NH_3-N present at the beginning and the end of the experiment in the synthetic N-emitting solution. The mass of NH_3-N recovered (in mg NH_3-N) was determined as the amount of NH_3-N present at the end of the experiment in the capture solution. To calculate the NH_3-N capture efficiency of the different extraction solutions (NH_3-N efficiency, %), the mass of NH_3-N recovered was divided by the amount of NH_3-N emitted. The overall ammonia removal efficiency was calculated by comparing the amount of NH_3-N recovered in the trapping solution with the initial NH_3-N concentration of the emitting solution. Finally, the NH_3-N mass flux across the membrane (in mg NH_3-N$\cdot$cm$^{-2}\cdot$d^{-1}), produced as a consequence of the gas concentration gradient across the membrane, was estimated by considering the average mass of N captured per day and the membrane surface area.

2.5. Statistical Analysis

First, the data were tested for homogeneity and homoscedasticity using the Shapiro-Wilk and Levene tests. A two-way analysis of variance (ANOVA) was then conducted, followed by a post hoc comparison of means using Tukey's test ($p < 0.05$). R statistical software [41] was used for the statistical analyses.

2.6. Economic Analysis of the Different NH_3 Capture Solutions

To compare the different capture solutions used, an economic analysis was carried out for each of them. To this end, operation costs (associated with the electric power consumption of the capture solution recirculation pump) and chemical costs (quantity of acid necessary to carry out the captures obtained in this study or quantity of injected CO_2) were considered, which are summarized in Table S2 for the summer and winter scenarios, respectively. Reagent prices were obtained from various chemical suppliers and should be taken as estimates, given that actual values fluctuate with the market.

In the experiments referred herein, which lasted 7 days, the saturation of the trapping solutions was not reached in any case, so the compensations offered by the final products obtained were not relevant and were not considered in the economic analysis.

3. Results and Discussion

3.1. pH and Electrical Conductivity Evolution of NH_3 Trapping Solutions

The pH and electrical conductivity initial and final values (after 7 days) of the eight trapping solutions (sulfuric, phosphoric, nitric, carbonic, acetic, citric, and maleic acids, and water), at the two temperatures tested (25 and 2 °C), are shown in Table 1.

The pH of the trapping solutions increased in all cases at the end of the experiment, indicating that all solutions were capable of trapping NH_3. The smallest differences between the pH measured at the end and the beginning of the experiment were found for the solutions based on strong acids (i.e., sulfuric and nitric), while the largest increases in pH occurred for water at both temperatures (6.7→8.6 and 6.0→8.7), acetic acid at 2 °C (2.2→3.8), citric acid at both temperatures (1.7→3.0) and phosphoric acid at both temperatures (1.3→2.0).

The increase in pH in the water-only trapping solution can be attributed to the high solubility of NH_3 in water [42]. In the case of the 2 °C acetic acid trapping solution, the higher pH increase (compared to the 25 °C scenario) could be due to lower volatilization of acetic acid at lower temperatures, which would have favored NH_3 capture [33].

As for the EC values in the different trapping solutions, a strong decrease was observed in the sulfuric and nitric acid solutions at the end of the experiment. In this regard, strong acids are totally ionized in an aqueous solution, showing a very high EC and yielding many H^+ ions to react with NH_3. As NH_3 capture occurs, the conductivity of the solution decreases because the acidic permeate solution with high conductivity slowly changes to an NH_4SO_4/NH_4NO_3 solution (which has lower conductivity than the previous acidic medium) [32]. In the case of phosphoric acid, this behavior is not so marked because, although it is an inorganic acid, it has a weaker character.

In the case of weak organic acids, the conductivity remained stable or increased, as their starting conductivity is low and that of the formed salts is higher.

Table 1. pH and EC values at the beginning and the end of the experiment for the different NH_3 trapping solutions at two temperatures, 25 and 2 °C, representative of the winter and summer scenarios, respectively.

Stripping Solution	Parameter	T (°C)	Experimental Time	
			Day 1	Day 7
Sulfuric Acid	pH	25	0.6 ± 0.0	0.7 ± 0.0
		2	0.3 ± 0.1	0.5 ± 0.2
	CE	25	207.3 ± 0.6	170.5 ± 3.1
		2	289.7 ± 19.3	206.7 ± 5.5
Phosporic Acid	pH	25	1.3 ± 0.1	2.0 ± 0.0
		2	1.3 ± 0.1	2.0 ± 0.1
	CE	25	23.2 ± 0.3	18.6 ± 0.4
		2	22.9 ± 0.6	18.2 ± 0.9
Nitric acid	pH	25	0.5 ± 0.0	0.6 ± 0.1
		2	0.4 ± 0.0	0.6 ± 0.0
	CE	25	224.7 ± 1.2	174.3 ± 14.5
		2	234.7 ± 1.2	181.2 ± 2.6
Carbonic acid	pH	25	6.0 ± 0.6	6.6 ± 0.0
		2	5.9 ± 1.3	6.5 ± 0.1
	CE	25	0.2 ± 0.0	4.4 ± 0.2
		2	0.3 ± 0.1	3.5 ± 0.3
Acetic acid	pH	25	2.4 ± 0.0	2.9 ± 0.2
		2	2.1 ± 0.1	3.8 ± 0.0
	CE	25	1.5 ± 0.1	1.9 ± 0.5
		2	1.6 ± 0.2	9.3 ± 0.6
Citric acid	pH	25	1.7 ± 0.1	3.0 ± 0.1
		2	1.7 ± 0.1	3.0 ± 0.0
	CE	25	5.8 ± 0.1	11.2 ± 1.0
		2	5.7 ± 0.1	12.3 ± 0.4
Maleic acid	pH	25	1.3 ± 0.0	1.5 ± 0.1
		2	1.1 ± 0.0	1.7 ± 0.1
	CE	25	31.9 ± 0.1	26.5 ± 0.1
		2	30.8 ± 0.3	25.2 ± 0.1
H_2O	pH	25	6.7 ± 0.2	8.6 ± 0.1
		2	6.0 ± 0.2	8.7 ± 0.0
	CE	25	0.7 ± 0.1	12.4 ± 1.8
		2	0.4 ± 0.2	9.6 ± 0.1

All values are expressed as mean ± s.d. of n = 3.

3.2. NH_3-N Recovery Comparison at Different Temperatures

3.2.1. Evolution of NH_3-N Recovery

Figure 1 shows the evolution of the amount of NH_3-N recovered by the eight trapping solutions over the 7-day experimental period at the two temperatures (25 and 2 °C).

The accumulation of NH_3-N in the capture solutions followed a linear trend at both temperatures, with R^2 values above 0.93 and 0.91 at 25 and 2 °C, respectively, consistent with the behavior observed in other studies for sulfuric acid-based trapping solutions conducted over 14-day and 240-day periods [26,36].

At 25 °C, the average capture rates were 397 ± 12, 295 ± 43, 352 ± 20, 49 ± 1, 25 ± 13, 295 ± 53, 252 ± 12, and 285 ± 41 mg NH_3-N·d^{-1} for sulfuric acid, phosphoric acid, nitric acid, carbonic acid, acetic acid, citric acid, maleic acid, and water capture solutions, respectively. At 2 °C, the capture rates were 407 ± 19, 362 ± 27, 376 ± 17, 53 ± 5, 238 ± 25, 369 ± 34, 354 ± 34, and 338 ± 12 mg NH_3-N·d^{-1} for sulfuric acid, phosphoric acid, nitric acid, carbonic acid, acetic acid, citric acid, maleic acid, and chilled water, respectively.

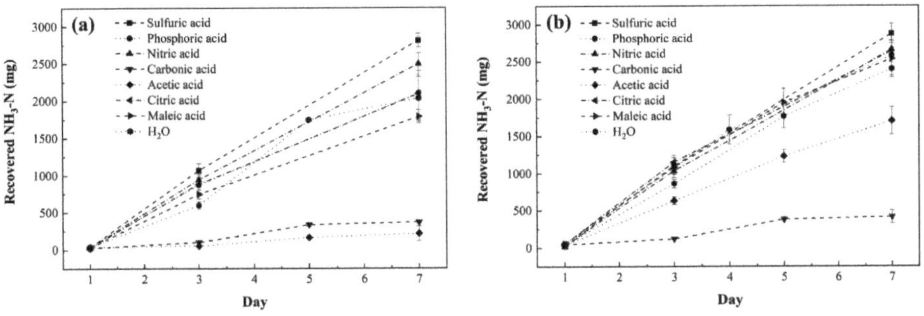

Figure 1. Evolution of the amount of TAN captured in the different trapping solutions during the 7-day experiments at (**a**) 25 °C and (**b**) 2 °C. All values are expressed as mean ± s.d. of n = 3.

3.2.2. NH_3-N Recovery Comparison at 25 °C

Figure 2 shows the total amount of NH_3-N recovered by the eight trapping solutions at the two temperatures (25 and 2 °C) at the end of the experiment.

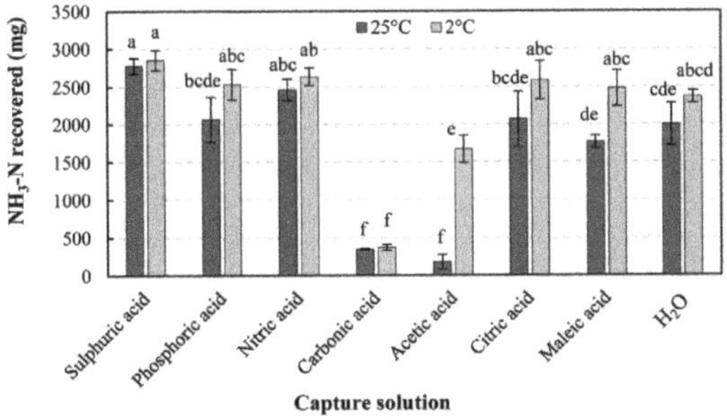

Figure 2. Mass of NH_3-N recovered by the different trapping solutions at two temperatures after 7 days. Values with different letters are significantly different at $p \leq 0.05$ according to Tukey's HSD test. All values are expressed as means of n = 3.

When the trapping solutions were at 25 °C (summer scenario), the sulfuric and nitric acid solutions showed the best response in terms of NH_3 trapping, followed by phosphoric acid, citric acid, deionized water, maleic acid, carbonic acid, and acetic acid.

Such higher recovery observed for strong inorganic acids is due to the stronger interaction between ammonium and the anion resulting from acid dissociation [32], and the results are consistent with those obtained by Damtie et al. [32], who used hydrophobic hollow fiber membrane contactors to recover NH_3 from human urine at room temperature, using sulfuric acid, nitric acid, phosphoric acid, and water (as a control) as capture solutions. They found that sulfuric acid was slightly better at NH_3 capture than other acids, but without significant differences. In this regard, it should be clarified that in our study there were no statistically significant differences between sulfuric acid and nitric acid, but there were significant differences between sulfuric acid and phosphoric acid. Such differences may be tentatively attributed to the fact that ammonium phosphate is less soluble in water than ammonium sulfate, with solubility values of 42.9 kg·L^{-1} for [$(NH_4)_2PO_4$] and 70.6 for [$(NH_4)_2SO_4$] at 0 °C, and of 68.6 vs. 75.4 kg·L^{-1} at 20 °C, respectively [31].

Concerning the organic acids, the NH_3 trapping results at 25 °C may be deemed as good (except for acetic acid): citric acid was significantly different from sulfuric acid, but not from nitric and phosphoric acids, and maleic acid was also comparable to phosphoric acid. It is worth noting that citric acid has three H^+ ions per acid molecule and is stronger than the other two assayed organic acids, which explains why better NH_3 capture results were obtained. As for maleic acid, authors such as Starmans and Melse [43] agree that, among the organic acids, it is the second-best alternative to sulfuric acid after citric acid.

The observed behavior differs from that reported by Jamaludin et al. [33] for an anaerobic digestate NH_3 stripping system using citric and acetic acids, cold water, and mineral salts (Epsom and gypsum). These authors found that citric acid achieved a similar stripping performance to sulfuric acid (with an NH_3 recovery of 561.3 mg versus 607.5 mg, respectively), but at the cost of using twice as much citric acid as sulfuric acid (due to its partial dissociation characteristics). In this study, the NH_3 recovery for both solutions was significantly different (2065 mg for citric acid versus 2776 mg for sulfuric acid).

As for the low NH_3 recovery attained using acetic acid as the trapping solution, it may be due to its lower acid strength ($K_a = 1.75 \times 10^{-5}$) compared to citric ($7.45 \times 10^{-4}$) or maleic ($9.1 \times 10^{-3}$) acids [33], and to its high Henry's volatility constant ($K_H = 2.50 \times 10^{-2}$ m^3 Pa mol^{-1}). This is >10^{12} times higher than, for example, those of sulfuric (7.69×10^{-14} m^3 Pa mol^{-1}) or citric acid (3.33×10^{-17} m^3 Pa mol^{-1}) [44]. Using acetic acid, Jamaludin et al. [33] obtained recoveries of 204.3, 421, and 536.1 mg NH_3 for scrubbing temperatures of 67–70, 36–40, and 15–18 °C, respectively, compared to 561.3 mg NH_3 when using citric acid.

In the solution generated by bubbling CO_2 in water (carbonic acid), very low NH_3 recovery values were also observed (343 mg NH_3). This result can be tentatively attributed to the fact that the CO_2 that is not absorbed in water, in the gas phase, can generate a competitive occupation of the membrane pores, reducing the NH_3 flux inside the membrane, which in turn leads to a reduction of NH_3 capture. This phenomenon was also observed by authors such as Zhang et al. [45] in CapAmm extraction systems using carbonated water as an adsorbent.

In relation to the water solution, NH_3 amounts similar to those obtained with weak acids were recovered. In this sense, it must be taken into consideration that NH_3 has a high solubility in water due to its polarity, forming hydrogen bonds with water molecules [46], and is retained in the solution provided that pH 9.2 is not exceeded [32,45]. However, the good NH_3 uptake results can also be attributed to the joint diffusion of NH_3 and CO_2 through the membrane, in such a way that, in addition to ammonium hydroxide and NH_3(aq), ammonium bicarbonate is also present in the uptake solution. Such presence of bicarbonate in the water samples was confirmed by laboratory analysis by titration with 0.1N HCl (not shown). This would be in good agreement with the results obtained by Vanotti et al. [47]: in a study on NH_3 recovery from poultry litter, they found that the amount of nitrogen recovered in the tank after 37 days was higher using distilled water as the trapping solution than using 1N sulfuric acid (4012 vs. 1156 mg N·L^{-1}, respectively). Based on analyses of NH_3 and carbonate concentrations in the recirculated water, they showed that gaseous NH_3 was recovered as ammonium bicarbonate salt and concluded that CO_2 also permeated through the e-PTFE membrane (similar to the one used in this study) and drove NH_3 fixation with water.

3.2.3. NH_3-N Recovery Comparison at 2 °C

NH_3 recovery at 2 °C increased in all trapping solutions compared to the values reported above at 25 °C (Figure 2), although differences were only statistically significant in the case of acetic acid and maleic acid. Being aqueous solutions, the observed improvement may be mainly attributed to the increase in NH_3 solubility in water with decreasing temperature [48] (ca. 90 g/100 mL at 0 °C vs. ca 32 g/100 mL at 25 °C). However, differences associated with the acid present in the capture solution cannot be ruled out. For instance, in the case of sulfuric acid, Kurtén et al. [49] reported that NH_3:H_2SO_4 ratios of nucleating clusters varied as a function of temperature and NH_3 concentration. At a 1 ppm NH_3 concentration, they estimated that $(H_2SO_4)_2 \cdot NH_3$ clusters' relative concentration would decrease from 50.2% to 4.2% when the temperature was decreased from 25 to 0 °C, whereas $(H_2SO_4)_2 \cdot (NH_3)_2$ clusters' relative concentration would increase from 49.8 to 95%. Hence, a higher capture ratio should be possible at 0 °C for the same amount of acid.

3.2.4. Differences in NH_3 Flux, NH_3-N Capture, and NH_3-N Removal Efficiencies

The NH_3 flux rate, NH_3-N capture efficiency, and overall NH_3-N removal efficiency values for the eight capture solutions at 25 and 2 °C are summarized in Table 2.

Table 2. NH_3 flux rates (N flux), NH_3-N capture efficiencies, and overall NH_3-N removal efficiencies for the different trapping solutions at the two assayed temperatures, 25 and 2 °C, representative of the winter and summer scenarios, respectively.

Capture Solution	T (°C)	N Flux (mg N·cm^{-2}·d^{-1})	NH_3-N Capture Efficiency (%)	NH_3-N Removal Efficiency (%)
Sulfuric acid	25	2.4 ± 0.1 a	87	46
	2	2.5 ± 0.1 a	89	48
Phosphoric acid	25	1.8 ± 0.3 bcde	64	34
	2	2.2 ± 0.2 abc	79	42
Nitric acid	25	2.2 ± 0.1 abc	77	41
	2	2.3 ± 0.1 ab	82	44
Carbonic acid	25	0.3 ± 0.0 f	11	6
	2	0.3 ± 0.0 f	12	6
Acetic acid	25	0.2 ± 0.1 f	6	3
	2	1.5 ± 0.2 e	52	28
Citric acid	25	1.8 ± 0.3 bcde	65	34
	2	2.3 ± 0.2 abc	81	43
Maleic acid	25	1.5 ± 0.1 de	55	29
	2	2.2 ± 0.2 abc	77	41
H_2O	25	1.7 ± 0.3 cde	62	33
	2	2.1 ± 0.1 abcd	74	39

Values followed by the same letter are not significantly different at $p \leq 0.05$ according to Tukey's HSD test. All values are expressed as mean ± s.d. of n = 3.

Most of the calculated NH_3 fluxes across the e-PTFE membrane were higher than those reported by Zhang et al. [45] (0.56–1.21 mg·cm^{-2}·d^{-1}), which should be mainly ascribed to the use of a much more concentrated NH_3-emitting synthetic solution (43 mg NH_4^+-N vs. 6000 mg NH_3-N·L^{-1}). Soto-Herranz et al. [36] reported flux rates of 2.14 ± 0.2 mg N·cm^{-2}·d^{-1} using sulfuric acid as the NH_3 trapping solution at 25 °C, for an NH_3 emitting solution concentration, membrane surface area, and flux rate similar to those used in this study, so the results reported herein (2.4 ± 0.1 mg N·cm^{-2}·d^{-1}) would be comparable.

The NH_3 capture yield at 25 °C ranged from 62% (water) to 87% (sulfuric acid), except for acetic acid (6%) and carbonic acid (11%) solutions. As discussed above, this could be due to the volatile character of acetic acid [44] and the competitive nature of CO_2 molecules in the membrane pores—which would impair NH_3 transfer—, respectively.

In the winter scenario, at 2 °C, the NH_3 capture efficiency barely changed for strong acids (+2% and +5% for sulfuric and nitric acid, respectively), but substantially improved for the other trapping solutions: phosphoric acid (64→79%), citric acid (65→81%), maleic acid (55→77%) and water (62→74%). Regarding acetic acid, although it significantly improved its capture yield up to 52%, it was still lower than that of the aforementioned capture solutions. As for the CO_2 saturated solution (H_2CO_3), the yield continued to be very low (12%).

Comparing the values obtained with those reported in the literature, the NH_3 capture efficiencies for sulfuric, phosphoric, and nitric acid obtained in this study were similar to those attained by other authors [40,45], and the differences may be attributed to the fact that the capture system was not the same. Riaño et al. [50], using a submerged GPM system with a sulfuric acid capture solution, reported capture efficiencies of up to 79.7%; Zhang et al. [45], using a CapAmm system, obtained NH_3 capture efficiencies of 75% for phosphoric acid and 73% for sulfuric acid; and Reig et al. [40], using hollow fiber liquid-liquid membrane contactors (HF-LLMC) combined with ion exchange, attained NH_3 capture efficiencies of 74% and 85% with nitric and phosphoric acid, respectively.

Using citric acid, Jamaludin et al. [33] obtained NH_3 recoveries from digestate of up to 90%, higher than that achieved in this study at low temperature (81%).

On the other hand, the NH_3 capture yields achieved using water as the capture solution (62–74%) were higher than the values referred by authors such as Jamaludin et al. [33] and Zhang et al. [45], who obtained efficiencies in the 35–61% range.

Using carbonic acid as the NH_3 capture solution, recovery efficiencies of 11–12% were obtained, lower than those reported by Zhang et al. [45].

Concerning the overall NH_3-N removal efficiencies, they were in the 34–48%, 29–43% and 33–39% range for mineral acids, citric and maleic acid, and water, respectively. These values were approximately half of the NH_3-N capture efficiencies, given that suspended GPM systems only recover nitrogen from the headspace (i.e., the TAN retained in the emitting solution cannot be recovered) and the average NH_3-N emission in the reported experiments was ca. 53%. For comparison purposes, in the work by Samani-Majd et al. [25], in which a hybrid GPM system (combining a submerged GPM system with a suspended GPM system) with a sulfuric acid trapping solution was used for ammonia recovery from free-stall dairy manure, an overall removal efficiency of 48% was reached; and in the work by Riaño et al. [50], focused on ammonia recovery from raw swine manure with a submerged GPM system using a sulfuric acid capture solution, the overall removal efficiency was in the 14.3–49.5% range.

3.3. *Comparison of Economic Costs Associated with the Different NH_3 Capture Solutions on a Laboratory Scale*

The costs associated with the use of each capture solution in the summer (25 °C) and winter (2 °C) scenarios are shown in Figure 3.

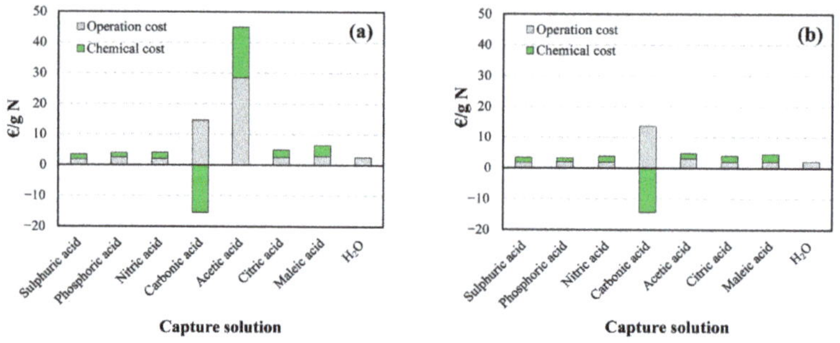

Figure 3. Economic analysis of the suspended GPM-based NH_3 capture process using eight different types of trapping solutions in the (**a**) summer (25 °C) and (**b**) winter (2 °C) scenarios.

In the summer scenario (25 °C), estimated operation costs would be lower for strong inorganic acids (1.82 and 2.05 €/g N for sulfuric acid and nitric acid, respectively) than for weak acids (2.44, 2.44, and 2.86 €/g N for phosphoric, citric, and maleic acid, respectively) or water (2.52 €/g N). Due to the low recovery yields, the operation costs for acetic acid and carbonic acid would be much higher (28.51 and 14.68 €/g N, respectively) and their use would not be advisable.

In the winter scenario (2 °C), operation costs would decrease in all cases, but not in the same proportion. Differences between those associated with the use of strong acids (1.77 and 1.91 €/g N for sulfuric and nitric, acid respectively) and those resulting from the weak acid-based alternatives (1.99, 1.95, and 2.04 €/g N for phosphoric, citric and maleic acid, respectively) or water (2.13 €/g N) would be noticeably smaller.

Zhang et al. [45] also estimated lower operating costs (1.02–1.30 €/kg N) when using inorganic acid sorbents (sulfuric, phosphoric, nitric, and hydrochloric acid) than weak sorbents (1.7 and 2.5 €/kg N for carbonic acid and water, respectively). However, the estimated costs are only comparable within each study, due to the differences in energy and reagent costs between the study by Zhang et al. [45] and the present study.

Regarding chemical costs, in the summer scenario (25 °C), those associated with the use of organic acids (2.52 and 3.52 €/g N for citric and maleic acid, respectively) would be noticeably higher than those of strong acids (1.61 and 2.04 €/g N for sulfuric and nitric acid, respectively). Interestingly, the use of phosphoric acid would be the cheapest option (1.49 €/g N), if water is excluded. These differences would again be substantially reduced in the winter scenario: while chemical costs reduction would be smaller than 6.5% for strong acids (1.57 and 1.91 €/g N for sulfuric and nitric acid, respectively), it would be in the 18.6–28.9% range for weak acids (1.21, 2.01, and 2.51 €/g N for phosphoric acid—again the cheapest option—, citric acid, and maleic acid, respectively). In the case of the H_2CO_3 solution, since ammonia extraction contributed to CO_2 fixation, the expenditure on reagents could be considered negative (for instance, Zhang et al. [45] considered earnings of \$0.37 kg^{-1} N).

If the total costs associated with the use of each trapping solution in both scenarios (summer and winter) are considered, the most viable option for NH_3 capture in both cases would be water, with a total cost of 2.52 €/g N in summer and 2.13 €/g N in winter, a result consistent, for example, with that obtained by Zhang et al. [45] for a CapAmm system. Although the NH_3 capture efficiency using this capture solution was not the highest, the savings in reagents make it a very interesting option as an NH_3 capture solution.

Citric or phosphoric acid solutions would also be viable options for NH_3 capture, with total associated costs of 3.96 and 3.20 €/g N, respectively (compared to 3.33 €/g N for sulfuric acid) in the winter scenario, and 4.96 and 3.93 €/g N, respectively (compared to 3.43 €/g N for sulfuric acid) in the summer scenario. The use of citric acid may be interesting to produce organic fertilizers suitable for organic farming [51], which cannot be produced from mineral acids. However, double-dosing would be required, due to its partial dissociation [33]. The phosphoric acid choice would be supported by economic considerations and safety considerations (in comparison with sulfuric and nitric acids).

Although in the study presented in Figure 3 the profit contributed by the final products obtained has not been included, it should be taken into consideration that, although the NH_3 capture efficiencies with water (62–74%) were lower than those achieved by the rest of the capture solutions, the final product has a higher market value than those derived from other fertilizers such as sulfuric acid (25% ammoniacal solutions have a selling price of approximately 1.30 €/kg N, compared to 0.69 €/kg N for 21% ammonium sulfate salt). As for weak acid end products, market prices for $(NH_3)_2HPO_4$ and ammonium acetate (ca. 1.10 and 1.06 €/kg N, respectively) are also higher than that of $(NH_4)_2SO_4$. The additional revenue from these end products would make water and phosphoric acid trapping solutions even more attractive and could help to offset the higher costs associated with the use of citric acid.

3.4. Applicability and Limitations of the Study

3.4.1. Applicability of the Study

The use of distilled (or tap) water as the capture solution would be the most recommendable option, as it has a moderate/high NH_3 capture performance regardless of the operating temperature used and the associated costs would be the lowest. In this case, the final product obtained (ammonia water) could be used, for instance, in waste incinerators for NO_x capture by selective (non-catalytic) reduction [33].

Concerning the use of weak acids, as noted above, phosphoric acid would be cheaper than strong acids in winter conditions. However, despite being more expensive, the use of citric acid as an alternative capture solution may also be advantageous in winter. In this regard, there is consensus that there is a significant increase in NH_3 emissions in the summer months compared to the winter months [52], as the increase in outside temperature in the summer months causes an increase in the temperature inside the farm, which accelerates the NH_3 release rate from the slurry/manure [53] and increases the concentration of gases in the air. However, the above reasoning ignores the impact of ventilation. Guo et al. [54] studied the concentration ranges of air pollutants in pig and poultry houses equipped with forced ventilation systems and found that the concentrations greatly varied depending on the season: for pig houses, the NH_3 concentration ranged from 27 ppm in winter to 0.8 ppm in summer, and for poultry houses, from 25 ppm in winter to 2 ppm in summer.

Since the capture efficiency is directly proportional to the NH_3 concentration in the livestock housing environment, the winter period would be the most favorable to apply GPM technology. In fact, in a pilot-scale NH_3 capture study using suspended GPM in pig and poultry houses [26], the NH_3 capture efficiency rate in winter was 2 to 4 times higher than in summer. Taking this factor into consideration, and in view that the total cost associated with the use of citric acid trapping solutions in winter conditions would be 18.8% and 3.6% higher than those resulting from the use of strong acids (sulfuric and nitric acid, respectively), their replacement with this safer and more environmentally benign alternative would be supported. Furthermore, as indicated above, the resulting salt is considered a biofertilizer, which has a higher market value than salts formed from inorganic acids [33] and does not contain sulfur, making it suitable for soils with excess sulfur problems [55]. In addition, it improves the bioavailability of phosphorus in the soil for plant uptake, as citrate ions form complexes with metal cations, thus solubilizing precipitated and adsorbed phosphorus [56].

3.4.2. Limitations of the Study

In this work, the performance of different solutions for NH_3 gas capture at a laboratory scale was studied to select the most promising ones. However, under such strictly controlled working conditions, the reported capture performances are expected to be higher than those achievable under real farm conditions. In this regard, a previous study [26] showed that capture yields were reduced by the management practices of the farms where the pilot plant based on a suspended GPM system was installed, atmospheric conditions, etc. Therefore, studies under real conditions in different types of livestock housing facilities, locations, and over longer periods are necessary to confirm the preliminary conclusions obtained in the laboratory.

In addition, since the behavior of the capture solutions was only evaluated at two temperatures (2 and 25 °C), a more detailed investigation of the performance of the NH_3 capture solutions over a wider range of temperatures and using smaller intervals would be needed to extrapolate the results to the varied atmospheric conditions that may occur in livestock facilities.

It is also important to note that the economic study presented in Section 3.3 has been carried out in an economic context marked by uncertainty and high price volatility (due to the conflict in Ukraine), so it is not possible to make cost comparisons with previous studies.

4. Conclusions

This study investigated the use of different trapping solutions at two temperatures (2 and 25 °C) in a GPM system with suspended e-PTFE membranes to recover NH_3 released from a synthetic solution of 6000 mg $N \cdot L^{-1}$. At 25 °C (summer scenario), strong acids (sulfuric and nitric) were the most efficient, followed by weak acids (phosphoric, citric, and maleic) and water, with efficiencies of 87 and 77%, 55–65%, and 62%, respectively. Despite their advantages in terms of lower reagent costs, trapping solutions based on acetic acid and carbonated water (carbonic acid) had to be discarded due to the high volatility of the former and possible back-diffusion of CO_2 and NH_3 through the membrane when the latter was used. In winter conditions (2 °C), the NH_3 capture efficiencies of all solutions improved, with particularly noticeable increases for phosphoric, citric, and maleic acids (+15%, +16%, and +22%, respectively). Bringing together technical and economic considerations, water may be put forward as the most advisable trapping solution, given that it offers moderate-high recovery yields in both summer and winter conditions and 24–38% lower total costs than traditionally used strong inorganic acids. Concerning weak acids, given that the capture efficiencies of GPM systems under real operating conditions are higher in winter and that cost differences would be small, it may also be advisable to opt for citric acid or phosphoric acid in winter. In addition to a lower environmental impact, the use of these three alternative capture solutions would have the advantage of obtaining end products with a higher market value, which would partly offset the costs associated with the implementation of GPM technology.

Supplementary Materials: The following supporting information can be downloaded at: https://www.mdpi.com/article/10.3390/membranes12060572/s1, Figure S1. Scheme of the NH_3 capture process by the suspended gas-permeable membrane system in a closed circuit using different trapping solutions at 25 °C; Figure S2. Scheme of the NH_3 capture process by the suspended gas-permeable membrane system in a closed circuit using different trapping solutions at 2 °C; Figure S3. Scheme of the NH_3 capture process by the suspended gas-permeable membrane system in a closed circuit using carbonic acid as the ammonia trapping solution at 2 °C; Table S1. Characteristics of the e-PTFE membrane used in the experiments; Table S2. Summary of operation and chemical costs, as well as end-product market prices, for the eight ammonia trapping solutions under analysis.

Author Contributions: Conceptualization, M.S.-B. and J.M.A.-R.; data curation, M.S.-H.; formal analysis, M.S.-H., J.M.A.-R. and P.M.-R.; funding acquisition, M.S.-B.; investigation, M.S.-H., M.S.-B., J.M.A.-R. and P.M.-R.; methodology, M.S.-B. and J.M.A.-R.; resources, M.S.-B.; supervision, M.S.-B.; validation, J.M.A.-R., M.S.-B. and P.M.-R.; visualization, M.S.-H.; writing—original draft, M.S.-H., M.S.-B., J.M.A.-R. and P.M.-R.; writing—review and editing, M.S.-H. and P.M.-R. All authors have read and agreed to the published version of the manuscript.

Funding: This work was funded by the European Union in the framework of the LIFE Project Green Ammonia (LIFE20 ENV/ES/000858, 'Market technology based on membranes for the reduction of ammonia in livestock farms').

Institutional Review Board Statement: Not applicable.

Informed Consent Statement: Not applicable.

Data Availability Statement: The data presented in this study are available on request from the corresponding author. The data are not publicly available due to their relevance as part of an ongoing Ph.D. thesis.

Conflicts of Interest: The authors declare no conflict of interest. The funders had no role in the design of the study; in the collection, analyses, or interpretation of data; in the writing of the manuscript, or in the decision to publish the results.

References

1. Vanotti, M.B.; García-González, M.C.; Szögi, A.A.; Harrison, J.H.; Smith, W.B.; Moral, R. Removing and Recovering Nitrogen and Phosphorus from Animal Manure. *Anim. Manure Prod. Charact. Environ. Concerns Manag.* **2020**, *67*, 275–321. [CrossRef]
2. Bouwman, L.; Goldewijk, K.K.; Van Der Hoek, K.W.; Beusen, A.H.W.; Van Vuuren, D.P.; Willems, J.; Rufino, M.C.; Stehfest, E. Erratum: Exploring global changes in nitrogen and phosphorus cycles in agriculture induced by livestock production over the 1900–2050 period. *Proc. Natl. Acad. Sci. USA* **2013**, *110*, 21196. [CrossRef] [PubMed]
3. Leip, A.; Billen, G.; Garnier, J.; Grizzetti, B.; Lassaletta, L.; Reis, S.; Simpson, D.; Sutton, M.A.; De Vries, W.; Weiss, F.; et al. Impacts of European livestock production: Nitrogen, sulphur, phosphorus and greenhouse gas emissions, land-use, water eutrophication and biodiversity. *Environ. Res. Lett.* **2015**, *10*, 115004. [CrossRef]
4. Vanotti, M.; Szogi, A.; Pilar Bernal, M.; Martinez, J. Livestock waste treatment systems of the future: A challenge to environmental quality, food safety, and sustainability. OECD Workshop. *Bioresour. Technol.* **2009**, *100*, 5371–5373. [CrossRef]
5. Vanotti, M.B.; Dube, P.J.; Szogi, A.A.; García-González, M.C. Recovery of ammonia and phosphate minerals from swine wastewater using gas-permeable membranes. *Water Res.* **2017**, *112*, 137–146. [CrossRef]
6. Bonmatí, A.; Flotats, X. Air stripping of ammonia from pig slurry: Characterisation and feasibility as a pre- or post-treatment to mesophilic anaerobic digestion. *Waste Manag.* **2003**, *23*, 261–272. [CrossRef]
7. Shin, C.; Szczuka, A.; Jiang, R.; Mitch, W.A.; Criddle, C.S. Optimization of reverse osmosis operational conditions to maximize ammonia removal from the effluent of an anaerobic membrane bioreactor. *Environ. Sci. Water Res. Technol.* **2021**, *7*, 739–747. [CrossRef]
8. Das, P.; Prasad, B.; Singh, K.K.K. Applicability of Zeolite Based Systems for Ammonia Removal and Recovery from Wastewater. *Water Environ. Res.* **2017**, *89*, 840–845. [CrossRef]
9. Schwarzwälder Sprovieri, J.A.; Octavio de Souza, T.S.; Contrera, R.C. Ammonia removal and recovery from municipal landfill leachates by heating. *J. Environ. Manag.* **2020**, *256*, 109947. [CrossRef]

10. Miles, A.; Ellis, T.G. Struvite precipitation potential for nutrient recovery from anaerobically treated wastes. *Water Sci. Technol.* **2001**, *43*, 259–266. [CrossRef]
11. Riaño, B.; García-González, M.C. On-farm treatment of swine manure based on solid-liquid separation and biological nitrification-denitrification of the liquid fraction. *J. Environ. Manag.* **2014**, *132*, 87–93. [CrossRef] [PubMed]
12. Magrí, A.; Béline, F.; Dabert, P. Feasibility and interest of the anammox process as treatment alternative for anaerobic digester supernatants in manure processing—An overview. *J. Environ. Manag.* **2013**, *131*, 170–184. [CrossRef] [PubMed]
13. Gude, V.G. Wastewater treatment in microbial fuel cells—An overview. *J. Clean. Prod.* **2016**, *122*, 287–307. [CrossRef]
14. Cai, T.; Park, S.Y.; Li, Y. Nutrient recovery from wastewater streams by microalgae: Status and prospects. *Renew. Sustain. Energy Rev.* **2013**, *19*, 360–369. [CrossRef]
15. Zhi, R.; Cao, K.; Zhang, G.; Zhu, J.; Xian, G. Zero excess sludge wastewater treatment with value-added substances recovery using photosynthetic bacteria. *J. Clean. Prod.* **2020**, *250*. [CrossRef]
16. Garcia-González, M.C.; Vanotti, M.B. Recovery of ammonia from swine manure using gas-permeable membranes: Effect of waste strength and pH. *Waste Manag.* **2015**, *38*, 455–461. [CrossRef]
17. Lorick, D.; Macura, B.; Ahlström, M.; Grimvall, A.; Harder, R. Effectiveness of struvite precipitation and ammonia stripping for recovery of phosphorus and nitrogen from anaerobic digestate: A systematic review. *Environ. Evid.* **2020**, *9*, 1–20. [CrossRef]
18. Xiang, S.; Liu, Y.; Zhang, G.; Ruan, R.; Wang, Y.; Wu, X.; Zheng, H.; Zhang, Q.; Cao, L. New progress of ammonia recovery during ammonia nitrogen removal from various wastewaters. *World J. Microbiol. Biotechnol.* **2020**, *36*, 1–20. [CrossRef]
19. Abeliovich, A.; Azov, Y. Toxicity of ammonia to algae in sewage oxidation ponds. *Appl. Environ. Microbiol.* **1976**, *31*, 801–806. [CrossRef]
20. González-García, I.; Riaño, B.; Molinuevo-Salces, B.; Vanotti, M.B.; García-González, M.C. Improved anaerobic digestion of swine manure by simultaneous ammonia recovery using gas-permeable membranes. *Water Res.* **2021**, *190*. [CrossRef]
21. Dube, P.J.; Vanotti, M.B.; Szogi, A.A.; García-González, M.C. Enhancing recovery of ammonia from swine manure anaerobic digester effluent using gas-permeable membrane technology. *Waste Manag.* **2016**, *49*, 372–377. [CrossRef] [PubMed]
22. Shi, X.; Zuo, J.; Zhang, M.; Wang, Y.; Yu, H.; Li, B. Enhanced biogas production and in situ ammonia recovery from food waste using a gas-membrane absorption anaerobic reactor. *Bioresour. Technol.* **2019**, *292*, 121864. [CrossRef] [PubMed]
23. Munasinghe-Arachchige, S.P.; Cooke, P.; Nirmalakhandan, N. Recovery of nitrogen-fertilizer from centrate of anaerobically digested sewage sludge via gas-permeable membranes. *J. Water Process Eng.* **2020**, *38*, 101630. [CrossRef]
24. Rothrock, M.J.; Szögi, A.A.; Vanotti, M.B. Recovery of ammonia from poultry litter using flat gas permeable membranes. *Waste Manag.* **2013**, *33*, 1531–1538. [CrossRef]
25. Samani Majd, A.M.; Mukhtar, S. Ammonia recovery enhancement using a tubular gas-permeable membrane system in laboratory and field-scale studies. *Trans. ASABE* **2013**, *56*, 1951–1958. [CrossRef]
26. Soto-Herranz, M.; Sánchez-Báscones, M.; Antolín-Rodríguez, J.M.; Martín-Ramos, P. Pilot plant for the capture of ammonia from the atmosphere of pig and poultry farms using gas-permeable membrane technology. *Membranes* **2021**, *11*, 859. [CrossRef]
27. Tao, W.; Ukwuani, A.T. Coupling thermal stripping and acid absorption for ammonia recovery from dairy manure: Ammonia volatilization kinetics and effects of temperature, pH and dissolved solids content. *Chem. Eng. J.* **2015**, *280*, 188–196. [CrossRef]
28. Guštin, S.; Marinšek-Logar, R. Effect of pH, temperature and air flow rate on the continuous ammonia stripping of the anaerobic digestion effluent. *Process Saf. Environ. Prot.* **2011**, *89*, 61–66. [CrossRef]
29. Evans, T.D. Recovering ammonium and struvite from digested sludge dewatering liquors. In Proceedings of the IWA Specialist Conference Moving Forward Wastewater Biosolids Sustainability, Jeju Island, Korea, 20–23 May 2007.
30. Licon Bernal, E.E.; Maya, C.; Valderrama, C.; Cortina, J.L. Valorization of ammonia concentrates from treated urban wastewater using liquid-liquid membrane contactors. *Chem. Eng. J.* **2016**, *302*, 641–649. [CrossRef]
31. Minocha, V.K.; Rao, A.V.S.P. Ammonia removal and recovery from urea fertilizer plant waste. *Environ. Technol. Lett.* **1988**, *9*, 655–664. [CrossRef]
32. Damtie, M.M.; Volpin, F.; Yao, M.; Tijing, L.D.; Hailemariain, R.H.; Bao, T.; Park, K.D.; Shon, H.K.; Choi, J.S. Ammonia recovery from human urine as liquid fertilizers in hollow fiber membrane contactor: Effects of permeate chemistry. *Environ. Eng. Res.* **2021**, *26*, 1–9. [CrossRef]
33. Jamaludin, Z.; Rollings-Scattergood, S.; Lutes, K.; Vaneeckhaute, C. Evaluation of sustainable scrubbing agents for ammonia recovery from anaerobic digestate. *Bioresour. Technol.* **2018**, *270*, 596–602. [CrossRef] [PubMed]
34. Ferraz, F.M.; Povinelli, J.; Vieira, E.M. Ammonia removal from landfill leachate by air stripping and absorption. *Environ. Technol.* **2013**, *34*, 2317–2326. [CrossRef] [PubMed]
35. Zhang, C.; Ma, J.; Waite, T.D. Ammonia-Rich Solution Production from Wastewaters Using Chemical-Free Flow-Electrode Capacitive Deionization. *ACS Sustain. Chem. Eng.* **2019**, *7*, 6480–6485. [CrossRef]
36. Soto-Herranz, M.; Sánchez-Báscones, M.; Antolín-Rodríguez, J.M.; Vanotti, M.B.; Martín-Ramos, P. Effect of acid flow rate, membrane surface area, and capture solution on the effectiveness of suspended gpm systems to recover ammonia. *Membranes* **2021**, *11*, 538. [CrossRef] [PubMed]
37. Sürmeli, R.; Bayrakdar, A.; Çalli, B. Removal and recovery of ammonia from chicken manure. *Water Sci. Technol.* **2017**, *75*, 2811–2817. [CrossRef] [PubMed]
38. Rodriguez, M.R.; Losada, E.; Besteiro, R.; Arango, T.; Velo, R.; Ortega, J.A.; Fernandez, M.D. Evolution of NH3 concentrations in weaner pig buildings based on setpoint temperature. *Agronomy* **2020**, *10*, 107. [CrossRef]

39. Salgado, I.G.; Guigui, C.; Sperandio, M. Transmembrane chemical absorption technology for ammonia recovery from wastewater: A critical review. *Chem. Eng. J.* **2022**, *444*, 136491. [CrossRef]
40. Reig, M.; Vecino, X.; Gibert, O.; Valderrama, C.; Cortina, J.L. Study of the operational parameters in the hollow fibre liquid-liquid membrane contactors process for ammonia valorisation as liquid fertiliser. *Sep. Purif. Technol.* **2021**, *255*, 117768. [CrossRef]
41. R Core Team. *R: A Language and Environment for Statistical Computing*; R Foundation for Statistical Computing: Vienna, Austria, 2020.
42. Wilhelm, E.; Battino, R.; Wilcock, R.J. Low-Pressure Solubility of Gases in Liquid Water. *Chem. Rev.* **1977**, *77*, 219–262. [CrossRef]
43. Starmans, D.A.J.; Melse, R.W. *Alternatives for the Use of Sulphuric Acid in Air Scrubbers*; Report 385; Wageningen UR Livestock Research: Lelystad, The Netherlands, 2011. (In Dutch)
44. Sander, R. Compilation of Henry's law constants (version 4.0) for water as solvent. *Atmos. Chem. Phys.* **2015**, *15*, 4399–4981. [CrossRef]
45. Zhang, C.; Ma, J.; Waite, T.D. The impact of absorbents on ammonia recovery in a capacitive membrane stripping system. *Chem. Eng. J.* **2020**, *382*, 122851. [CrossRef]
46. Palakodeti, A.; Azman, S.; Rossi, B.; Dewil, R.; Appels, L. A critical review of ammonia recovery from anaerobic digestate of organic wastes via stripping. *Renew. Sustain. Energy Rev.* **2021**, *143*, 110903. [CrossRef]
47. Vanotti, M.B.; Sanchez Bascones, M.; Millner, P.; Szogi, A.A.; Hashem, F. Development of treatment modules for capturing ammonia emmisions from poultry manures and recovering the nitrogen. In Proceedings of the IIV Symposium on Agricultural and Agroindustrial Waste Management, Rio de Janeiro, Brazil, 5–7 May 2015; pp. 7–10.
48. Hales, J.M.; Drewes, D.R. Solubility of ammonia in water at low concentrations. *Atmos. Environ.* **1979**, *13*, 1133–1147. [CrossRef]
49. Kurtén, T.; Torpo, L.; Sundberg, M.R.; Kerminen, V.M.; Vehkamäki, H.; Kulmala, M. Estimating the NH3:H2so4 ratio of nucleating clusters in atmospheric conditions using quantum chemical methods. *Atmos. Chem. Phys.* **2007**, *7*, 2765–2773. [CrossRef]
50. Riaño, B.; Molinuevo-Salces, B.; Vanotti, M.B.; García-González, M.C. Ammonia recovery from digestate using gas-permeable membranes: A pilot-scale study. *Environments* **2021**, *8*, 133. [CrossRef]
51. De Dobbelaere, A.; De Keulenaere, B.; De Mey, J.; Lebuf, V.; Meers, E.; Ryckaert, B.; Schollier, C.; Van Driessche, J. *Small-Scale Anaerobic Digestion: Case Studies in Western Europe*; Inagro: Rumberke, Belgium, 2015.
52. Cao, Y.; Bai, Z.; Misselbrook, T.; Wang, X.; Ma, L. Ammonia emissions from different pig production scales and their temporal variations in the North China Plain. *J. Air Waste Manag. Assoc.* **2021**, *71*, 23–33. [CrossRef]
53. Arogo, J.; Zhang, R.H.; Riskowski, G.L.; Christianson, L.L.; Day, D.L. Mass transfer coefficient of ammonia in liquid swine manure and aqueous solutions. *J. Agric. Eng. Res.* **1999**, *73*, 77–86. [CrossRef]
54. Guo, L.; Zhao, B.; Jia, Y.; He, F.; Chen, W. Mitigation Strategies of Air Pollutants for Mechanical Ventilated Livestock and Poultry Housing—A Review. *Atmosphere* **2022**, *13*, 452. [CrossRef]
55. Brienza, C.; Sigurnjak, I.; Michels, E.; Meers, E. Ammonia Stripping and Scrubbing for Mineral Nitrogen Recovery. In *Biorefinery of Inorganics: Recovering Mineral Nutrients from Biomass and Organic Waste*; Meers, E., Velthof, G., Michels, E., Rietra, R., Eds.; John Wiley & Sons, Inc.: Hoboken, NJ, USA, 2020; pp. 95–106. [CrossRef]
56. Menezes-Blackburn, D.; Paredes, C.; Zhang, H.; Giles, C.D.; Darch, T.; Stutter, M.; George, T.S.; Shand, C.; Lumsdon, D.; Cooper, P.; et al. Organic Acids Regulation of Chemical-Microbial Phosphorus Transformations in Soils. *Environ. Sci. Technol.* **2016**, *50*, 11521–11531. [CrossRef]

membranes

MDPI

Article

Membrane Dialysis for Partial Dealcoholization of White Wines

José Ignacio Calvo [1,2,*], Jaime Asensio [3], Daniel Sainz [3], Rubén Zapatero [3], Daniel Carracedo [3], Encarnación Fernández-Fernández [3], Pedro Prádanos [2,4], Laura Palacio [2,4] and Antonio Hernández [2,4]

1 Institute of Sustainable Processes (ISP), Dr. Mergelina s/n, 47071 Valladolid, Spain
2 Departamento de Física Aplicada, ETSIIAA, Universidad de Valladolid, Avenida de Madrid 57, 34004 Palencia, Spain; ppradanos@uva.es (P.P.); laura.palacio@uva.es (L.P.); antonio.hernandez@uva.es (A.H.)
3 Área de Tecnología de los Alimentos, Escuela Técnica Superior de Ingenierias Agrarias, Universidad de Valladolid, Avenida de Madrid 50, 34004 Palencia, Spain; jaimeasensioo@gmail.com (J.A.); dsainze@gmail.com (D.S.); rd.dr91@gmail.com (R.Z.); carracedoesguevillas@gmail.com (D.C.); encarnacion.fernandez@uva.es (E.F.-F.)
4 Departamento de Física Aplicada, Facultad de Ciencias, Universidad de Valladolid, Paseo de Belén 7, 47011 Valladolid, Spain
* Correspondence: joseignacio.calvo@uva.es

Abstract: Membrane dialysis is studied as a promising technique for partial dealcoholization of white wines. The performance of three membrane processes applied for the partial dealcoholization of white wines of the Verdejo variety has been studied in the present work. Combination of Nanofiltration with Pervaporation, single step Pervaporation and, finally, Dialysis, have been applied to white wines from same variety and different vintages. The resulting wines have been chemically and sensorially analyzed and results have been compared with initial characteristics of the wines. From the results obtained, we can conclude that all procedures lead to significant alcohol content reduction (2%, 0.9% and 1.23% v/v respectively). Nevertheless, the best procedure consists in the application of Dialysis to the wines which resulted in a reasonable alcohol content reduction while maintaining organoleptic properties (only 14 consumers were able to distinguish the filtered and original wines, with 17 consumers needed to be this differences significant) and consumer acceptability of the original wine. Therefore, membrane dialysis, as a method of partial dealcoholization of white wines, has undoubted advantages over other techniques based on membranes, which must be confirmed in subsequent studies under more industrial conditions. This work represents the first application of Dialysis for the reduction of alcohol content in wines.

Keywords: wine dealcoholization; nanofiltration; pervaporation; dialysis; white wine

1. Introduction

In recent years, most world winemaking regions have found that climatic change, leading to generalized global warming and important reduction in total rainfall, resulted in severe problems for the production of quality wines, among which: difficulties to achieve a correct balance between phenolic and technological grape maturation; grapes raisin before reaching a good maturation; accumulation of aromas in the white grape berry, more frequently in cold climates [1] and finally, resulting musts which have lower acidity, thus an excessive pH, and higher sugar contents with variation in the concentration of polyphenols [2].

This increase of sugar in must lead to wines with higher alcoholic content [3,4], which also increases the perception of a health risk from consumers, [5], a sensitivity that also could be supported by the restrictive legislation on alcohol rates allowed when driving. An excessive alcohol content also affects the taxes associated with the exportation of wines, making them difficult to compete in markets that are already very saturated and to consumers who traditionally have little appetite for high alcohol content beverages.

Of course, premature grape harvest and winemaking could reduce the potential alcohol content of the berries. Nevertheless, this would affect the final wine quality, since the phenolic maturity would not be yet fully achieved leading to more acid and less colored wines, at least in the case of red wines [6].

To reduce the alcoholic content of wines, different techniques can be used. The points to be met in order to carry out partial dealcoholization are, [7]:

The decrease in the acquired alcoholic strength by volume cannot be greater than 20% vol of original wine and final product must meet the definition of wine.

The treated wines must be suitable for direct human consumption (not having organoleptic defects).

If operations have been carried out to increase the alcoholic degree in one of the winemaking input products, the alcohol could not be eliminated.

Studies developed in the field of viticulture show strategies to act on the techniques of wine cultivation (irrigation or plant canopy management to control photosynthesis), thus avoiding the excessive accumulation of sugar [8]. There are even techniques that are performed before alcoholic fermentation, such as the addition of glucose-oxidase and catalase enzymes [9].

Many studies have also evaluated the suitability of non-*Saccharomyces* and *Saccharomyces cerevisiae* yeast strains that could prevent ethanol formation in alcoholic fermentation [10–15].

Several publications have successfully studied a pre-fermentation approach by membrane Nanofiltration (NF) for sugar control in grape must [6,16–19]. Results showed that the best NF technique was a two-stage NF process [18], which promoted higher recovery of grape must chemical properties and fewer volume losses. However, sensory evaluation of the wines obtained showed that they were less persistent in mouth and had lower flavor in comparison to control wines. This aroma depletion was attributed to the possible loss of volatile compounds (primary aroma compounds) during grape must nanofiltration.

In order to minimize those losses, primary aroma compounds could be recovered, before NF, from the grape juice and added to the filtered must before fermentation. Traditional aroma recovery processes such as distillation, adsorption and solvent extraction are not useful since the operation at high temperature [20] could deteriorate original juice and its aromas. In view of their intrinsic characteristics, namely high selectivity and possibility of operation at moderate temperatures, Pervaporation (PV) is a membrane process that seems highly appropriate for the separation of dilute species in liquid solutions [21–23]. Specifically, organophilic PV membranes have a high potential for recovering natural and natural-like aroma compounds, highly diluted in complex aqueous media, [24].

On the other hand, non-alcoholic (or reduced alcoholic beverages) can be produced by eliminating ethanol from a fully fermented beverage (post-fermentation approach). The most common separation techniques for the dealcoholization of beverages are thermal treatment or membrane-based processes [25]. The heat treatment processes include evaporation and distillation or steam separation, in both cases under vacuum conditions.

Other techniques as osmotic distillation or rotating cone columns have been successfully used [26–30].

Membrane-based processes include Electrodialysis (ED) mainly used for tartaric stabilization, acidification, and de-acidification of musts and wines, [31]. While the membrane processes used for alcohol reduction of beverages include Reverse Osmosis (RO) [32,33], NF [34], and PV [35].

Among membrane separation processes, Dialysis (D) offers interesting features which suggest that its application for the partial reduction of alcohol in wines could be highly beneficial. For example, it does not use any pressure or temperature gradient, which could result in sensorial worsening of the resulting product. It has been successfully applied for alcohol reduction of beers [36]. Nevertheless, to the best of our knowledge, no study has been done to apply Dialysis to wine dealcoholization [37].

Therefore, the scope of the present research is to compare the feasibility of several approaches for the reduction of alcohol content in a white wine. Three distinct approaches (from more conventional Nanofiltration to less aggressive Pervaporation or Dialysis) have been tested on similar white wines and features and differences of the resulting wines will be analyzed in terms of alcohol reduction, chemical properties, and organoleptic descriptors.

2. Materials and Methods

2.1. Dealcoholization Procedures

Three procedures of membrane-based wine dealcoholization will be compared in this work: (a) The first approach consisted, as previously commented, in using a NF membrane to reduce the alcohol content of the treated wine. NF permeate resulted in a significant reduction of alcohol content but also an important decrease of aromatic compounds. Therefore, permeate was further pervaporated to recover some of the aromas lost during NF. From the retentate of NF and the aromas recovered by PV, the starting wine is reconstructed, from which a sensible reduction of the alcohol content is expected, as well as an aromatic and organoleptic composition, as close as possible to the original wine. (b) A second approach consisted in a one-step filtration. Wine was subjected now to a PV step, from which part of the alcohol of the wine was recovered as permeate. Also, some aromas of the wine were pervaporated and passed through the membrane. Since most aroma compounds are more volatile than alcohol, the initial fraction of PV permeate was rich on aromas and less in alcohol. Consequently, this first fraction was recovered and added to the original wine once filtration finished. The rest of the PV permeate, mostly composed of alcohol, was discarded to get the maximum alcohol degree reduction. (c) The third approach consisted also in a single step filtration, by means of a D setup. Wine and pure distilled water were recirculated from both sides of an appropriated membrane by means of two peristaltic pumps in no transmembrane pressure conditions. Two transport mechanisms acting simultaneously (osmotic flow of water to the wine side and diffusion of volatile compounds of wine, mainly alcohol, to the water side) lead to a resulting wine with lower alcohol content.

Consequently, three experimental devices were designed to perform the NF, PV, and D stages, respectively—devices that will be described below.

2.2. Filtration Devices

2.2.1. Nanofiltration Experimental Set Up

NF was carried out in an automated pilot plant (see Figure 1), designed at our Research Lab (SMAP), and consisting of a 30 L stainless steel feeding tank with a cold jacket (1) for temperature control through a Neslab (ThermoFischer, Madrid, Spain) refrigerator bath (2). Feed fluid was recirculated with a Hydra-Cell D10 pump (Wanner Eng., Minneapolis, MN, USA) (3) to a flat membrane module (4) with transverse flow, housing the membrane (5), with an active membrane area $A = 6.6 \times 10^{-3}$ m^2. Two digital pressure gauges located before and after the membrane module measured the inlet (P_i) and outlet (P_o) pressure. In order to manually adjust the pressure inside the module, a needle valve was placed at the output of the unit. The transverse flow was measured with a Tecfluid (Barcelona, Spain) flow meter (measuring range 0–10 L/min) and automatically adjusted to avoid the pressure exceeding a critical level (settled at 40 bar), so that no breakage occurred in the membrane. Tangential speed, inlet and outlet pressure, and permeability measurements are completely automatized.

Feed stream temperature was kept constant at 6 °C during filtration, while pressure was kept controlled at 30 $\pm$ 1 Bar and the feed rate was adjusted to 6.0 $\pm$ 0.2 L/min. NF permeate was collected in a graduated 2 L beaker to estimate the volume yet permeated.

After finishing the NF step, permeate was kept in the fridge up to complete the PV step.

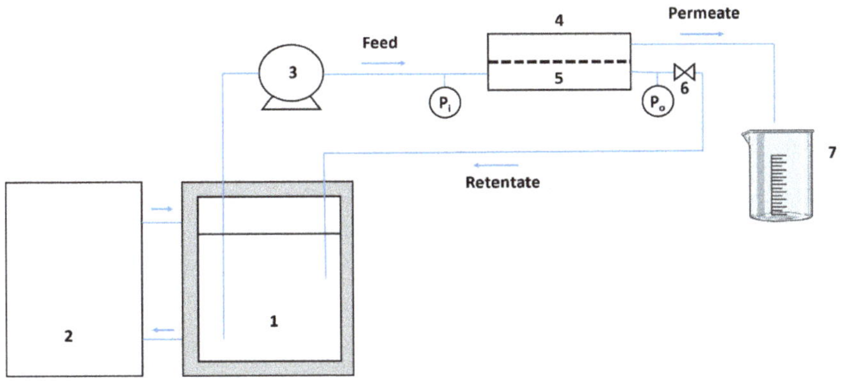

Figure 1. Nanofiltration (NF) experimental system.

2.2.2. Pervaporation Experimental Set Up

The recovery of the aroma compounds of NF permeate (Case 1) or the aroma recovery and alcohol elimination (Case 2) was carried out in a laboratory cell equipped with a PV flat sheet. The installation built for the experiments is schematized in Figure 2.

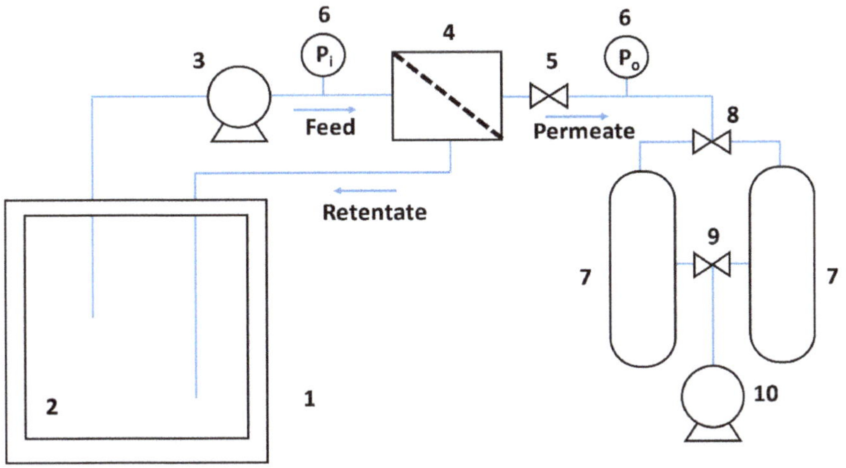

Figure 2. Pervaporation (PV) experimental system.

The pervaporation device contains a pervaporation membrane, placed in a flat sheet cell (4) that provides an active area of 6.6 × 10^{-3} m^2. A Flojet (Xylem, Hertfordshire, UK) electric pump (3), model R2100-232, extracts the wine to be pervaporated from the feed tank (2) of 3 litre capacity and equipped with a cold jacket (1) controlling temperature by a Julabo model F12 (Iberfluid, Madrid, Spain) circulation bath. This fluid circulates tangentially over the membrane to limit the effects of concentration polarization. Two nanoscale pressure gauges (6) are placed before and after the cell containing the membrane to avoid overpressures in the membrane; which can be controlled by a needle valve (5), placed after the outer meter. In order to control the flow rate of the fluid through the circuit, a flowmeter (Parker, Cleveland, OH, USA) of the rotameter type is placed. On the other hand, two nitrogen traps (7) are placed at the exit of the pervaporation membrane cell, prepared to collect permeate. Prior to these traps, a three-way valve (8) is placed to direct permeate towards a liquid nitrogen trap or another. Permeate is then condensed in those

2 liquid nitrogen traps alternately. While one of the traps is tempered, the other continues collecting permeate that passes through the membrane. Between the nitrogen traps and the vacuum pump, another three-way valve is used (9), which allows the suction of this pump to be diverted to one or the other trap. This pump (10) is of the vacuum displacement type of the brand Agilent Technologies (Madrid, Spain) model SH 110, which achieves a minimum pressure of 2 mmHg. The liquid nitrogen traps are kept inside a Dewar vessel in which the nitrogen is poured, delaying the evaporation of it. Food fat was also used in order to facilitate the closure of the flask where permeate is collected, and a 20 mL volumetric pipette to extract permeate without touching the fat.

2.2.3. Dialysis Set Up

The equipment used to perform D step is schematized in Figure 3. It consists of an Eheim (Leadfluid, Hebei, China) peristaltic pump, model 1048 (5) that aspires the wine (1.4 L approximately) from the 2 L Erlenmeyer (1) to the cell (3) containing the dialysis membrane (4), with an effective membrane area of 4.4×10^{-3} m^2. Then it returns to the initial Erlenmeyer flask, so that it recirculates continuously at a given flow rate.

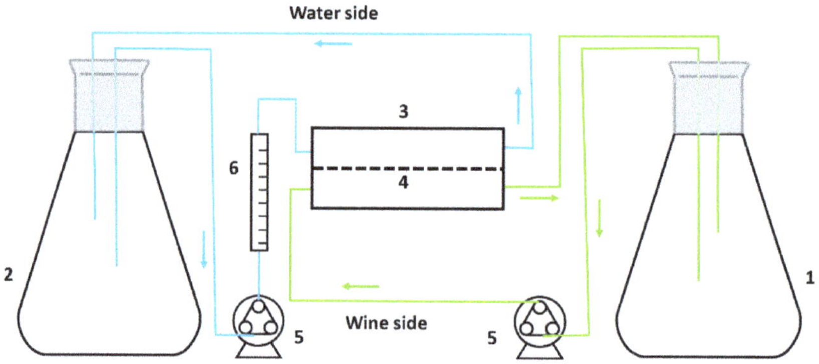

Figure 3. Dialysis (D) experimental system.

On the other hand, another Erlenmeyer flask (2) contains 1.4 L of a complementary liquid (in our case Milli-Q water, from Merck, Mollet del Vallés, Spain) that is aspirated by a peristaltic pump equal to the one described above (5). It passes in a co-current direction through the membrane cell and collects the dialyzed alcohol from the wine. Finally, the flow rate is controlled by a rotameter type Parker flowmeter (6).

Both Erlenmeyer flasks are covered with Parafilm® to prevent oxidation and loss of aromas and alcohol from the synthetic and real wine. In this way, we reduce the error due to losses that may occur during the process.

2.3. Membranes

The NF membrane used in the first case was DK (DK1812C-28D, from Lenntech, Delfgauw, The Netherlands), from the D series, an asymmetric membrane made by phase inversion, from General Electric (GE). The DK membrane has a minimum rejection of 98% over 2000 ppm of $MgSO_4$ at 25 °C (77 °F) and 110 psi operating pressure.

The PV membrane used in cases 1 and 2 consisted in a PDMS based organophilic membrane made and commercialized by Pervatech (Rijssen, The Netherlands). Previous work in our lab (23) showed that PDMS based membranes give nice performance for aroma recovery through PV of natural grape must.

Regarding D membranes, to be sure of the optimal alcohol recovery of the process, three suitable membranes were studied for the Dialysis system:

Membrane type PLGC 15005: Marketed by Merck (Madrid, Spain) under the commercial name of Ultracel, it is a UF membrane composed of low binding regenerated cellulose.

It has a nominal molecular weight value of 10,000 Da and is presented in the form of discs with a diameter of 150 mm.

X25 SPECTRA/POR membranes: In this work 2 different membranes of this type, supplied by Fisher Scientific (Madrid, Spain), were tested. These membranes come in the form of flat sheets composed of regenerated cellulose and they are aimed for use in dialysis experiments. They present good compatibility with dilute strong acids and bases or concentrated weak acids and bases, many alcohols, and organic compounds. The two different types of membranes tested are: X25 SPECTRA/POR 1, 6-8KD, MWCO 240 × 240 mm sheet, with a nominal molecular cut-off weight (MWCO) of between 6000–8000 Da, and presented as flat sheets with a length of 240 mm and a width of 240 mm; and X25 SPECTRA/POR 2, 12–14KD, MWCO 200 × 200 mm sheet, having a MWCO between 12,000–14,000 Da, also in the form of flat sheets of 200 mm in length and 200 mm in width.

Main nominal characteristics of the used membranes are summarized in Table 1.

Table 1. Nominal characteristics of membranes used.

Process	Membrane	Manufacturer	pH Range	Max. Temperature /(°C)	Max. Pressure /(bar)	MWCO /(kDa)
NF	DK-28D	General Electric	3–9	80	30	0.15–3
PV	PDMS	Pervatech	1–12	70	5	-
D	PLG15005	Merck	2–12	121	5	10
D	SpectraPor1	Fisher Sci.	2–12	121	-	6–8
D	SpectraPor2	Fisher Sci.	2–12	121	-	12–14

2.4. Chemical Analysis

After each filtration procedure, the analysis of the main oenological parameters were carried out on both wines, the original wine and the dealcoholized wine of all tests, nanofiltration, pervaporation, or dialysis.

Chemical parameters as pH, total and volatile acidity, alcoholic degree, total polyphenol index (TPI), total and free sulphur dioxide, and color were determined according to the principles and methods established by the International Organization of Vine and Wine [38]. All the chemical analysis were carried out in duplicate for each wine, being both results averaged. From the data obtained from the chemical analyses, a statistical treatment was carried out by means of analysis of variance (ANOVA). For this, the Statgraphics Centurion XVIII program (StatPoint Technologies, Inc., Warrenton, VA, USA) has been used.

Finally, the time evolution of the PV and D processes was followed by the aid of a differential refractometer, from Atago (Saitama, Japan). Regularly, small quantities (5 mL) were obtained from the wine side and their refractive index measured in that device. Then, using a calibration curve, refractive index measured data were converted in contents of alcohol.

2.5. Synthetic Wine

While, for the case of NF or PV, we had previous experience on the suitability of the membranes selected and the expected duration of the experiments, [24], for the case of D, some previous experiments were needed aimed to select the most appropriated membrane.

Then, for the third case, a synthetic wine was prepared and used to test the best performance membranes. Synthetic wine was composed of alcohol at 96% vol. and mostly distilled water (88.6% vol., for an approximate final graduation of 11.4% vol.).

2.6. Verdejo White Grape Wine

The wine used in this study consisted in an experimental Verdejo wine elaborated in the Agricultural Engineering School (University of Valladolid, Palencia, Spain). The variety, with origin in the region of Castilla y León, produces full flavored white wines (Verdejo appellation). Different amounts of such experimental wines (coming from different years and harvests) were used in all case studies. For each case, a different number of wine bottles was used.

To homogenize the different bottles used and prevent membrane clogging since the original wines were bottled unfiltered, a similar procedure was followed in all cases: The number of bottles necessary for each experience was selected. Then, all the wine was homogenized in an appropriated tank to facilitate the sedimentation of the thickest particles. Finally, the resulting homogeneous wine was again bottled in 0.75 L standardized bottles and used for the filtration experiments designed for each case. In all cases enough number of bottles remained stored to be used as a reference sample to carry out the subsequent wine analyses and tastings

Case 1: (NF + PV)

2014 Vintage. The initial alcohol content of the wine used in Case 1 was 11.2% *v*/*v*

Case 2: (PV)

2015 Vintage. For this case, initial alcohol content was 11.5% *v*/*v*.

Case 3: (D)

Finally, the wine used in this case (from 2018 Vintage) presented an initial alcohol content of 9.90% *v*/*v*.

In all three cases, the wine was elaborated just after grape harvesting, while filtration procedure, chemical analysis, and sensorial analysis were accomplished in the following year (2015, 2016, and 2019, respectively).

2.7. Sensory Analysis

Sensory evaluation of the wines was conducted in three sessions (one for each case of study), using in all cases triangle tests to demonstrate the existence or not of a noticeable difference between products [39], and check-all-that-apply (CATA) questions [40] to characterize organoleptic properties of treated and untreated wines.

The check-all-that-apply (CATA) questionnaire contained 15 terms related to the sensory characteristics of the Verdejo wines grouped in two visual terms (clean, rusty), eight olfactory terms (fruity, citric, tropical, herbaceous, balsamic, reduced, aniseed, aroma intensity), four terms in the mouth (bitter, acid, mouth volume persistence), and finally consumers were asked to give a general evaluation of the wine by stating whether the wines were acceptable for them.

Consumers were asked to check all the terms that they considered appropriate to describe each wine. The terms were selected based on published data [41] considering the descriptors selected by trained assessors in preliminary studies, along with the typical characteristics of Verdejo wines.

A total of eighty-six people participated in the study. The first session was conducted with twenty-six consumers, while twenty-seven participated in the second testing, and thirty-three tested the wines from third case. In all cases, consumers were students and professors of the Oenology degree in the Agricultural Engineering School with experience in wine tasting and ages between 18 and 50.

In all sessions, the samples were served as 25 mL aliquots in standardized wineglasses [42], which were coded with 3-digit numbers randomly generated. The serving temperature of the samples was 10 ± 1 °C. All these sensory evaluations were carried out at the Sensory Science Laboratory of the School of Agricultural Engineering, at the University of Valladolid, Palencia (Spain), in individual booths designed in accordance with ISO 8589 [43].

For the triangle test, the significance level was determined according to the number of tasters who correctly distinguished between the tested samples, as well as the proportion of the population that the sample can distinguish [38].

Frequency of use of each one of the terms of the CATA question was determined by counting the number of consumers that used that term to describe each sample. Cochran's Q test [44] was carried out to identify significant differences among samples for each of the sensory terms using IBM SPSS Statistics software, version 24.0.

3. Results and Discussion

3.1. Filtration

3.1.1. Case 1 (NF + PV)

Case 1 consisted in two filtration steps (NF and PV) separately, and a further reconstruction of the wine, so each step will be now commented.

Nanofiltration

For the NF process, departing from an initial effective volume of 8 L of wine, after 10 h of nanofiltration, 1700 mL of permeate were collected in a graduated beaker (total capacity 2.0 L). Said permeate was stored in a refrigerator at the end of the NF stage to assure its best conservation until PV step was carried out. During NF filtration, periodic measurements of permeate flux were done showing that membrane permeability was constant during all the process. So, we can conclude that membrane was not appreciably affected by fouling.

Pervaporation

Regarding the pervaporation process, it was started with the initial volume of 1700 mL of permeate resulting from the nanofiltration step and, after 520 min, 91 mL of pervaporate were obtained distributed in 12 extractions as shown in Figure 4.

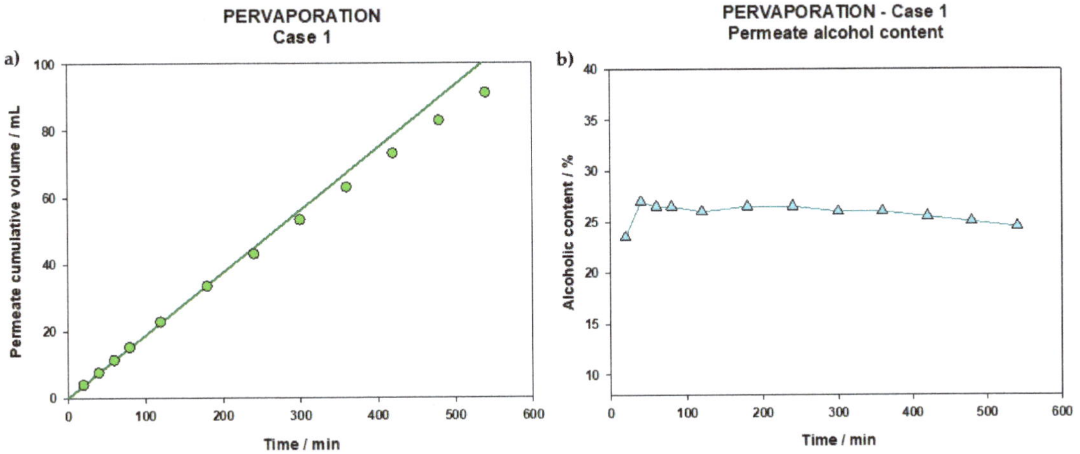

Figure 4. (**a**) Cumulative volume of permeate collected during PV step of Case 1. (**b**) Alcohol content of PV permeate (Case 1).

In Figure 4a it can be observed that the slope of the volume of permeate collected versus time was reasonably constant over time; indicating that the membrane at no time was filled and only vapor transport occurred (something otherwise expected when working with gaseous and liquid phases on both sides of the membrane). In fact, the figure also presents a straight line corresponding to the linear fit of the first points. It can be seen that, in the initial moments, the flow is constant due to the passage of aromas and a greater presence of ethanol in the liquid phases, but afterwards, a slight decrease in flow is observed due to the elimination of the most volatile (and organophilic) compounds due to the reduction of alcohol in the liquid phase.

A refractometry follow-up of the alcoholic content of permeate collected can be seen in Figure 4b. The first extraction, which corresponds to the aromas according to their volume [23], corresponds to a concentration of 23.5%. The next extraction clearly increases this alcohol content, followed by a slow decrease up to a final value of 24.5% (after 540 min of PV filtration). The fact that the first point has a lower value may be related to the time-lag of the membrane until the stationary profile is defined, since the refractometer will not differentiate alcohol from other volatiles. These concentrations of ethanol in the permeate are lower than another study in which a wine was dealcoholized by PV [35]; this may be due to the use of a different type of membrane or a starting ethanol concentration of the pervaporated fluid.

Since the content of alcohol in permeate is always higher than original wine, we can be sure that PV retentate alcohol content will decrease during the process. That is the reason why the alcohol content of the permeate extractions is also decreasing due to the decrease in alcohol of retentate which makes the alcohol extraction process less effective with time. This is in line with the decrease in flow observed in Figure 4, which, as we have said, is due to the decrease in the concentration of alcohol in the feed. The same reduction of components collected in PV was observed in another study [45] as the PV process progressed.

Taking into account the alcoholic content of all resulting fractions (11% *v*/*v* for the NF retentate and 6.9% *v*/*v* for PV retentate) the wine was reconstructed to achieve the maximum alcohol reduction permitted. So, 1.47 L of retained NF and 1.36 L of retained PV were mixed and added the 3.8 mL of aromas collected during initial steps of PV. Finally, 2.83 L of dealcoholized wine were obtained with a final alcohol content of 9.2% *v*/*v*.

3.1.2. Case 2 (PV)

Next, filtration results coming from the second approach will be discussed.

Pervaporation

For this process, a total volume of 2.5 L of the control wine was filtered. The PV process lasted for 526 min (8.8 h) and after that a volume of 21.8 mL of PV permeate was obtained, distributed in 15 extractions, whose cumulative volumes can be seen in the following graph (Figure 5a).

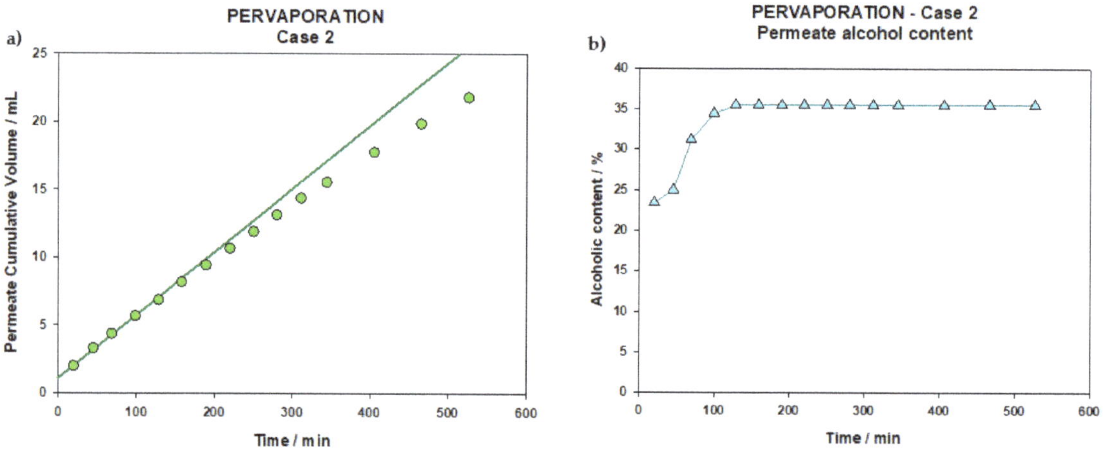

Figure 5. (**a**) Cumulative volume of permeate collected during PV (Case 2). (**b**) Alcohol content of PV permeate (Case 2).

Similarly to that arisen in the PV step of Case 1, this figure allows for the assurance that the membrane was never liquid filled. The hydrophobicity of the PDMS membrane, succeeded in preventing the passage of aqueous solutions. So, only the volatile substances present in the medium, favored by the vacuum created on the permeate side, vaporize and pass the membrane, being collected in the nitrogen traps at a constant rate.

The first extraction corresponds to the aromas according to its volume [17,18,23], and the fact is that it presents the lowest refractive index (1.345, which corresponds to an alcoholic degree of 23.4% vol.). Therefore, the content of this first extraction will be added later to the final wine. As can be seen in the Figure 5b, the values for alcohol content are increasing until reaching the fifth extraction, in which the highest value is reached, 35.5% vol. Afterwards the permeate graduation stabilizes, remaining almost constant during the rest of the process. In any case the behavior is very similar to the previous case (Figure 4a,b), but in this case alcohol values in the permeate are sensibly higher than in Case 1 and a longer membrane stabilization time is needed, possibly because the membrane is in contact with wine rather than with an NF permeate that should only contain small molecules and a sensibly lower amount of alcohol.

Wine Reconstruction

Once PV is finished, a partially dealcoholized wine (retentate) is obtained. The 2 mL that were obtained from the first PV permeate extraction, which should be the fraction with the highest aromatic content as previously mentioned, were added to this retentate. A total of 2.25 L of treated wine were obtained, which were bottled instantly, avoiding possible oxidations.

3.1.3. Case 3 (D)

In Case, 3, prior to the filtration step, a study was conducted to select the most appropriate membrane for the Dialysis process.

Membrane Selection

As commented previously, for this case, several membranes were tested to check their performance in the D process. Same synthetic wine was subjected to D filtration for more than 21 h, with extractions (5 mL) from each side flask (wine and water) every hour to control (by refractometry) the content of alcohol extracted from wine side and which part of it passed to the water side. After 4 h, the process continued all night and a final extraction was made on the filtration end.

Time evolution of the alcohol contents in both wine side and water side flasks are compared in Figure 6 for all membranes studied. Here, it can be clearly seen that Ultracel membrane presents the best results, with a clearly faster alcohol reduction which allows us to obtain the expected reduction of alcohol degree in sensibly lower times.

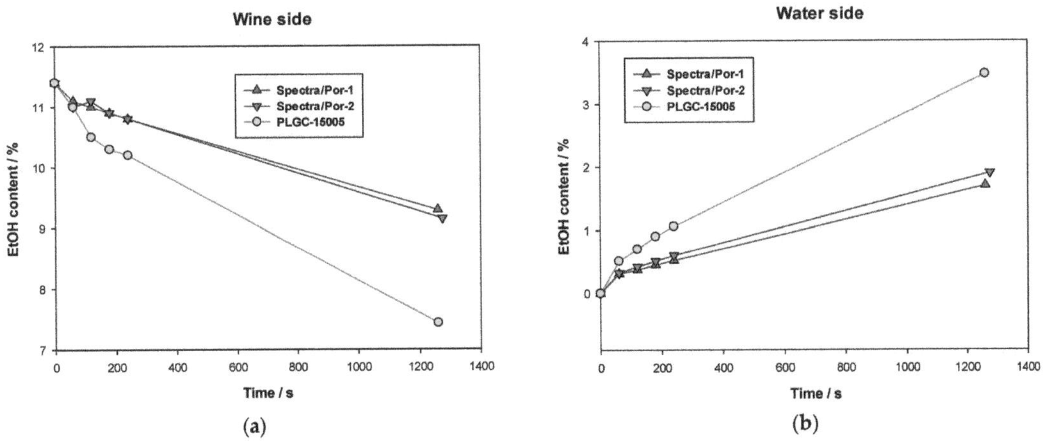

Figure 6. Alcohol content evolution during D for each membrane tested: (**a**) Wine side, (**b**) Water side.

Filtration of Real Wine

According to the results of the previous study, it seems clear that, among the membranes tested, PLGC15005 offers the best performance, at least in terms of alcohol recovery.

Therefore, this membrane was selected to perform the D filtration of the actual white wine. For such purpose, and taking into account that OIV does not allows alcohol reductions over 20% of initial value (which means 2% degrees for a wine initially around 10% vol.), the filtration time was reduced to approximately 8 h.

To get enough filtered wine for the sensory testing, D was performed in three sessions with an initial amount of 1.4 L of the original white wine, filtered in each session. All sessions lasted for 8 h, and the resulting wines of the 3 sessions were homogenized and bottled again for further chemical analysis and sensorial testing

3.2. Chemical Analysis

The reconstructed wine and the original one were analyzed for the parameters described in Materials and Methods Section and for all 3 cases of study. Corresponding results are presented in Table 2 for all Cases.

Table 2. Mean values and standard deviations of wines chemical composition (for all 3 cases).

	Case 1 (2014 Vintage)		Case 2 (2015 Vintage)		Case 3 (2018 Vintage)	
	Original Wine	**NF + PV**	**Original Wine**	**PV**	**Original Wine**	**D**
pH	3.17 ± 0.03 [a]	3.22 ± 0.01 [a]	2.90 ± 0.03 [a]	3.00 ± 0.01 [a]	3.24 ± 0.02 [a]	3.04 ± 0.01 [b]
Total acidity (g/L tartaric acid)	4.90 ± 0.01 [a]	4.02 ± 0.02 [b]	4.40 ± 0.01 [a]	4.40 ± 0.01 [a]	4.6 ± 0.06 [a]	4.2 ± 0.00 [b]
Volatile acidity (g/L acetic acid)	0.16 ± 0.01 [b]	0.22 ± 0.01 [a]	0.16 ± 0.01 [a]	0.13 ± 0.01 [b]	0.34 ± 0.01 [a]	0.30 ± 0.02 [a]
Color (AU)	0.103 ± 0.02 [a]	0.122 ± 0.01 [a]	0.097 ± 0.02 [b]	0.105 ± 0.01 [a]	0.094 ± 0.002 [a]	0.084 ± 0.003 [a]
Alcohol content (% *v*/*v*)	11.20 ± 0.10 [a]	9.20 ± 0.10 [b]	11.50 ± 0.10 [a]	10.60 ± 0.20 [b]	9.90 ± 0.00 [a]	8.70 ± 0.00 [b]
Free sulfur dioxide (mg/L)	nd	nd	3 ± 0.0 [a]	3 ± 0.0 [a]	43 ± 2 [a]	29 ± 6 [a]
Total sulfur dioxide (mg/L)	75 ± 0.02 [a]	66 ± 0.06 [b]	109 ± 0.0 [a]	89 ± 0.06 [b]	98 ± 2.1 [a]	103 ± 4.0 [a]
TPI (Total Polyphenol Index)	4 ± 0.1 [a]	4 ± 0.1 [a]	4 ± 0.0 [a]	4 ± 0.0 [a]	5 ± 0.05 [a]	4 ± 0.16 [b]

nd: not detected. [a,b]: different letters indicate significant differences among samples for each case ($p < 0.05$, Tukey test).

Firstly, a consideration must be made about the analysis of the original wines. Since they correspond to different harvests and years, they are expected to have slightly different compositions, reflecting the atmospheric and maturation history of each vintage.

Referring to the change of the parameters from original to filtered wines we can note several aspects as follows. pH remains almost unchanged for all cases, with a slight increase in Cases 1 and 2 and slight decrease in Case 3. Otherwise, total acidity decreased significantly for Cases 1 and 3, and remained constant for Case 2. For Case 1, this decrease may be due to selectivity of the NF membrane to tartaric acid (majority within a wine), as was also observed in previous works [46]. This selectivity to tartaric acid could also be the explanation of the slight reduction of total acidity for Case 3. Normally a reduction in total acidity is accompanied by a certain increase of pH, which did not happen for Case 3. However, it should be considered that pH and total acidity refer to different contributions. While pH accounts for the proton concentration, total acidity is expressed in terms of tartaric acid, one of the acid components of wine, along with malic, citric, and lactic. Then, different behavior of pH and total acidity can be experimentally found. On the other side, volatile acidity increased slightly for Case 1, with a very low decrease in Case 2 and a bit higher for Case 2. Regarding the volatile acidity, a significant increase in the dealcoholized wine can be seen; this may be due to the excessive recirculation (especially during NF step), that caused a high aeration, and to the high temperature at which the nanofiltration retention was maintained during pervaporation. Previous studies [46], showing an increase in the content of acetaldehyde, acetic acid precursor, were observed in wines that underwent a dealcoholization using membranes. As in the case of total acidity, in studies carried out partially removing the sugars from a must of Verdejo, to later elaborate wine with it [47], the volatile acidity of the resulting wine was equal and even lower when made with NF or UF. As demonstrated in other studies [48], the temperature and lighting to which the wine was subjected during the NF and PV processes could favor the increase of the volatile acidity of our wine that is dealcoholized.

The content of free sulfur dioxide in both samples is low for Cases 1 and 2 (not detected for Case 1) while a bit higher for Case 3. This relatively low values of free sulfides is connected with the time passed (from initial bottling to the present study), which is clearly longer in the Cases 1 or 2 (2014 and 2015 vintage, respectively), and also with the homogenization and sedimentation that caused small losses of free sulfides that could have been in the wine. Regarding the total sulfur dioxide, it fell significantly for Cases 1 and 2, remaining almost constant for Case 3.

The most important chemical parameter here is the alcohol content, whose reduction is the objective of this work. For Case 1, alcohol content reduces from 11.20% in the original wine to the final 9.20% in the reconstructed one, thus achieving nearly the maximum allowed reduction. However, this wine was reconstructed with appropriated additions of NF retentate and PV permeate, so the alcohol reduction was forced. Second Case leaded to a lower alcohol reduction (0.9%), while D achieved a gentle alcohol reduction (1.2%) which could be higher, if necessary, just by increasing slightly the filtration time.

Here are insignificant changes on TPI after filtration in Cases 1 and 2, with significant reduction for Case 3. While color increases sensibly in Case 1 and, to a minor extent, in Case 2, Case 3 led to almost equal values of color, which means that the wine does not suffer from oxidation or browning after processing.

3.3. *Sensory Analysis*

3.3.1. Triangular Test of Differences

For all results of triangular test, hits means that the judge was able to discriminate which of the 3 wines presented was different while wrong answers correspond to a failure. Table 3 presents results of a triangular test for all Cases, with the following number of judges: 26 for Case 1, 27 for Case 2, and 33 for Case 3. For the sake of clarity, the vintage of each case must also be included, and finally the number of hits needed to consider both wines (filtered and original), distinguishable with a $p < 5\%$ level of significance.

Table 3. Data obtained in the triangular test (for all Cases).

	Case 1 (26 Judges)	Case 2 (27 Judges)	Case 3 (33 Judges)
Hits	23	14	14
Failures	3	13	19
Vintage	2014	2015	2018
Significant number of answers	14	14	17

In the first two cases it was found that the tasters successfully distinguished between both wines at a significance level of $p \leq 0.05$, however in Case 3 the tasters were not able to distinguish between the two wines and, therefore there are no perceptible differences between the samples ($p \leq 0.05$). As can be seen in last row of Table 3, only 14 consumers were able to distinguish both wines, while 17 were needed to be sure of such distinguishability.

In addition, the confidence interval on the proportion of the population that can distinguish the sample was calculated. In Case 1 the upper confidence limit is 0.99 and the lower one is 0.67; in Case 2 the upper confidence limit is 0.51 and the lower one is 0.04; finally, regarding Case 3 the upper confidence limit is 0.34 and the lower one has a negative value. From the statistical analysis of data, we can conclude that the actual proportion of the population capable of distinguishing the samples is between 99 and 67% in Case 1, between 51 and 4% in Case 2 and, between 0 and 34% in Case 3, at a confidence level of 95%. Therefore, the organoleptic differences in Case 3, if any, must be a little noticeable.

3.3.2. Check-All-That-Apply (CATA)

Now, the data obtained during the CATA test will be presented and commented (Table 4).

Table 4. Frequency of citations obtained in the CATA test used by consumers to describe each sample for all cases.

	CASE 1 (2014 Vintage)			CASE 2 (2015 Vintage)			CASE 3 (2018 Vintage)		
		Samples			Samples			Samples	
Number	Terms	Original Wine	NF + PV	Terms	Original Wine	PV	Terms	Original Wine	D
1	Clean **	12	23	Clean **	10	20	Clean ns	24	21
2	Rusty ns	3	2	Rusty ns	6	5	Rusty ns	1	3
3	Fruity ***	17	4	Fruity ns	15	17	Fruity ns	14	12
4	Citric ns	14	11	Citric ns	10	12	Citric ns	16	20
5	Tropical **	11	3	Tropical ns	5	4	Tropical ns	5	8
6	Herbaceous ns	6	9	Herbaceous ns	9	10	Herbaceous ns	12	9
7	Balsamic ns	8	5	Balsamic ns	7	4	Balsamic ns	4	4
8	Reduced ns	3	4	Reduced ns	5	4	Reduced ns	3	5
9	Aniseed ns	7	4	Aniseed ns	6	8	Aniseed ns	6	7
10	Aroma intensity **	11	2	Aroma intensity *	14	10	Aroma intensity ns	15	10
11	Bitter ns	14	12	Bitter ns	13	12	Bitter ns	10	13
12	Acid ns	18	17	Acid **	16	9	Acid ns	19	23
13	Mouth volume **	12	4	Mouth volume *	12	7	Mouth volume ns	5	5
14	Persistence **	15	5	Persistence *	15	10	Persistence ns	9	10
15	Acceptable **	16	7	Acceptable ns	16	13	Acceptable ns	10	10

***: indicates significant differences at $p < 0.001$. **: indicates significant differences at $p < 0.01$. *: indicates significant differences at $p < 0.05$. ns: indicates no significant differences ($p > 0.05$) according to Cochran's Q test.

Case 1: As can be seen from the previous Table, filtered wine is considered as cleaner than original wine, but this is usual after any filtration procedure. On the other hand, the great majority of the aromatic attributes of the dealcoholized wine, minus the herbaceous aroma, were less significant than in the case of the original wine. Likewise, the overall aromatic intensity of the dealcoholized wine was also valued as being considerably lower than that of the original wine. Also, important tasting parameters as volume in mouth and persistence are registered by judges in much lower extent for filtered wine. Finally, the overall valuation of the wines also is less favorable to the treated wine, so we can conclude that the procedure applied in Case 1 results in a worse wine in aromatic, gustatory, and overall aspects.

Contrary to what we have found in our study for yet developed wines, when the dealcoholization of the wines has been carried out by means of a partial reduction of the sugars of the starting musts [47], the differences have been much less sensorial. So, from both sensory tests we can conclude that, in Case 1, treated wine presents important organoleptic differences with original wine which makes this option not very useful for industrial applications.

Case 2: Regarding the sensory attributes related to the visual phase, the control wine has a greater limpidity (expected after filtration), and a less oxidized appearance than the control wine.

In the olfactory phase, the dealcoholized wine seems to have lost part of its aromatic qualities, but not as strongly as in Case 1. Anyway, this leads to sensible damage of its aromatic intensity. These changes resulted in a lower acceptability of the filtered wine with respect to the original one.

Case 3: According to the Cochran test for results of Case 3, none of the attributes presented significant differences between original and treated wine. However, some comments should be made.

It is a bit surprising that the descriptor clean obtained higher attributions for the original wine that for the treated one, since normally filtration results in cleaner wines. In any case, the differences are so small that they could be neglected. Also, Dialysis treated wine is found as more oxidized than the original wine but both have such small punctuations of rust that these differences can be considered as likely experimental errors.

Referring to olfactory parameters, it seems Dialysis filtration has not resulted in a sensible loss of aromatic aspects; even most parameters present some increase. However, whole D wine is reported as having a less intense aroma.

Regarding gustatory aspects, filtered wine has very similar valuations as the original one, and finally this is also reflected in an exactly equal acceptability. It is worth noting that the wine of this year apparently was considered lower quality even in origin, as it only reached 30% acceptance as compared with almost 60% for the other two Cases. It must be remembered that each case corresponded to a different vintage and different winemaking, which reflects the sensitivity of the wine to the initial quality of the grapes.

3.3.3. CATA Comparison of All Cases

For a better discussion of the influence of both membrane-based approaches on final oenological parameters, a comparison of the changes induced for both treatments in the resulting wines is presented in Figure 7. In this figure, the difference between the original and the treated wines is plotted for each parameter. There, a positive answer for a given parameters means that (in the opinion of the judge) the tested sample possesses such a descriptor; the number of positive answers were recorded for each descriptor and for both original and treated wines in the two cases in the study. Then, we can define the change observed in each descriptor according to the following expression:

$$Ch_i\ (\%) = 100 \times (Tr_i - Or_i)/Or_i \quad (1)$$

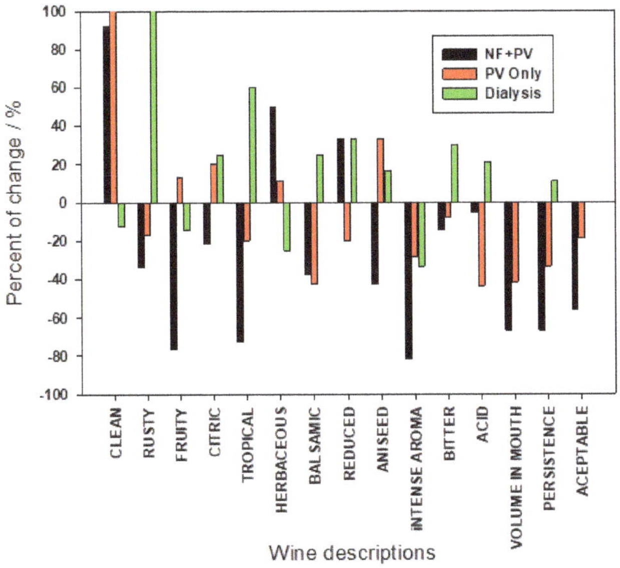

Figure 7. Differences obtained for all parameters tested in the CATA test (in percentage) with the original wines, for all cases studied.

Here Ch_i is the change observed for the i-th descriptor (from those presented in Table 4), Tr_i the value recorded for such descriptor (number of hits) for the treated wine, and Or_i the similar value obtained for original wine.

These percentage values are plotted in Figure 7 for each descriptor and for the three cases studied.

As can be seen, there are positive and negative differences indicating descriptors which appear more clearly in the treated wine (positive values) while others do not appear or appear with less frequency (negative values).

Clearly for treatments applied in Cases 1 and 2, the resulting wines were associated to the cleanness descriptor. Surprisingly, Case 3 led to an almost negligible change in limpidity, which could indicate this treatment is less aggressive to treated wines.

On the other hand, descriptors associated to a well-made and rounded wine as: volume in mouth, intensity of aroma or persistence, suffered important changes that clearly can be associated to a loss of aroma compounds during the filtration process. Anyway, this loss is lower in the case of the dialyzed wine (which in fact increases some aromatic perceptions) being these differences clearly higher for the case of pervaporated wine and clearly much higher when NF followed by PV was applied. It is clear that the loss of wine descriptors is mostly associated with the pumping needed during filtration, much more intense in the NF stage, while PV, which is a low-pressure process, needs less power of pumping, and finally D, where there is no transmembrane pressure, needs even less aggressive pumping.

Finally, the overall acceptability of the resulting wines is clearly related with the process suffered. In that sense, NF resulted in a strong fall of acceptability (clearly related with sensible losses in most tasting parameters), while PV also reduced acceptability of the wine but in a sensibly lower extent. Finally D treated wine showed same acceptability of the original wine.

4. Conclusions

Concerning the three procedures tested for the partial dealcoholization of white Verdejo wines, we can conclude the following:

NF process resulted the most aggressive treatment for the wine, due to the need of pumping (at relatively high pressures) resulting in sensible losses of its organoleptic qualities, still unrecovered after the PV application. This resulted in a reconstructed wine presenting notable sensorial differences with original wine, clearly reflected in the triangular test, where only 3 judges from 26 were unable to distinguish filtered and original wines.

The use of only PV, clearly less aggressive in terms of pumping, lead to a resulting wine with reasonability good chemical comparison with original wine along with sensory analyses better than NF. Certain loss of aroma, persistence, and volume in mouth, of filtered wine, made it less acceptable than the original. Triangular test confirmed this fact, with enough consumers able to distinguish both wines.

Finally, the use of Dialysis for the partial dealcoholization of the wine resulted in reasonably low chemical differences (except alcohol reduction). However, the most interesting feature is that for most sensorial characters tested, filtered wine presented similar marks as original one. On the other side, triangular test allows us to assure both wines are almost indistinguishable from organoleptic point of view.

As a final conclusion, we can state that, in the experimental conditions of our study, and applied to white wine of the Verdejo variety, the Membrane Dialysis process presents notable advantages over PV or combined NF + PV. This clearly better organoleptic behavior could envisage the utility of such process at industrial scale for partial dealcoholization of this kind of white wines, with scarce loss of sensorial and organoleptic properties of the wine and assuring the acceptability of them for potential consumers.

It is important to recall on a point that has so far been overlooked. Once demonstrated the feasibility of Dialysis for alcohol reduction of Verdejo white it is clear that application of this process only makes sense in the case of wines with natural alcohol contents considered excessive for the type of wine or the Appellation of Origin (AO) concerned, which was not the particular case for our wines. In general, the OIV does not allow for the name of wine to be given to products containing less than 8.5% alcohol (which was not the case in any of the treatments here studied). However, the different AOs can enforce more restrictive conditions to their wines. In particular, the Rueda appellation (which covers the different types of white wines made from the Verdejo grape in the region of Castilla y León) only grants this appellation to wines containing a minimum of 11%, so that none of the examples included here could be marketed under this appellation. It is therefore up to the oenologist of each winery to decide whether the starting wine has sufficient alcohol content to merit the treatment discussed here.

Finally, the influence of the alcohol reduction on the stability of the resulting wines needed to be tested in a longer study. These studies would ensure the availability of wines treated by dialysis for subsequent maturation in barrels and the consequent increase in their sensory characteristics and the resulting added value.

Author Contributions: J.A., D.S., R.Z. and D.C. performed the experiments. J.I.C. and P.P. designed the filtration experiments. E.F.-F. directed the organoleptic tests and the chemical analysis. J.I.C. and E.F.-F. realized the data analysis and interpretation, and the writing of the draft. A.H., P.P. and L.P. revised critically the manuscript and collaborated in the final version writing. All authors have read and agreed to the published version of the manuscript.

Funding: This research was funded by the Regional Government of Castilla y León (Projects: CLU2017-09, UIC082 and VA088G19) and the Spanish "Ministerio de Ciencia, Innovación y Universidades" (projects: PID2019-109403RB-C21/AEI/10.13039/501100011033 and PID2019-104129GB-I00/MCIN/AEI/10.13039/501100011033). The APC was founded by project: PID2019-109403RB-C21/AEI/10.13039/501100011033.

Institutional Review Board Statement: Not applicable.

Data Availability Statement: The data presented in this study are available on request from the corresponding author.

Acknowledgments: Silvia Teresa Gallego (from SMAP) and Miguel Ángel García Esteban (from ETSIIAA) are acknowledged for their help with chemical and refractometric analysis.

Conflicts of Interest: All authors declare hereby that there exist no conflict of interest related with this manuscript and the research in it presented.

References

1. Schneider, C. Grapevine and climatic changes: A glance at the situation in Alsace. *Agronomie* **2005**, *25*, 93–99. [CrossRef]
2. Jones, G.V.; White, M.A.; Cooper, O.R.; Storchmann, K. Climate Change and Global Wine Quality. *Clim. Chang.* **2005**, *73*, 319–343. [CrossRef]
3. Godden, P. Persistent wine instability issues. *Aust. Grapegrower Winemaker* **2000**, *443*, 10–14.
4. De Orduña, R.M. Climate change associated effects on grape and wine quality and production. *Food Res. Int.* **2010**, *43*, 1844–1855. [CrossRef]
5. Bucher, T.; Deroover, K.; Stockley, C. Low-Alcohol Wine: A Narrative Review on Consumer Perception and Behaviour. *Beverages* **2018**, *4*, 82. [CrossRef]
6. García-Martín, N.; Perez-Magariño, S.; Ortega-Heras, M.; González-Huerta, C.; Mihnea, M.; González-Sanjosé, M.L.; Palacio, L.; Prádanos, P.; Hernández, A. Sugar reduction in white and red musts with nanofiltration membranes. *DESALINATION Water Treat.* **2011**, *27*, 167–174. [CrossRef]
7. OIV, OIV-OENO Resolution 394B-2012. 2012. Available online: https://www.oiv.int/public/medias/1432/oiv-oeno-394b-2012-en.pdf (accessed on 10 April 2022).
8. Novello, V.; De Palma, L. Viticultural strategy to reduce alcohol levels in wine. In Proceedings of the 1st OENOVITI International Symposium: Alcohol Level Reduction in Wine, Bordeaux, France, 6 September 2013.
9. Röcker, J.; Schmitt, M.; Pasch, L.; Ebert, K.; Grossmann, M. The use of glucose oxidase and catalase for the enzymatic reduction of the potential ethanol content in wine. *Food Chem.* **2016**, *210*, 660–670. [CrossRef]
10. Ehsani, M.; Fernández, M.R.; Biosca, J.A.; Julien, A.; Dequin, S. Engineering of 2,3-Butanediol Dehydrogenase To Reduce Acetoin Formation by Glycerol-Overproducing, Low-Alcohol Saccharomyces cerevisiae. *Appl. Environ. Microbiol.* **2009**, *75*, 3196–3205. [CrossRef]

11. Kutyna, D.R.; Varela, C.; Henschke, P.A.; Chambers, P.J.; Stanley, G.A. Microbiological approaches to lowering ethanol concentration in wine. *Trends Food Sci. Technol.* **2010**, *21*, 293–302. [CrossRef]
12. Varela, C.; Kutyna, D.R.; Solomon, M.R.; Black, C.A.; Borneman, A.; A Henschke, P.; Pretorius, I.; Chambers, P.J. Evaluation of Gene Modification Strategies for the Development of Low-Alcohol-Wine Yeasts. *Appl. Environ. Microbiol.* **2012**, *78*, 6068–6077. [CrossRef]
13. Gonzalez, R.; Quirós, M.; Morales, P. Yeast respiration of sugars by non-Saccharomyces yeast species: A promising and barely explored approach to lowering alcohol content of wines. *Trends Food Sci. Technol.* **2013**, *29*, 55–61. [CrossRef]
14. Contreras, A.; Hidalgo, C.; Henschke, P.A.; Chambers, P.J.; Curtin, C.; Varela, C. Evaluation of Non-Saccharomyces Yeasts for the Reduction of Alcohol Content in Wine. *Appl. Environ. Microbiol.* **2014**, *80*, 1670–1678. [CrossRef] [PubMed]
15. Tilloy, V.; Ortiz-Julien, A.; Dequin, S. Reduction of Ethanol Yield and Improvement of Glycerol Formation by Adaptive Evolution of the Wine Yeast Saccharomyces cerevisiae under Hyperosmotic Conditions. *Appl. Environ. Microbiol.* **2014**, *80*, 2623–2632. [CrossRef] [PubMed]
16. Kiss, I.; Vatai, G.; Bekassy-Molnar, E. Must concentrate using membrane technology. *Desalination* **2004**, *162*, 295–300. [CrossRef]
17. Salgado, C.M.; Palacio, L.; Prádanos, P.; Hernández, A.; González-Huerta, C.; Pérez-Magariño, S. Comparative study of red grape must nanofiltration: Laboratory and pilot plant scales. *Food Bioprod. Process.* **2014**, *94*, 610–620. [CrossRef]
18. Salgado, C.M.; Fernández-Fernández, E.; Palacio, L.; Hernández, A.; Prádanos, P. Alcohol reduction in red and white wines by nanofiltration of musts before fermentation. *Food Bioprod. Process.* **2015**, *96*, 285–295. [CrossRef]
19. Versari, A.; Ferrarini, R.; Parpinello, G.P.; Galassi, S. Concentration of Grape Must by Nanofiltration Membranes. *Food Bioprod. Process.* **2003**, *81*, 275–278. [CrossRef]
20. Bengtson, G.; Pingel, H.; Crespo, J.P.S.G. Recovery of aroma compounds from a wine-must fermentation by organophilic pervaporation. *Biotechnol. Bioeng.* **1999**, *62*, 412–421. [CrossRef]
21. Karlsson, H.O.; Tragardh, G. Applications of pervaporation in food processing. *Trends Food Sci. Technol.* **1996**, *7*, 78–83. [CrossRef]
22. Pereira, C.C.; Ribeiro, C.P.; Nobrega, R.; Borges, C. Pervaporative recovery of volatile aroma compounds from fruit juices. *J. Membr. Sci.* **2006**, *274*, 1–23. [CrossRef]
23. Salgado, C.M.; Fernández-Fernández, E.; Palacio, L.; Carmona, F.J.; Hernandez, A.; Prádanos, P. Application of pervaporation and nanofiltration membrane processes for the elaboration of full flavored low alcohol white wines. *Food Bioprod. Process.* **2016**, *101*, 11–21. [CrossRef]
24. Brazinha, C.; Alves, V.D.; Viegas, R.M.; Crespo, J.G. Aroma recovery by integration of sweeping gas pervaporation and liquid absorption in membrane contactors. *Sep. Purif. Technol.* **2009**, *70*, 103–111. [CrossRef]
25. Brányik, T.; da Silva, D.P.; Baszczyňski, M.; Lehnert, R.; Silva, J.B.D.A.E. A review of methods of low alcohol and alcohol-free beer production. *J. Food Eng.* **2012**, *108*, 493–506. [CrossRef]
26. Pickering, G.J. Low- and Reduced-alcohol Wine: A Review. *J. Wine Res.* **2000**, *11*, 129–144. [CrossRef]
27. Aguera, E.; Bes, M.; Roy, A.; Camarasa, C.; Sablayrolles, J.-M. Partial Removal of Ethanol during Fermentation to Obtain Reduced-Alcohol Wines. *Am. J. Enol. Vitic.* **2010**, *61*, 53.
28. Schmidtke, L.M.; Blackman, J.; Agboola, S.O. Production Technologies for Reduced Alcoholic Wines. *J. Food Sci.* **2011**, *77*, R25–R41. [CrossRef]
29. Corona, O.; Liguori, L.; Albanese, D.; Di Matteo, M.; Cinquanta, L.; Russo, P. Quality and volatile compounds in red wine at different degrees of dealcoholization by membrane process. *Eur. Food Res. Technol.* **2019**, *245*, 2601–2611. [CrossRef]
30. Liguori, L.; Russo, P.; Albanese, D.; Di Matteo, M. Effect of Process Parameters on Partial Dealcoholization of Wine by Osmotic Distillation. *Food Bioprocess Technol.* **2013**, *6*, 2514–2524. [CrossRef]
31. El Rayess, Y.; Mietton-Peuchot, M. Membrane Technologies in Wine Industry: An Overview. *Crit. Rev. Food Sci. Nutr.* **2015**, *56*, 2005–2020. [CrossRef]
32. Labanda, J.; Vichi, S.; Llorens, J.; López-Tamames, E. Membrane separation technology for the reduction of alcoholic degree of a white model wine. *LWT* **2009**, *42*, 1390–1395. [CrossRef]
33. Pilipovik, M.; Riverol, C. Assessing dealcoholization systems based on reverse osmosis. *J. Food Eng.* **2005**, *69*, 437–441. [CrossRef]
34. Verhoef, A.; Figoli, A.; Leen, B.; Bettens, B.; Drioli, E.; Van der Bruggen, B. Performance of a nanofiltration membrane for removal of ethanol from aqueous solutions by pervaporation. *Sep. Purif. Technol.* **2008**, *60*, 54–63. [CrossRef]
35. Takács, L.; Vatai, G.; Korány, K. Production of alcohol free wine by pervaporation. *J. Food Eng.* **2007**, *78*, 118–125. [CrossRef]
36. Petkovska, M.; Leskošek, I.; Nedović, V. Analysis of Mass Transfer in Beer Diafiltration with Cellulose-Based and Polysulfone Membranes. *Food Bioprod. Process.* **1997**, *75*, 247–252. [CrossRef]
37. Sam, F.E.; Ma, T.-Z.; Salifu, R.; Wang, J.; Jiang, Y.-M.; Zhang, B.; Han, S.-Y. Techniques for Dealcoholization of Wines: Their Impact on Wine Phenolic Composition, Volatile Composition, and Sensory Characteristics. *Foods* **2021**, *10*, 2498. [CrossRef]
38. OIV. *Compendium of International Methods of Wine and Must Analysis*, 2021 ed.; International Organisation of Vine and Wine: Paris, France, 2021. Available online: https://www.oiv.int/public/medias/7907/oiv-vol1-compendium-of-international-methods-of-analysis.pdf (accessed on 10 April 2022).
39. *ISO 4120:2004*; Sensory Analysis-Methodology-Triangle Test. International Organization for Standardization: Geneva, Switzerland, 2004.
40. Ares, G.; Varela, P. *Comparison of Novel Methodologies for Sensory Characterization in Novel Techniques in Sensory Characterization and Consumer Profiling*; CRC Press, Taylor and Francis Group: Boca Raton, FL, USA, 2016.

41. Rodríguez-Nogales, J.M.; Fernández-Fernández, E.; Vila-Crespo, J. Characterisation and classification of Spanish Verdejo young white wines by volatile and sensory analysis with chemometric tools. *J. Sci. Food Agric.* **2009**, *89*, 1927–1935. [CrossRef]
42. *ISO-3591:1977*; Sensory Analysis. Apparatus. Wine-Tasting Glass, ISO 3591. International Standardization Office: Geneva, Switzerland, 1977. Available online: https://www.iso.org/standard/9002.html (accessed on 10 April 2022).
43. *ISO Standard 8589:2007*; Sensory Analysis—General Guidance for the Design of Test Rooms. International Organization for Standardization: Geneva, Switzerland, 2007.
44. Mckean, J.W.; Manoukian, E.B. Mathematical Nonparametric Statistics. *J. Am. Stat. Assoc.* **1988**, *83*, 1217. [CrossRef]
45. She, M.; Hwang, S.-T. Recovery of key components from real flavor concentrates by pervaporation. *J. Membr. Sci.* **2006**, *279*, 86–93. [CrossRef]
46. Catarino, M.; Mendes, A. Dealcoholizing wine by membrane separation processes. *Innov. Food Sci. Emerg. Technol.* **2011**, *12*, 330–337. [CrossRef]
47. García-Martín, N.; Palacio, L.; Prádanos, P.; Hernández, A.; Ortega-Heras, M.; Pérez-Magariño, S.; González-Huerta, D. Evaluation of several ultra- and nanofiltration membranes for sugar control in winemaking. *Desalination* **2009**, *245*, 554–558. [CrossRef]
48. Pérez-Coello, M.; González-Viñas, M.; García-Romero, E.; Díaz-Maroto, M.; Cabezudo, M. Influence of storage temperature on the volatile compounds of young white wines. *Food Control* **2003**, *14*, 301–306. [CrossRef]

membranes

MDPI

Article

Flat PVDF Membrane with Enhanced Hydrophobicity through Alkali Activation and Organofluorosilanisation for Dissolved Methane Recovery

Ramón Jiménez-Robles [1], Beatriz María Moreno-Torralbo [2], Jose David Badia [2], Vicente Martínez-Soria [1] and Marta Izquierdo [1,*]

[1] Research Group in Environmental Engineering (GI2AM), Department of Chemical Engineering, School of Engineering, University of Valencia, Avda. Universitat s/n, 46100 Burjassot, Spain; ramon.jimenez@uv.es (R.J.-R.); vmsoria@uv.es (V.M.-S.)

[2] Research Group in Materials Technology and Sustainability (MATS), Department of Chemical Engineering, School of Engineering, University of Valencia, Avda. Universitat s/n, 46100 Burjassot, Spain; bmmortor@itq.upv.es (B.M.M.-T.); jose.badia@uv.es (J.D.B.)

* Correspondence: marta.izquierdo-sanchis@uv.es; Tel.: +34-963-543-737; Fax: +34-963-544-898

Abstract: A three-step surface modification consisting of activation with NaOH, functionalisation with a silica precursor and organofluorosilane mixture (FSi_T), and curing was applied to a poly(vinylidene fluoride) (PVDF) membrane for the recovery of dissolved methane (D-CH_4) from aqueous streams. Based on the results of a statistical experimental design, the main variables affecting the water contact angle (WCA) were the NaOH concentration and the FSi_T ratio and concentration used. The maximum WCA of the modified PVDF ($mPVDF_{max}$) was >140° at a NaOH concentration of 5%, an FSi_T ratio of 0.55 and an FSi_T concentration of 7.2%. The presence of clusters and a lower surface porosity of mPVDF was detected by FESEM analysis. In long-term stability tests with deionised water at 21 L h^{-1}, the WCA of the mPVDF decreased rapidly to around 105°, similar to that of pristine nmPVDF. In contrast, the WCA of the mPVDF was always higher than that of nmPVDF in long-term operation with an anaerobic effluent at 3.5 L h^{-1} and showed greater mechanical stability, since water breakthrough was detected only with the nmPVDF membrane. D-CH_4 degassing tests showed that the increase in hydrophobicity induced by the modification procedure increased the D-CH_4 removal efficiency but seemed to promote fouling.

Keywords: anaerobic digestion; hydrophobicity; membrane stability; methane recovery; organofluorosilane; PVDF; surface modification

1. Introduction

Methane (CH_4) produced as biogas in anaerobic digesters is partially dissolved in the digestate, so the anaerobic reactor effluents can present high amounts of dissolved CH_4 (D-CH_4), the discharge of which causes fugitive emissions of this important greenhouse gas and biofuel. The use of membrane contactors as desorption devices for the recovery of D-CH_4 from anaerobic reactor effluents has attracted the strong interest of researchers in recent years [1–5]. As well as CH_4 degassing from aqueous streams, membrane contactors can also be used in other stripping and gas absorption processes such as CO_2 capture from flue gas or biogas streams or gas removal from aqueous solutions using transmembrane chemical absorption [6,7]. The most common industrial configuration is the hollow-fibre membrane contactor (HFMC), characterised by a high specific surface area, compactness, ease of scaling, and cost-effectiveness. Nevertheless, HFMC is easy to damage and susceptible to pore blocking owing to the stress from the liquid flux and/or the fouling phenomenon [6,8,9], and it cannot be easily repaired or refreshed which can dramatically increase the maintenance and operation costs. Although many studies have focused on HFMCs, flat-sheet membrane modules (FMs) present several advantages related to their simplicity, versatility, and ease of cleaning and replacement, which make them attractive for use in some applications and also in investigations at a laboratory scale [8]. FMs can be especially useful where fouling or membrane stability are the focus of the research, or to facilitate the performance comparison of membrane materials with different characteristics [8,10]. The most typically used polymers for commercial membrane contactors include polydimethylsiloxane (PDMS), polysulfone (PSF), polyacrylonitrile (PAN), polypropylene (PP), polytetrafluoroethylene (PTFE), and poly(vinylidene fluoride) (PVDF). PVDF membranes have been intensively studied in recent years [11] due to their high performance and thermal, mechanical, and chemical resistance [5,12]. In addition, they present a lower raw material cost and ease of synthesis, casting, and extrusion to form different membrane configurations [12] as well as having a good thermodynamic affinity with various fluoropolymers [13].

Despite the numerous advantages of polymeric membrane contactors relative to traditional separation units, two major shortcomings are apparent when operating with membranes: wetting and fouling phenomena. In microporous membranes, the liquid pressure over the membrane surface can induce the partial or total filling of the membrane pores with the liquid phase when a critical transmembrane pressure is surpassed [5,14]. This membrane wetting increases the mass transfer resistance and, consequently, reduces the separation efficiency. Only 5% of membrane pores occupied by the liquid phase can induce an increase of around 20% in the overall mass transfer resistance [15]. The wetting resistance is influenced by membrane properties such as the pore size and distribution, roughness and, especially, the hydrophobicity of the membrane material [5,16,17]. In operation with anaerobic reactor effluents, the fouling agents contained in the stream can adhere to the membrane surface [18]. This fouling cake adds restrictions for mass transfer and can also physically and chemically deteriorate the membrane [9]. The tendency of the membrane to be fouled depends mainly on issues such as membrane hydrophobicity and roughness, the concentration of the foulants and ionic strength of the solution, and membrane module configuration, among other factors [9,18,19].

In order to overcome these shortcomings, polymer membranes can be enhanced at the synthesis step by controlling different operational conditions or by using additives to add new functionalities or change the membrane morphology [9,20]. The surface modification of existing membranes has gained a great deal of attention in recent years due to the versatility offered by the different techniques [9,12]. Surface modification techniques can be divided into physical and chemical treatments. Chemical treatments commonly involve grafting modifying agents onto the membrane by means of covalent bonding, thus achieving a stronger adhesion force. Consequently, the modifiers are less susceptible to removal from the membrane by liquid flux. Recently, surface chemical modification has focused on increasing the hydrophobicity of the membrane surface to avoid the wetting phenomenon [10,21,22]. Thus, fluorinated organic compounds and siloxanes have been developed and evaluated as modifying agents, since they are well-known to reduce surface free energy and, consequently, to increase hydrophobicity [5]. Indeed, the superhydrophobic degree (water contact angle, WCA > 150°) has been reported after surface modification with organosilanes [23]. In the case of PVDF, PTFE, or PDMS, with poor reactivity, a step prior to grafting the modifiers is needed to activate the membranes [5,12]. Different techniques, such as hydroxylation by NaOH or plasma treatment, have been evaluated, in which the membrane is subjected to an alkaline or oxygen-radical environment, respectively, in order to form active sites on the membrane (e.g., hydroxide or peroxide groups) to react further with the functionalisation agents that confer enhanced surface performance [17,24,25].

The long-term stability of the membrane during operation plays a critical role in the design of the surface modification [26]. The stress and dragging effect of a continuous liquid flux over the modified membranes in long-term operations can remove the coating in the case of physical modification or alter the surface morphology, producing a change in the surface properties such as the hydrophobicity [8,19,27] and reducing the membrane lifetime. Reports into the evaluation of the structural and chemical stability of modified membranes after long-term operations for D-CH_4 recovery are very scarce but would provide useful preview information about the economic viability of applying membrane modification procedures at the industrial scale. The absorption of CO_2 using membrane contactors has been more widely investigated, and higher CO_2 fluxes and stability in long-term tests have been reported with modified membranes compared to non-modified membranes under different modification procedures [25,28–31].

In this context, the general aim of this work was to evaluate the effect of a chemical surface modification applied over a commercial PVDF membrane for the recovery of D-CH_4 from aqueous streams. Firstly, a statistical experimental design of a three-step modification procedure (activation, functionalisation, and curing) was proposed to optimise the surface hydrophobicity of the membrane, measured by the WCA. Then, the stability and the membrane performance in D-CH_4 recovery by modified PVDF membranes were evaluated and benchmarked with non-modified PVDF. These membranes were tested in an FM module under several operational conditions with deionised water (DW) and real anaerobic reactor effluent (AE).

2. Materials and Methods

2.1. Membrane Surface Modification Procedure

Flat-sheet PVDF membranes were supplied by Dorsan Filtration (Spain). The membrane was composed of hydrophobic PVDF supported on a polyester (PET) non-woven support, resulting in a microporous structure. The main characteristics of the membrane are reported in Table 1.

Table 1. Characteristics of the flat-sheet PVDF membrane.

Property	Value
Structure	Microporous
Support	Polyester (PET)
Thickness, μm	159 ± 2 [a]
Pore diameter, μm	0.2 [b]
Porosity, %	62 ± 3 [c]
Liquid entry pressure, bar	1.8 [b]
Static water contact angle, °	103.4 ± 1.6 [a]

[a] Measured. [b] Provided by the manufacturer (Dorsan Filtration, Spain). [c] Determined by gravimetric analysis [32].

The surface modification of the PVDF was carried out in a three-step procedure: (1) activation, (2) functionalisation, and (3) curing.

In the activation step, the PVDF was hydroxylated with 2.5 mL of sodium hydroxide (NaOH) solution per cm^2 of membrane in an orbital shaker at 150 rpm. According to preliminary experiments (Supplementary Material S1), the temperature and time of the activation step were fixed at 50 °C and 1 h, respectively, and the NaOH concentration ($NaOH\%_{wt}$) ranged from 1 to 6%, as suitable for the purposes of activation. Afterwards, the activated membranes were immediately rinsed with milliQ water to stop the reaction and the excess of water on the sample surface was gently removed. In the hydroxylation reaction, the OH^- ions replaced the F atoms from the PVDF chains, which promoted a decrease in its hydrophobicity, mainly by a defluorination mechanism [24]. The detailed reactions proposed for the modification of PVDF are shown in the Supplementary Material S2.

In the functionalisation step, the organofluorosilane 1H,1H,2H,2H-perfluorooctyltriethoxysilane (Dyn; Dynasylan® F8261, Evonik GmbH, Hanau-Wolfgang, Germany) and tetraethyl orthosilicate (TEOS; ≥99%, Sigma-Aldrich, Saint Louis, MI, USA) were used as modifying agent and silica precursor, respectively, to carry out the grafting of additional fluoride chains onto the PVDF. A mixture of 2-propanol (IPA; HPLC grade, VWR

Chemicals, Radnor, PA, USA) and milliQ water in a molar ratio of 57:1 was used as solvent (labelled as IPA/H_2O). For the preparation of the functionalisation solution (FS), an initial mixture of the organofluorosilane and silica precursor was prepared and then the IPA/H_2O solvent was added dropwise [27]. The presence of water in the FS was needed in order to hydrolyse the silica precursor, as reported by several authors [27,31,33,34]. Once the solvent was fully added, in order to promote the hydrolysis of the TEOS [27,34], the pH of the FS was adjusted with concentrated HCl (37 wt %, reagent grade, VWR Chemicals, Radnor, PA, USA) to a value of 2–3. Finally, the FS remained under magnetic stirring at 200 rpm at room temperature for 24 h in order to achieve the complete hydrolysis of the TEOS. The mixture of Dyn and TEOS in the FS was labelled 'FSi_T', where 'F' denotes the fluorine modifying agent Dyn and 'Si_T' the silica precursor TEOS. The volumetric ratio of the modifying agent in the FSi_T mixture (FSi_T ratio = Dyn volume/FSi_T mixture volume) ranged from 0.17 to 0.94, and the FSi_T volumetric percentage in the solvent (FSi_T%$_v$) ranged from 0.8 to 9.2%. The ranges of both parameters were set from the preliminary experiments shown in Supplementary Material S1. In the functionalisation step, the membrane samples were immersed in 2 mL of the FS per cm^2 of the membrane for 1 h at 100 rpm and room temperature and then rinsed with IPA. The reactions involved in the functionalisation are detailed in Supplementary Material S2.

Finally, for the curing step, the membrane was kept in an oven at 60 °C overnight on a Petri dish. Functionalisation and curing times were also established based on the preliminary experiments. The modified PVDF was labelled 'mPVDF' and the non-modified PVDF as 'nmPVDF'. A scheme of the membrane modification procedure is shown in Figure 1.

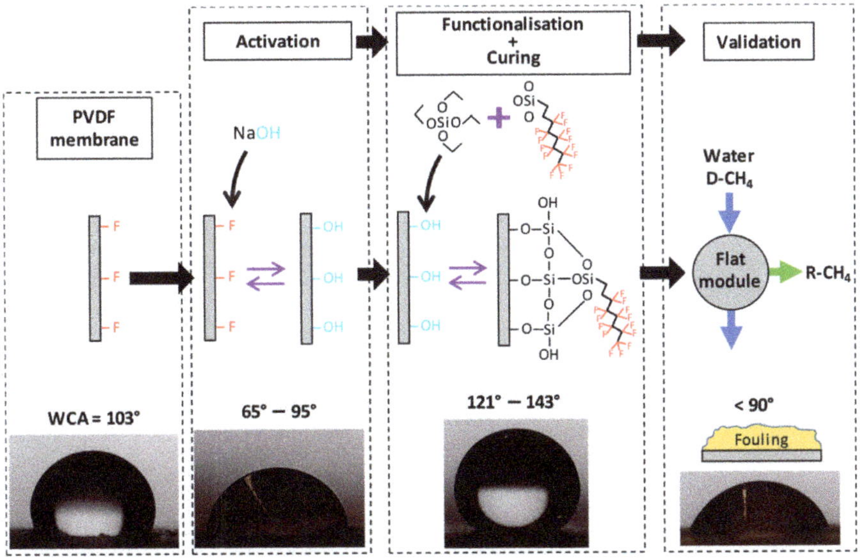

Figure 1. Scheme of the modification procedure of poly(vinylidene fluoride) (PVDF) membranes and the validation step. Water contact angle (WCA): static water contact angle; D-CH_4: methane dissolved in liquid streams; R-CH_4: methane recovered in gaseous form.

2.2. Design of Experiments and Statistical Analysis

The effect of the modification procedure on the hydrophobicity of PVDF was initially evaluated on 2 × 2-cm membrane specimens. The main variables involved in the modification procedure were analysed by means of a statistical experimental design in order to maximise the static WCA of the membrane surface. After preliminary experiments in which the activation time (1 h) and temperature (50 °C), and functionalisation and curing times (1 h and >12 h, respectively) were fixed, the main variables affecting the final hydrophobicity of the PVDF were attributed to FSi_T ratio, $FSi_T\%_v$, and $NaOH\%_{wt}$. These variables (factors) were evaluated at the values (levels) shown in Table 2. A 3^3 complete factorial design with the central point and corner levels was carried out in order to identify the significant effects of the factors and their interactions on the response variable, i.e., the WCA. Then, statistical analysis based on multiple linear regression and a response surface was carried out for the determination of the optimum values of the factors leading to the maximum response. This analysis included the additional axial points specified in Table 2. The different runs of the set of experiments were carried out randomly in duplicate to avoid systematic errors. The statistical software Minitab® (Minitab, LLC., State College, PA, USA) was used for the design of the experiments and the statistical analysis.

Table 2. Factors and levels used in the design of the experiments and statistical analysis of the modification procedure.

Independent Variables (Factors)		Levels				
		Axis (−)	Corner (−)	Central Point	Corner (+)	Axis (+)
A	FSi_T ratio	0.17	0.29	0.55	0.79	0.94
B	$FSi_T\%_v$, %	0.8	2.4	4.6	7.2	9.2
C	$NaOH\%_{wt}$, %	1.0	2.0	3.5	5.0	6.0

2.3. *Evaluation of the Modified Membrane Performance*

The performance of recovery of D-CH_4 from liquid streams and the long-term stability of both nmPVDF and mPVDF were evaluated in degassing tests and long-term tests, respectively. Experiments were carried out using DW (<1 μS cm^{-1}) and an effluent stream from an anaerobic digester for comparison purposes. The anaerobic effluent (AE) was collected from the anaerobic reactor of the urban Quart-Benager II wastewater treatment plant (Valencia, Spain) and filtered with a 10–20 μm filter before use in the tests. The pH, conductivity, alkalinity, chemical oxygen demand, and total and volatile suspended solids of the filtered AE were 7.84, 8.61 mS cm^{-1}, 2102 mg $CaCO_3$ L^{-1}, 414 mg O_2 L^{-1}, 75.75 mg L^{-1}, and 67.50 mg L^{-1}, respectively. A circular FM made of stainless steel with an effective contact area of 17.3 cm^2 was used for both performance and long-term stability tests.

2.3.1. Dissolved Methane Recovery from Water Streams

The laboratory-scale system with the FM used for the non-steady-state degassing tests was extensively described in our previous work [8] and is depicted in Figure 2. For the degassing test, a 2 L glass bottle acting as the liquid feed tank was filled initially with the liquid (DW or AE). In the first step, the liquid was saturated with 99.5$\%_v$ CH_4 (Carburos Metálicos, Barcelona, Spain) in a packed-bed absorption column in counter-current flux. The liquid stream flowed in a closed loop through the system using a peristaltic pump (Watson-Marlow Fluid Technology Solutions, Falmouth, UK), keeping the gas ports of the module closed (dashed line, Figure 2). This liquid stream reached a concentration of 20.7 ± 1.1 mg CH_4 L^{-1}. In the second step, the saturated liquid was pumped from the liquid feed tank to the FM and then recirculated continuously to the tank operating in a closed loop in order to recover the D-CH_4 (continuous line, Figure 2). The total duration of these non-steady-state experiments was 5 h, and the liquid flow rates (Q_L) were fixed at 3.5 and 21.0 L h^{-1} (liquid velocity ranging from 2.8 to 16.9 cm s^{-1}) based on our previous work [8]. All experiments were carried out in the sweep-gas mode, using a N_2 flow rate (Q_{N2}) of 4 L h^{-1} (gas velocity of 0.3 cm s^{-1}). The liquid and sweep gas flowed in a cross-flow configuration inside the FM and the experiments were conducted at room temperature (24 ± 1 °C). Removal of D-CH_4 was evaluated from liquid samples of 5 mL collected at different times during the degassing test to determine the change in D-CH_4 concentration measured by gas chromatography. Duplicates of liquid samples were collected at the outlet of the liquid feed tank. The results reported in this paper are the removal efficiency (RE, %) and the CH_4 flux (J_{CH4}) obtained after 5 h of experimentation. As an example, the profiles of the D-CH_4 concentration, RE, and J_{CH4} obtained at a Q_L of 21 L h^{-1} are shown in Supplementary Material S3.

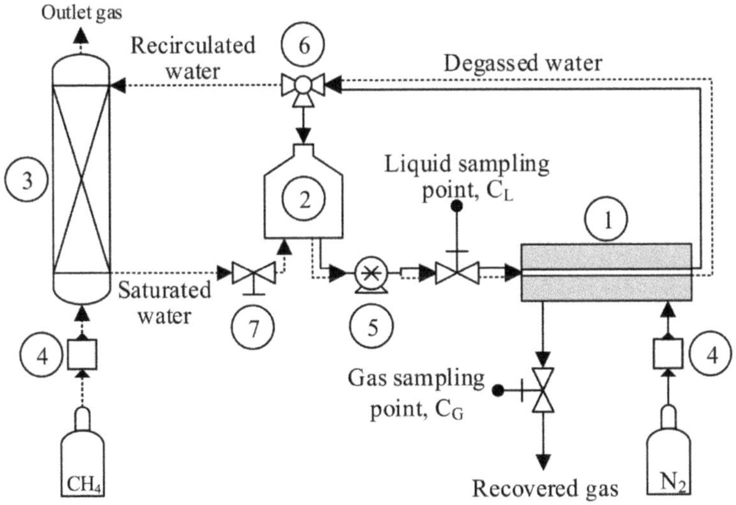

Figure 2. Scheme of the experimental system for dissolved methane degassing tests from aqueous streams with a flat-sheet membrane module. (1) membrane module, (2) liquid feed tank, (3) saturation column, (4) mass flow controller, (5) peristaltic pump, (6) 3-way valve, and (7) on/off valve.

2.3.2. Long-Term Surface Stability Tests

Experiments using DW and AE were carried out to evaluate the behaviour of the PVDF membranes during long-term operation using the former laboratory-scale system and FM. The long-term tests were carried out to evaluate the stability of the hydrophobicity of the membrane surface under a continuous liquid flux with different types of aqueous stream. Initially, the liquid feed tank was filled with DW or AE, the saturation column disconnected, and the gas ports of the FM closed for this type of test. Then, a constant water flux was applied through the liquid side of the FM in the closed loop defined by the continuous line in Figure 2. Based on our previous study [8], the long-term tests with DW were operated at the highest value of Q_L of 21 L h^{-1} until a constant WCA was reached. The long-term tests with AE were carried out at the lowest Q_L tested, 3.5 L h^{-1}, in order to promote the potential fouling deposition on the membrane for up to around 830 h. In the experiments with AE, the anaerobic water inside the system was replaced up to three times. The membrane was extracted periodically to measure its WCA in order to monitor the surface hydrophobicity with time of use, as reported in the literature [8,27,33]. Similarly, for the membranes treating the AE, the membrane performance for D-CH_4 degasification was evaluated at different times of use using AE saturated with CH_4 as the liquid stream.

2.4. Analytical Methods

The surface hydrophobicity of the membranes was evaluated by means of the WCA as mentioned above. WCA measurements were conducted by the sessile drop technique [35] of depositing a water drop of 5.5 $\pm$ 0.1 µL onto the membrane surface using a syringe pump (KF Technology s.r.l., Rome, Italy) at room temperature (~25 °C). An image of the water drop profile was captured at 15 s with a digital microscope (Handheld Digital Microscope Pro, Celestron LLC., Torrance, CA, USA) under white light (Philips HUE Lamp,

Koninklijke Philips NV, Amsterdam, The Netherlands). ImageJ software was used for image processing using the *Contact Angle Plug-in* based on the ellipse approximation. The WCA was evaluated at different positions on the membrane and a mean value obtained from at least four measurements. The WCA values of the wet membranes from the long-term stability tests were measured after removing excess water and evaporation of moisture at room temperature for approximately 1 h with forced aeration.

The morphology of the membrane surface and cross-section, as well as its chemical composition, were analysed by field emission scanning electron microscopy (FESEM) equipped with energy dispersive X-ray spectrometer (EDX) with an accelerating voltage of 20 kV (Hitachi S4800, Hitachi Ltd., Tokyo, Japan). For image acquisition, the samples were placed on a metal holder and then coated with a fine layer of Au/Pd by sputtering in vacuum for 1 min. FESEM was also applied to measure the membrane thickness and the surface porosity by means of the abovementioned ImageJ software [36].

The liquid samples collected from the degassing tests were analysed by the adapted head-space method described elsewhere [37–40]. Briefly, 5 mL liquid samples were collected in sealed 16 mL vials prefilled with air. These vials were allowed to reach equilibrium between phases in an orbital shaker at 200 rpm and 25 °C for 30 min. After this, 0.5 mL of head-space gas was injected into a gas chromatograph (GC 7820 A, Agilent Technologies Spain S.L., Madrid, Spain) equipped with Agilent HP-PLOT/U and Agilent HP-MOLESIEVE columns and a thermal conductivity detector to measure the CH_4 concentration in the gas phase. The CH_4 concentration in the liquid phase was calculated by applying Henry's Law, and then both the gas and liquid phase concentrations were used to calculate the concentration in the liquid sample (C_L, mg L^{-1}).

The RE of D-CH_4 (%) and CH_4 flux (J_{CH4}, g s^{-1} m^{-2}) through the membrane were used to evaluate the performance of the D-CH_4 recovery process in the degassing tests, defined as follows:

$$RE = \frac{C_{L0} - C_{Lt}}{C_{L0}} \times 100 \tag{1}$$

$$J_{CH4} = \frac{V_T \times 10}{A} \times \frac{dC_L}{dt} \tag{2}$$

where C_{L0} and C_{Lt} (mg L^{-1}) are the CH_4 concentrations in the liquid at the beginning and at time on stream (t) of the degassing test from the liquid feed tank, respectively; V_T is the total liquid volume in the system (2.35 L); and A (cm^2) is the effective membrane area. The variation in liquid volume flow rate was assumed to be negligible for the application of Equation (1).

3. Results and Discussion

3.1. Maximisation of Membrane Hydrophobicity

A statistical experimental design evaluating the main parameters that affect the membrane surface hydrophobicity was conducted to maximise the response variable, i.e., the WCA, of the modified membranes. After the preliminary experiments (Supplementary Material S1), the variables FSi_T ratio, $FSi_T\%_v$, and $NaOH\%_{wt}$ were found to be the main factors to be optimised. Initially, a 3^3 complete factorial design was carried out in order to determine the effects of the different factors, including individual effects and those of the 2-way and 3-way interactions on the response. The effects of the factors and their interactions can be observed in the Pareto diagram of the factorial design shown in Figure 3. The critical standardised effect (2.05) was calculated from the Student's T distribution ($t_{\alpha/2}$,ν) with a significance level (α) of 0.05 and 27 degrees of freedom (ν), associated with the error of the factorial design. As observed in the Pareto diagram, the standardised effect of the different factors and their interactions were higher than the critical value, except for the interaction between FSi_T ratio and $FSi_T\%_v$. Thus, the interaction of these two factors did not seem to affect the WCA of the mPVDF and their effects could be ignored.

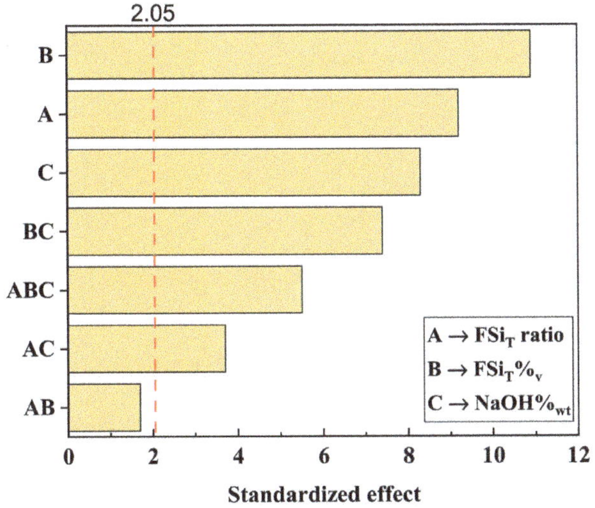

Figure 3. Pareto diagram of the standardised effects of the factors analysed in the 3^3 complete factorial design. The dashed red line represents the critical standardised effect for a significance level of 0.05.

The individual factors presented the highest effects on the response, the most important being those involved in the functionalisation step (FSi$_T$ ratio and FSi$_T$%$_v$). This result can be understood, since the FSi$_T$ ratio and FSi$_T$%$_v$ determined the number of O–Si bonds on the PVDF chains and the total F content at the end of the functionalisation process, these resulting in higher WCA values as the F content on the surface increased. Additionally, the increase in O–Si bonds grafted onto the PVDF chains allowed reaction with higher numbers of Dyn molecules, increasing the total F content of the surface of the modified membrane. The NaOH%$_{wt}$ also showed a high effect on the response, since this factor determined the number of active sites (OH^-) on the PVDF chains where the silanol groups from TEOS were grafted, leading to O–Si bonds [23,25]. Consequently, the higher the NaOH%$_{wt}$ applied, the higher the number of F atoms from the PVDF chains that were substituted by OH^- groups, allowing a higher number of O–Si bonds and Dyn molecules to be grafted. This was in accordance with the significant standardised effect of the interaction of FSi$_T$ ratio and FSi$_T$%$_v$ with NaOH%$_{wt}$ (ABC) observed in the Pareto diagram.

The effects of the individual factors on the response are depicted in the main-effects plot shown in Figure 4. A WCA higher than that of the nmPVDF ($103.4 \pm 1.6°$) was obtained for all the modified membranes used, with an overall mean value of 131°, showing that the functionalisation procedure proposed in this work was suitable to increase the PVDF membrane hydrophobicity. The WCA clearly increased with the FSi$_T$%$_v$ and NaOH%$_{wt}$, in accordance with the previous discussion, although NaOH%$_{wt}$ should not be higher than 5% in order to avoid membrane degradation. In contrast, a maximum WCA was found with the variation of the FSi$_T$ ratio, as reported by other authors [23,31]. In that case, the WCA increased initially with the FSi$_T$ ratio until a maximum value was reached at a FSi$_T$ ratio of around 0.55. An increase in FSi$_T$ ratio caused a higher number of Dyn molecules to be grafted onto the PVDF chains. In contrast, a higher FSi$_T$ led to a decrease in WCA, which could be explained by taking into account that the amount of TEOS decreased as Dyn increased, and therefore fewer O–Si bonds were formed for the subsequent grafting of Dyn molecules. Thus, a value of around 0.55 seemed to provide an appropriate number of O-Si bonds to graft the available Dyn onto the FS and avoid a large excess of unreacted Dyn.

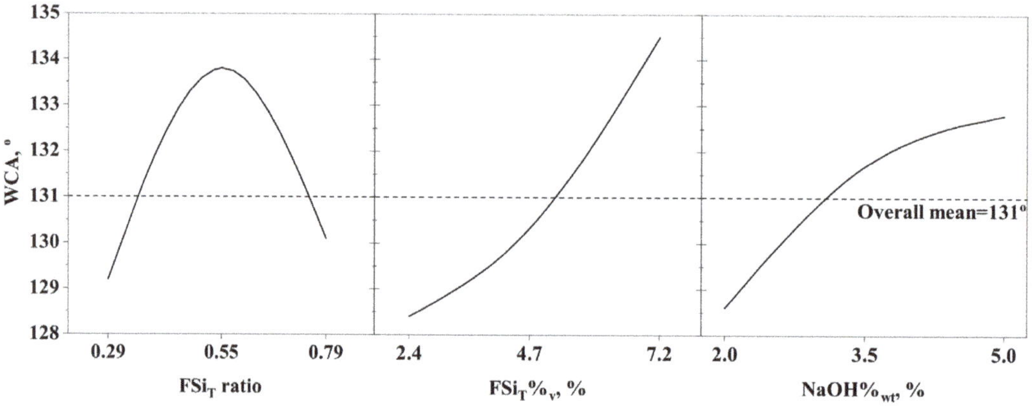

Figure 4. Main-effects plot of the factors analysed in the 3^3 complete factorial design.

After determining the main effects of the factors and their interactions with the 3^3 factorial design, a new set of experiments was conducted at the axis levels of the three factors. A statistical analysis based on response surface methodology was carried out on all the data obtained at the central point, corner, and axis levels. All these data were fitted by means of a multiple linear regression model including the linear (A, B, C), square (AA, BB, CC), 2-way interaction (AB, AC, BC), and 3-way interaction (ABC) terms. The 2-way interaction between FSi$_T$ ratio and FSi$_T$%$_v$ was also included in the model. Since large effects were observed for the individual factors (A, B, and C, Figure 3), their square terms could also present high significance in the model and were therefore included. The resulting model is represented in the form of a response surface and contour plot at a NaOH%$_{wt}$ of 5% in Figure 5.

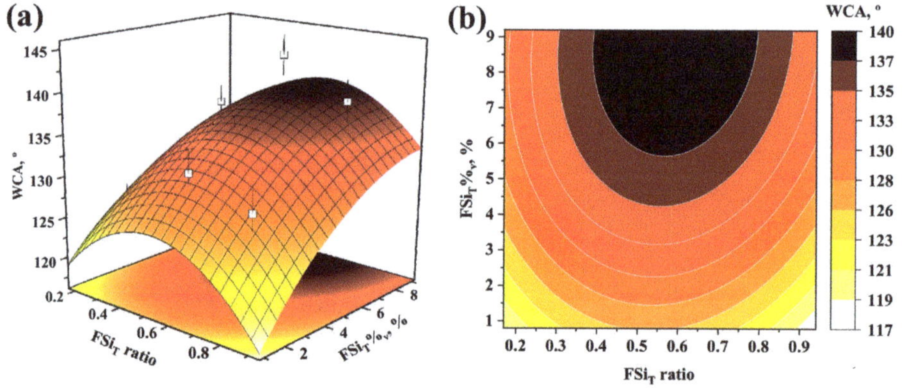

Figure 5. (**a**) Response surface and (**b**) contour plot obtained by linear regression from the overall data of the design of the experiments with a NaOH concentration of 5%. Experimental values are depicted as open symbols in the response surface plot and the error bars denote the standard deviation.

Based on the model explained above, the WCA initially increased with FSi$_T$%$_v$ up to a value around 6% and then remained nearly constant for a fixed FSi$_T$ ratio. For example, the model predicted a WCA of 137–140° for FSi$_T$%$_v$ ranging from 6 to 9.2% at a constant FSi$_T$ ratio of 0.55 with a level of confidence of 95%. Regarding the effect of the FSi$_T$ ratio on the WCA, the predicted trend showed a maximum value of WCA at intermediate values of FSi$_T$ ratio, as reported in the literature [23,31]. Thus, the maximum WCA values were achieved for a FSi_T ratio of around 0.55 at any $FSi_T\%_v$ value. This trend was in accordance with the previous analysis of effects and the experimental results shown in Figure 5a (open symbols). Thus, the maximum experimental WCA obtained was 143.2 ± 2.5°, while the model predicted a WCA of 138.8 ± 2.2° at an FSi_T ratio of 0.55, $FSi_T\%_v$ of 7.2%, and $NaOH\%_{wt}$ of 5%. Likewise, the maximum WCA value predicted by the model was 140.5 ± 6.6° at an FSi_T ratio of 0.59, $FSi_T\%_v$ of 9.2%, and $NaOH\%_{wt}$ of 6%. These results led to the conclusion that increases in FSi_T ratio, $FSi_T\%_v$, and $NaOH\%_{wt}$ over 0.55, 7.2%, and 5%, respectively, would not significantly increase the WCA and that they could be set as the optimal experimental conditions to achieve the highest hydrophobicity. The PVDF sample modified under the optimal conditions is termed hereafter as '$mPVDF_{max}$'.

For the sake of benchmarking the results, the maximum WCA obtained in this work (143.2 ± 2.5°) represented a 39% increase from the initial value of the nmPVDF (103.4 ± 1.6°) and was higher than most values found in the literature for comparable surface modification approaches. The maximum WCA was obtained at a low $NaOH\%_{wt}$ (5%) with IPA as solvent, unlike the other hazardous solvents such as toluene, hexane and tetrahydrofuran reported in the literature. Similar results of WCA increase have been reported by other authors. Sairiam et al. [25] reported a maximum WCA of 119.5° (from an initial value of 68.9°) with the same modifying agent as used in this work, using hexane as solvent. They applied longer treatments, and no WCA increase for the modified membranes was reported with $NaOH\%_{wt}$ higher than 10%. Wongchitphimon et al. [31] found a maximum WCA of 127.8° for modified PVDF (from an initial value of 95.5°) at a $NaOH\%_{wt}$, activation time, and temperature of 10%, 1 h, and 60 °C, respectively, and functionalised with the commercial organofluorosilane Fluorolink® S10 (FS10) at a FSi_T concentration of 5% and a ratio of FS10 to TEOS of 3:2 using toluene as solvent. In addition, they suggested destruction of the membrane integrity, since they reported a final WCA decrease and an increase in pore size at activation times higher than 1 h and a $NaOH\%_{wt}$ of 10%. Moreover, Sethunga et al. [21] obtained a modified PVDF with a maximum WCA of 111.7° (from an initial value of 84.5°) with a somewhat lower $NaOH\%_{wt}$ (pH = 10) and activation and functionalisation times of 0.5 and 1.5 h using Fluorolink® S10 and toluene as modifying agent and solvent, respectively. They also reported a final WCA decrease with an increase in $NaOH\%_{wt}$ and activation time, suggesting a conversion of the hydroxide groups to carbonyl groups that hindered the attachment of TEOS [21]. The maximum value was reported by Zheng et al. [23], with a maximum WCA of 157° (from an initial value of 89°) under more severe activation conditions with a $NaOH\%_{wt}$ of 30% and an activation time of 3 h and using organosilanes free of fluorine groups at a concentration of around 25% in the FS and toluene as solvent.

3.2. Structure and Chemical Composition of the Modified PVDF

An inspection of the surface and cross-section of nmPVDF and $mPVDF_{max}$ was carried out by means of a FESEM analysis. The nmPVDF presented a homogenous surface with a high surface porosity of around 55% (Figure 6a). The effect of membrane modification resulted in a heterogeneous surface with porous zones and zones with dense and smooth clusters (Figure 6b) and less surface porosity of around 27%. Similar observations have been reported by other authors [23,25]. This can be attributed mainly to a coating-like effect of the polymerisation/condensation of the FSi_T components (TEOS and Dyn) on the surface leading to a higher hydrophobicity. Thus, these observations led to the conclusion that the homogeneous distribution and availability of the OH^- groups on the PVDF chains are critical factors for the grafting and condensation of the FSi_T in order to promote a homogeneous surface, since the OH^- groups act as active sites [21].

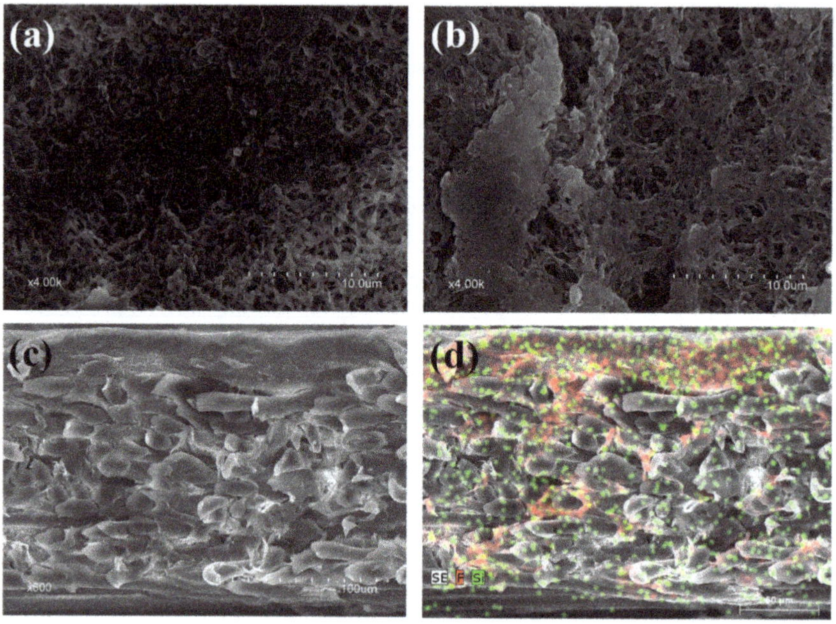

Figure 6. FESEM images of (**a**) non-modified PVDF (nmPVDF) surface, (**b**) modified PVDF ($mPVDF_{max}$) surface, (**c**) modified PVDF ($mPVDF_{max}$) cross-section, and (**d**) EDX mapping of modified PVDF ($mPVDF_{max}$) cross-section (red dots for F and green dots for Si).

The grafting of the FSi_T onto the membrane surface was also verified by measuring the fluorine and silicon content of both nmPVDF and $mPVDF_{max}$ by EDX spectroscopy. The $mPVDF_{max}$ showed a slightly higher F content than that obtained for the nmPVDF (55% and 53%, respectively) indicating that the Dyn had been grafted on the surface. Additionally, the $mPVDF_{max}$ presented a Si content of 0.5% due to the grafting and condensation of the silanols from TEOS over the surface as well as the grafting of Dyn.

From the cross-section inspection, the overall thickness of the $mPVDF_{max}$ was slightly greater than that of the nmPVDF, with values of 164 ± 1 μm and 159 ± 2 μm, respectively. A dense and thick top layer was observed for the $mPVDF_{max}$ with a thickness between 15 and 37 μm (Figure 6c), composed of the residual PVDF chains and especially of the grafted Dyn, since EDX analysis showed a high F content distributed homogeneously throughout this layer (red dots in Figure 6d). An FESEM image of the nmPVDF cross-section is shown in Supplementary Material S4 as an example, and a detailed analysis can be found in our previous work [8]. The Dyn seemed to be grafted both at the surface level and inside the membrane pores, reaching the interface between the PVDF and the PET membrane support and increasing the thickness of that top layer. The presence of this top layer was probably responsible for the increase in hydrophobicity and likely to improve the wetting resistance of the membrane. The EDX analysis also showed the presence of Si (green dots in Figure 6d) from the silica precursor TEOS, located mainly in the top layer, but Si was also found in all the sections but at a lower concentration, as also reported in the literature [41]. Therefore, the condensation of silanols occurred mainly at the surface level [24]. The Si content detected inside the membrane could be attributed to a deposition of the residual FSi_T after solvent evaporation in the curing step, since no reactions between silanols and the PET support were expected [21].

Degradation of the mPVDF activated at $NaOH\%_{wt}$ levels above 6% was confirmed by FESEM-EDX analyses (Supplementary Material S1 and S4). Circular holes were observed on the membrane surface showing a chemical degradation of the active layer, since no F was detected inside the holes. This degradation was accompanied with a visual colour change of orange stains to the membrane surface. The high concentration of NaOH degraded the membrane by the loss of PVDF chains affecting the membrane integrity [21,31]. The increase in NaOH$\%_{wt}$ or activation time increased the extent of defluorination, leading to a severe alteration to the surface chemical structure. For instance, this reaction under severe conditions could convert the PVDF to polyacetylene chains followed by detaching from the surface [24]. Other authors observed a similar colour change in PVDF treated with an alkaline solution, and they established that the F content was lower as the colour became darker [42,43]. A decrease in the mechanical properties of the PVDF after NaOH treatment has been reported, especially for NaOH$\%_{wt} \geq 10\%$, leading to a severe reduction in the degree of crystallinity of the PVDF structure [43]. The PET support could also be degraded, since PET is hydrolysed under severe alkali concentrations and temperature, so the membrane support can play a critical role in the selection of the activation technique and conditions.

3.3. *Performance of the PVDF Membranes in Stability in Long-Term Operation Tests*

The stability of PVDF in long-term operation was evaluated by monitoring its WCA as a useful indicator of changes or alterations occurring on the membrane surface [8,27,33]. The results of the variation in WCA during use of the nmPVDF and mPVDF$_{max}$ are shown in Figures 7a,b for tests with DW and AE, respectively.

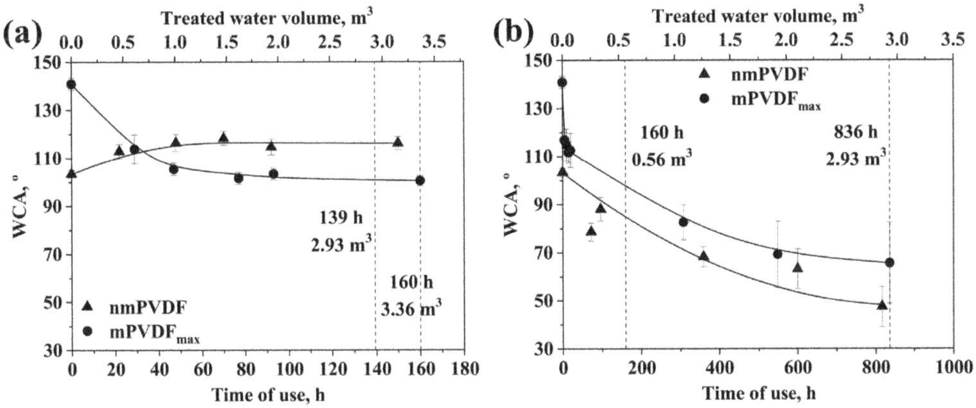

Figure 7. Water contact angle versus time of use (bottom axis) and volume of treated water (top axis) for non-modified PVDF (nmPVDF) and PVDF modified under optimal conditions (mPVDF$_{max}$) with (**a**) deionised water at a liquid flow rate of 21 L h^{-1} and (**b**) anaerobic reactor effluent at a liquid flow rate of 3.5 L h^{-1}. Data on nmPVDF with deionised water from [8]. The error bars denote the standard deviation.

The WCA of the mPVDF$_{max}$ decreased continuously from 140.9 $\pm$ 2.5° at 0 h to 100.5 $\pm$ 2.2° at 160 h in the test with DW (Figure 7a). The WCA decrease was more pronounced in the first 47 h, reaching a value of 105.4 $\pm$ 2.5° and then staying almost constant; the final WCA of mPVDF$_{max}$ was similar to the initial value of the nmPVDF (100.5 $\pm$ 2.2° and 103.4 $\pm$ 1.6°, respectively). The WCA decrease could be due to the reduction of F content detected in the EDX analysis from 55% to 53%, similar to the pristine nmPVDF. This F loss could be attributed to the stress and dragging effect of the liquid flux removing the polymeric chains and/or clusters weakly linked to the surface. Nevertheless, Si content increased up to values of 2% suggesting that the most external layer was composed of the Dyn molecules and the silica precursor was strongly linked to the membrane surface. The WCA after the activation step was of 88.5 $\pm$ 5.4°, lower than that of the mPVDF$_{max}$ at 160 h (100.5 $\pm$ 2.2°), indicating that siloxane chains and Dyn molecules remained on the surface at the end of the long-term experiment with a moderate liquid flow rate of 21 L h^{-1}. Future research would be needed to improve the fixation of macromolecules onto the membrane.

At the end of the long-term test with DW, a permanent plastic deformation at surface level was observed for the $mPVDF_{max}$ from the FESEM images (Supplementary Material S4) in which the surface porosity declined at values lower than 10%. Similar observations were found for the nmPVDF resulting in a dense-like surface [8]. Moreover, the functional top layer formed onto the membrane appeared to have been substantially reduced and the membrane thickness decreased to a value of 148 ± 12 μm, indicating that part of this surface layer was removed.

Some authors have monitored membrane stability and wetting resistance in long-term operations by means of the gas flux change instead of WCA. Using tap water, Sethunga et al. [21] reported a stable CH_4 flux of around 36 × 10^{-5} g m^{-2} s^{-1} over 240 h at a liquid flow rate of 0.6 L h^{-1} using a modified PVDF, and Wongchitphimon et al. [44] also reported a stable CH_4 flux of around 66.7 × 10^{-5} g m^{-2} s^{-1} over 300 h at a liquid flow rate of 5.2 L h^{-1} using polyimide membranes functionalised with organofluorosilanes.

Regarding the results with AE at a liquid flow rate of 3.5 L h^{-1} (Figure 7b), the WCA of the nmPVDF decreased continuously from 103.4 ± 1.6° at the beginning to 47.4 ± 8.4° after 816 h, showing a reduction of 54% from the initial value. A gradual darkening of the membrane was apparent during the tests with AE (Supplementary Material S5), indicating that the entire surface was covered by a fouling cake, which could explain the WCA decrease due to the hydrophilisation promoted by some fouling agents [45]. The WCA trend was similar for both the nmPVDF and $mPVDF_{max}$, but the final WCA of the $mPVDF_{max}$ was higher, with a value of 65.6 ± 16.9° after 836 h, representing a reduction of 53% from the initial value (140.9 ± 2.5°) in spite of the $mPVDF_{max}$ surface being darker at the end of the test. In addition, the presence of a fouling cake on the membrane could be confirmed by two additional observations: (1) the WCA initially decreased faster with AE than with DW despite the lower liquid flow rate in both the nmPVDF and $mPVDF_{max}$ membranes (dashed line at 160 h in Figure 7) and (2) the final WCA obtained with AE was lower than with DW at the same volume of treated water and with lower flow rate (dashed line at 2.93 m^3 of treated water in Figure 7). Regarding observation number (1), the relatively sudden WCA decline of the $mPVDF_{max}$ in the first 50 h could be attributed to a fast attachment of organic/biological matter in AE with more affinity with hydrophobic surfaces [46]. All these results indicated that the membrane surface was covered by the deposition of fouling agents, obscuring the hydrophobic nature of the membrane surface; the hydrophobic level of the surface was dominated by the nature and amount of fouling agents present.

A water breakthrough was detected on the nmPVDF at 816 h, indicating mechanical or chemical deterioration of the membrane surface. Interestingly, a longer useful lifetime was achieved for the $mPVDF_{max}$ since no water breakthrough was observed. Thus, the modification procedure seemed to improve the stability of the PVDF in long-term operations with real AEs. In this regard, the surface modification should be taken into account as a strategy to increase the useful lifetime of the membrane since a greater chemical or mechanical resistance could be induced. In order to compare the viability of using both non-modified and modified membranes, a suitable economic analysis should also consider the additional cost of the modification procedure and its effect on the membrane performance and useful lifetime.

3.4. Performance of the PVDF Membranes for the Recovery of Dissolved Methane from Water

The performance of the $mPVDF_{max}$ in degassing tests for D-CH_4 recovery was evaluated for both DW and AE and compared with the performance of the nmPVDF.

Results obtained with DW are presented in Table 3, in which the D-CH_4 RE and the methane flux (J_{CH4}) at different liquid flow rates (Q_L) for both materials are compared. An RE of 42% and 21% were obtained for $mPVDF_{max}$ at a Q_L of 21.0 and 3.5 L h^{-1}, respectively. A slight improvement in membrane performance was achieved after the membrane modification process, especially when a Q_L higher than 3.5 L h^{-1} was applied. Thus, the RE at a Q_L of 21.0 L h^{-1} was slightly higher than that seen with the nmPVDF (39%). No significant differences were observed in the RE at the lowest Q_L of 3.5 L h^{-1} for both membranes, this being attributed mainly to the limiting mass transfer resistance in the liquid phase, especially at low liquid flow rates and velocities [8,37,47].

Table 3. Effect of the time of use on the dissolved CH_4 removal efficiency (RE) and CH_4 flux (J_{CH4}) in degassing tests (at 5 h of time on stream) with non-modified PVDF (nmPVDF) and modified PVDF ($mPVDF_{max}$) at different liquid flow rates (Q_L) of distilled water and anaerobic reactor effluent.

Membrane	Q_L, L h^{-1}	Time of Use, h	RE, %	$J_{CH4} \times 10^5$, g s^{-1} m^{-2}
		(a) With deionised water		
nmPVDF	3.5	0	19 ± 1 [a]	27 [a]
	21.0	0	39 ± 1 [a]	46 [a]
$mPVDF_{max}$	3.5	0	21 ± 2	29
	21.0	0	42 ± 1	52
		(b) With anaerobic effluent		
nmPVDF	3.5	0	19 ± 1	27
	3.5	720	19 ± 2	26
	21.0	0	39 ± 1	46
	21.0	744	34 ± 2	39
$mPVDF_{max}$	3.5	0	22 ± 3	29
	3.5	528	19 ± 1	24
	21.0	0	42 ± 1	52
	21.0	672	20 ± 1	26

[a] Previous work [8].

The CH_4 flux was also calculated for the degassing tests, and the results are shown in Table 3. The maximum CH_4 flux was observed for the $mPVDF_{max}$ at the highest Q_L of 21 L h^{-1} with a value of 52 × 10^{-5} g s^{-1} m^{-2} since this test showed the maximum RE value. The CH_4 fluxes obtained in this work under unsteady-state conditions were similar to those reported by other authors operating under steady-state conditions. Wongchitphimon et al. [44] obtained CH_4 fluxes of 42 × 10^{-5} to 80 × 10^{-5} g s^{-1} m^{-2} for polyimide membranes modified with organofluorosilanes at liquid flow rates ranging from 2.6 to 5.8 L h^{-1}. Sethunga et al. [21] reported CH_4 fluxes of 17 × 10^{-5} to 39 × 10^{-5} g s^{-1} m^{-2} for PVDF modified with organofluorosilanes with Q_L ranging from 0.2 to 0.5 L h^{-1}. Another work of Sethunga et al. [48] used polydimethylsiloxanes as modifying agents, and CH_4 fluxes between 16 × 10^{-5} and 48 × 10^{-5} g s^{-1} m^{-2} were reported with Q_L ranging from 0.05 to 3.0 L h^{-1}. All these abovementioned works conducted degassing tests under steady-state conditions with tap water and initial D-CH_4 concentrations between 12 and 15 mg L^{-1}.

The $mPVDF_{max}$ showed a greater mass transport of CH_4 than did the nmPVDF under the same liquid and gas hydraulic conditions and at the highest Q_L (21.0 L h^{-1}), since the obtained RE and CH_4 flux were higher with the $mPVDF_{max}$. Thus, these results indicated that the mass transfer resistance of the $mPVDF_{max}$ membrane was lower than that of nmPVDF, which could be due to the prevention of membrane wetting associated with the increase in WCA obtained after the modification of the membrane. The lower surface porosity and slightly higher thickness in the $mPVDF_{max}$ seemed not to significantly affect to the overall mass transfer resistance, being that the hydrophobicity increase was the predominant factor in the decrease of the mass transfer resistance. The higher hydrophobicity observed on the surface of the $mPVDF_{max}$ involved an increase in the liquid entry pressure; therefore, the membrane pores were less susceptible to being wetted or filled with the liquid phase. It is well-known that membrane wetting leads to an increase in membrane mass transfer resistance for CH_4 transport from the liquid phase to the gas phase, since CH_4 presents a higher diffusion coefficient in air than in water [15,49]. Similarly, an increase in degree of wetting with liquid flow rate has been reported, showing a huge influence on membrane resistance, especially for porous membranes [14]. In addition, the reduction in surface porosity observed after modification can also increase wetting resistance, as previously explained.

Likewise, a RE value of 41% was obtained at a Q_L of 21.0 L h^{-1} with $mPVDF_{max}$ after 20 h, when its WCA was reduced to 122°. This RE was similar to the initial value of 42% obtained for the $mPVDF_{max}$ with a WCA of ~140°. Thus, it can be concluded that the modification procedure improved the stability of the PVDF in terms of membrane performance, in contrast to those structural and performance changes previously observed for the nmPVDF [8].

Both the nmPVDF and $mPVDF_{max}$ used in the long-term stability tests (Section 3.3) with AE were evaluated in degassing tests at different times of use in order to determine the effect on D-CH_4 recovery of an actual stream and level of hydrophobicity. The AE previously saturated with CH_4 was used as the liquid stream. The results in terms of RE and J_{CH4} at different times of use and at liquid flow rates of 3.5 and 21.0 L h^{-1} for the nmPVDF and $mPVDF_{max}$ are shown in Table 3. Similar performances were obtained at a Q_L of 3.5 L h^{-1} for the pristine membranes (time of use of 0 h), with nmPVDF at 720 h and $mPVDF_{max}$ at 528 h, with values of RE and J_{CH4} of around 20% and 26×10^{-5} g s^{-1} m^{-2}, respectively. Thus, the potential effect on membrane performance of the foulants contained in the liquid stream could be considered negligible for both membranes at low liquid flow rates, at least, for a time of use less than 800 h. In contrast, lower RE values were obtained at the higher Q_L of 21.0 L h^{-1} for the used nmPVDF and $mPVDF_{max}$ membranes at 744 and 672 h, respectively. For example, REs of 39% and 34% were obtained for the pristine and used nmPVDF, respectively (decline in J_{CH4} of 17%). This RE decline was more severe for the $mPVDF_{max}$ in that it decreased from 42% to 20% for the pristine and used membrane, respectively, at a liquid flow rate of 21.0 L h^{-1} (decline in J_{CH4} of 51%). In this case, the deposition of foulants onto the membrane involved an additional and non-negligible mass transfer resistance, since the liquid resistance decreased at higher liquid flow rates. Similar declines in CH_4 flux, and even a severe reduction of up to 90% after only 40 h, have been reported by other authors using modified membranes to treat AEs [10,48,50], while a stable flux was maintained when using tap water [21,44]. The results allowed the conclusion that the modification of PVDF promoted a higher additional mass transfer resistance due to the deposition of foulants contained in the AE hindering D-CH_4 recovery, even though greater stability and resistance were achieved, since no water breakthrough was observed during the long-term test. The apparent higher fouling deposition on the mPVDF could be due to the increase in membrane hydrophobicity which favoured the deposition of hydrophobic foulants, and/or the change in the surface charge which could promote an increase of the electrostatic attraction between foulants and the modified membrane [46,51]. Further investigations should be focused on the mechanisms and characterisation of fouling in order to propose strategies for its control and prevention.

4. Conclusions

The hydrophobicity of a PVDF membrane surface has been maximised by chemical grafting in a three-step procedure (activation, functionalisation, and curing) in order to increase the membrane performance in the recovery of dissolved methane from aqueous streams. The main variables in the modification procedure that significantly affected the water contact angle were found to be the NaOH concentration in the activation step and the modifying agent/precursor (FSi_T) ratio and concentration in the functionalisation step. The contact angle of the modified PVDF increased continuously with NaOH and

FSi_T concentration, although the concentration of NaOH was limited by the degradation observed on the membrane surface at concentrations of >5%. In addition, no significant increase in contact angle was observed at an FSi_T concentration of >7.2%. In contrast, a maximum contact angle was observed for an FSi_T ratio of 0.55. Thus, these values were established as the optimal conditions in order to maximise the hydrophobicity of the membrane surface, achieving a contact angle of >140°. This value represented an increase of >36% with respect to the initial contact angle of the non-modified PVDF (103°).

Regarding the stability of the membrane under long-term operation, the modified PVDF experienced a continuous decrease in contact angle to similar values as the pristine non-modified PVDF in a distilled water stream. In contrast, the contact angle of the modified PVDF was always higher when real anaerobic reactor effluent was used as the liquid stream. However, the final contact angle was quite low, at values of around 70° and 50° for the modified and non-modified PVDF, respectively, after ~800 h. In addition, the non-modified PVDF was strongly degraded at ~800 h since a water breakthrough occurred. Those outcomes allowed the conclusion that the modified PVDF presented greater chemical resistance and stability under operational conditions.

With regard to methane recovery, at the highest liquid flow rate tested (21 L h^{-1}), the modified PVDF showed a slightly higher performance than the non-modified PVDF. However, when real AE was used, the performance of the modified PVDF membrane decreased faster, suggesting that modification could favour fouling. In contrast, at the lowest liquid flow rate (3.5 L h^{-1}) no significant differences were observed between the modified and non-modified membranes, which could be explained by taking into account the high and limiting mass transfer resistance of the liquid phase under this operational condition, and so the effect of the additional resistances of the membrane could be ignored.

Supplementary Materials: The following supporting information can be downloaded at: https://www.mdpi.com/article/10.3390/membranes12040426/s1: Figure S1. (a) Effect of the NaOH concentration ($NaOH\%_{wt}$) (activation time of 1 h at 60 °C) and (b) the activation time (5% NaOH at 50 °C) on the water contact angle (WCA) after activation step. Table S1. Effect of the composition of the functionalisation solution on the water contact angle (WCA) after the curing step of the modified PVDF (activation: 5% NaOH, 1 h, 50 °C; functionalisation: 1 h, room temperature; curing: overnight, 60 °C), Figure S2. Effect of the functionalisation time on the water contact angle (WCA) after the curing step (activation: 5% NaOH, 50 °C, 1 h; functionalisation: $FSi_T\%_v$ = 3.6%, FSi_T ratio = 0.41; curing: overnight, 60 °C), Figure S3. Effect of the curing time on water contact angle (WCA) after the curing step (activation: 5% NaOH, 50 °C, 1 h; functionalisation: $FSi_T\%_v$ = 3.6%, FSi_T ratio = 0.41, 1 h; curing: 60 °C), Figure S4. Hydroxylation reaction of the poly(vinylidene fluoride) (PVDF) with sodium hydroxide in the activation step, Figure S5. Hydrolysis reaction of the tetraethyl orthosilicate (TEOS), Figure S6. Condensation reaction of the silanols on the hydroxylated PVDF in the functionalisation step, Figure S7. Condensation reaction of the 1H,1H,2H,2H-perfluorooctyltriethoxysilane (Dynasylan F8261®) onto the membrane surface in the functionalisation step, Figure S8. Examples of the profiles of (a) dissolved methane concentration (C_L) and removal efficiency (RE) and (b) CH_4 flux (J_{CH4}) versus time on stream for degassing tests with the modified PVDF with the highest hydrophobicity ($mPVDF_{max}$) at a liquid and N_2 flow rate of 21 L h^1 and 4 L h^{-1}, respectively, Figure S9. Cross section FESEM image of the non-modified PVDF, Figure S10. (a) FESEM image of a modified PVDF surface focused on a degraded zone (membrane activated with a $50\%_{wt}$ NaOH solution). (b) Fluor distribution on the membrane surface around a degraded zone, Figure S11. FESEM images of the (a) surface and (b) cross section of the modified PVDF ($mPVDF_{max}$) surface after 160 h of use with deionized water at a liquid flow rate of 21 L h^{-1}, Figure S12. Images of the non-modified and modified PVDF membrane surface (nmPVDF and $mPVDF_{max}$, respectively) during the long-term tests with an anaerobic reactor effluent at a liquid flow rate of 3.5 L h^{-1}. References [23–25,27,29–31,33,34,42,43] are cited in supplementary materials.

Author Contributions: Conceptualization, M.I., J.D.B. and V.M.-S.; methodology, R.J.-R., B.M.M.-T., J.D.B., V.M-S. and M.I.; validation, R.J.-R., B.M.M.-T., J.D.B. and M.I.; formal analysis, R.J.-R.; investigation, R.J.-R. and B.M.M.-T.; resources, J.D.B.; data curation, R.J.-R. and B.M.M.-T.; writing—original draft preparation, R.J.-R.; writing—review and editing, R.J.-R., J.D.B., V.M.-S. and M.I.; visualization,

R.J.-R.; supervision, M.I., J.D.B. and V.M.-S.; project administration, M.I., J.D.B. and V.M.-S.; funding acquisition, M.I. and V.M.-S. All authors have read and agreed to the published version of the manuscript.

Funding: This research was funded by the Conselleria d'Innovació, Universitats, Ciència i Societat Digital—Generalitat Valenciana (project GV/2019/149). The PhD grant of R. Jiménez-Robles was funded by the Ministerio de Universidades, Spain (Beca de Formación de Profesorado Universitario FPU19/02478).

Institutional Review Board Statement: Not applicable.

Informed Consent Statement: Not applicable.

Data Availability Statement: The data presented in this study are available on request from the corresponding author.

Conflicts of Interest: The authors declare no conflict of interest.

References

1. Crone, B.C.; Garland, J.L.; Sorial, G.A.; Vane, L.M. Significance of dissolved methane in effluents of anaerobically treated low strength wastewater and potential for recovery as an energy product: A review. *Water Res.* **2016**, *104*, 520–531. [CrossRef] [PubMed]
2. Heile, S.; Chernicharo, C.; Brandt, E.; McAdam, E. Dissolved gas separation for engineered anaerobic wastewater systems. *Sep. Purif. Technol.* **2017**, *189*, 405–418. [CrossRef]
3. Mora, E.C.; Chernicharo, C.D.L. Use of membrane contactors for removing and recovering dissolved methane from anaerobic reactors effluents: State-of-the-art, challenges, and perspectives. *Rev. Environ. Sci. Bio/Technol.* **2020**, *19*, 673–697. [CrossRef]
4. Rongwong, W.; Lee, J.; Goh, K.; Karahan, H.E.; Bae, T.-H. Membrane-based technologies for post-treatment of anaerobic effluents. *npj Clean Water* **2018**, *1*, 21. [CrossRef]
5. Xu, Y.; Goh, K.; Wang, R.; Bae, T.-H. A review on polymer-based membranes for gas-liquid membrane contacting processes: Current challenges and future direction. *Sep. Purif. Technol.* **2019**, *229*, 115791. [CrossRef]
6. Zhang, N.; Pan, Z.; Zhang, Z.; Zhang, W.; Zhang, L.; Moreno, F.M.B.; Lichtfouse, E. CO2 capture from coalbed methane using membranes: A review. *Environ. Chem. Lett.* **2019**, *18*, 79–96. [CrossRef]
7. Klaassen, R.; Feron, P.; Jansen, A. Membrane Contactors in Industrial Applications. *Chem. Eng. Res. Des.* **2005**, *83*, 234–246. [CrossRef]
8. Jiménez-Robles, R.; Gabaldón, C.; Badia, J.; Izquierdo, M.; Martínez-Soria, V. Recovery of dissolved methane through a flat sheet module with PDMS, PP, and PVDF membranes. *Sep. Purif. Technol.* **2021**, *282 Pt B*, 120057. [CrossRef]
9. Rana, D.; Matsuura, T. Surface Modifications for Antifouling Membranes. *Chem. Rev.* **2010**, *110*, 2448–2471. [CrossRef]
10. Sethunga, G.; Karahan, H.E.; Wang, R.; Bae, T.-H. Wetting- and fouling-resistant hollow fiber membranes for dissolved methane recovery from anaerobic wastewater treatment effluents. *J. Membr. Sci.* **2020**, *617*, 118621. [CrossRef]
11. Liu, F.; Hashim, N.A.; Liu, Y.; Moghareh Abed, M.R.; Li, K. Progress in the production and modification of PVDF membranes. *J. Membr. Sci.* **2011**, *375*, 1–27. [CrossRef]
12. Kang, G.; Cao, Y. Application and modification of poly(vinylidene fluoride) (PVDF) membranes—A review. *J. Membr. Sci.* **2014**, *463*, 145–165. [CrossRef]
13. Al-Gharabli, S.; Mavukkandy, M.O.; Kujawa, J.; Nunes, S.; Arafat, H.A. Activation of PVDF membranes through facile hydroxylation of the polymeric dope. *J. Mater. Res.* **2017**, *32*, 4219–4231. [CrossRef]
14. Henares, M.; Ferrero, P.; San-Valero, P.; Martínez-Soria, V.; Izquierdo, M. Performance of a polypropylene membrane contactor for the recovery of dissolved methane from anaerobic effluents: Mass transfer evaluation, long-term operation and cleaning strategies. *J. Membr. Sci.* **2018**, *563*, 926–937. [CrossRef]
15. Wang, R.; Zhang, H.; Feron, P.; Liang, D. Influence of membrane wetting on CO2 capture in microporous hollow fiber membrane contactors. *Sep. Purif. Technol.* **2005**, *46*, 33–40. [CrossRef]
16. Velasco, P.; Jegatheesan, V.; Thangavadivel, K.; Othman, M.; Zhang, Y. A focused review on membrane contactors for the recovery of dissolved methane from anaerobic membrane bioreactor (AnMBR) effluents. *Chemosphere* **2021**, *278*, 130448. [CrossRef]
17. Miller, D.J.; Dreyer, D.R.; Bielawski, C.W.; Paul, D.R.; Freeman, B.D. Surface Modification of Water Purification Membranes. *Angew. Chem. Int. Ed.* **2017**, *56*, 4662–4711. [CrossRef]
18. Al-Juboori, R.; Yusaf, T. Biofouling in RO system: Mechanisms, monitoring and controlling. *Desalination* **2012**, *302*, 1–23. [CrossRef]
19. Zhao, S.; Liao, Z.; Fane, A.; Li, J.; Tang, C.; Zheng, C.; Lin, J.; Kong, L. Engineering antifouling reverse osmosis membranes: A review. *Desalination* **2020**, *499*, 114857. [CrossRef]
20. Wongchitphimon, S.; Wang, R.; Jiraratananon, R.; Shi, L.; Loh, C.H. Effect of polyethylene glycol (PEG) as an additive on the fabrication of polyvinylidene fluoride-co-hexafluropropylene (PVDF-HFP) asymmetric microporous hollow fiber membranes. *J. Membr. Sci.* **2011**, *369*, 329–338. [CrossRef]

21. Sethunga, G.; Rongwong, W.; Wang, R.; Bae, T.-H. Optimization of hydrophobic modification parameters of microporous polyvinylidene fluoride hollow-fiber membrane for biogas recovery from anaerobic membrane bioreactor effluent. *J. Membr. Sci.* **2018**, *548*, 510–518. [CrossRef]
22. Hashemifard, S.A.; Ismail, A.F.; Matsuura, T.; DashtArzhandi, M.R. Performance of silicon rubber coated polyetherimide hollow fibers for CO2 removal via a membrane contactor. *RSC Adv.* **2015**, *5*, 48442–48455. [CrossRef]
23. Zheng, Z.; Gu, Z.; Huo, R.; Luo, Z. Superhydrophobic poly(vinylidene fluoride) film fabricated by alkali treatment enhancing chemical bath deposition. *Appl. Surf. Sci.* **2009**, *256*, 2061–2065. [CrossRef]
24. Ross, G.; Watts, J.; Hill, M.; Morrissey, P. Surface modification of poly(vinylidene fluoride) by alkaline treatment1. The degradation mechanism. *Polymer* **2000**, *41*, 1685–1696. [CrossRef]
25. Sairiam, S.; Loh, C.H.; Wang, R.; Jiraratananon, R. Surface modification of PVDF hollow fiber membrane to enhance hydrophobicity using organosilanes. *J. Appl. Polym. Sci.* **2013**, *130*, 610–621. [CrossRef]
26. Badia, J.; Gil-Castell, O.; Ribes-Greus, A. Long-term properties and end-of-life of polymers from renewable resources. *Polym. Degrad. Stab.* **2017**, *137*, 35–57. [CrossRef]
27. Oldani, V.; Sergi, G.; Pirola, C.; Sacchi, B.; Bianchi, C. Sol-gel hybrid coatings containing silica and a perfluoropolyether derivative with high resistance and anti-fouling properties in liquid media. *J. Fluor. Chem.* **2016**, *188*, 43–49. [CrossRef]
28. Zhang, Y.; Wang, R.; Yi, S.; Setiawan, L.; Hu, X.; Fane, A.G. Novel chemical surface modification to enhance hydrophobicity of polyamide-imide (PAI) hollow fiber membranes. *J. Membr. Sci.* **2011**, *380*, 241–250. [CrossRef]
29. Zhang, Y.; Wang, R. Novel method for incorporating hydrophobic silica nanoparticles on polyetherimide hollow fiber membranes for CO2 absorption in a gas–liquid membrane contactor. *J. Membr. Sci.* **2013**, *452*, 379–389. [CrossRef]
30. Zhang, Y.; Wang, R. Fabrication of novel polyetherimide-fluorinated silica organic-inorganic composite hollow fiber membranes intended for membrane contactor application. *J. Membr. Sci.* **2013**, *443*, 170–180. [CrossRef]
31. Wongchitphimon, S.; Wang, R.; Jiraratananon, R. Surface properties of fluorinated hybrid coatings(PVDF–HFP) hollow fiber membrane for membrane gas absorption. *J. Membr. Sci.* **2011**, *381*, 183–191. [CrossRef]
32. Liu, L.; Shen, F.; Chen, X.; Luo, J.; Su, Y.; Wu, H.; Wan, Y. A novel plasma-induced surface hydrophobization strategy for membrane distillation: Etching, dipping and grafting. *J. Membr. Sci.* **2015**, *499*, 544–554. [CrossRef]
33. Fabbri, P.; Messori, M.; Montecchi, M.; Pilati, F.; Taurino, R.; Tonelli, C.; Toselli, M. Surface properties of fluorinated hybrid coatings. *J. Appl. Polym. Sci.* **2006**, *102*, 1483–1488. [CrossRef]
34. Huang, X.; Zhang, J.; Wang, W.; Liu, Y.; Zhang, Z.; Li, L.; Fan, W. Effects of PVDF/SiO2 hybrid ultrafiltration membranes by sol–gel method for the concentration of fennel oil in herbal water extract. *RSC Adv.* **2015**, *5*, 18258–18266. [CrossRef]
35. Hebbar, R.S.; Isloor, A.M.; Ismail, A.F. Contact Angle Measurements. In *Membrane Characterization*, 1st ed.; Hilal, N., Ismail, A., Matsuura, T., Oatley-Radcliffe, D., Eds.; Elsevier: Amsterdam, The Netherlands, 2017; pp. 219–255. [CrossRef]
36. Abbasgholipourghadim, M.; Mailah, M.; Darus, I.Z.M.; Ismail, A.F.; Dashtarzhandi, M.R.; Abbasgholipourghadim, M.; Khademi, S. Porosity and Pore Area Determination of Hollow Fiber Membrane Incorporating Digital Image Processing. In *Recent Advances in Mechanics and Mechanical Engineering, Proceedings of the 6th International Conference on Fluid Mechanics and Heat & Mass Transfer (FLUIDSHEAT '15), Proceedings of the 1st International Conference on Mechanical and Transportation Engineering (ICMTE '15), Kuala Lumpur, Malaysia, 23–25 April 2015*; WSEAS Press: Athens, Greece, 2015; pp. 118–123.
37. Jiménez-Robles, R.; Gabaldón, C.; Martínez-Soria, V.; Izquierdo, M. Simultaneous application of vacuum and sweep gas in a polypropylene membrane contactor for the recovery of dissolved methane from water. *J. Membr. Sci.* **2020**, *617*, 118560. [CrossRef]
38. Cookney, J.; Cartmell, E.; Jefferson, B.; McAdam, E.J. Recovery of methane from anaerobic process effluent using poly-di-methyl-siloxane membrane contactors. *Water Sci. Technol.* **2012**, *65*, 604–610. [CrossRef]
39. Bandara, W.M.; Satoh, H.; Sasakawa, M.; Nakahara, Y.; Takahashi, M.; Okabe, S. Removal of residual dissolved methane gas in an upflow anaerobic sludge blanket reactor treating low-strength wastewater at low temperature with degassing membrane. *Water Res.* **2011**, *45*, 3533–3540. [CrossRef]
40. Henares, M.; Izquierdo, M.; Marzal, P.; Martínez-Soria, V. Demethanization of aqueous anaerobic effluents using a polydimethyl-siloxane membrane module: Mass transfer, fouling and energy analysis. *Sep. Purif. Technol.* **2017**, *186*, 10–19. [CrossRef]
41. Sethunga, G.; Lee, J.; Wang, R.; Bae, T.-H. Influences of operating parameters and membrane characteristics on the net energy production in dense, porous, and composite hollow fiber membrane contactors for dissolved biomethane recovery. *J. Membr. Sci.* **2020**, *610*, 118301. [CrossRef]
42. Ross, G.; Watts, J.; Hill, M.; Morrissey, P. Surface modification of poly(vinylidene fluoride) by alkaline treatment Part 2. Process modification by the use of phase transfer catalysts. *Polymer* **2001**, *42*, 403–413. [CrossRef]
43. Hashim, N.A.; Liu, Y.; Li, K. Stability of PVDF hollow fibre membranes in sodium hydroxide aqueous solution. *Chem. Eng. Sci.* **2011**, *66*, 1565–1575. [CrossRef]
44. Wongchitphimon, S.; Rongwong, W.; Chuah, C.Y.; Wang, R.; Bae, T.-H. Polymer-fluorinated silica composite hollow fiber membranes for the recovery of biogas dissolved in anaerobic effluent. *J. Membr. Sci.* **2017**, *540*, 146–154. [CrossRef]
45. Zarebska, A.; Amor, C.; Ciurkot, K.; Karring, H.; Thygesen, O.; Andersen, T.; Hägg, M.-B.; Christensen, K.; Norddahl, B. Fouling mitigation in membrane distillation processes during ammonia stripping from pig manure. *J. Membr. Sci.* **2015**, *484*, 119–132. [CrossRef]
46. Costa, F.C.; Ricci, B.C.; Teodoro, B.; Koch, K.; Drewes, J.E.; Amaral, M.C. Biofouling in membrane distillation applications—A review. *Desalination* **2021**, *516*, 115241. [CrossRef]

47. Henares, M.; Izquierdo, M.; Penya-Roja, J.; Martínez-Soria, V. Comparative study of degassing membrane modules for the removal of methane from Expanded Granular Sludge Bed anaerobic reactor effluent. *Sep. Purif. Technol.* **2016**, *170*, 22–29. [CrossRef]
48. Sethunga, G.; Karahan, H.E.; Wang, R.; Bae, T.-H. PDMS-coated porous PVDF hollow fiber membranes for efficient recovery of dissolved biomethane from anaerobic effluents. *J. Membr. Sci.* **2019**, *584*, 333–342. [CrossRef]
49. Sethunga, G.; Lee, J.; Wang, R.; Bae, T.-H. Influence of membrane characteristics and operating parameters on transport properties of dissolved methane in a hollow fiber membrane contactor for biogas recovery from anaerobic effluents. *J. Membr. Sci.* **2019**, *589*, 117263. [CrossRef]
50. Rongwong, W.; Goh, K.; Sethunga, G.; Bae, T.-H. Fouling formation in membrane contactors for methane recovery from anaerobic effluents. *J. Membr. Sci.* **2018**, *573*, 534–543. [CrossRef]
51. Guo, J.; Farid, M.U.; Lee, E.-J.; Yan, D.Y.-S.; Jeong, S.; An, A.K. Fouling behavior of negatively charged PVDF membrane in membrane distillation for removal of antibiotics from wastewater. *J. Membr. Sci.* **2018**, *551*, 12–19. [CrossRef]

membranes

MDPI

Article

A Novel Methodology to Obtain the Mechanical Properties of Membranes by Means of Dynamic Tests

Antonia Lima-Rodriguez [1,*], Jose Garcia-Manrique [1], Wei Dong [2] and Antonio Gonzalez-Herrera [1]

[1] Department of Civil Engineering, Materials and Manufacturing, School of Engineering, University of Malaga, C/Arquitecto Francisco Peñalosa 6, 29071 Malaga, Spain; josegmo@uma.es (J.G.-M.); agh@uma.es (A.G.-H.)

[2] VA Loma Linda Healthcare System and Department of Otolaryngology—Head & Neck Surgery, Loma Linda University Health, 11234 Anderson St., Loma Linda, CA 92354, USA; wei.dong@va.gov

* Correspondence: tlima@uma.es

Abstract: A new, non-destructive methodology is proposed in this work in order to determine the mechanical properties of membrane using vibro-acoustic tests. This procedure is based on the dynamic analysis of the behavior of the membrane. When the membrane is subjected to a sound excitation it responds by vibrating based on its modal characteristics and this modal parameter is directly related to its mechanical properties. The paper is structured in two parts. First, the theoretical bases of the test are presented. The interaction between the sound waves and the membrane (mechano-acoustic coupling) is complex and requires meticulous study. It was broadly studied by means of numerical simulations. A summary of this study is shown. Aspects, such as the position of the sound source, the measuring points, the dimensions of the membrane, the frequency range, and the magnitudes to be measured, among others, were evaluated. The validity of modal analysis curve-fitting techniques to extract the modal parameter from the data measures was also explored. In the second part, an experimental test was performed to evaluate the validity of the method. A membrane of the same material with three different diameters was measured with the aim of estimating the value of the Young's modulus. The procedure was applied and satisfactory results were obtained. Additionally, the experiment shed light on aspects that must be taken account in future experiments.

Keywords: membrane characterization; experiment; finite element model

1. Introduction

Thin film materials are elements present in multitude of engineering problems, such as the study of microdevices or biomechanics, and their characterization is essential to predicting and simulating their behavior. Nevertheless, the determination of the mechanical properties of a material is a difficult problem to solve when it is in its membrane form. These materials present a small thickness and low stiffness. Traditional experimental tests present inaccuracies in the characterization of some of these materials and mostly involve the destructive testing of the material.

One of the best-known tests to determine mechanical properties is the tensile test. With this test, the Young's modulus is determined, among other mechanical characteristics, and it is a destructive test. Many authors have used this test to determine the Young's modulus of thin film materials of different compositions. In [1], both the tensile test and resonance techniques were used to mechanically characterize boron-doped silicon films with thicknesses of 1 to 4 μm. In [2], the Young's modulus of fetal preterm and term amniotic membranes were measured by using a stress-strain apparatus. In [3], thin micromembranes were subjected to compressive stresses, with or without initial deflections, to determine the Young's modulus and the residual stress of the micromembranes for use in microdevices.

Other works are based on the relationship between deflection and pressure. In [4], this relationship was used to obtain the Young's modulus and internal stress of composite rectangular membranes of 2 mm by 8 mm.

Resonance methods are used in the literature to mechanically characterize thin film materials [5–9]. In [5,6,9], cantilever beams made from thin films were forced to vibrate. Young's moduli were calculated by measuring the resonance frequencies of the beams. In [7], the influence on Young's modulus of aging phenomena in heavily doped micromachined silicon cantilever beams was studied. In [8], the first and second symmetric vibration modes of a circular diaphragm thin film were measured to calculate Young's modulus and the residual stress of the film, based on the resolution of the Rayleigh–Ritz equation, analytically.

Focusing on the field of interest of the authors, the study of the tympanic membrane, there are quite a few studies carried out to determine the Young's modulus. The function of this membrane is very important in the hearing process, since it is responsible for transmitting the sound waves captured from the external auditory canal into the ear. One of the most important parameters to obtain a physical model of this membrane is the elastic modulus. Authors have usually carried out experiments with samples cut from the tympanic membrane to characterize them mechanically. In [10], the Young's modulus of a strip of human tympanic membrane was measured using a longitudinal vibration technique, while ref. [11] performed a beam bending test and determined the Young's modulus and ref. [12] reported results of uniaxial tests of the human eardrum. In [13,14], the tensile test was applied to eardrum specimens from fresh human cadavers. In [13], the Young's modulus was determined through mechanical tension tests and a t test was used to evaluate it statistically. In [14], the Ogden hyperelastic model and the digital image correlation method were used to analyze the experimental results. Recently, ref. [15] compared the acoustic behavior of different types of membranes for use in the repair of chronic tympanic membrane perforations. Membranes of different thicknesses (10–100 μm) were tested in a model with a simulated ear canal to determine the frequency response function (FRF) and their resistance to pressure loads. The tensile test was used to obtain the mechanical rigidity to pressure loads.

Although these measurement methods can be useful on certain occasions, they also have certain limitations. As the tympanic membrane is a biological material, it is not always possible to have enough samples to carry out a destructive test, and measurements cannot be carried out in situ. Furthermore, another aspect to consider is that the results are subject to significant uncertainties due to the heterogeneous composition of the tympanic membrane and its anisotropic properties. In fact, the range of Young's modulus results for the tympanic membrane is very broad, even of different orders of magnitude in many cases, due to the alteration of the material during the experiment.

There are interesting works focused on nano- or micro-indentations for the mechanical characterization of tympanic membranes [16–20]. In [19], the material was also subjected to a uniaxial tensile test to validate the results obtained, and in [20], the tissue indentation technique and an inverse finite element analysis were used to estimate the Young's modulus of the tympanic membrane. Other authors have also used this technique to calculate the elastic modulus for other types of film-type materials [21,22].

The nano- or micro-indentation method presents good results to obtain the strain–stress curve of the membrane. Nonlinear material behavior with a large strain can be captured, but less information is obtained at the low strain range. The problem is that it is precisely at that low strain range where most membranes behave in vibrations due to acosutic waves.

Another technique used to calculate the mechanical properties of materials is AFM force spectroscopy [23–25]. In these works, the calculation of Young's modulus is focused on microscopic materials. In these cases, it may be interesting to monitor the state of a sample during the test, to be sure that unwanted failures do not occur in the material [26,27].

There are other works that study the mechanical behavior of the tympanic membrane using resonance methods [28,29]. In [28], in addition to this dynamic method, used a constitutive model to estimate the properties based on known stiffness. After that, the existing experimental data were reinterpreted using the classical theory of composite lamination.

In [29], a new Hopkinson tension bar was developed to investigate the mechanical behavior of the tympanic membrane at high strain rates.

The ASTM E 756-05 standard [30] presents an experiment that enables the determination dynamic properties, such as natural frequencies and the damping of the material, and then establishes analytical relationships with mechanical properties, such as Young's modulus (E) or the transverse modulus (G). The tests were based on the analysis of the behavior of an excited bracket by means of a vibration in a frequency range of 50–5000 Hz at different temperatures. In the case of non-self-supporting materials, such as the case described here, it uses a configuration where it is attached to a base material with sufficient rigidity. In these cases, the test is repeated with the composite piece and with the base material, to establish relationships between the damping of the second versus the set. The need to join both materials, the inclusion of adhesives and the simplifications assumed imply precision levels that the standard establishes at 20%. Although they are a starting point, the properties of materials change when they are deployed in film or membrane forms.

Due to the need for a test to determine the mechanical characteristics of thin film materials, a non-destructive methodology is proposed to determine the Young's modulus using vibro-acoustic tests, based on the dynamic analysis of the behavior of the membrane. The tested sample remains intact and can be reused in other configurations or tests in the future. A membrane of a material subjected to a sound excitation responds by vibrating in a range of small deformations and small displacements, where bending is the main mechanical behavior. This vibration depends on the dynamic characteristics of the membrane in terms of resonance frequencies and vibration modes. The main purpose of the methodology is to relate the expected response to a controlled excitation of a specimen based on its mechanical properties, specifically to obtain the Young's modulus information.

However, this measurement is not easy to obtain. The interaction between the sound waves and the response of the membrane (mechanical acoustic coupling) is complex and requires a meticulous study. The response is affected by a multitude of factors whose influence must be studied. The position of the sound source, the measuring points, the dimensions of the membrane, the sound frequencies or the magnitudes to be measured, among others, must be considered.

The methodology proposed by the authors is based on a controlled experiment where the range of sound pressure on the membrane guarantees the vibrations exhibit small-strain behavior and present a linear and elastic response. In order to avoid sources of uncertainty as much as possible, the main parameters were studied by previous tests or numerical simulations with Ansys software. Once the experiment procedure was established, different strategies of data post-processing were analyzed, including direct observation and Nyquist diagrams [31]. The experiment configuration (circular specimens) enables the use of a known analytical solution to describe the dynamic response, which has been widely covered in the literature for extreme cases of plate and membrane behaviors [32,33] and for intermediate situations [34,35]. Finally, experimental displacement measurement at strategic point of the specimen identified a group of main frequencies that, introduced in the analytical formulation, determined the mechanical properties.

The paper is structured as follows. In the materials and methods section, the general description of the methodology proposed is presented. The main dynamic response of a membrane subject to sound pressure is introduced. In addition, a numerical study to evaluate the potentialities of this method is described. The main parameters influencing the tests are analyzed based on these numerical simulations. In the results and discussion sections, the methodology is applied to the analysis of a membrane similar to the tympanic membrane. Finally, the main conclusions are outlined.

2. Materials and Methods

The methodology of the experiment was developed to evaluate mechanical characteristics (Young's modulus, pretension on membranes, damping, etc.) through the experimental and analytical correlation with a model test.

The configuration of the experiment must fulfill a series of requirements:

- The specimen has no added material.
- The test is non-destructive and repeatable with the same specimen.
- The specimen is subjected to a sound pressure that causes a vibration under the regime of small deformations.
- The configuration of the elements (source of sound, capture device, specimen, etc.) and the post-processing strategy must enable the detection of enough resonant frequencies to apply correlations.
- The test must reproduce a dynamic behavior in the specimen with a contrasted analytical solution.
- The experiment will be easy to reproduce with media currently available in standard laboratories.

2.1. Experiment Configuration

On Figure 1 we can see a schematic description of the experimental test proposed. The simplest configuration of the experiment was chosen: a sound source at 90°, and measurement of the center of the membrane. The sound source was placed at a distance d perpendicular to, and aligned with, the center of one of the faces of the membrane (angle of 90°). The harmonic experiment applies a stimulus through pressure to the membrane. The membrane specimen was placed into a baffle device with a free circular surface.

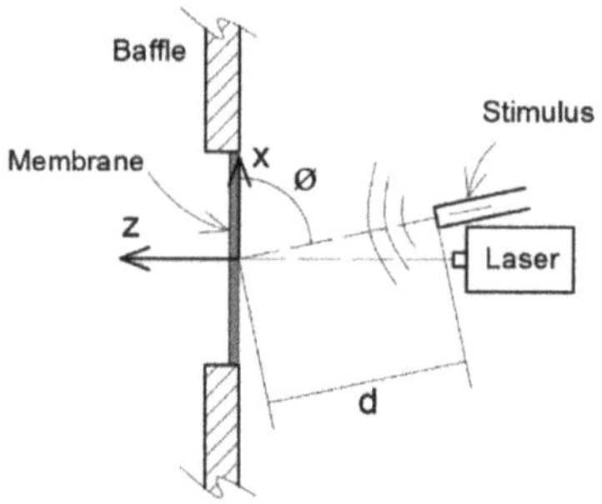

Figure 1. Configuration setup.

With this configuration, the axial symmetry vibration modes (1, 4, 9, etc.) can be obtained experimentally. With the data of the vibration modes, the Young's modulus of the material can be calculated, which is the objective of the experiment.

2.2. The Shape of the Specimen. Analytical Correlation

The selected shape for specimens was circular. This is a common shape which is easy to reproduce with a device that serves as a support and leaves the circular portion of the specimen free.

There is also a theoretical, well-known relation between the frequencies of resonance and the mechanical properties of the material (Young's modulus, Poisson's ratio, density and geometry) for a circular membrane with clamped boundary conditions and no applied load. The equation to calculate the natural frequencies is [33]:

$$f = \frac{\alpha}{2\pi a^2}\sqrt{\frac{gD}{\gamma h}} = \frac{\alpha}{2\pi a^2}\sqrt{\frac{Eh^2}{12(1-\upsilon^2)\rho}} \quad (1)$$

where f is the natural frequency in Hz, a is the radio of the membrane, $\gamma h/g$ is the mass per unit area, D is the flexural rigidity, E is the Young's modulus, h is the membrane thickness, υ is the Poisson's ratio, ρ is the density and α is a constant that depends on the number of nodal diameters n and the number of nodal circles s of the mode considered. Figure 2 reproduce the nodal diameters (n) and nodal circles (s) for the first nine modes of vibration of a circular membrane.

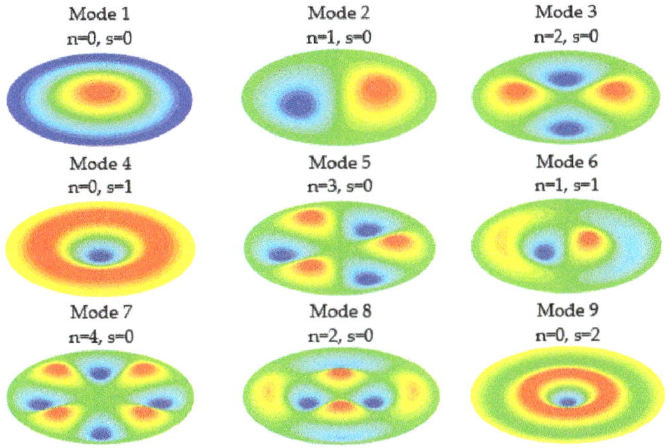

Figure 2. Modal shapes of the first 9 vibration modes.

To identify frequencies, the experiment measured a selected number of points of displacement using a Laser Doppler Vibrometer (LDV, Polytec), then calculated the harmonic response and post-processed that information with an automatic Matlab software routine.

The different parameters that influences the resonant frequencies of the membrane, such as the dynamic response of an excitation, were studied. These parameters included the geometry of the membrane, damping and the position and arrangement of the excitation source, among others.

2.3. Measurement Objetive Region and Position of the Source. Numerical Analysis

The deformed shape of the membrane was different in each vibration mode. Figure 2 shows the shape of the first nine modes of vibration of a circular membrane. There were points of the membrane with large displacements compared to others. This issue must be taken into account in the choice of measurement points. The choice of which points to measure determines the modes that can be detected. In the experiment it was required to be able to conveniently excite the vibration modes we wanted to detect.

To study the influence of the parameters on the vibration modes, an extensive numerical study was carried out with the finite element software ANSYS. This methodology allowed us to determine the most suitable measurement points in the test, their position and the angle of the sound source.

Simulations of the harmonic response of a membrane with a diameter of 1 cm and different thicknesses were carried out. The mechanical properties of the membrane were: density 1200 kg/m^3, Young's modulus 2 GPa and Poisson's ratio 0.35. Two positions of the source were considered: 90° and 45° between the axis of the source and the plane membrane. The distance from the sound source was 5 cm for a membrane of 40 µm thickness and 1 cm from the center of a membrane of 60 µm thickness. The harmonic response was compared with the modal response to analyse the influence of the combined parameters.

Figures 3 and 4 present the results for the membrane of 40 µm thickness, measured from two points of the membrane, at the center of the diameter and the point of coordinates y = 0 and x = radio/2, for a frequency range between 0.5 and 15 kHz, and for sound sources at 90° and 45°. The excitation that caused the response of the membrane was a sound wave originating from the source with a sound pressure of 1 Pa.

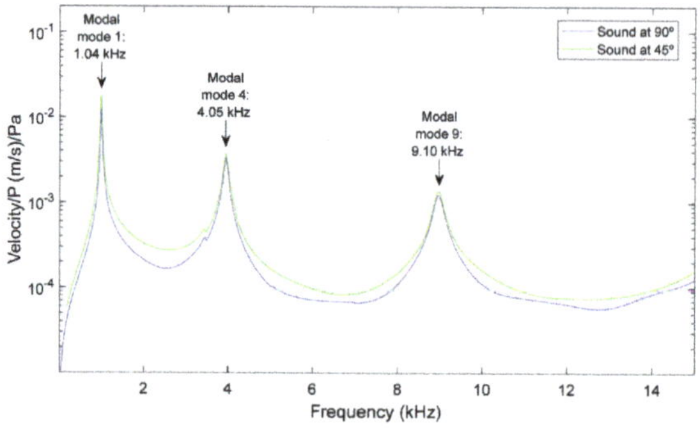

Figure 3. Center point of a membrane of 40 μm. Velocity magnitude at different positions of the source: (a) 90° and (b) 45°.

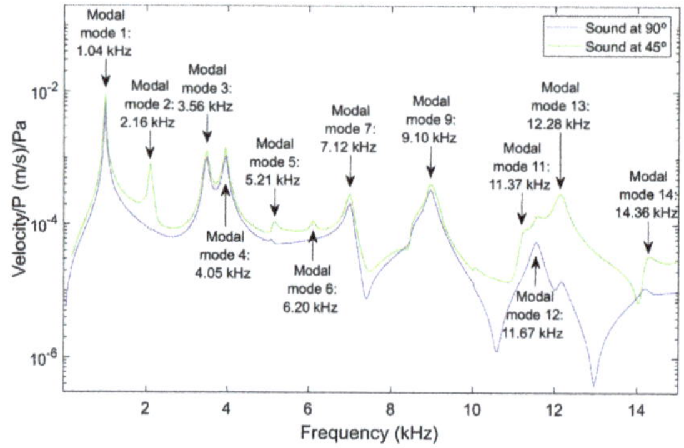

Figure 4. Point y = 0 x = radio/2 for a membrane of 40 μm. Velocity magnitude at different positions of the source: (a) 90° and (b) 45°.

For the center of the membrane and the two positions of the source, it was observed that only the axial symmetry modes appeared, namely modes 1, 4 and 9. At this point, there was no significant influence from the position of the source. In the case of the point at y = 0 and x = radio/2, more modes of vibration were present in the response of the membrane. In the case of the position of the source at 45° almost all modal modes were observed. The selection of the position of the source in this controlled geometry allow us to involve more or fewer modes, depending on the data we needed to correlate properties of each specimen.

The influence of damping was also studied. Figure 5 shows data from one of the cases analyzed: the response of the membranes with thicknesses of 60 μm and an angle of the sound source of 90° from the central point of the membrane and damping values of 1%, 5%, 10%, 20%, 30% and 40%. Only the modes with axial symmetry were observed, as in Figure 3.

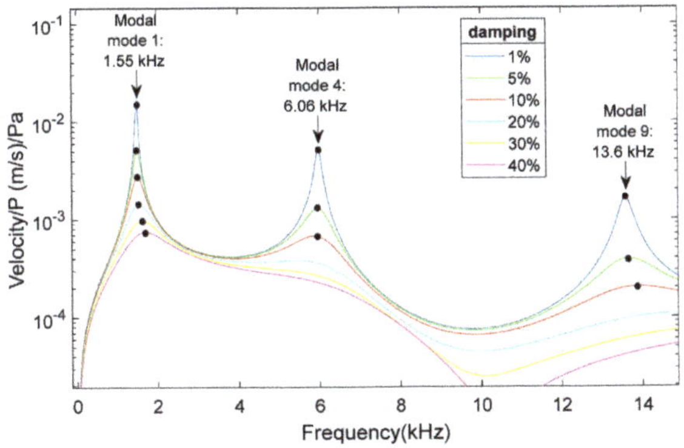

Figure 5. Center point for membranes of 60 μm. Velocity magnitude for the positions of the source at 90° for different damping values: 1%, 5%, 10%, 20%, 30% and 40%.

It was observed that, for larger values of damping, the peaks become softer, as expected. However, the peak values, which correspond to the resonant frequency for each case, generally moved to the right, increasing their values so the resonant frequency increases with increasing damping. For a high enough damping value, it is not possible to determinate the location of the peak of the graph. These simulations helped us to anticipate the number of potential main frequencies available for each type of specimen. In Figure 5, we detected three modes under 20% of damping. In higher-damping situations, we would recourse to reconfigure the test (position of the source, pressure level, etc).

Another important aspect to determine was the feasibility to detect the movement of the selected points. Figure 6 presents numerical results from some points of the surface of a membrane of 40 μm, with 90° source angle at 1 Pa. Velocity was selected as variable to evaluate the sensibility needed from the Laser Doppler Vibrometer, depending on the point analyzed. A value of $2{\cdot}10^{-5}$ m/s was established as the minimum velocity for a good detection of movement by a standard LDV. Figure 7 presents the points of the surface that met the condition of being accurately detectable by LDV in the simulation configurations for the first six modes that could be observed with these configurations: modes 1, 3, 4, 7, 8 and 9. This routine allowed us to adjust the combined pressure level and LDV sensibility in each experiment.

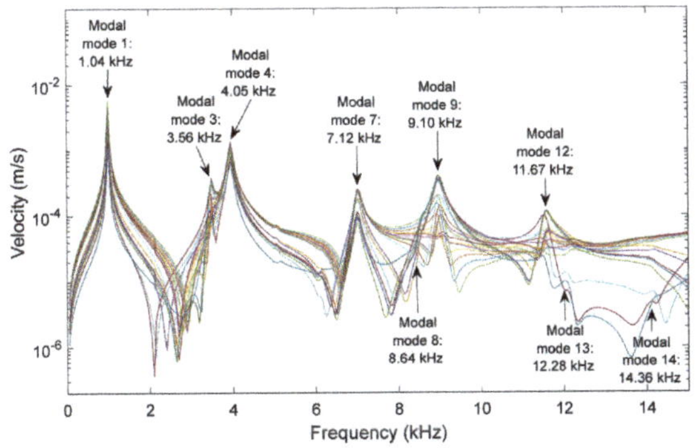

Figure 6. Surface responses of some representative points for a membrane of 40 μm, 90° source angle and 1 Pa of pressure in the source.

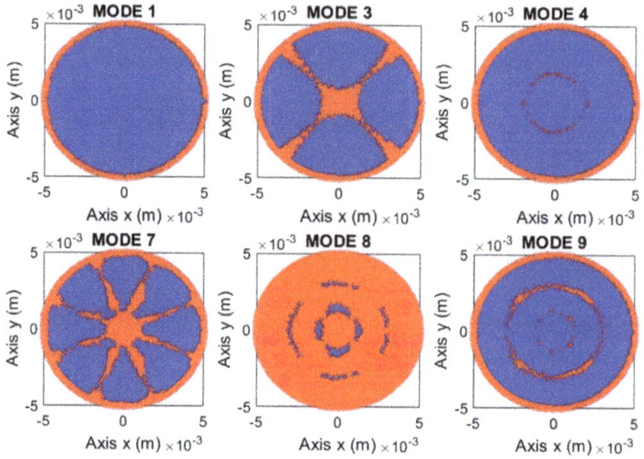

Figure 7. Candidate points from the surface of the membrane for a membrane with a thickness of 40 microns and with an angle of 90°. Modes 1, 3, 4, 7, 8 and 9.

2.4. Circle Fit Method

The correlation among the natural frequencies and the FRF was based on the visual observation of the maximum peaks of the response. This was an intuitive procedure, which is valid when numerical data were used with a low level damping, especially for illustrative purposes. Nevertheless, experimental data showed higher damping and complexity, and we resorted to numerical curve-fitting techniques to identify the resonance frequencies.

In this particular case, we used the classical circular fit method, which is used for the calculation of vibration modes and damping [31] and is based on the use of the Nyquist diagram. Additional FRF curve-fitting techniques were explored, but this simple method was proven to be valid for the purpose of this work.

In any case, it must be stated that the use of the classic FRF curve-fitting techniques implies the admission of certain assumptions. These techniques are used in the field of experimental modal analysis and their validity in this case is not strictly justified. They are based on the curve-fitting of FRF, which is considered as the mobility response (normally in terms of acceleration) to a force applied. Both parameters were required to be measured in the experiment.

In this specific problem, the output was easily measured but the input could not be obtained because it was a spatial pressure distribution along the membrane. The parameter that we could measure in this experiment was the sound pressure at a certain point close to the sound source. In this paper we refer to FRF as the ratio of the mobility output to the sound pressure (instead of mobility output and force applied). As we will see, this assumption was acceptable.

The circular fit method was based on the Nyquist diagram where the presence of a natural mode of vibration can be seen because the function takes the form of a circumference. Its size and position depend on the different parameters that characterize the mode (natural frequency, damping coefficient, etc.), as well as on the rest of the existing modes in that function.

The method consists of fitting a circle to a complex plane using the least squares model. For the method to be valid, several requirements must be met as that the damping is small and that the modes are decoupled.

Equation (2) represents the displacement frequency response function, where νk is the natural frequency, αk is related to the damping coefficient, $U_k + iV_k$ is the residual of this mode, and $R_k + iI_k$ represents the influence of the rest of the modes.

$$H(\omega) = \frac{U_k + iV_k}{i\omega - \alpha_k - i\nu_k} + R_k + iI_k \tag{2}$$

Damping can be calculated with Equation (3) [36]. In this equation, ρ is the angle formed by a point on the circumference (with a frequency w), with the point of the modal frequency (ν_k), provided that the angle is not greater than 90°.

$$\alpha_k = \frac{\omega - \nu_k}{tg\frac{\rho}{2}} \tag{3}$$

In Figure 8 the Nyquist diagrams for the points obtained in the experimental test with two examples of circular fit are presented.

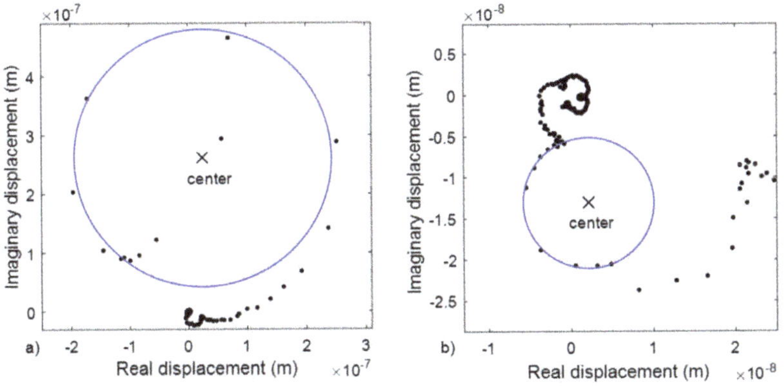

Figure 8. Nyquist diagrams. Application of the circular fit method for two resonance frequencies (**a**,**b**).

3. Experimental Results

The method used to correlate mechanical properties and frequency parameter based on numerical data was straightforward. Nevertheless, its ability remains to be proven based on real experimental data. In experimental tests, many aspects come into play, especially when the manipulation of small samples is necessary, the magnitude to measure is small and the setup presents many parameters to consider. The main goal of the present work was to check the validity in practice of the assumptions made.

3.1. Experiment Setup

According to the experiment setup described in the previous section, several experiments were implemented. They were performed in the VA Loma Linda Healthcare System, CA, USA. Figure 9 shows the setup of these experiments. The membrane was supported by a structure that acted as a baffle. The membrane was in a horizontal position and the speaker was placed below the membrane at a distance of 1 cm. The velocity was measured using an LDV placed on the opposite side. The figure shows both spots used in this paper: the center of the membrane and the external border. Two Sokolich ultrasound microphones were used, one to calibrate the sound pressure at the speaker and to monitor the sound pressure level at the speaker output (mic 1) and another to capture sound transmission (mic 2). The first microphone was used as the input signal while the second was not used in this present work.

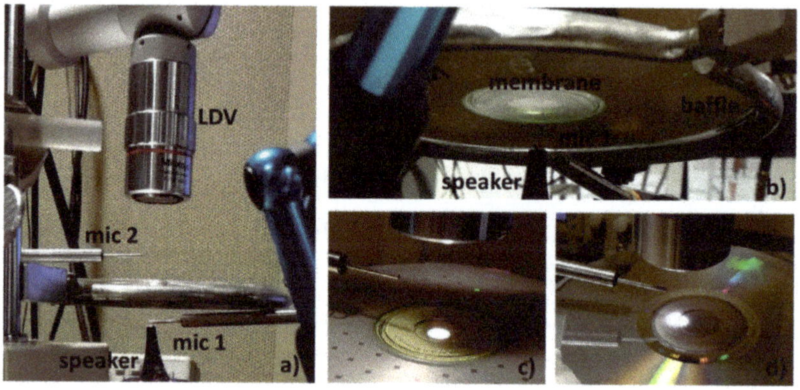

Figure 9. Disposition of the different elements in the experiment (**a**,**b**). Measurement at the center of the membrane (**c**). Measurement at a support point (**d**).

Sound-induced membrane velocities were measured from two perpendicular lines across the center of the membrane with many locations along the line and a location resolution of 0.2 mm, using a Laser-Doppler Vibrometer (LDV, OFV-5000, Polytec Inc., Germany). Stimulus and acquisition programs were written in Matlab and TDT Visual Design Studio. The sampling frequency of the TDT system was 200 kHz. Pure tones (1–2 s duration) were generated by a Tucker Davis Technology (TDT) system III. Data were stored following the removal of the first 4,096 points of the response waveform to avoid the transient, and after time-averaging the remaining waveform, typically in 50 time-locked segments. The responses were later analyzed by Fourier transform in Matlab. The frequencies were swept from 0.5 to 15 kHz in 500 Hz steps.

The material used was parafilm. It is a membrane material commonly used in laboratories to seal recipients. Its thickness and mechanical properties make it ideal for this experiment due to some similarity to the tympanic membrane. According to the manufacturer information the membrane thickness was approximately 130 microns, and its density was 992 kg/m^3. No information was provided about is elastic modulus.

Three different samples were prepared with circular membranes with diameters of 5, 10 and 15 mm. The membranes were clamped between two discs with a hole corresponding to the desired diameter. The discs acted as baffles, such that the sound pressure only pushed one of the membrane sides.

The velocity of the central point of the membrane was measured. The support system was also measured as control. Sound pressure was applied at an approximate distance of 1 cm, with an angle Φ of 90°. This corresponded to the simplest setup possible, leading to a theoretical axial symmetric response. The excitation frequencies ranged from 0 and 10 kHz for the 10 and 15 mm diameter membranes, and from 0 and 15 kHz for the 5 mm membrane due to its higher natural frequencies. The frequency sample was $\Delta f = 47.7$ Hz for all the cases.

3.2. *Experiment Results*

The results obtained for the different cases are plotted in Figure 10. They corresponds to the transfer function of the velocity at the membrane divided by the speaker sound pressure (input, mic 1). Results corresponding to the support system (border) were also plotted and it will be of great interest to interpret the results as it will be explained later. It was magnified ($\times 10$) to facilitate comparison.

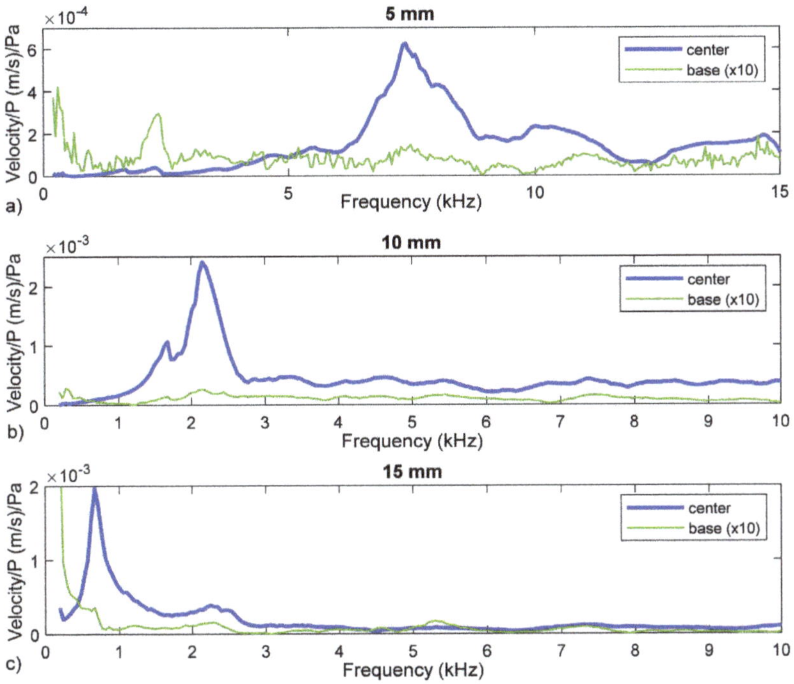

Figure 10. Membrane velocity transfer function. Responses of membranes with diameters of 5 mm (**a**), 10 mm (**b**), and 15 mm (**c**) for the center location (blue curve) and edge location (green curve). Responses at the edge location are scaled up to 10 x.

Visually, we could identify some maximum that we could interpret as being linked to the natural frequencies. For each diameter, a clear first resonance was observed (at 7360, 2124 and 667 Hz, respectively). Nevertheless, it was difficult to distinguish some additional maximum at higher frequencies. So, we resorted to evaluating these frequencies by means of curve fit methods, particularly by means of the circular fit method. Additionally, as will be shown, the visually detected frequencies were proven to be erroneous.

3.3. Circular Fit Results

As described in previous section, the circular fit method is based on the Nyquist diagram, where the presence of a natural mode of vibration can be identified as a circumference on the FRF. This characteristic is correct in the case of a mechanical system where the FRF is defined as displacement divided by force applied. In our case, we cannot measure the force applied, we only have the sound pressure at a certain distance (mic 1). Thus, the use of this methodology could be questionable. As we saw in this experiment, this limitation can be overcome, and we obtained resonant frequencies with great accuracy.

In Figure 11, the results shown in Figure 10 are plotted in terms of displacements in the Nyquist diagram. Circular shapes can be clearly observed in all the cases, even for the base data.

A semi-automatic sweep procedure was applied to identify circular fits for the data shown in Figure 11. Tables 1 and 2 show the most significant values obtained for each case, at the center of the membrane and at the border, respectively.

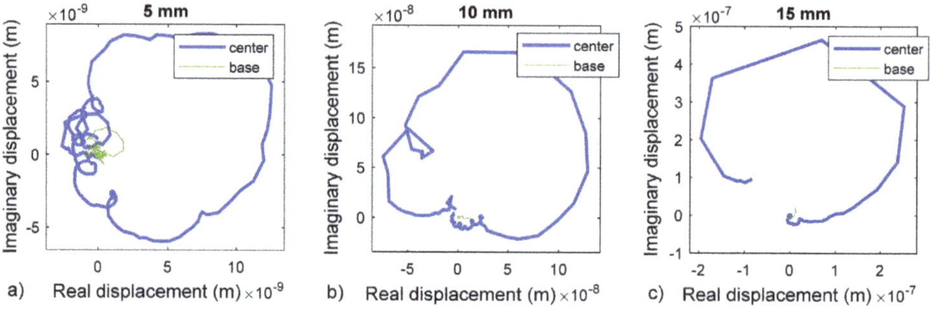

Figure 11. Nyquist diagram for membranes with diameters of 5 mm (**a**), 10 mm (**b**), and 15 mm (**c**) for the center location (blue curve) and edge location (green curve).

Table 1. Resonant frequencies detected by circular fit method at the center of the membrane (Hz).

5 mm	10 mm	15 mm
2312	1552	648.5
4602	2124	2545
6652	3322	3210
7366	4087	5031
9985	4649	7934
10848	5415	8179
12709	5973	8373
13472	6132	8757
14662	6649	
	7366	
	7612	
	8173	
	8685	
	9272	

Table 2. Resonant frequencies detected by circular fit method at the border (base) of the membrane (Hz).

5 mm	10 mm	15 mm
2439	2124	2240
4492	3933	4594
5396	4599	5315
7116	7366	7320
9669	8179	8173
10751	8798	8752
12771	9853	9225
13572		9798
14677		

Many frequencies were detected on the border of the support device. Some of these frequencies were present at the center of the membrane and some were not. In Figure 12, the results obtained are plotted in terms or acceleration. This magnitude provides higher values at higher frequencies, allowing us to observe the phenomenon present at a higher range of frequencies. In this figure, vertical lines were added at the frequencies calculated by circular fit method, for the center of the membrane (red) and for the border (grey).

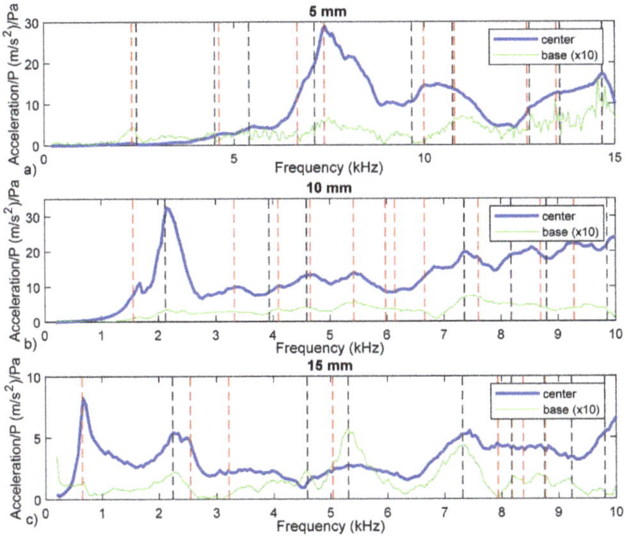

Figure 12. Membrane acceleration transfer function. Responses of membranes with diameters of 5 mm (**a**), 10 mm (**b**), and 15 mm (**c**) for the center location (blue curve) and edge location (green curve). Resonant frequencies added at the center (red line) and at the border (grey line).

4. Discussion

Analyzing the data plotted in Figure 12, we determined the capabilities and limitation of the methodology proposed. Among others, we can highlight the following issues:

- Many frequencies were detected, more than the number of expected frequencies, considering the numerical model of the problem (Figure 2).
- Some of the frequencies were coincident approximately with the observed peak, both at the membrane and at the border.
- Some other frequencies appeared where no clear observable maximum was present.
- When the frequencies were close to peaks, they were not the same values.

To understand all these aspects, we evaluated each of them in order to extract the data useful for our purpose.

The first point to comment on is the presence of resonant frequencies on the support of the membrane. The design of the experiment did not expect significative values at this location. Actually, the magnitude measured was very low (on the figures it was magnified ten times). Some peaks can be seen but it was expected that their influence on the vibration of the membrane could be negligible. Nevertheless, one finding was that they have influence on the response of the membrane.

In Figure 12, it can be observed that the same frequencies (or very close frequencies) are present in the response of the membrane and the border (close magenta and cyan lines). This means that the membrane vibrates in response to a mechanical input (the displacement of the base) instead of the acoustical input.

It can be seen that a group of frequencies was present in the three experiments with slightly similar frequencies: 2200, 4500, 7300 and 9750 Hz (see Table 2). This was probably due to the structural dynamics behavior of the whole experimental system being related to natural frequencies of this system. This effect is not desirable, and future experiments should be designed to eliminate it.

In any case, it is not a real limitation, because we can ignore these frequencies and consider as valid natural frequencies of the membrane (the goal of the experiment) only the remainder of the frequencies shown in Table 1.

Thus, the first frequency for each case would be 6652 Hz (5 mm), 1552 Hz (10 mm) and 648 Hz (15 mm). These frequencies are not exactly the maximum peaks observed in Figure 10 or Figure 12. In the 15 mm case, the difference was small (648 compared to 667 Hz) and was due to the influence of damping, as stated in the previous section. In the other cases, the resonant frequencies of the structural system are masking the membrane's natural frequencies and can lead to a wrong interpretation of the data. As a rule, for future tests, the measurement of the supporting structure should be required to obtain valid results.

The ability of the circular fit method to detect properly these values and many others impossible to detect visually was remarkable. More complex curve fitting methods that we are currently evaluating exist, but the results provided with this simple method are very satisfactory.

Now that we have the natural frequencies for each diameter (those marked with an asterisk in Table 1), we can use them to evaluate the mechanical properties of the material. In this experiment, we have an acceptably accurate knowledge of the main parameters of the material and the membrane (density, diameter and thickness). The only parameter left to identify is the Young's modulus (E). As the material is the same, the value calculated must be the same for the three diameters.

For this purpose, we calculated the natural frequencies for membranes with those diameters, but changing the value of the Young's modulus (in a range from 2.6 to 4.4 GPa). The results obtained are shown in Figure 13. It can be seen how the natural frequencies increase with E (black lines).

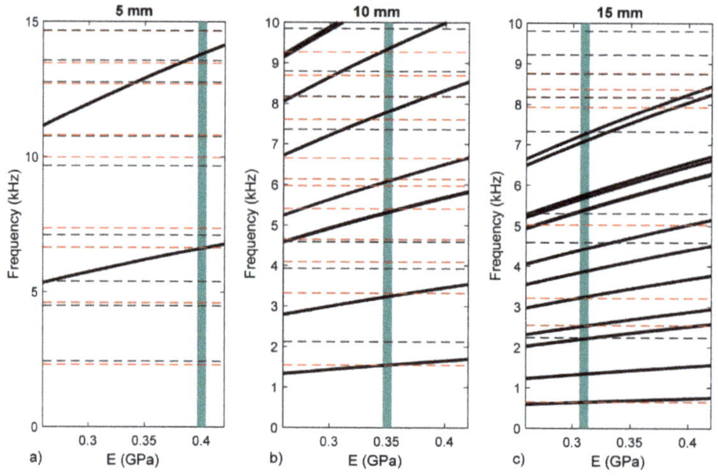

Figure 13. Young's modulus determination for membranes with diameters of 5 mm (**a**), 10 mm (**b**), and 15 mm (**c**).

Then, we can plot the frequencies obtained experimentally as horizontal lines. Crossing these lines with the numerical results, we can obtain the value of E coherent with the experiment results.

In the first case, 5 mm, we see that we only have one measurement by which to identify the value of E. The first natural frequency provided a value of E = 0.4 GPa. A second crossing point with the second natural frequency at 13.4 KHz appeared, but we could not consider it valid as it also appeared on the base movement. In this case, we can estimate a value for E, but we do not have a second point to validate. The high stiffness of this small membrane provides very high frequencies, difficult to measure from a practical point of view.

For the 10 mm case, we have more frequencies in the range of the experiment. Therefore, we have more valid crossing point. The experimental results cut f_1, f_2, f_3, f_4, f_5 and f_6. The exact point was slightly different for each frequency, but we can consider that the value E = 0.35 GPa produces less than a 5% error. Due to the axial symmetry configuration of the experiment, we only expected to obtain f_1 and f_4. As this condition is difficult to reproduce in practice, we obtained more information than expected. In summary, this proved to be the best experimental setup to obtain the value of E.

Finally, in the 15 mm case, we had valid cuts with f_1, f_4 and f_5, and a value of E = 0.31 GPa was obtained. In this case, fewer frequencies were detected in the low frequency range, probably the experiment was centered more properly, and we had no information from f_2 and f_3. Many higher frequencies were detected that were difficult to distinguish from the base frequencies and correlate with the higher modes.

The variability on the results obtained was interesting. Young's modulus was obtained in a broad range (0.31−0.4 GPa). One of the reasons for this variability is that probably there is a certain heterogeneity of the material (parafilm) and different samples can lead to different responses. If necessary, this aspect can be clarified in future with additional tests.

Nevertheless, the more interesting at this point is to analyze where these discrepancies are coming from and how can we limit them in future experiments.

Not all the experiments had the same validity. The results obtained with the 10 mm membrane proved to be the best suited for this test.

In the case of the 5 mm membrane, we only had one measurement by which to estimate E; if there was some small deviation, we could not detect it. The range of frequencies of the experiment was short. Additionally, we must consider the influence of small uncertainties in the known parameters, such as geometry, density, etc. We will focus on the diameter parameters. If we consider the numerical model to compare, there was another fact that could have altered the result. In the numerical model, we can introduce an exact diameter and ideal boundary conditions (fixed ends). Both conditions are intimately linked, and were difficult to distinguish in the real experiment, especially for smaller sizes. With a smaller diameter, small errors on the diameter size estimation leads to a broad range of E uncertainty. Suppose a 0.5 mm error in the three cases under study, the range of uncertainty would be a 10% in the 5 mm case and 3% in the 15 mm case, and the natural frequencies are inversely proportional to the quadratic value of this parameter (Equation (1)). A total of 0.5 mm of uncertainty in this experiment is a credible estimate, considering the practical difficulties of affixing the sample on the support.

Another aspect to consider is the frequency sample of the data. It was the same for all the cases (47.7 Hz). If we calculate the ratio $f_1/\Delta f$ we would have an idea of the discretization level of the FFT, which would be the worst in the case of the 15 mm membrane. This can be observed in the Nyquist diagram in Figure 11, where the circles are more easily identified with the lower diameter membranes than with the 15 mm membrane.

Based on this discussion and the experiment results, we concluded that a range of values from 0.31 to 0.35 GPa was more reliable in this case, probably closer to 0.35 GPa.

5. Conclusions

A new methodology to obtain the mechanical properties of a membrane was shown. The numerical bases were stated and a particular experiment was used to check the validity of the procedure.

The three different configurations of the experiment provided information to prove the utility of the method and to evaluate key points for improvement.

The Young's modulus was identified for the material tested with three different setups. A certain variability was observed in the results, but the general procedure was considered valid. The difficulties shown during the experimental tests have revealed key aspects that must be considered in the design of future experiments.

Among others, the influence of the sample size on the uncertainty of the procedure or the vibration of the support system that contaminated the output signal were highlighted and explained.

It was proven that the use of classic FRF curve-fitting techniques is applicable to this experiment even when the input measured is the sound pressure from a certain distance instead of the force applied to a structure. The circular method proved its utility. Other FRF curve-fitting techniques are being investigated to improve the results.

Author Contributions: Conceptualization, A.G.-H. and J.G.-M.; methodology and software, A.L.-R. and A.G.-H.; validation, A.L.-R., A.G.-H. and J.G.-M.; formal analysis, A.L.-R.; investigation, A.L.-R. and A.G.-H.; writing—original draft preparation, all authors.; writing—review and editing, all authors; supervision and project administration, W.D., A.G.-H. and J.G.-M.; funding acquisition, J.G.-M. and A.G.-H. All authors have read and agreed to the published version of the manuscript.

Funding: This research was funded by the FEDER program and the University of Malaga, grant number UMA FEDERJA-214, by the grant PRX19/00139 of the Government of Spain and by U.S. Department of Veterans Affairs Merit Review award of Rehabilitation Research & Development Service (RR&D) C2296-R (W Dong).

Institutional Review Board Statement: Not applicable.

Informed Consent Statement: Not applicable.

Data Availability Statement: The data presented in this study are available on request from the corresponding author. The data are not publicly available due to needs the permission of principal investigators of each research project involved.

Acknowledgments: We also would like to thank Mario Gutierrez, who performed velocity measurements at VA Loma Linda Healthcare System, CA, USA.

Conflicts of Interest: The authors declare no conflict of interest. The funders had no role in the design of the study; in the collection, analyses, or interpretation of data; in the writing of the manuscript, or in the decision to publish the results.

References

1. Ding, X.; Ko, W.H.; Mansour, J.M. Residual stress and mechanical properties of boron-doped p+-silicon films. *Sens. Actuators A Phys.* **1990**, *23*, 866–871. [CrossRef]
2. Benson-Martin, J.; Zammaretti, P.; Bilic, G.; Schweizer, T.; Portmann-Lanz, B.; Burkhardt, T.; Zimmermann, R.; Ochsenbein-Kölble, N. The Young's modulus of fetal preterm and term amniotic membranes. *Eur. J. Obstet. Gynecol. Reprod. Biol.* **2006**, *128*, 103–107. [CrossRef] [PubMed]
3. Ye, X.Y.; Zhang, J.H.; Zhou, Z.Y.; Yang, Y. Measurement of Young's modulus and residual stress of micromembranes. In Proceedings of the MHS'96 Seventh International Symposium on Micro Machine and Human Science, Nagoya, Japan, 2–4 October 1996; pp. 125–129.
4. Tabata, O.; Kawahata, K.; Sugiyama, S.; Igarashi, I. Mechanical property measurements of thin films using load-deflection of composite rectangular membrane. *Micromech. MEMS Class. Semin. Pap. 1990* **1997**, *20*, 667–671. [CrossRef]
5. Petersen, K.E.; Guarnieri, C.R. Young's modulus measurements of thin films using micromechanics. *J. Appl. Phys.* **1979**, *50*, 6761–6766. [CrossRef]
6. Kiesewetter, L.; Zhang, J.M.; Houdeau, D.; Steckenborn, A. Determination of Young's moduli of micromechanical thin films using the resonance method. *Sens. Actuators A Phys.* **1992**, *35*, 153–159. [CrossRef]
7. Tabib-Azar, M.; Wong, K.; Ko, W. Aging phenomena in heavily doped (p+) micromachined silicon cantilever beams. *Sens. Actuators A Phys.* **1992**, *33*, 199–206. [CrossRef]
8. Ma, S.; Wang, S.; Iacopi, F.; Huang, H. A resonant method for determining the residual stress and elastic modulus of a thin film. *Appl. Phys. Lett.* **2013**, *103*, 031603. [CrossRef]
9. Dos Santos, E.A.; Rehder, G.P. Study of the young's modulus in microstructures through the resonance frequency technique for applications in commercial CMOS processes. *Mater. Res.* **2017**, *20*, 81–84. [CrossRef]
10. Kirikae, I. *The Structure and Function of Middle Ear*; Tokyo University Press: Tokyo, Japan, 1960.
11. Von Békésy, G. *Experiments in Hearing*; McGraw-Hill: New York, NY, USA, 1960.
12. Decraemer, W.F.; Maes, M.A.; Vanhuyse, V.J. An elastic stress-strain relation for soft biological tissues based on a structural model. *J. Biomech.* **1980**, *13*, 463–468. [CrossRef]
13. Zahnert, T.; Hüttenbrink, K.-B.; Mürbe, D.; Bornitz, M. Experimental investigations of the use of cartilage in tympanic membrane reconstruction. *Otol. Neurotol.* **2000**, *21*, 322–328. [CrossRef]

14. Cheng, T.; Dai, C.; Gan, R.Z. Viscoelastic Properties of Human Tympanic Membrane. *Ann. Biomed. Eng.* **2007**, *35*, 305–314. [CrossRef]
15. Allardyce, B.J.; Rajkhowa, R.; Dilley, R.J.; Xie, Z.; Campbell, L.; Keating, A.; Atlas, M.D.; von Unge, M.; Wang, X. Comparative acoustic performance and mechanical properties of silk membranes for the repair of chronic tympanic membrane perforations. *J. Mech. Behav. Biomed. Mater.* **2016**, *64*, 65–74. [CrossRef] [PubMed]
16. Huang, G.; Daphalapurkar, N.P.; Gan, R.Z.; Lu, H. A method for measuring linearly viscoelastic properties of human tympanic membrane using nanoindentation. *J. Biomech. Eng.* **2008**, *130*, 014501. [CrossRef] [PubMed]
17. Daphalapurkar, N.P.; Dai, C.; Gan, R.Z.; Lu, H. Characterization of the linearly viscoelastic behavior of human tympanic membrane by nanoindentation. *J. Mech. Behav. Biomed. Mater.* **2009**, *2*, 82–92. [CrossRef]
18. Buytaert, J.A.N.; Aernouts, J.E.F.; Dirckx, J.J.J. Indentation measurements on the eardrum with automated projection moiré profilometry. *Opt. Lasers Eng.* **2009**, *47*, 301–309. [CrossRef]
19. Aernouts, J.; Couckuyt, I.; Crombecq, K.; Dirckx, J.J.J. Elastic characterization of membranes with a complex shape using point indentation measurements and inverse modelling. *Int. J. Eng. Sci.* **2010**, *48*, 599–611. [CrossRef]
20. Hesabgar, S.M.; Marshall, H.; Agrawal, S.K.; Samani, A.; Ladak, H.M. Measuring the quasi-static Young's modulus of the eardrum using an indentation technique. *Hear. Res.* **2010**, *263*, 168–176. [CrossRef]
21. Wu, D.; Chen, Y.; Manna, S.; Talley, K.; Zakutayev, A.; Brennecka, G.L.; Ciobanu, C.V.; Constantine, P.; Packard, C.E. Characterization of Elastic Modulus Across the (Al1-xScx)N System Using DFT and Substrate-Effect-Corrected Nanoindentation. *IEEE Trans. Ultrason. Ferroelectr. Freq. Control* **2018**, *65*, 2167–2175. [CrossRef]
22. Poilane, C.; Delobelle, P.; Lexcellent, C.; Hayashi, S.; Tobushi, H. Analysis of the mechanical behavior of shape memory polymer membranes by nanoindentation, bulging and point membrane deflection tests. *Thin Solid Film.* **2000**, *379*, 156–165. [CrossRef]
23. Gaboriaud, F.; Parcha, B.S.; Gee, M.L.; Holden, J.A.; Strugnell, R.A. Spatially resolved force spectroscopy of bacterial surfaces using force-volume imaging. *Colloids Surf. B Biointerfaces* **2008**, *62*, 206–213. [CrossRef]
24. Longo, M.; De Santo, M.P.; Esposito, E.; Fuoco, A.; Monteleone, M.; Giorno, L.; Jansen, J.C. Force spectroscopy determination of Young's modulus in mixed matrix membranes. *Polymer* **2018**, *156*, 22–29. [CrossRef]
25. Et-Thakafy, O.; Guyomarc'h, F.; Lopez, C. Young modulus of supported lipid membranes containing milk sphingomyelin in the gel, fluid or liquid-ordered phase, determined using AFM force spectroscopy. *Biochim. Biophys. Acta Biomembr.* **2019**, *1861*, 1523–1532. [CrossRef]
26. Tng, K.H.; Antony, A.; Wang, Y.; Leslie, G.L. 11—Membrane ageing during water treatment: Mechanisms, monitoring, and control. In *Advances in Membrane Technologies for Water Treatment*; Woodhead Publishing Series in Energy; Basile, A., Cassano, A., Rastogi, N.K., Eds.; Woodhead Publishing: Oxford, UK, 2015; pp. 349–378. ISBN 978-1-78242-121-4.
27. Du, S.; Qiang, Y.; Liu, J. Erythrocyte Membrane Failure by Electromechanical Stress. *Appl. Sci.* **2018**, *8*, 174. [CrossRef] [PubMed]
28. Fay, J.; Puria, S.; Decraemer, W.F.; Steele, C. Three approaches for estimating the elastic modulus of the tympanic membrane. *J. Biomech.* **2005**, *38*, 1807–1815. [CrossRef] [PubMed]
29. Luo, H.; Dai, C.; Gan, R.Z.; Lu, H. Measurement of young's modulus of human tympanic membrane at high strain rates. *J. Biomech. Eng.* **2009**, *131*, 064501. [CrossRef]
30. ASTM. *Standard Test Method for Measuring Vibration-Damping Properties of Materials*; American Society for Testing and Materials: West Conshohocken, PA, USA, 2005.
31. Ewins, D.J.; Saunders, H. Modal Testing—Theory and Practice. *J. Vib. Acoust.* **1986**, *108*, 109–110. [CrossRef]
32. Rayleigh, J.W.S. *The Theory of Sounds*, 2nd ed.; Dover Publications Inc.: London, UK, 1998; ISBN 978-0486602929.
33. Timoshenko, S. *Vibration Problems in Engineering*; Wiley: New York, NY, USA, 1928.
34. Thein, W. Vibration of Circular Plates. *J. Acoust. Soc. Am.* **1962**, *34*, 275–281. [CrossRef]
35. Lima-Rodriguez, A.; Gonzalez-Herrera, A.; Garcia-Manrique, J. Study of the dynamic behaviour of circular membranes with low tension. *Appl. Sci.* **2019**, *9*, 4716. [CrossRef]
36. Carrascosa, L.I.; Busturia, J.M.; Gimenez, J.G. Global Experimental Modal Analysis. A Comparison of Different Methods. In Proceedings of the 3rd International Modal Analysis Conference, San Sebastian, Spain, 24–29 June 1985; Volume 1, pp. 311–321.

membranes

MDPI

Article

New Insights on the Effects of Water on Polymer Inclusion Membranes Containing Aliquat 336 Derivatives as Carriers

Clàudia Fontàs [1,*], Ruben Vera [1], Enriqueta Anticó [1], María del Valle Martínez de Yuso [2], Enrique Rodríguez-Castellón [3] and Juana Benavente [4]

1 Department of Chemistry, University of Girona, 17003 Girona, Spain; ruben.vech@gmail.com (R.V.); enriqueta.antico@udg.edu (E.A.)
2 Laboratorio de Espectroscopía de Rayos-X, Servicios Centrales de Apoyo a la Investigación de la Universidad de Málaga, 29071 Málaga, Spain; mvyuso@uma.es
3 Departamento de Química Inorgánica, Facultad de Ciencia, Universidad de Málaga, 29071 Málaga, Spain; castellon@uma.es
4 Departamento de Física Aplicada I, Facultad de Ciencias, Universidad de Málaga, 29071 Málaga, Spain; j_benavente@uma.es
* Correspondence: claudia.fontas@udg.edu

Abstract: Surface characterization of polymer inclusion membranes (PIMs) using the polymers cellulose triacetate and polyvinyl chloride, containing different ionic liquids (ILs) as carriers, has been performed. Three different ILs have been tested: commercial trioctyl methylammonium chloride (Aliquat 336–AlqCl$^-$) and two derivatives bearing the counter anion NO_3^- or SCN^- ($AlqNO_3$ and AlqSCN, respectively). Surface analysis was performed by scanning electron microscopy (SEM) and X-ray photoelectron spectroscopy (XPS) for both dry membranes and PIMs immersed for 4 days in ultrapure water to investigate the effect of the interaction of water with the membrane's morphology and composition. XPS analysis of the PIMs revealed that immersion in ultrapure water causes a decrease in the atomic concentration percentage (A.C.%) of the specific IL atoms (Cl, S, and N) when compared with dry samples. Moreover, SEM images of the PIMs containing the IL $AlqNO_3$ showed an alteration in the morphology of the membrane due to water contact at surface level, whereas no changes were observed at a bulk level. These changes in the surface composition of the water equilibrated PIMs may be associated with the solubilization of the IL in the water solution, which, therefore, may affect the reactivity of the membrane's surface. To better understand this effect, PIMs containing both AlqCl and $AlqNO_3$ as carriers were used for arsenic (V) transport. It was found that AlqCl was the most effective IL and that the effectivity of the PIM on As(V) removal was not affected after five cycles of the membrane's reuse.

Keywords: polymer inclusion membranes (PIMs); X-ray photoelectron spectroscopy (XPS); scanning electron microscopy (SEM); ionic liquids; Aliquat 336; As(V)

1. Introduction

Polymer inclusion membranes (PIMs) have emerged in recent years as a separation technology that is able to provide a highly selective, efficient, and cost-effective process for the removal of pollutants from aqueous matrices [1–3]. PIMs, which are usually considered to be a type of liquid membrane [4], basically consist of a polymeric matrix that incorporates a carrier (or extractant) which facilitates the transport of the chemical species of interest from the feed phase to the stripping phase, where the analytes are released. A plasticizer is also sometimes added to improve the mechanical properties of this kind of membrane. However, it has already been proved that the use of ionic liquids (ILs) as carriers favors the plasticity of the membrane, allowing the elimination of plasticizers and, consequently, simplifying the formulation of the PIMs [5].

Ionic liquids (ILs) are low vapor pressure salts of an organic cation and an organic or inorganic anion [6]. They also have limited solubility, good electrical conductivity, and thermal stability, as well as high ion mobility and viscosity [6]. Their low vapor pressure allows ILs to be used in a large number of different applications, such as solvents for organic reactions and catalysis, electrodeposition, extraction, and CO_2 capture devices [7–9]. ILs may be used as substitutes for volatile organic solvents and in new processes, since the variety of possible chemical compositions allows their properties to be tailored to specific applications [10]. ILs have also been considered as good candidates to improve the behavior of membranes for fuel cell applications [11–14].

Among the different ILs, trioctylmethylammonium chloride (commercially available as Aliquat 336 –AlqCl-) is widely used in the preparation of PIMs, whereas cellulose triacetate (CTA) and polyvinylchloride (PVC) are commonly considered as polymer supports [3,5,15–18]. However, the great majority of studies using PIMs evaluate their use in transport systems without considering basic aspects regarding the composition of PIMs or solution interactions, despite such information being of significant interest in selecting the most appropriate membrane. In our previous studies [5,19,20], we demonstrated that both physicochemical and electrochemical characteristics of PIMs were affected by the amount of Aliquat 336 content in addition to the nature of the counter-anion (Cl^-, NO_3^- or SCN^-). It was established from contact angle measurements that the hydrophobicity of the PIMs with different IL derivatives followed the trend AlqSCN > $AlqNO_3$ > AlqCl, which was related to the strength of the ion pair formed between the cationic and anionic part of the IL, and closely related to the stability of the membrane.

Taking into account that the surface of a membrane is the first flow barrier, its characterization is of great importance, since it can significantly influence the properties of the membrane itself. Scanning electron microscopy (SEM), atomic force microscopy (AFM), X-ray photoelectron spectroscopy (XPS), and contact angle measurement are commonly used to determine morphology (dense or porous), structure (symmetric or composite), topography (pore-size distribution), chemical composition, and the wettability of membranes [21–24]. Other techniques such as tangential streaming potential (TSP), phase-contrast scanning force microscopy (SFM), and spectroscopic ellipsometry (ES) are used for more specific membrane characterization (surface charge, material viscoelastic behavior, or layer thickness in the case of non-uniform or composite membranes) [25–28]. In particular, XPS is nowadays one of the most popular surface analytical techniques due not only to the exceptional combination of compositional and chemical information, but also to the operational simplicity of this high vacuum technique and the availability of commercial equipment. XPS can be used to determine or confirm the empirical formula of a material and the quantity of the elements present on or near the surface (up to 10 nm of depth) and also for fouling studies in the case of membranes used in different separation processes [26,29].

In this study, both morphological and chemical surface analysis for PIMs prepared with three ILs (commercial AlqCl and the two derivatives, $AlqNO_3$ and AlqSCN) are described. XPS investigations of PIMs after water contact were performed and the results are compared with those previously obtained for dry PIMs to estimate possible hydration effects. The effect of the base polymer is also considered by comparing AlqCl–CTA and AlqCl–PVC samples. SEM images were also obtained for pristine PIMs containing the different ILs at different contents in a CTA matrix, and for the PIM containing $AlqNO_3$, before and after water contact. Finally, PIMs containing both AlqCl and $AlqNO_3$ have been evaluated for the transport of As(V).

2. Materials and Methods

2.1. Reagents and Solutions

Aliquat 336 was obtained from Sigma-Aldrich (USA), whereas the polymers CTA and PVC were purchased from Fluka (Bern, Switzerland). Chloroform (for CTA) and tetrahydrofuran, THF (for PVC), both from Panreac (Barcelona, Spain), were used to dissolve the polymer and the ILs. NaSCN and $NaNO_3$ (Panreac, Barcelona, Spain) were used as the anion source to prepare the Aliquat 336 derivatives.

Stock solution (100 mg L^{-1}) of As(V) was prepared from solid $Na_2HAsO_4 \cdot 7H_2O$ purchased from Merck (Darmstadt, Germany). Working solutions of arsenate in ultrapure water were prepared by dilution of the corresponding stock solution. Sodium chloride, obtained from Fluka (Bern, Switzerland), was used to prepare the stripping solution.

All reagents and solvents were of analytical reagent grade. Ultrapure water obtained from a Milli-Q Plus water purification system (Millipore Ibérica, Madrid, Spain) was used.

2.2. IL Preparation

Commercial Aliquat 336 was used for the preparation of the two other ILs, $AlqNO_3$ and AlqSCN, by exchanging the chloride anion present in the formulation of Aliquat 336 by nitrate or thiocyanate anions, respectively (see their chemical structure in Figure 1a) [19]. For this, 5 g of Aliquat 336 were transferred into a separation funnel and dissolved in 50 mL of chloroform. Different sodium salts ($NaNO_3$ and NaSCN) were slowly added to the solution in an amount that was 30% greater than the amount of Aliquat 336. The mixture was vigorously stirred for 5 h at room temperature. Afterwards, it was rinsed with water to remove both the excess sodium salt as well as the sodium chloride that was formed. Washing was repeated until no chloride was detected in the water sample (by the addition of silver nitrate) and the organic solution was dried with magnesium sulphate. Finally, chloroform was evaporated under reduced pressure until a brownish viscous liquid was obtained.

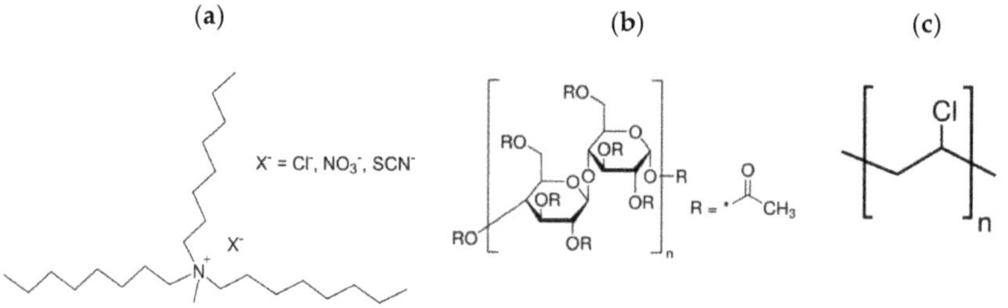

Figure 1. Chemical structure of: (**a**) ILs; (**b**) cellulose triacetate (CTA); and (**c**) polyvinyl chloride (PVC).

2.3. PIMs Preparation

Preparation of PIMs using cellulose triacetate (CTA) or polyvinyl chloride (PVC) as base polymers (see Figure 1b,c for the chemical structure of the polymers) was previously detailed [5,19]. Briefly, 200 mg of CTA or 400 mg of PVC were dissolved in 20 mL of chloroform or in THF, respectively; then, the corresponding amount of the ionic liquid was added and magnetically stirred for 2 h. Finally, the solution was poured into a 9.0 cm diameter flat bottom glass Petri dish, which was set horizontally and loosely covered. The solvent was allowed to evaporate overnight at room temperature and the resulting film was peeled off the Petri dish.

In this study, membranes with different ILs:polymer ratios have been used to reveal specific changes or effects more clearly. The amount of IL incorporated in the PIM is described in % (mass), and the content was 60% and 30% for the characterization studies for all three ILs in CTA. Membranes with AlqCl and PVC were also investigated. Consequently, the resulting membranes were named as follows: (60%AlqCl, 60%$AlqNO_3$, or 60%AlqSCN)–40%CTA and 60%AlqCl–40%PVC and (30%AlqCl, 30%$AlqNO_3$, or 30%AlqSCN)–70%CTA. Furthermore, a PIM made of 50%$AlqNO_3$–50%CTA was also prepared for the investigation of the effect of water using SEM and for transport studies.

2.4. XPS Measurements

In order to evaluate the effect of water contact on the surface of the membrane, pieces of the PIMs were submerged in ultrapure water for 4 days (50 mL of ultrapure water and a piece of membrane of 4 cm^2). These PIMs were labelled as "water equilibrated" samples and identified with the letter w, whereas dry samples (pristine PIMs) were identified by the letter d.

Chemical surface characterization of the PIMs was performed by a Physical Electronics spectrometer (PHI 5700) with X-ray Mg K_α radiation (300 W, 15 kV, 1253.6 eV) as the excitation source. High-resolution spectra were recorded at a given take-off angle of 45° by a concentric hemispherical analyzer operating in the constant pass energy mode at 29.35 eV, using a 720 µm diameter analysis area. Accurate ±0.1 eV binding energies were determined with respect to the position of the adventitious C *1s* peak at 285.0 eV, and the residual pressure in the analysis chamber was maintained below 5×10^{-7} Pa during data acquisition.

Membranes were mounted on a sample holder and kept overnight at high vacuum in the preparation chamber before being transferred to the analysis chamber of the spectrometer for testing. When analyzing the pure IL, a drop was deposited on a glass slide. Each spectral region was scanned several times until a good signal-to-noise ratio was observed. The PHI ACCESS ESCA-V6.0 F software package was used for acquisition and data analysis. A Shirley-type background was subtracted from the signals. Recorded spectra were always fitted using Gauss–Lorentz curves to accurately determine the binding energy (BE) of the different element core levels, as has previously been described in detail [30]. Atomic concentration percentages (A.C.%) of the characteristic elements found on the surfaces of the analyzed samples were determined, taking into account the corresponding area sensitivity factor [30] of the different measured spectral regions.

2.5. Scanning Electron Microscopy

SEM images were performed with a scanning electron microscope FE-SEM Hitachi, S-4100 (Tokyo, Japan), and the samples were placed on a stub and coated with carbon (K950 turbo evaporator, Emitech, Germany). For cross-section observations, the membranes were frozen and broken under liquid nitrogen. Digital images were collected and processed by the Quartz PCI software. This technique was used to obtain micrographs of the different PIM formulations as well as of dry and water-equilibrated membranes (after being immersed in ultrapure water) in the case of the PIM made of 50%AlqNO_3–50%CTA.

2.6. Transport Experiments for As(V)

Transport experiments were carried out using both 50%AlqNO_3–50 CTA and 50%AlqCl–50%CTA membranes in a two-compartment permeation cell containing 190 mL of both feed (consisting of 10 mg L^{-1} As at pH 7) and stripping (0.1 M NaCl) solutions, well stirred, and a PIM with an exposed area of 11.5 cm^2. Samples from feed and stripping solutions were taken after 24 h for subsequent metal content analysis. All experiments were carried out at room temperature of 22 ± 1 °C. Transport efficiency (TE) of As was determined by using Equation (1):

$$TE(\%) = \frac{[\mathrm{As(V)}]\mathrm{strip(t)}}{[\mathrm{As(V)}]\mathrm{feed(0)}} \times 100 \quad (1)$$

where $[\mathrm{As(V)}]_{\mathrm{strip(t)}}$ denotes As concentration in the stripping compartment at an elapsed time t and $[\mathrm{As(V)}]_{\mathrm{feed(0)}}$ is the initial As concentration in the feed phase. An inductively coupled plasma emission spectrometer (ICP-AES) (Liberty RL, Varian Australia, Mulgrave, Vic., Australia) was used for the analysis of arsenic in the feed and stripping phases at 228.8 nm.

3. Results and Discussion

3.1. Chemical Surface Characterization of PIMs by XPS

Taking into account that PIMs are commonly used in separation processes for analytes present in aqueous solutions, it is of interest to establish possible chemical changes associated with water–IL–polymer interactions. In fact, the XPS technique has been employed for the characterization of PIMs prepared with the IL Aliquat 336 and its derivatives in dry state in earlier studies [5,19]. Table 1 shows the average A.C.% values (determined from measurements performed on both faces of the membranes) of the elements found on the water-equilibrated PIMs surfaces. For comparison purposes, data for dry state from [19] are also included. Common atoms in all PIMs made of CTA are, on the one hand, N (from the cationic part of the IL –trioctyl metil ammonium cation-) and, on the other hand, O and C, both from the polymer and/or the IL. A certain percentage of silicon, a non-characteristic atom which did not originate from either the IL or the polymer, was also detected, which is associated with SiO_2 and attributed to the membrane preparation. Focusing on the A.C.% of the common atoms, it can be observed in Table 1 that water-equilibrated samples show lower C% than dry ones, resulting from the solubilization of the components of the membrane containing aliphatic carbon [31,32]. In the case of O% and Si%, the effect is the opposite, since their content is higher for PIM-w samples. This fact can be associated with a more superficial re-orientation of the hydrophobic silicon compounds in water-equilibrated samples, as was found in the analysis of regenerated cellulose membranes reported in [33].

Table 1. Average values of the A.C.% of the elements found on the surfaces of water-equilibrated (w) and dry (d) PIMs (data for dry state from [20]). *nd: not detected.*

PIM Composition		<C> (%)	<O> (%)	<N> (%)	<Cl> (%)	<S> (%)	<Si> (%)
60%AlqCl–40%CTA	w	78.6 ± 1.9	16.5 ± 0.4	1.8 ± 0.2	1.5 ± 0.1	*nd*	1.3 ± 0.8
	d	86.2 ± 0.8	8.2 ± 1.1	2.7 ± 0.1	2.5 ± 0.2	*nd*	0.4 ± 0.2
60%AlqNO$_3$–40%CTA	w	76.3 ± 2.8	15.4 ± 0.7	4.1 ± 0.3	*nd*	*nd*	2.6 ± 1.2
	d	81.1 ± 1.3	13.5 ± 0.9	4.0 ± 0.2	*nd*	*nd*	1.2 ± 0.6
60%AlqSCN–40%CTA	w	78.3 ± 2.1	14.1 ± 0.8	3.4 ± 0.4	*nd*	1.6 ± 0.3	1.7 ± 0.9
	d	86.0 ± 1.5	6.2 ± 1.1	4.5 ± 0.3	*nd*	4.5 ± 0.3	0.8 ± 0.4
60%AlqCl–40%PVC	w	89.7 ± 0.5	2.4 ± 0.5	2.8 ± 0.1	4.0 ± 0.1	*nd*	0.7 ± 0.3
	d	89.0 ± 0.3	4.0 ± 0.3	3.1 ± 0.1	3.7 ± 0.1	*nd*	0.2 ± 0.1

In order to better evaluate the changes caused by water contact to the surface of the different PIMs, we need to take into account the changes in atomic content (%) of the specific anions: chlorine in the case of the PIM 60%AlqCl–40%CTA and sulphur for the PIM 60%AlqSCN–40%CTA. It is important to bear in mind that PIMs containing AlqNO$_3$ do not possess a specific atom characteristic of the IL, since N is present in both the anion and the cation of the IL. Moreover, chlorine is not a suitable atom to evaluate the possible changes in the PIM made of PVC, since Cl is also present in both the polymer and the IL.

As can be observed in Table 1, chlorine is not present in any PIMs containing either AlqSCN or AlqNO$_3$, confirming that a complete anion exchange has taken place in the preparation of these derivatives. Focusing on the PIM 60%AlqCl–40%CTA, a reduction of 40% of chlorine was found when compared with the dry sample. The reduction of S for PIM 60%AlqSCN–40%CTA was 36%, slightly lower than the value for the PIM with AlqCl. These results show a significant loss of IL on the surface of the PIMs.

In order to estimate changes to the membrane surface associated with the hydration state for 60%AlqNO$_3$–40%CTA and 60%AlqCl–40%PVC membranes, it is necessary to analyze core lever signals of N and Cl, respectively.

Figure 2 shows the comparison of N *1s* core level signals obtained for dry (Figure 2a) and water-equilibrated (Figure 2b) samples of the four membranes that were studied. As can be observed, the spectra show different signals depending on which IL is present in the membrane, but also a new peak at a B.E. at approximately 399.5 eV. To evaluate the origin of this new peak, a film made only of the polymer CTA was also immersed in ultrapure water and the XPS spectrum was recorded. This spectrum is presented in Figure 3 (100% CTA-w) together with the spectrum of the 60% AlqCl–40% CTA-w PIM, for comparison purposes. As can be seen, the new peak is due to the polymer, and might be associated with the hydration of impurities in the CTA matrix.

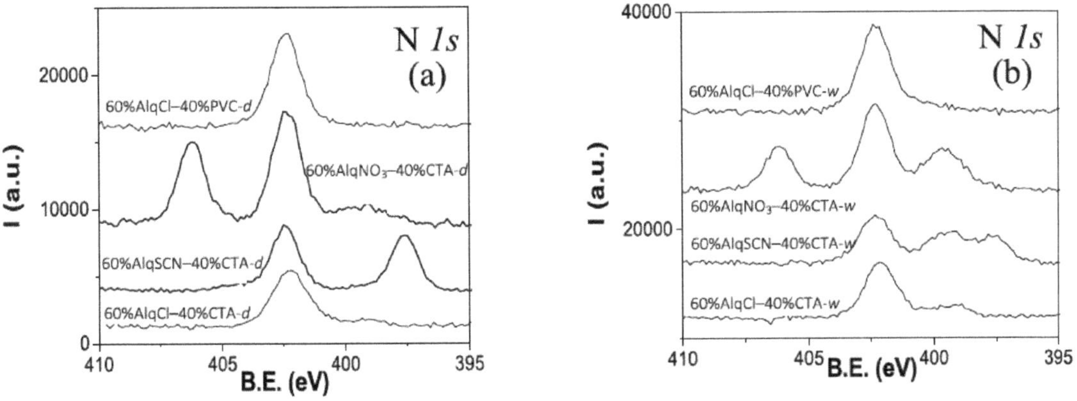

Figure 2. N *1s* core level spectra for dry (**a**) and water-equilibrated (**b**) samples of the studied PIMs.

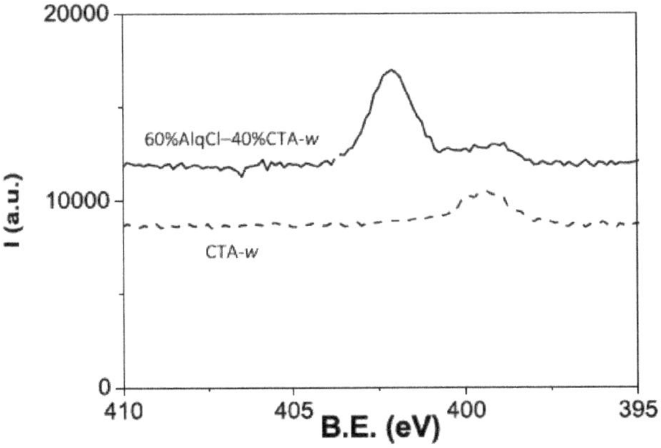

Figure 3. N *1s* core level spectra for water-equilibrated samples of 100% CTA and a 60% AlqCl–40% CTA PIM.

The percent areas and corresponding A.C.% of the different nitrogen contributions were obtained by deconvolution of the spectra shown in Figure 2, and their values are indicated in Table 2. N_1 at 406.4 eV corresponds to NO_3^- anion; N_2 at 402.4 eV to the N of the cationic part of the IL; and N_3 at 397.7 eV to SCN^- anion. In all cases, a clear reduction in the areas of nitrogen contributions exists when dry and water-equilibrated samples of membranes are compared. Taking into account the values indicated in Tables 1 and 2, the following ratios for the water-equilibrated samples were obtained: $N_2/Cl = 1.1$ (60%AlqCl–40%CTA-w membrane), $N_2/S = 0.94$ and $N_2/N_3 = 1.9$ (60%AlqSCN–40%CTA-w), and $N_2/N_1 = 1.9$ (60%AlqNO$_3$–40%CTA-w); however, these last two ratios would be reduced if the new peak contributions (at 399.5 eV) were also added ($N_2/N_3{}^+$new peak = 0.9 and N_2/N_{1+}new peak = 1.1), in agreement with the compounds' stoichiometry. Moreover, a reduction of approximately 21% in the percentages of N_1 for water-equilibrated

60%$AlqNO_3$–40%CTA when compared with dry samples was obtained, which demonstrates the greater stability of this PIM formulation (in the case of PIMs with AlqCl and AlqSCN the reduction was 40% and 36%, respectively, as stated above).

Table 2. Area percentages (between brackets in italics) and A.C.% after deconvolution of nitrogen signals obtained for dry and water-equilibrated PIMs.

PIM Composition		N_1, 406.4 eV		N_2, 402.4 eV		N_3, 397.7 eV	
60%AlqCl–40%CTA	d			(91%)	2.5%		
	w			(86.0%)	1.6%		
60%$AlqNO_3$–40%CTA	d	(35.1%)	1.4%	(55.4%)	2.2%		
	w	(27.2%)	1.1%	(50.5%)	2.1%		
60%AlqSCN–40%CTA	d			(50.7%)	2.3%	(44.1%)	2.0%
	w			(45.4%)	1.6%	(25.1%)	0.85%
60%AlqCl–40%PVC	d			(100%)	3.1%		
	w			(100%)	2.8%		

Figure 4 shows the Cl 2*p* spectrum for dry and water-equilibrated samples of the two PIMs obtained with AlqCl and the two polymers (60%AlqCl–40%CTA (Figure 4a) and 60%AlqCl–40%PVC (Figure 4b)). Both dry samples as well as the 60%AlqCl–40%CTA-w membrane seem to be completely covered by the ionic liquid (which is also represented in Figure 4 by a dotted line), and they show the typical signal for ionic chlorides [34]; however, upon hydration of the PVC-based sample, the presence of the covalent chloride of PVC is clearly observed in Figure 4b. These results provide extra evidence that the hydration process removes part of the AlqCl IL from the PIM surfaces, leaving the polymer structure exposed.

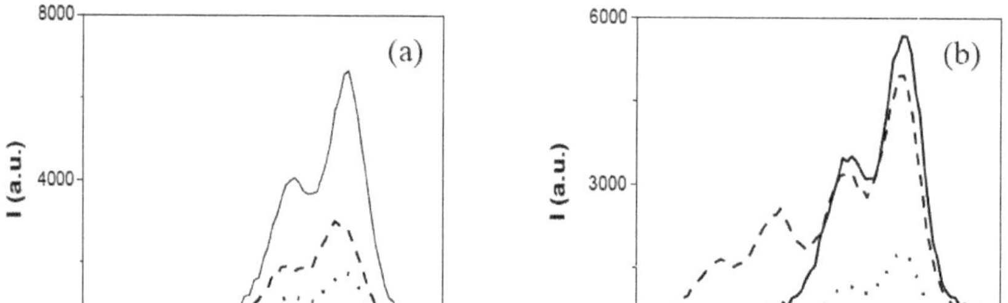

Figure 4. Cl 2*p* core level spectra for (**a**) 60%AlqCl–40%CTA-d (solid line) and 60%AlqCl–40%CTA-w (dashed line); (**b**) 60%AlqCl–40%PVC-d (solid line); and 60%AlqCl–40%PVC-w (dashed line). Dotted line in (**a**,**b**) corresponds to pure IL AlqCl.

Similar results were obtained for PIMs with 30% of IL. Figure 5 shows a comparison of nitrogen core level spectra for 30%AlqCl–70%CTA, 30%$AlqNO_3$–70%CTA, and 30%AlqSCN–70%CTA PIMs. Due to the higher CTA content of these membranes, the peak at ~399.5 eV is clearly observed even in dry samples, with the water-equilibrated samples contributing more nitrogen.

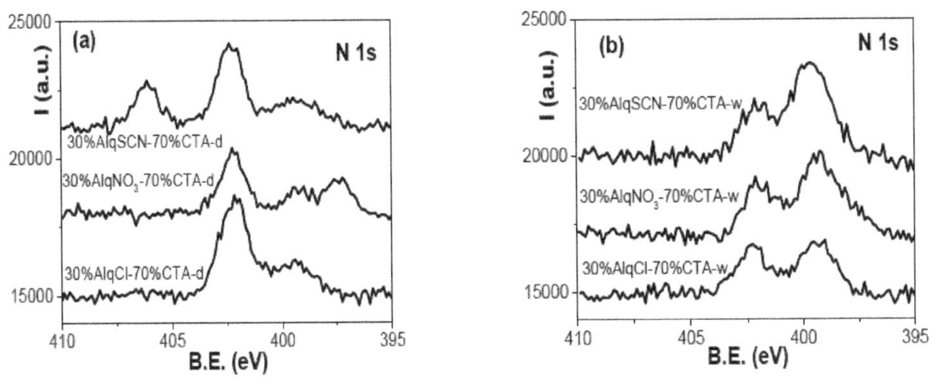

Figure 5. N *1s* core level spectra for 30%AlqCl–70%CTA, 30%$AlqNO_3$–70%CTA, and 30%AlqSCN–70%CTA PIMs. (**a**) Dry samples and (**b**) water-equilibrated samples.

Consequently, the XPS analysis performed with water-equilibrated samples of different PIMs demonstrates a certain loss of the ILs used for membrane fabrication, in agreement with reported results obtained from weight loss measurements [35,36], but also the modifications of the PIMs associated with water contact, which depend on both the IL and the polymer.

3.2. Morphological Characterization with SEM

Possible PIM modifications depending on the type of IL, the concentration, and the contact with water were investigated using SEM. It is significant to note the importance of IL content on the mechanical stability and manageability of PIMs, since brittle samples are obtained when low IL content is used (less than 20%, approximately), but the plasticity of membranes at high IL percentages (higher than 70%) may be too great [5]. It was anticipated that the IL would be physically inserted within the polymer chains, giving a nano-scale structuration, as reported by other authors [37]. As expected, the dry membranes obtained were dense and with no apparent porosity, as can be observed in the SEM micrographs for the PIMs with the lowest IL concentration (i.e., 30%) (Figure 6).

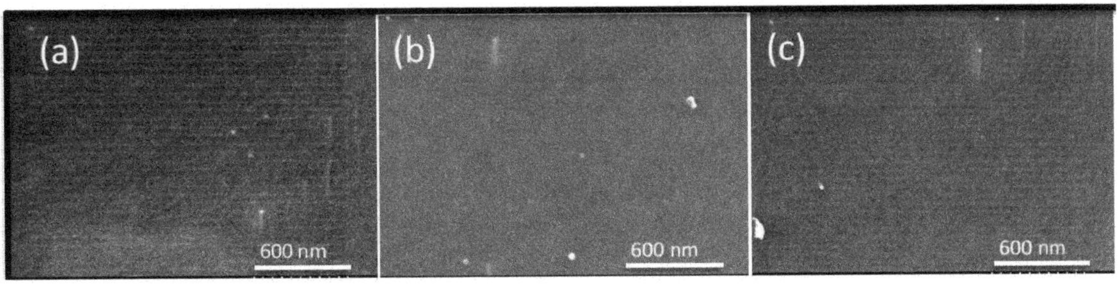

Figure 6. Surface SEM images of dry PIMs with a composition of 30%IL–70%CTA. (**a**) AlqCl; (**b**) $AlqNO_3$; and (**c**) AlqSCN.

However, in membranes with a higher IL content (60%), the morphology of the PIMs varies depending on the IL incorporated (see Figure 7). As can be observed in Figure 7a, in the case of AlqCl, the characteristics of the PIM are similar to those of the PIM containing 30% of IL. However, in the case of both PIMs that include the derivatives $AlqNO_3$ and AlqSCN, they present a surface with nano-spheres that are uniformly distributed along the surface. The presence of such small nano-spheres can be attributed to the formation of ionic aggregates of IL due to poorer miscibility and/or dispersion of these derivatives in the matrix. The formation of such ionic aggregates in the presence of an excess of IL has previously been determined by other authors [38] and confirmed by dynamic mechanical analysis [39].

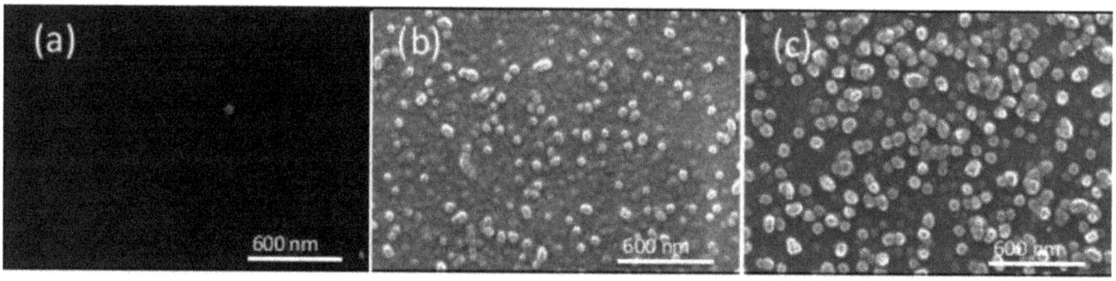

Figure 7. Surface SEM images of dry PIMs with a composition of 60%IL–40%CTA. (**a**) AlqCl; (**b**) $AlqNO_3$; and (**c**) AlqSCN.

Interestingly, the cross-section images (see Figure 8) show that the morphology of the bulk membrane is completely homogeneous, without the presence of aggregates of the IL, thus demonstrating the significance of the surface characterization.

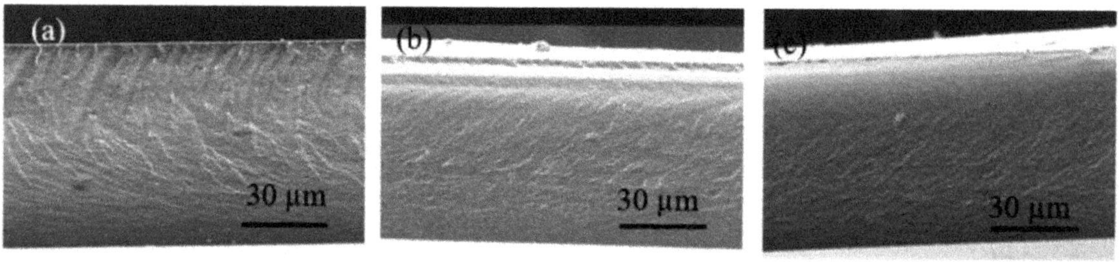

Figure 8. Cross-section SEM images of dry PIMs with a composition of 60%IL–40%CTA. (**a**) AlqCl; (**b**) $AlqNO_3$; and (**c**) AlqSCN.

Importantly, from our results, we can hypothesize that the miscibility of the IL with the polymer is lower for $AlqNO_3$ and AlqSCN than for AlqCl. Other authors have pointed to the existence of a miscibility limit between the pre-polymer and the ionic liquid in an epoxy matrix [37,39].

As it was previously indicated, the formulations used for PIM characterization by XPS and SEM techniques (60% IL/40% polymer or 30%IL/70% polymer) were selected to highlight IL or polymer characteristics, as the PIMs were almost at the limit of adequate mechanical properties. Consequently, a PIM containing an intermediate amount of $AlqNO_3$ (50%$AlqNO_3$–50%CTA) was also studied, and the results are shown in Figure 9. SEM images were recorded before and after immersion of the PIM in ultrapure water, and both the surface and cross-section images were analyzed. As can be observed, the nanospheres in the surface of the dry PIM disappear when the PIM is contacted with water. A comparison of cross-section images for water-equilibrated and dry samples shows practically no modification of the membrane bulk. Consequently, it seems that water contact causes morphological changes to the surface of PIMs but not to the bulk part.

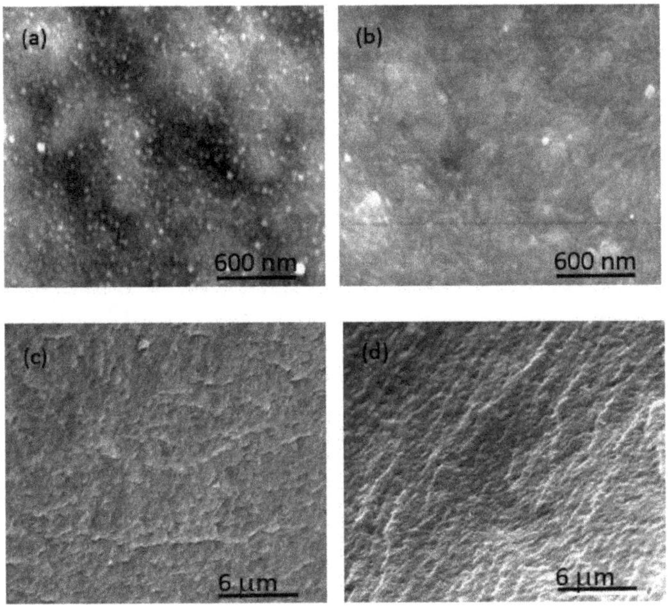

Figure 9. SEM images from a pristine 50%$AlqNO_3$–50%CTA PIM: surface (**a**) and cross-section (**c**). The same membrane after immersion in ultrapure water: surface (**b**) and cross-section (**d**).

3.3. Comparision of IL Performance in Transport Experiments

Taking into account the results from the characterization of the membranes, we wanted to investigate whether the stability of the membrane was related to its effectivity as a separation system. For that, we chose the transport of As(V) as the case study. The extraction and transport of arsenic species by means of AlqCl incorporated in a membrane have been successfully reported in the literature [3,40]. However, arsenic transport has not been investigated using the nitrate derivative as a carrier. At neutral pH (~7), the arsenate ion is present mainly as $H_2AsO_4^-$ and $HAsO_4^{2-}$, which are in equal proportion in the solution. This fact allows the As(V) species to be extracted through an ion-exchange mechanism according to the following stoichiometric equation:

$$HAsO_4^{2-} + 2(AlqX^-) \leftrightharpoons [(Alq^+)_2HAsO_4^{2-}] + 2X^- \quad (2)$$

where Alq^+ represents the cationic part of Aliquat 336 molecule and X^- represents the counter-anion chloride or nitrate.

An alternative which reverses this extraction reaction can be based on the use of chloride at high concentration, as it is expected that arsenic species will be released due to the dissociation of the ion pair formed in the organic phase.

Therefore, we evaluated PIMs (average thickness of 60 µm) consisting of 50% CTA as the base polymer and 50% AlqCl or $AlqNO_3$ as the carriers; the first one is the IL that is easily released from the PIM surface, while the second is the IL that provides more stable membranes.

The transient As(V) concentration profiles in both the feed and the stripping solutions are shown in Figure 10. As can be seen, the more stable PIM (made of CTA and $AlqNO_3$) results in the least efficiency in terms of As(V) transport. This fact may be due to the great interaction between Alq+ and the anion NO_3^-, which form a stable ionic pair. Therefore, nitrate is hardly exchanged by arsenate, resulting in poor transportation (about 10%). However, in the case of the PIM containing AlqCl, despite the possible alteration on the surface due to the IL loss, it results in a very effective membrane to remove arsenate.

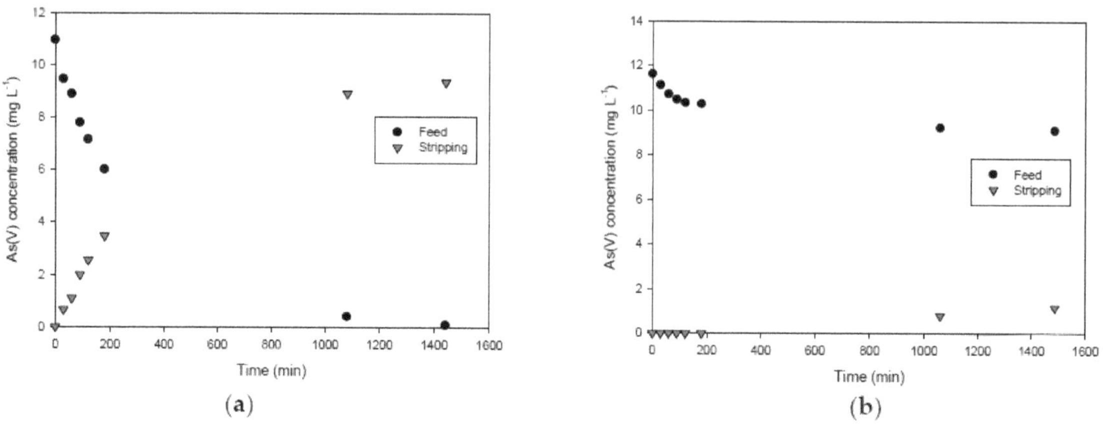

Figure 10. Transient concentration curves in As(V) transport experiments involving 50%CTA–50%AlqCl (**a**) and 50%CTA–50%AlqNO$_3$ (**b**). (Experimental conditions: feed solution: 10 mg L^{-1} As(V), pH = 7; stripping solution: 0.1 M NaCl in (**a**) and 0.1 M NaNO$_3$ in (**b**)).

Moreover, to examine the possible long-term stability of PIMs, we used the same membrane (50%CTA–50%AlqCl) repeatedly in successive experiments transporting As(V) from the aqueous solution at the same concentration (5 cycles of 24 h each). The results obtained in terms of As(V) transport efficiency are plotted in Figure 11 and it can be seen that only slightly variations were measured, with average transport efficiency of <TE> = (79 ± 4)%, without signs of structural weakening.

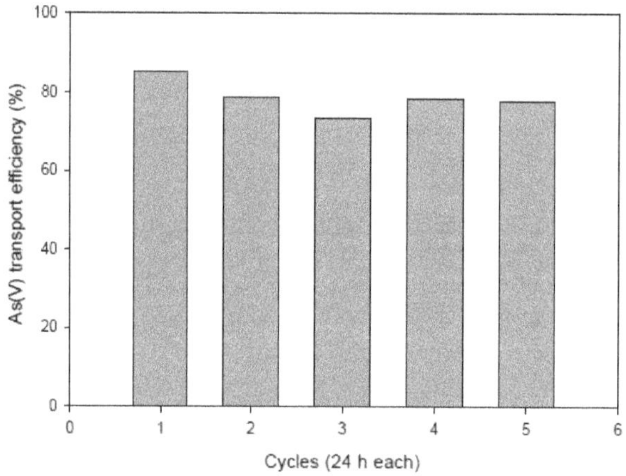

Figure 11. Transport efficiency of the PIM system after five cycles of 24 h each. Feed solution: [As(V)] = 10 mg L^{-1}, pH 7. Stripping solution: 0.1 M NaCl.

Then, we can presume that, for the particular formulation of the PIM tested, surface modification does not have a dramatic effect on the transport efficiency since the bulk concentration of the IL is sufficient to ensure the transportation. However, in other systems where the surface of the membrane has an active role, for example when the chemical reaction taking place in the interface acts as a limiting factor or when the membrane itself acts as a catalyst, the modifications occurring at the membrane surface, such as those determined by XPS analysis, could affect the transport, which helps us to understand the behavior of the system.

4. Conclusions

Surface characterization by XPS and SEM techniques for different polymer inclusion membranes has been performed with the samples in dry and water-equilibrated states to estimate the possible loss of IL and/or their chemical surface modifications depending on both IL-anion and polymer support. The PIMs studied here were fabricated with three ILs, AlqCl, $AlqNO_3$, and AlqSCN, as carriers and cellulose triacetate (CTA) or polyvinyl chloride (PVC) as the support matrix at two different formulations (60% IL-40% polymer or 30%IL-70% polymer). XPS results show the partial loss of the IL in the water-equilibrated samples, which is more significant for the AlqCl–CTA PIM; approximately 40% is calculated from the deconvolution of the XPS spectrum. PIMs made of $AlqNO_3$-CTA show the lower loss of the IL from the surface. The nature of the IL-anion also affects the structuration of the polymer as observed in SEM images. It was observed that AlqCl is more miscible with the polymer matrix, whereas, for the other anions, nano-spheres are formed on the surface of the PIM while no changes at bulk level are observed. These nano-spheres disappear upon the PIM coming into contact with water. Moreover, the 50%AlqCl-50%CTA membrane shows good transport efficiency to As(V) (approximately 80%) as well as adequate stability. This latter point was established by considering PIM reusability studies, which demonstrate that the PIM does not practically modify its transport ability after 5 days in contact with aqueous solutions.

Author Contributions: Conceptualization, J.B., C.F. and E.A.; methodology, M.d.V.M.d.Y. and E.R.-C.; investigation, R.V.; writing—original draft preparation, J.B., C.F. and E.A.; writing—review and editing, C.F., E.A. and J.B.; supervision, C.F.; project administration, C.F.; funding acquisition, C.F. and E.A. All authors have read and agreed to the published version of the manuscript.

Funding: This research was funded by the Ministerio de Ciencia, Innovación, y Universidades (MCIU), grant number PID2019-107033GB-C22/AEI/10.13039/501100011033 and RTI2018-099668-B-C22. Research Group FQM 258 (Junta de Andalucía, Spain) is also acknowledged.

Conflicts of Interest: The authors declare no conflict of interest. The funders had no role in the design of the study; in the collection, analyses, or interpretation of data; in the writing of the manuscript, or in the decision to publish the result.

References

1. Nghiem, L.; Mornane, P.; Potter, I.; Perera, J.; Cattrall, R.; Kolev, S. Extraction and transport of metal ions and small organic compounds using polymer inclusion membranes (PIMs). *J. Membr. Sci.* **2006**, *281*, 7–41. [CrossRef]
2. Casadellà, A.; Schaetzle, O.; Nijmeijer, K.; Loos, K. Polymer Inclusion Membranes (PIM) for the Recovery of Potassium in the Presence of Competitive Cations. *Polymers* **2016**, *8*, 76. [CrossRef]
3. Güell, R.; Anticó, E.; Kolev, S.D.; Benavente, J.; Salvadó, V.; Fontàs, C. Development and characterization of polymer inclusion membranes for the separation and speciation of inorganic As species. *J. Membr. Sci.* **2011**, *383*, 88–95. [CrossRef]
4. Parhi, P.K. Supported Liquid Membrane Principle and Its Practices: A Short Review. *J. Chem.* **2013**, *2013*, 618236. [CrossRef]
5. Vázquez, M.I.; Romero, V.; Fontàs, C.; Anticó, E.; Benavente, J. Polymer inclusion membranes (PIMs) with the ionic liquid (IL) Aliquat 336 as extractant: Effect of base polymer and IL concentration on their physical-chemical and elastic characteristics. *J. Membr. Sci.* **2014**, *455*, 312–319. [CrossRef]
6. Bennet, M.D.; Leo, D.J. Ionic Liquids as stable solvents for ionic polymer transducers. *Sens. Actuators A* **2004**, *115*, 79–90. [CrossRef]
7. Welton, T. Room-Temperature Ionic Liquids. Solvents for synthesis and catalysis. *Chem. Rev.* **1999**, *99*, 2071–2083. [CrossRef]
8. Armand, M.; Endres, F.; MacFarlane, D.; Ohno, H.; Scrosati, B. Ionic-liquid Materials for the Electrochemical Challenges of the Future. *Nat. Mater.* **2009**, *8*, 621–629. [CrossRef]
9. Kogelnig, D.; Stojanovic, A.; Galanski, M.; Groessl, M.; Jirsa, F.; Krachler, M.; Keppler, B. Greener synthesis of new ammonium ionic liquids and their potential as extracting agents. *Tetrahedron Lett.* **2008**, *49*, 2782–2785. [CrossRef]
10. Mikkola, J.-P.; Virtanen, P.; Sjöholm, R. Aliquat 336®—A versatile and affordable cation source for an entirely new family of hydrophobic ionic liquids. *Green Chem.* **2006**, *8*, 250–255. [CrossRef]
11. Neves, L.A.; Benavente, J.; Coelhoso, I.M.; Crespo, J.G. Design and characterization of Nafion membranes with incorporated ionic liquid cations. *J. Membr. Sci.* **2010**, *347*, 42–53. [CrossRef]
12. Gao, J.; Liu, J.; Liu, W.; Li, B.; Xin, Y.; Yin, Y.; Gu, J.; Zou, Z. Proton exchange membrane fuel cell working at elevated temperature with ionic liquid as electrolyte. *Int. J. Electrochem. Sci.* **2011**, *6*, 611–616.

13. Díaz, M.; Ortiz, A.; Ortiz, I. Progress in the use of ionic liquids as electrolyte membranes in fuel cells. *J. Membr. Sci.* **2014**, *469*, 379–396. [CrossRef]
14. Neves, L.A.; Coelhoso, I.; Crespo, J.G. Methanol and gas crossover through modified nafion membranes by incorporation of ionic liquid cations. *J. Membr. Sci.* **2010**, *360*, 363–370. [CrossRef]
15. Gherasim, C.-V.I.; Bourceanu, G.; Olariu, R.-I.; Arsene, C. Removal of lead(II) from aqueous solutions by a polyvinyl-chloride inclusion membrane without added plasticizer. *J. Membr. Sci.* **2011**, *377*, 167–174. [CrossRef]
16. Pont, N.; Salvadó, V.; Fontàs, C. Selective transport and removal of Cd from chloride solutions by polymer inclusion membranes. *J. Memb. Sci.* **2008**, *318*, 340–345. [CrossRef]
17. Matsumoto, M.; Murakami, Y.; Minamidate, Y.; Kondo, K. Separation of lactic acid through polymer inclusion membranes containing ionic liquids. *Sep. Sci. Technol.* **2012**, *47*, 354–359. [CrossRef]
18. Nasser, I.I.; Amor, F.I.E.H.; Donato, L.; Algieri, C.; Garofalo, A.; Drioli, E.; Ahmed, C. Removal and recovery of Ag(CN)2- from synthetic electroplating baths by polymer inclusion membranes containing Aliquat 336 as a carrier. *Chem. Eng. J.* **2016**, *295*, 207–217. [CrossRef]
19. Vera, R.; Gelde, L.; Anticó, E.; de Yuso, M.V.M.; Benavente, J.; Fontàs, C. Tuning physicochemical, electrochemical and transport characteristics of polymer inclusion membrane by varying the counter-anion of the ionic liquid Aliquat 336. *J. Membr. Sci.* **2017**, *529*, 87–94. [CrossRef]
20. Benavente, J.; Romero, V.; Vázquez, M.I.; Anticó, E.; Fontàs, C. Electrochemical Characterization of a Polymer Inclusion Membrane Made of Cellulose Triacetate and Aliquat 336 and Its Application to Sulfonamides Separation. *Separations* **2018**, *5*, 5. [CrossRef]
21. Kallioinen, M.; Nyström, M. Membrane Surface Characterization. In *Advanced Membrane Technology and Applications*; Li, N.N., Fane, A.G., Ho, W.S.W., Matsuura, T., Eds.; Chapter 32; Wiley Online Library: Hoboken, NJ, USA, 2008; pp. 841–877.
22. Xu, Z.; Wan, L.; Huang, X. Techniques for Membrane Surface Characterization. In *Surface Engineering of Polymeric Membranes*; ATSTC Book Series; Springer: Berlin/Heidelberg, Germany, 2009; Chapter 2; pp. 5–63.
23. Ariza, M.J.; Castellón, E.R.; Rico, R.; Benavente, J.; Muñoz, M.; Oleinikova, M. X-ray Photoelectron Spectroscopy Analysis of Di-(2-ethylhexyl) Phosphoric Acid Activated Membranes. *J. Colloid Interface Sci.* **2000**, *226*, 151–158. [CrossRef]
24. Ariza, M.J.; Benavente, J.; Rodríguez-Castellón, E.; Palacio, L. Effect of Hydration of Polyamide Membranes on the Surface Electrokinetic Parameters: Surface Characterization by X-Ray Photoelectronic Spectroscopy and Atomic Force Microscopy. *J. Colloid Interface Sci.* **2002**, *247*, 149–158. [CrossRef]
25. Szymczyk, A.; Fatin-Rouge, N.; Fievet, P. Tangential streaming potential as a tool in modeling of ion transport through nanoporous membranes. *J. Colloid Interface Sci.* **2007**, *309*, 245–252. [CrossRef]
26. de Yuso, M.V.M.; Calderón, A.B.; Romero, V.; Cuberes, M.T.; Benavente, J. Chemical and inhomogeneity surface changes of a Nafion membrane associated to its doping with the room temperature ionic liquid AliquatCl. *Surf. Interf. Anal.* **2016**, *48*, 561–565. [CrossRef]
27. Ogielo, W.; Wormeester, H.; Wessling, M.; Benes, N.E. Spectroscopic ellipsometry analysis of a thin film composite membrane consisting of polysulphone on a porous a-alumina support. *ACS Appl. Mater. Interfaces* **2012**, *4*, 935–943. [CrossRef]
28. Palacio, L.; Prádanos, P.; Hernández, A.; Ariza, M.J.; Benavente, J.; Nyström, M. Phase-contrast scanning force microscopy and chemical heterogeneity of GR polysulphone ultrafiltration membranes. *Appl. Phys. A* **2001**, *73*, 555–560. [CrossRef]
29. Buechner, C.; Gericke, S.M.; Trotochaud, L.; Karslıoğlu, O.; Raso, J.; Bluhm, H.; Karslıoğlu, O. Quantitative characterization of a desalination membrane model system by X-ray photoelectron spectroscopy. *Langmuir* **2019**, *35*, 11315–11321. [CrossRef]
30. Briggs, D.; Seah, M.P. *Practical Surface Analysis*, 2nd ed.; Vol I: Auger and X-ray Photoelectron Spectroscopy; John Wiley & Sons: Chichester, UK, 1995.
31. Ariza, M.J.; Benavente, J.; Rodríguez-Castellón, E. The capability of X-ray Photoelectron Spectroscopy in the Characterization of Membranes: Correlation between Surface Chemical and Transport Properties in Polymeric Membranes. In *Handbook of Membrane Research: Properties, Performance and Applications*; Gorley, S.V., Ed.; Nova Science Publishers, Inc.: New York, NY, USA, 2009; pp. 257–290, ISBN 978-1-60741-638-8.
32. Keurentjes, J.T.F.; Harbrecht, J.G.; Brinkman, D.; Hanemaaijer, J.H.; Stuart, M.A.C.; van't Riet, K. Hydrophobicity measurements of microfiltration and ultrafiltration membranes. *J. Membr. Sci.* **1989**, *47*, 333–344. [CrossRef]
33. Ramos, J.D.; Milano, C.; Romero, V.; Escalera, S.; Alba, M.C.; Vázquez, M.I.; Benavente, J. Water effect on physical-chemical and elastic parameters for a dense cellulose regenerated membrane. Transport of different aqueous electrolyte solutions. *J. Membr. Sci.* **2010**, *352*, 153–159. [CrossRef]
34. Moulder, J.F.; Stickl, W.F.; Sobol, P.E.; Bomben, K.D. *Handbook of X-Ray Photoelectron Spectroscopy*; Perkin-Elmer: Eden Prairie, MN, USA, 1992.
35. Kagaya, S.; Ryokan, Y.; Cattrall, R.W.; Kolev, S. Stability studies of poly(vinyl chloride)-based polymer inclusion membranes containing Aliquat 336 as a carrier. *Sep. Purif. Technol.* **2012**, *101*, 69–75. [CrossRef]
36. Tomás-Alonso, F.; Rubio, A.M.; Giménez, A.; Rios, A.P.D.L.; Salar-García, M.J.; Ortiz-Martínez, V.M.; Hernández-Fernández, F.J. Influence of ionic liquid composition on the stability of polyvinyl chloride-based ionic liquid inclusion membranes in aqueous solution. *AIChE J.* **2016**, *63*, 770–780. [CrossRef]
37. Livi, S.; Silva, A.A.; Thimont, Y.; Nguyen, T.K.L.; Soares, B.G.; Gérard, J.-F.; Duchet-Rumeau, J. Nanostructured thermosets from ionic liquid building block–epoxy prepolymer mixtures. *RSC Adv.* **2014**, *4*, 28099–28106. [CrossRef]

38. Livi, S.; Gérard, J.-F.; Duchet-Rumeau, J. Ionic liquids: Structuration agents in a fluorinated matrix. *Chem. Commun.* **2011**, *47*, 3589–3591. [CrossRef]
39. Zhang, X.; Sun, H.; Yang, C.; Zhang, K.; Yuen, M.M.F.; Yang, S. Highly conductive polymer composites from room-temperature ionic liquid cured epoxy resin: Effect of interphase layer on percolation conductance. *RSC Adv.* **2013**, *3*, 1916–1921. [CrossRef]
40. Vera, R.; Anticó, E.; Fontàs, C. The Use of a Polymer Inclusion Membrane for Arsenate Determination in Groundwater. *Water* **2018**, *10*, 1093. [CrossRef]

membranes

MDPI

Article

Nitrate Removal by Donnan Dialysis and Anion-Exchange Membrane Bioreactor Using Upcycled End-of-Life Reverse Osmosis Membranes

Amaia Lejarazu-Larrañaga [1,2,*], Juan M. Ortiz [1], Serena Molina [1,*], Sylwin Pawlowski [3], Claudia F. Galinha [3], Vanessa Otero [4,5], Eloy García-Calvo [1,2], Svetlozar Velizarov [3] and João G. Crespo [3]

1 IMDEA Water Institute, Avenida Punto Com, 2, Alcalá de Henares, 28805 Madrid, Spain; juanma.ortiz@imdea.org (J.M.O.); eloy.garcia@imdea.org (E.G.-C.)
2 Chemical Engineering Department, University of Alcalá, Ctra. Madrid-Barcelona Km 33.600, Alcalá de Henares, 28871 Madrid, Spain
3 LAQV-REQUIMTE, Department of Chemistry, NOVA School of Science and Technology, FCT NOVA, Universidade NOVA de Lisboa, 2829-516 Caparica, Portugal; s.pawlowski@fct.unl.pt (S.P.); cf.galinha@fct.unl.pt (C.F.G.); s.velizarov@fct.unl.pt (S.V.); jgc@fct.unl.pt (J.G.C.)
4 LAQV-REQUIMTE, Department of Conservation and Restoration, NOVA School of Science and Technology, FCT NOVA, Universidade NOVA de Lisboa, 2829-516 Caparica, Portugal; van_otero@campus.fct.unl.pt
5 VICARTE, Department of Conservation and Restoration, NOVA School of Science and Technology, FCT NOVA, Universidade NOVA de Lisboa, 2829-516 Caparica, Portugal
* Correspondence: amaia.ortiz@imdea.org (A.L.-L.); serena.molina@imdea.org (S.M.)

Abstract: This work explores the application of Reverse Osmosis (RO) upcycled membranes, as Anion Exchange Membranes (AEMs) in Donnan Dialysis (DD) and related processes, such as the Ion Exchange Membrane Bioreactor (IEMB), for the removal of nitrate from contaminated water, to meet drinking water standards. Such upcycled membranes might be manufactured at a lower price than commercial AEMs, while their utilization reinforces the commitment to a circular economy transition. In an effort to gain a better understanding of such AEMs, confocal μ-Raman spectroscopy was employed, to assess the distribution of the ion-exchange sites through the thickness of the prepared membranes, and 2D fluorescence spectroscopy, to evaluate alterations in the membranes caused by fouling and chemical cleaning The best performing membrane reached a 56% average nitrate removal within 24 h in the DD and IEMB systems, with the latter furthermore allowing for simultaneous elimination of the pollutant by biological denitrification, thus avoiding its discharge into the environment. Overall, this work validates the technical feasibility of using RO upcycled AEMs in DD and IEMB processes for nitrate removal. This membrane recycling concept might also find applications for the removal and/or recovery of other target negatively charged species.

Keywords: membrane recycling; nitrate removal; Donnan Dialysis; Ion-Exchange Membrane Bioreactor; circular economy

1. Introduction

Reverse Osmosis (RO) desalination market is in constant growth [1]. However, due to RO performance deterioration, an annual membrane replacement rate ranging between 5% in brackish water RO desalination to 20% in sea water RO desalination is estimated [2]. To date, end-of-life RO membranes are commonly wasted in landfills, generating an increasing environmental concern [3]. Membrane recycling is a more sustainable alternative, and hence, it is gaining an increasing attention by researchers and industry stakeholders. Recently, the upcycling of end-of-life RO membranes for the preparation of Anion-Exchange Membranes (AEMs) has been proposed [4–6]. This alternative entails the deconstruction of the RO module, allowing for the individual management of flat sheet membranes and plastic components. The resulting membranes have a good permselectivity but a relatively high electrical resistance, which increases the energy consumption associated with their use in electrodialysis (ED) [4]. In addition, ED necessarily removes both cations and anions (i.e., leading to demineralization), and it might be less appropriate in certain applications, such as drinking water or beverages production, in which the ionic composition should be preserved as much as possible. In that respect, an electrochemical potential gradient driven process, such as Donnan Dialysis (DD) and its integration with biotransformation, referred to as Ion-Exchange Membrane Bioreactor (IEMB), could entail a more suitable application for such RO upcycled AEMs. Since DD, utilizes counter-diffusion of two or more ions through an ion-exchange membrane to achieve their separation, its energy demand is much lower in comparison to that of ED [7]. Indeed, owing to its simplicity of installation and low operating costs (under batch operation conditions DD can be operated even without pumping), DD has been presented as an appropriated technology for household water treatment in remote rural areas of developing countries (not connected to centralized water supply facilities) [8,9]. Furthermore, the IEMB has been reported as an efficient technology for removal and degradation of target ionic pollutants, such as nitrate, from drinking water supplies [10].

Nitrate is toxic for human health. The abusive use of nitrogen-rich fertilizers, uncontrolled sludge and slurry dumping, and a poor wastewater treatment (i.e., an uncompleted biological denitrification process), have risen its concentration in natural waters (specially in groundwaters) to critical levels worldwide [11,12]. The World Health Organization (WHO) has limited its concentration to a maximum of 50 mg L^{-1} in drinking water, along with the recommendation to keep it below 25 mg L^{-1} [13]. However, in drinking water production, nitrate is not eliminated by regularly used technologies (i.e., coagulation-flocculation, decantation, filtration, and disinfection), and frequently other methods, such as RO [14], ED [15], ion-exchange resins [16], and biological denitrification [17], have been employed for this purpose. These processes have several disadvantages such as: high economic and energy costs, the extensive use of chemicals, or the inception of secondary pollution problems in the treated water (biological contamination) [18].

For the removal of nitrate by DD, an anion-exchange membrane (AEM) is used to separate the feed solution (i.e., water polluted with nitrate), from the receiver solution containing a higher concentration of a "driving" nontoxic counter-ion (chloride, in this work). In such system, chloride is transported through the AEM to the diluted nitrate-containing solution by diffusion, due to its concentration gradient. As the AEM rejects the co-ions, an electrochemical potential gradient is established across the membrane, causing the coupled back-transport of the target counter ion (nitrate, in this work) to the receiver solution [7]. The continuous diffusive transport of the driving counter ion to the diluted compartment eventually leads to the transport of nitrate against its concentration gradient (uphill transport), thus allowing an almost complete removal of the target ionic pollutant from water and its consequent concentration in the receiver compartment. Hence, a post treatment is needed for the elimination of the nitrate accumulated in the receiver compartment. To overcome this issue, the IEMB combines the DD principle with a biological conversion process in the receiver compartment, providing the simultaneous transport and elimination of nitrate by conversion to molecular nitrogen, in a single device and in a single step [10]. In the bioreactor, a microbial culture of denitrifying bacteria oxidizes a carbon source (the electron donor) added to the receiver. In the absence of oxygen (anoxic/anaerobic environment), the nitrate acts as the electron acceptor, and it is sequentially reduced to NO_2, NO, N_2O, and finally N_2 [19]. Furthermore, the integration of a membrane in the bioreactor allows the physical separation of the microbial culture and other pollutants from the treated water (e.g., residual carbon source and by-products from microbial metabolism), thus preventing the major problems associated to the application of biological treatments to drinking water production [20]. In previous studies, IEMB process achieved a high efficiency in drinking water denitrification when using Neosepta ACS membrane [21]. However, the high cost of such membrane could compromise the practical implementation of the IEMB concept, and thus, the use of more affordable membranes is highly desirable [22]. In this context, the AEMs prepared from upcycled RO membranes could entail a more economic approach.

In the present work, the use of AEMs upcycled from end-of-life RO membranes in DD and IEMB processes for the removal of nitrate from a synthetic polluted water was studied and compared with the performance of a Ralex ® AMH-PES (a commercial AEM manufactured by MemBrain ® s.r.o., Stráž pod Ralskem, Czech Republic) selected due to its high durability, good mechanical and chemical stability and a relatively low commercial cost. Further, the membrane structure was analysed by confocal µ-Raman spectroscopy, and alterations in the membranes due to fouling and the chemical cleaning were evaluated by 2D fluorescence spectroscopy. Finally, a preliminary estimation of the material cost to manufacture the upcycled membranes was performed.

2. Materials and Methods

2.1. Chemical Reagents

Sodium hypochlorite (NaClO, 14%), tetrahydrofuran (THF), sodium sulphate (Na_2SO_4), and ammonium chloride (NH_4Cl) were purchased from Scharlab, S.L., Barcelona, Spain. Sodium chloride (NaCl), sodium nitrate ($NaNO_3$), di-potassium hydrogen phosphate anhydrous (K_2HPO_4), humic acids (HA), Amberlite® IRA-402, and Lewatit ® Sybron Ionac ® SR-7 anion-exchange resins were supplied by Merck KGaA, Darmstadt, Germany. Purolite°® A600/9413 anion-exchange resin was supplied by MemBrain ® s.r.o., Stráž pod Ralskem, Czech Republic. Potassium dihydrogen phosphate (KH_2PO_4) was purchased from Chem-lab NV, Zedelgem, Belgium; sodium hydrogen phosphate (NaH_2PO_4) from Panreac Química S.L.U., Barcelona, Spain; magnesium sulphate heptahydrate ($MgSO_4 \cdot 7H_2O$) from LabChem Inc., Zelienople, PA, USA; sodium hydroxide (NaOH) from Fisher Scientific Co. LLC., Pittsburgh, PA, USA, and citric acid from VWR chemicals, USA. Polyvinylchloride (PVC, Mw = 112,000 g mol^{-1}) was supplied by ATOCHEM, Madrid, Spain. MilliQ water was used throughout the experiments.

2.2. Membranes

Figure 1. illustrates the different steps carried out for the preparation of the RO upcycled AEMs. To complement this representation, Scanning Electron Microscopy (SEM) images of the membranes are included in Figure 1(a3,c3,d3). Due to the morphological differences between the surfaces of the membranes, different microscope magnifications were employed. The methodology for conducting the SEM analyses was described elsewhere [4,23]. An end-of-life thin film composite polyamide RO membrane (TM 720-400, Toray Industries, Inc., Osaka, Japan), discarded by a desalination plant after more than three years of exploitation, was upcycled as mechanical support to prepare AEMs (Figure 1a) [4]. First of all, the end-of-life RO module was treated by passive immersion in 9000 mg L^{-1} NaClO solution for 90 h, as described in [24,25]. Then, the spiral wound configuration of the module was opened (module autopsy), using a radial arm saw, and the membrane sheets were unrolled (Figure 1b). Membrane coupons of approximately 315 cm^2 were then taken out to be used as mechanical support in the preparation of AEMs. The high exposure dose of NaClO employed in the previous step (810,000 ppm·h) ensured the complete elimination of the fouling and of the polyamide (PA) layer, and leaded to a considerably homogeneous and microporous polysulfone (PSf) surface (Figure 1c), as it was demonstrated in previous studies carried out by our research group [26–28]. Thus, the membranes achieved ultrafiltration (UF)-like properties in terms of salt rejection and hydraulic permeability [26]. Afterwards, the AEMs were prepared by casting a polymeric mixture on the polysulfone (PSf) surface of the recycled membrane (Figure 1d). The polymeric mixture contained PVC dissolved into THF and a finely grounded anion-exchange resin (sieved down to less than 50 µm), in a blend ratio of 0.5 g:0.5 g:10 mL, respectively, ion-exchange resin (g) to PVC (g), to THF (mL). The solvent was evaporated at room temperature for 1 h, and then, the membrane was immersed in a water bath. Three different types of anion-exchange resins were tested: Purolite® A600/9149, Amberlite® IRA-402 and Lewatit® Sybron Ionac® SR-7 (see Table 1). The resulting membranes were named as Pur-RE, Amb-RE, and Lew-RE, henceforth. The methodology for the preparation of the membranes is described in more detail in [4]. The microstructure of such membranes was deeply characterized before [4,6] (see Figure 1(d3), and it is further analysed by confocal μ-Raman spectroscopy in the present work (Section 3.1). The prepared membranes show a considerably uniform distribution of the ion-exchange resin at the membrane surface, having a heterogeneous microstructure due to the presence of conductive (ion-exchange resin particles) and non-conductive regions (PVC, PET, and PSf). Therefore, a commercial membrane with a heterogeneous structure was used as reference membrane, in this case, Ralex ® AMH-PES from Mega a.s., which was selected due to its durability, high mechanical stability and relatively low commercial cost.

Figure 1. Steps for the preparation of RO upcycled Anion Exchange Membranes: (**a**) End-of-life RO membrane, (**a1**) schematic representation, (**a2**), surface image, (**a3**) surface SEM image. (**b**) Disassembled end-of-life RO module and separation of flat sheet membrane coupons; (**c**) recycled membrane support, (**c1**) schematic representation, (**c2**) surface image, (**c3**) surface SEM image. (**d**) Prepared AEM (**d1**) schematic representation, (**d2**), surface image, (**d3**) surface SEM image. PA, polyamide; PSf, Polysulfone, PET, polyester.

Table 1. Main characteristics of the membranes under study. Thickness, water content, Ion-Exchange Capacity (IEC), electrical resistance, and permselectivity of all the membranes were experimentally measured under the same conditions and previously reported (data from [6]). AMH-PES membrane composition is detailed by the provider (data from [29]).

Anion-Exchange Membrane	AMH-PES	Pur-RE	Amb-RE	Lew-RE
Mechanical support	Polyester (PET)	End-of-life RO	End-of-life RO	End-of-life RO
Polymer binder	Polyethylene (PE)	PVC	PVC	PVC
Ion-exchange resin	Unspecified	Purolite® A600/9149	Amberlite® IRA-402	Lewatit® Sybron Ionac® SR-7
Ion-exchange group	R—$(CH_3)_3N^+$	R—$(CH_3)_3N^+$	R—$(CH_3)_3N^+$	R—$(C_3H_7)_3N^+$
Membrane thickness (μm)	645 ± 5	190 ± 4	184 ± 7	182 ± 7
Water content (%) [a]	50 ± 0	23 ± 4	21 ± 4	19 ± 1
IEC (mmol g^{-1}) [b]	2.19 ± 0.09	0.75 ± 0.14	0.62 ± 0.04	0.39 ± 0.01
Permselectivity (%) [c]	81 ± 1	65 ± 9	83 ± 7	66 ± 4
Electrical resistance (Ω cm^2) [d]	19 ± 3	56 ± 7	129 ± 1	120 ± 11

[a] Gravimetric method. [b] Immersed in KNO_3 1 M for 24 h under stirring (100 rpm); immersed in NaCl 0.5 M for 24 h under stirring (100 rpm); UV-VIS spectrophotometric determination of the nitrate released in NaCl solution. [c] Measured in 0.1 and 0.5 M NaCl solutions. [d] Measured in 0.5 M NaCl solution.

2.3. Confocal Micro-Raman Spectroscopy

In this work, confocal μ-Raman spectroscopy was used to analyse the distribution of the polymeric composition and the ion-exchange sites, along the thickness of the Pur-RE membrane. Since the same membrane preparation methodology was employed to prepare all the membranes (Pure-RE, Amb-RE and Lew-RE) and only the type of the ion exchange resin was different, analogous structures are expected in all the prepared membranes. To assess that structure, Pur-RE membrane was selected for confocal micro-Raman spectroscopy characterisation since that membrane showed the best performance under DD and IEMB operating conditions (results shown in Section 3.2). A Labram 300 Jobin Yvon spectrometer from Horiba Ltd., Japan, was used, equipped with a solid-state laser operating at 532 nm. The laser beam was focused with a 50 × Olympus objective lens. The analysis was performed with a neutral density filter of 0.6 optical density, with 15-s exposure time and 20 scans. Spectra were recorded as an extended scan and presented with a break between 1025 and 1085 cm^{-1} to remove the contribution of bands attributed to nitrate compounds. For obtaining the cross-section, the membrane was frozen into liquid nitrogen and broken properly. The analyses were performed in 10 consecutive points in a row across the membrane section, from the coated side to the PET side of the membrane. The distance between the analysed points was about 20 μm.

2.4. Donnan Dialysis (DD) and Ion-Exchange Membrane Bioreactor (IEMB) Experiments

2.4.1. Model Solutions, Test Cell, and Analytical Methods

The model polluted water, referred as feed henceforth, consisted of deionized water supplemented with 80 mg L^{-1} of nitrate (added as $NaNO_3$). This concentration was selected as a model high nitrate concentration (>50 mg L^{-1}), as it was used before in [30]. The receiver aqueous solution contained chloride as a major counter-ion (3545 mg L^{-1} of Cl^-, added as NaCl) and a mixture of inorganic nutrients (1000 mg L^{-1} K_2HPO_4, 592 mg L^{-1} KH_2PO_4, 500 mg L^{-1} NaH_2PO_4, 233 mg L^{-1} NH_4Cl, and 1000 mg L^{-1} $MgSO_4{\cdot}7H_2O$ [31]) to support the microbial growth in the bioreactor (when operating as IEMB). In order to maintain the same ionic strength in DD and IEMB operation conditions, the nutrient enriched water media was used as a receiver solution also for DD experiments. The receiver compartment in the IEMB (referred to as biocompartment, hereafter) was additionally supplemented with microbial culture (as described in Section 2.4.3) and ethanol (at a rate of 400 mg L^{-1} day^{-1} EtOH) to ensure the presence of the carbon source in the IEMB experiments.

The same test cell (Figure 2) was used in DD and IEMB experiments. The membrane effective area was 11.3 cm^2 and the volume of each of the two compartments was 175 mL. The coated side of the membrane was faced to the feed compartment (polluted water).

During the experiments, the solutions were continuously stirred to minimize possible concentration polarization effects. Before the experiments, all tested membranes (listed in Table 1) were equilibrated by immersion in the receiver solution (i.e., the salt mixture, without the microbial culture and ethanol) for 2 days.

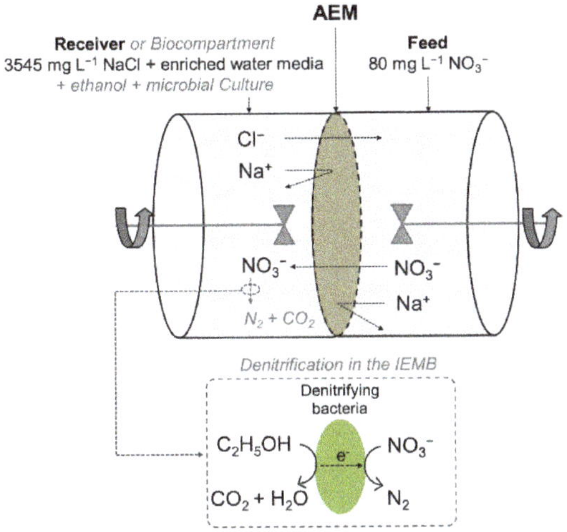

Figure 2. Schematic diagram of the test cell and representation of the main ionic transport in DD and IEMB experiments. Effective membrane area, 11.3 cm^2. Volume of each compartment, 175 mL. The modifications to operate as an IEMB are represented in grey colour and italic lettering. The receiver compartment under IEMB operation was referred to as biocompartment.

The experimental time was 96 h in the case of Pur-RE, Amb-RE, and Lew-RE membranes and 72 h in the case of AMH-PES (as nitrate was already completely removed from the feed after that period). The membranes were thoroughly rinsed with water before the chemical cleaning and after each step [32].

2.4.2. Operation as a Donnan Dialyzer

The time course concentration of nitrate was measured in both compartments during the experiments as described in Section 2.4.4. The membrane performance in DD operation was evaluated in terms of nitrate removal yield from the feed (%), and nitrate recovery in the receiver solution (%), as detailed in Equations (1) and (2), applicable to DD systems [33].

$$\text{Removal}_{NO_3^-(t)} = \frac{C_{NO_3^-,\text{ feed }(0)} - C_{NO_3^-,\text{ feed }(t)}}{C_{NO_3^-,\text{ feed }(0)}} \cdot 100\%, \quad (1)$$

$$\text{Recovery}_{NO_3^-(t)} = \frac{C_{NO_3^-,\text{ receiv. }(t)}}{C_{NO_3^-,\text{ feed }(0)}} \cdot 100\% \quad (2)$$

where $C_{NO_3^-}$ is the concentration of nitrate.

The conductivity was checked in both compartments during the experiments.

2.4.3. Operation as an Ion-Exchange Membrane Bioreactor

Microbial Culture

The activated sludge obtained from a laboratory bioreactor fed with real wastewater was used as inoculum. The inoculum was centrifuged for 30 min at 4000 rpm, under a temperature of 4 °C, for obtaining a pellet enriched in microorganisms and biomass. The pellet was resuspended in fresh nutrient enriched water media (composition detailed in

Section 2.4.1), 800 mg L^{-1} of ethanol (i.e., the carbon source in the IEMB) and 80 mg L^{-1} of nitrate (i.e., the electron acceptor under anoxic conditions, added as $NaNO_3$) were supplied to support the microbial activity and to ensure the acclimation of the bacteria to the experimental conditions. The culture was degasified with N_2 and CO_2 in order to preserve anaerobic conditions. Then, it was incubated at 37 °C for 5 days. The same procedure was repeated 3 times (centrifugation, resuspension in fresh enriched water media and addition of ethanol and nitrate). After that, the enriched microbial culture was placed in fresh enriched water media and stored at 4 °C, being ready to be used as inoculum in the IEMB.

Ethanol Permeation Studies

For the operation as an IEMB, it is of primary importance to avoid the permeation of the carbon source (ethanol, in this case), from the receiver solution to the feed, in order to prevent cross contamination issues (i.e., secondary pollution) in the treated water. The diffusion coefficients of ethanol across the analysed membranes were measured in a two compartments cell. (i.e., two glass bottles of 250 mL each, sealed with a lid, under stirring at 400 rpm). Each membrane under study, with an effective area of 4.52 cm^2, was placed between the compartments. One of the compartments was filled with an aqueous ethanol solution (1000 mg L^{-1} ethanol), and the other one was filled with MilliQ water. As long as ethanol was the only organic compound in the solutions, the Total Organic Carbon (TOC) concentration was analysed in both compartments during the experimental time course (15 days) and the concentration of ethanol was calculated. Then, the diffusion coefficients of ethanol through the membranes were calculated by the graphical representation of Fick's law Equation (3) [4].

$$J_D = -D \cdot \frac{dC}{dl} \tag{3}$$

where J_D (mol cm^{-2} s^{-1}) is the flux of ethanol through the membrane, D (cm^2 s^{-1}) is the diffusion coefficient, dC (mol cm^{-3}) is the concentration gradient (the driving force), and dl (cm) is the membrane thickness. In order to estimate the diffusion coefficient D, the linear region of the experiments was considered, where the slope of the curve C vs time has constant value, and assuming $dC \approx \Delta C$ and $dl \approx l$.

Operation as a Bioreactor

In the case of IEMB experiments, the receiver compartment (Figure 2) was filled with fresh enriched water media and inoculated with 15 mL of enriched microbial culture (the experimental procedure for obtaining the microbial culture has been described in this section in "*Microbial culture*" section). Before inoculation, the microbial culture was stored at room temperature for more than 48 h, for activation of the microbial metabolism. The anaerobic conditions in the receiver compartment were maintained during the experiment by continuously sparging nitrogen gas through the compartment headspace. The concentration of nitrate was measured in both compartments during the experimental time and nitrate removal yield was calculated using Equation (1). In addition, the concentration of nitrite was measured in both compartments, to verify that nitrite traces were not accumulated, ensuring that the complete bio-reduction of nitrate to nitrogen was attained. Nitrite is more toxic than nitrate, and its maximum allowed concentration is limited to 3 mg L^{-1} in drinking water [13]. In addition, the TOC in the feed (water) compartment was analysed to prove that the diffusion of ethanol from the biocompartment to the feed was avoided. The conductivity was checked in both compartments during the experiments.

2.4.4. Analytical Methods

Nitrate and nitrite were analysed based on the colorimetric cadmium reduction method, according with the standard methods [34], using a Skalar SAN++ CFA analyser (Skalar Analytical B.V., Breda, The Netherlands). The TOC was measured using a TOC-V CSH total organic carbon analyser (Shimadzu Corp., Kyoto, Japan), then the corresponding concentration of ethanol was calculated by a conversion factor. The conductivity of the solutions was measured using a Sension +EC7 instruments (from Hach Company, Loveland, Colorado, United States).

2.5. 2D Fluorescence Spectroscopy

2D fluorescence spectroscopy has been recently proposed for in situ monitoring bioreactors and ion-exchange membrane processes [35]. In this work, 2D fluorescence spectroscopy was used to analyse alterations in the membrane surface caused by (i) their use in the bioreactor experiments (as it is described in Section 2.4.3), (ii) an accelerated organic fouling test (on pristine membranes), and (iii) chemical cleaning (of pristine membranes and those used in IEMB).

The accelerated fouling test consisted of the passive immersion of the membranes in a 100 mg L^{-1} aqueous Humic Acid (HA) solution for 1 h, at room temperature [36].

The cleaning treatment involved the passive immersion of the membranes in alkali and acid solutions, and consisted of the following steps, according to [32]:

1. Rinse with DI water
2. Passive immersion in 5000 mg L^{-1} NaOH aqueous solution for 10 min, at room temperature.
3. Rinse with DI water
4. Passive immersion in a 20,000 mg L^{-1} citric acid aqueous solution for 30 min, at room temperature.
5. Rinse with DI water

A Varian Cary Eclipse fluorescence spectrophotometer, equipped with excitation and emission monochromators, and coupled to an optical fibre bundle probe was used. The optical fibre bundle is constituted by 294 optical fibres, each with a diameter of 200 μm and a length of 2 m. The Excitation–Emission Matrices (EEMs) of the membrane surface samples were obtained between 250 and 690 nm excitation and from 260 to 700 nm emission with 5 nm step and slits of 10 and 5 nm, respectively. EEMs of both coated (top) and PET (bottom) surfaces of all membranes were acquired when pristine, after used in IEMB, after accelerated fouling test and after chemical cleaning of pristine and used in IEMB. All measurements were performed in triplicate.

Before performing a principal component analysis (PCA) of obtained EEMs, spectral data were standardized (by subtracting averages and dividing by standard deviations) for each pair of excitation emission intensity. OCTAVE GNU software was used to implement the computational routines in PCA (through PARAFAC algorithm [37]). PCA was performed by using the following two different data combinations: (1) only EEMs of the prepared membranes (Pur-RE, Amb-RE, and Lew-RE, in a total of 91 spectra); (2) EEMs of the prepared membranes and commercial membrane together (Pur-RE, Amb-RE, Lew-RE, and AMH-PES in a total of 120 spectra). PARAFAC analysis of only modified membranes captured 81.76% of the variance by the first two principal components (PCs), while PARAFAC analysis of all membranes (including commercial) captured 83.84% variance by the first two PCs.

3. Results

3.1. Confocal Micro-Raman Spectroscopy

Confocal μ-Raman spectroscopy was used to understand better the polymeric composition/structure of the cross-section membrane Pur-RE (the best performing membrane under DD and IEMB operating conditions, as discussed below). Figure 3a represents the examined locations (from a to j) along the membrane cross-section, and Figure 3b shows the Raman spectra corresponding to these locations. The main characteristic Raman bands identifying the polymers in the membrane, as well as the Raman spectra of those polymers, were characterised according to the literature [38] and are presented as Supplementary material in Table S1 and Figure S1, respectively.

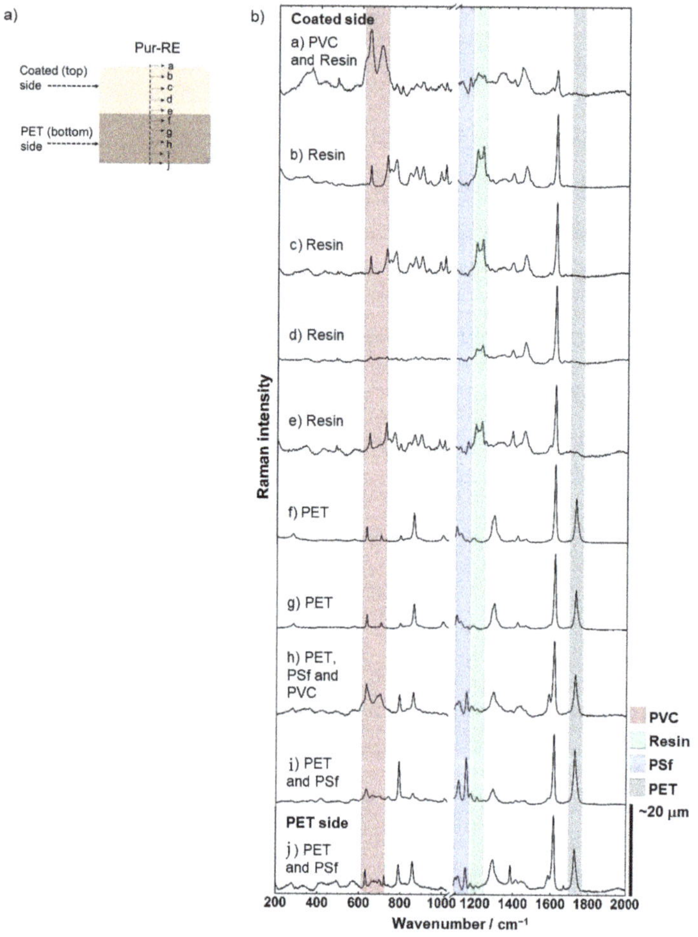

Figure 3. (**a**) Schematic representation of the analysed points in the membrane section. (**b**) μ-Raman spectra of the polymers in the membrane section, at the analysed points a to j. The main bands that allow for a clear identification of each polymer are highlighted.

At the outermost coated surface (top) of the membrane (point a), PVC is identified by its characteristic Raman bands at 638 and 695 cm^{-1} due to C-Cl stretching vibrations (highlighted with red in Figure 3b). The PVC was used as polymer binder in the coating solution, and thus, its presence in point a demonstrated an effective formation of a film on the coated membrane surface. At points a to e, the spectrum of the ion-exchange resin was found, particularly due to its characteristic Raman bands at 1190 and 1221 cm^{-1} attributed to ring C-H vibration modes (highlighted with green in Figure 3b), which verifies the successful deposition of the ion-exchange resin in the coated (top) surface of the membrane and, therefore, the presence of fixed charged functional groups. The lower intensity of the resin signal at the point a corresponds to an overshadowed effect by the PVC signal, whereas at points f to j (i.e., the PET bottom part of the membrane), the ion-exchange resin spectrum is no longer present. At these points (f to j), the spectra corresponding to PET and PSf, in different combinations, can be distinguished by their characteristic Raman bands at 1727 cm^{-1} for PET C =O stretching vibration (highlighted with grey in Figure 3b), and at 1150 and 1109 cm^{-1} for PSf C-O-C and SO_2 stretching vibrations, respectively (highlighted with blue in Figure 3b). In the recycled membrane support, the PSf layer conforms the outermost top surface of the membrane covering the PET layer (see Figure 1b), whereas in Pur-RE, the signal corresponding to PSf was detected deeply inside the PET layer (points h to j). These results indicate that the employed organic solvent in the coating mixture (i.e., THF) has dissolved the PSf layer of the recycled membrane support, causing the percolation of dissolved PSf inside the PET layer, while the solvation of the PSf layer promotes the penetration of the polymer binder (i.e., PVC) into the recycled membrane support, as it is demonstrated by the presence of the PVC signal at point h. As a result, the adhesion of the coated layer to the recycled membrane support is strengthened, providing a high mechanical stability to the membrane [4]. In contrast, the penetration of the ion-exchange resin in the membrane support is probably limited by the size of the ion-exchange resin particles. Overall, the absence of charged functional groups in combination with the presence of non-conductive polymers in this region of the membrane (points h to j) contributes to the high electrical resistance, low IEC, and low WC of the membrane (see Table 1) [6]. In conclusion, these results confirm successful upcycling of RO membranes into AEMs and the asymmetric distribution of the fixed charged functional groups along the thickness of the PU-RE membrane. In addition, since the only difference between the prepared membranes was the type of ion-exchange resin used during the preparation, analogous results are expected for Amb-RE and Lew-RE.

3.2. Donnan Dialysis (DD) and Ion-Exchange Membrane Bioreactor (IEMB) Experiments

3.2.1. Time Course Concentration of Nitrate in the Feed and in the Receiver Compartments

Figure 4 shows the time course concentration of nitrate in the feed and the receiver compartments and nitrate removal rates from the feed compartment during DD and IEMB experiments.

Figure 4a,b shows considerable differences in the performance of the membranes for reducing the concentration of the target ion to the maximum and the recommended levels (50 and 25 mg L^{-1}, respectively). Such differences are correlated to the membrane properties (reported in Table 1). Accordingly, the AMH-PES membrane clearly shows the best performance, achieving 96% and 98% removal within 24 h, under DD and IEMB operation conditions, respectively (Figure 4e). This membrane has a relatively uniform distribution of the ion-exchange resin particles across the membrane section (reported before, in [4]), together with the highest IEC and WC (2.19 $\pm$ 0.09 mmol g^{-1} and 50 $\pm$ 0%, respectively), a high permselectivity (81 $\pm$ 1%) and a relatively low electrical resistance (19 $\pm$ 3 Ω cm^2), in comparison with the prepared membranes.

In contrast, the asymmetric distribution of the ion-exchange functional groups in the prepared membranes (confirmed by μ-Raman spectroscopy, Section 3.1.) might reduce the counter-ion transport rate. Further, the relatively low ion-exchange capacity (IEC) of such membranes results in a lower number of fixed charged functional groups, meaning that a lower number of ion transport sites are available [39,40]. In addition, the low water content of the prepared membranes leads to a lower electrical conductivity [41]. As a result of the combination of the aforementioned properties, the prepared membranes have a high electrical resistance (see Table 1), as explained in detail in our previous works [4,6]. A high electrical resistance leads, under the same experimental conditions, to a slower effective transport of the ions (i.e., migration). Although, among the prepared membranes, Pur-RE shows the best performance, reaching a 57% and 56% removal within 24 h, under DD and IEMB operation conditions, respectively (Figure 4e). These results might be related to the higher IEC and the lower electrical resistance of the Pur-RE membrane (0.75 mmol g^{-1} and 56 Ω cm^2, respectively), in comparison with the IEC and the electrical resistance of Amb-RE (0.62 mmol g^{-1} and 129 Ω cm^2, respectively) and Lew-RE membranes (0.39 mmol g^{-1} and 120 Ω cm^2, respectively) [20].

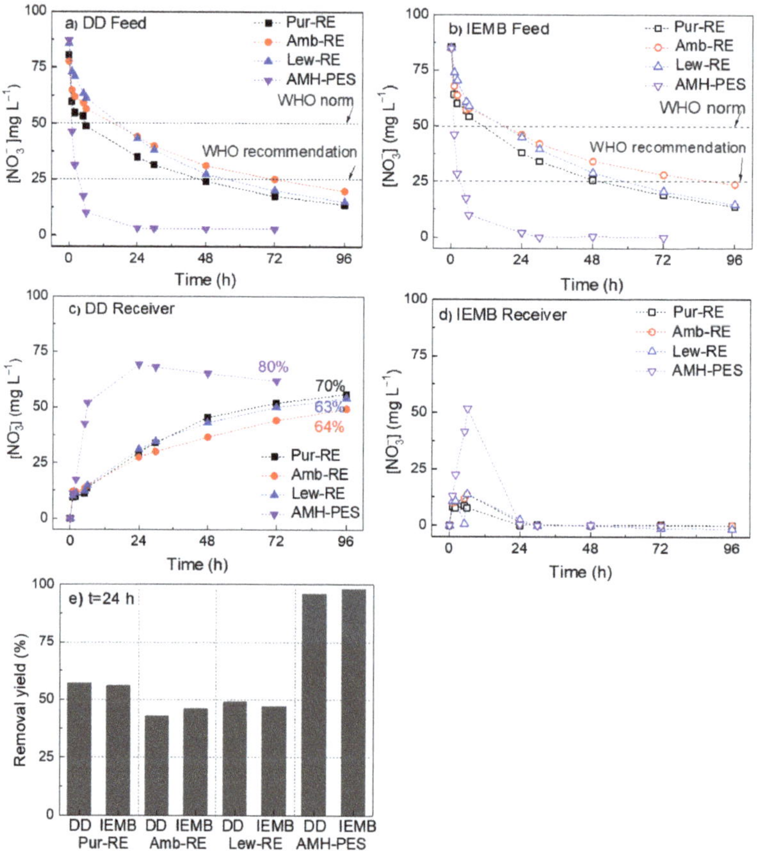

Figure 4. Time course concentration of nitrate ([NO_3^-]) in the feed compartment in (**a**) DD experiments and (**b**) IEMB experiments. Time course concentration of nitrate in the receiver compartment in (**c**) DD experiments (the numbers in % indicate the recovery yield of nitrate at the end of the experiment) and (**d**) IEMB experiments (where nitrate conversion to N_2 occurs). (**e**) Nitrate removal yield (%) from the feed within 24 h operation as DD and IEMB. Membranes: Pur-RE, Amb-RE, Lew-RE, AMH-PES.

Figure 4c shows the accumulation of nitrate in the receiver compartment during DD experiments. Among the prepared membranes, Pur-RE membrane shows again the best performance for the recovery of nitrate under DD operation (recovery of 70%), although the performance of AMH-PES is larger (recovery of 80%). Differently, in Figure 4d, corresponding to the IEMB operation, it can be observed that the integration of a bioreactor results in the successful elimination of nitrate from the receiver compartment by denitrifying bacteria in a single device and a single step. Moreover, the time course concentration analyses of nitrite (NO_2^-) showed that its accumulation was avoided in both compartments (results reported as Supplementary Material in Figure S2). The latter entails several advantages such as avoiding the discharge of nitrite into the environment, while enabling a prolonged use of the receiver solution. In addition, the concentration gradient of nitrate is effectively maintained during the process, which under a continuous operation would result in favoured nitrate transport in respect to that of other accompanying anions which are not bioreduced (not evident in the present case due to batch IEMB operating conditions and the use of a single anion (NO_3^-) in the feed) [42].

3.2.2. Co-Ion Leakage to the Feed Compartment

As membranes are not 100% permselective, co-ion leakage from the receiver to the feed compartment is expected. The co-ion leakage reduces the electrochemical potential difference established across the membrane, decreasing the back-transport of the target counter-ion to the receiver compartment. Moreover, it increases the conductivity in the feed compartment. In this respect, the WHO establishes a maximum conductivity of 1 mS cm^{-1} for water intended for drinking purposes [13]. Figure 5 shows the conductivity increase in the feed compartment after 72 h of experiment.

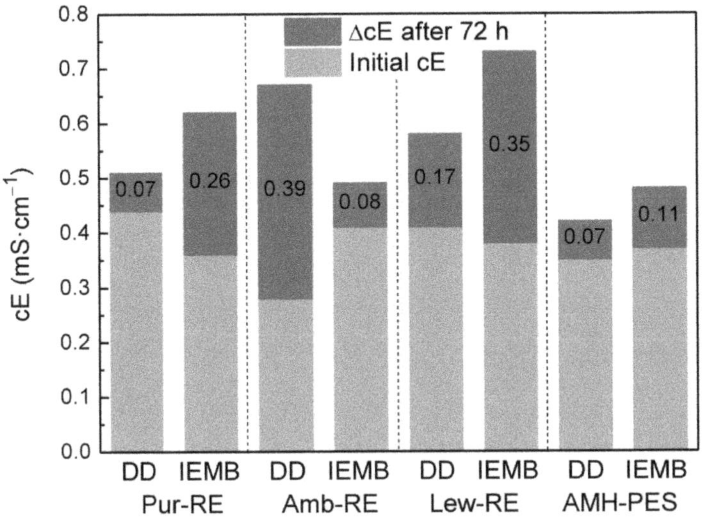

Figure 5. Increase of feed solution conductivity after 72 h of operation as a Donnan Dialyzer and as an IEMB.

As it can be observed in Figure 5, some co-ion leakage occurred in all experiments, resulting in a slight conductivity increase of the feed solution (e.g., a 0.07 and a 0.26 mS cm^{-1} increase using Pur-RE under DD and IEMB operating conditions, and a 0.07 and a 0.11 mS cm^{-1} increase using AMH-PES under the same conditions). However, the final conductivity in the feed compartment was maintained in all the cases below 1 mS cm^{-1}. Even if the permselectivity of Pur-RE (65%) is not as high as that of AMH-PES (81%), these small increases in conductivity demonstrated that co-ion leakage is not a problem in this process [20]. Indeed, the Pur-RE membrane achieved a better performance than the Amb-RE membrane (with a permselectivity of 83%), indicating that, in this case, other membrane properties may have a greater influence in the process performance, as it was discussed in the previous subsection.

3.2.3. Diffusion Permeability of the Carbon Source

The migration of the carbon source from the receiver to the feed compartment during IEMB operation might generate secondary pollution problems in the treated water and lead to the undesired growth of bacteria. Thus, for operation as a bioreactor, the membranes must assure a low permeability towards the employed carbon source, ethanol (selected as a non-charged electron donor) in this case. The TOC measurements performed during IEMB experiments showed that the migration of the carbon source and other metabolic by-products to the treated water was adequately avoided (i.e., TOC increase was not detected). To confirm this, additional tests on ethanol permeation were performed, and the results are shown in Figure 6.

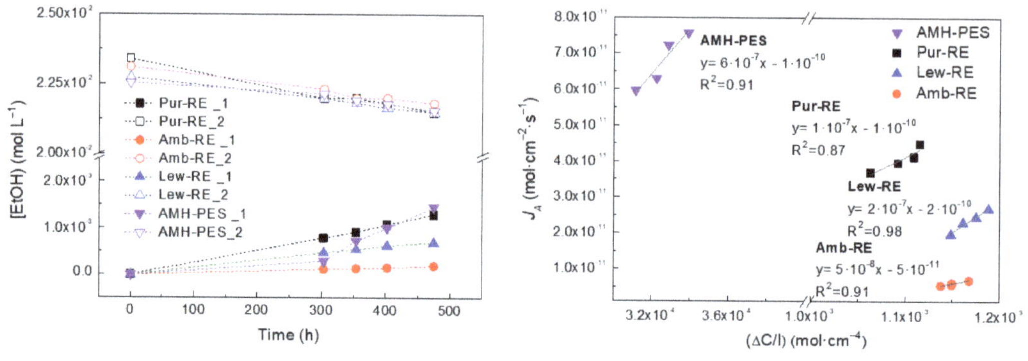

(**a**) (**b**)

Figure 6. Studies on ethanol permeation across the membranes. (**a**) Time course concentration change in the water compartment (1) and the ethanol compartment (2), (**b**) representation used for calculation of ethanol diffusion coefficients across the membranes and linear equations for each membrane.

From Figure 6, the diffusion coefficients of ethanol through the membranes are determined as $5{\cdot}10^{-8}$ cm^2 s^{-1} Amb-RE < $1{\cdot}10^{-7}$ cm^2 s^{-1} Pur-RE < $2{\cdot}10^{-7}$ cm^2 s^{-1} Lew-RE < $6{\cdot}10^{-7}$ cm^2 s^{-1} AMH-PES.

The advantageous lower diffusion coefficient of ethanol in Pur-RE, Amb-RE, and Lew-RE in respect to AMH-PES could be a consequence of the lower water content of the prepared membranes [6,43].

Overall, this work investigated for the first time the proof-of-concept validation of RO upcycled AEMs in DD and IEMB processes. Among the prepared membranes, Pur-RE shows the best performance in DD and IEMB operation as it combines the highest transmembrane flux of nitrate together with the best recovery yield in the receiver compartment (under DD operation), an adequately low co-ion leakage, and an undetectable transport of the carbon source and microbial metabolic by-products to the treated water compartment (under IEMB operation). Still, the lower transmembrane ion flux of Pur-RE in comparison with the reference commercial membrane could entail a limitation for its real application, and in this sense, future research should be dedicated to upgrading the performance of the membrane under DD and IEMB operating conditions. For that purpose, different strategies might be followed, either focused on improving membrane properties (e.g., by an activation treatment [5], using conductive nanomaterials as additives in the formulation [44,45], or increasing the selectivity towards the target counter ion [46]) or optimizing process parameters (e.g., increasing the relation between the membrane area per volume to be treated by using a plate and frame configuration [30], increasing the concentration of the stripping ion [47], or further avoiding co-ion leakage by choosing a divalent co-ion such as $CaCl_2$ [8]).

3.3. 2D Fluorescence Spectroscopy

In this work, 2D fluorescence spectroscopy has been employed to gain a better understanding about the stability of the membranes against fouling and chemical cleaning. Such properties might be related to the lifetime of the membranes in a real application. In addition, understanding the fluorescence response of the membranes could help the implementation of an in situ monitorization of the bioreactors [35].

Due to the complexity of the EEMs obtained from the 2D fluorescence spectroscopic analysis, a Principal Components Analysis (PCA) was conducted to reduce the dimensionality of the data, while capturing the principal interrelations of the original data [37]. The PCA performed using all the data (i.e., Pur-RE, Amb-RE, Lew-RE, and AMH-PES) is displayed as Supplementary Material in Figure S3. Due to significant differences among the PCs of the reference (AMH-PES) and the prepared membranes, another PCA was performed including only the spectral data of the prepared membranes (i.e., Pur-RE, Amb-RE, and Lew-RE, excluding AMH-PES), aiming at a more precise analysis of the relation between the PCs of the prepared membranes. Results corresponding to the PCA analysis of the coated side of prepared membranes are displayed in Figure 7, while results corresponding to the PET side of the membranes are displayed as Supplementary Material in Figure S4.

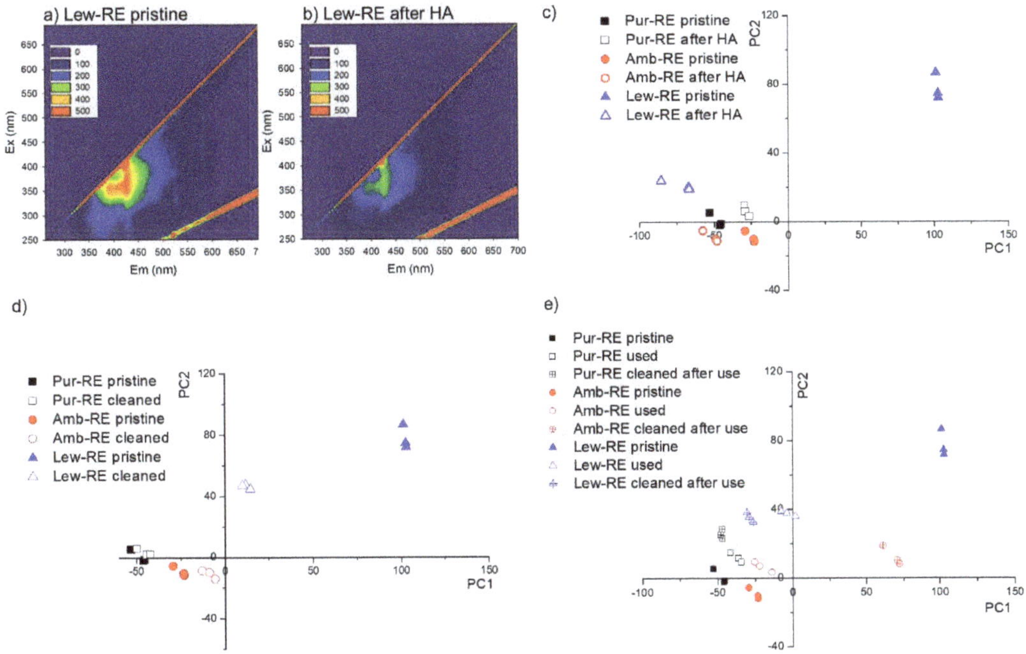

Figure 7. (**a**,**b**) Fluorescence spectra obtained from the coated (top) side of Lew-RE in pristine state (**a**) and after the accelerated fouling test (**b**); (**c**) scores for PC1 and PC2 obtained from the EEMs obtained from the coated (top) side of pristine membranes and membranes after the accelerated fouling test; (**d**) scores for PC1 and PC2 obtained from the EEMs obtained from the coated (top) side of pristine membranes and membranes after the chemical cleaning; (**e**) scores for PC1 and PC2 obtained from the EEMs of pristine membranes, used membranes (after the IEMB experiments), and cleaned membranes after their use in a first batch of experiments.

Figure 7a,b shows the EEMs of Lew-RE membrane before and after the accelerated fouling test. It can be observed that the pristine Lew-RE membrane displays a high fluorescence intensity at the wavelength (λ) region $\lambda_{excitation}$ = 320–420 nm/$\lambda_{emission}$ = 350–450 nm (Figure 7a) while, after the accelerated fouling test, the excitation emission intensity is reduced (Figure 7b). This reduction in the emission intensity might be attributed to the deposition of humic acids on the membrane surface, partially absorbing the light emitted by the membrane and decreasing the emission intensity [32]. In respect to pristine membranes, it can be observed by the PCA that the PCs of the Lew-RE membrane differs from the PCs of Pur-RE and Amb-RE membranes, indicating differences between their EEMs. These results could be related to the different ion-exchange resins used in the preparation of the membranes, as they differ in the substituents at the ion-exchange sites (see Table 1), having different hydrophilic properties [6]. In contrast, the PET sides of the membranes were found to be rather similar to each other (Figure S4a, Supplementary Materials), as the ion-exchange resin is barely present at the PET side (confirmed by μ-Ramn analyses).

In respect to the development of a fouling layer, the PCA analysis (Figure 7c) shows a considerable distance between the PCs of the pristine and fouled Lew-RE membrane, indicating this membrane is the one that suffers higher alterations due to organic fouling.

Regarding the effect of the chemical cleaning on pristine membranes (Figure 7d), it can be observed that the PC scores of the pristine and cleaned Lew-RE membrane are the most distant from each other, indicating that Lew-RE membrane seems to be significantly affected by the employed chemical cleaning. In contrast, the Pur-RE membrane shows a negligible alteration of its fluorescence spectra.

It can be observed in Figure 7e that, on the one hand, the fluorescence response of the Lew-RE membrane becomes significantly altered after its use in the IEMB. In contrast, Pur-RE and Amb-RE membranes show a small alteration after their use. It should be noticed that in these experiments the biocompartment was faced to the PET surfaces of the membranes, and thus, the coated side should not be affected by biofouling. Therefore, the alteration shown by the Lew-RE membrane could be related to the adsorption of nitrate at the membrane surface, facilitated by the higher affinity of this membrane towards nitrate ions [6], as it was demonstrated before that nitrate can induce a quenching effect on the fluorescence spectra [48]. On the other hand, it can be observed in Figure 7e that the "pristine state" of the used membranes was not fully recovered after chemical cleaning. Indeed, the PCs are even more distant from the original (pristine state) after chemical cleaning. In the case of the Amb-RE membrane, the alteration induced by chemical cleaning on the used membrane seems to be more significant than in the case of the pristine membrane (results shown in Figure 7d). Again, the Pur-RE membrane is the one that suffers a less significant alteration under the tested conditions. In contrast, the PET side of Lew-RE membrane shows the lowest alteration under these conditions (information detailed in Figure S4c, Supplementary Materials).

In conclusion, it can be stated that the variations in the excitation emission spectra are well captured by the first two PCs in the performed PCA analysis. Concerning the resistance to alterations caused by organic fouling and the exposure to acid and alkali solutions, the Lew-RE membrane shows to be the least resistant, while the Pur-RE membrane appears to be the most stable one, under the tested conditions.

As a complementary study, an Attenuated total reflectance–Fourier transform infrared spectroscopy (ATR-FTIR) analysis of the membranes in pristine state and after the accelerated fouling test and the chemical cleaning of pristine membranes was conducted (see Section S4 of the Supplementary Materials). Interestingly, differences in the functional groups of the membranes (after the accelerated fouling test and the chemical cleaning) were not detected, indicating that the mentioned conditions were not harsh enough to induce chemical alterations perceived by this technique. Accordingly, it is worthy to remark that, through the fluorescence response studies performed, it was possible to notice differences in the chemical composition of the membranes' surfaces, after the accelerated fouling test and the chemical cleaning, indicating therefore a high sensitivity of 2D fluorescence spectroscopy to detect subtle alterations at the membrane surfaces.

3.4. Preliminary Membrane-Associated Costs and Environmental Implications

The aim of this section is to provide a preliminary estimation, for the first time, of the approximated material cost of Pur-RE membrane. In addition, the economic affordability of Pur-RE membrane and reference commercial IEM (AMH-PES) were compared, taking into consideration the differences in ion transport performances between the membranes. In light of the preliminary nature of this proof-of-concept study, CAPEX and OPEX considerations in relation to the industrial application of DD and IEMB systems are out of the scope of this work. The cost of recycling an end-of-life brackish water RO membrane into UF-like properties, including capital expenditure (CAPEX) and the operational expenditure costs (OPEX), has been estimated in a previous study [49], as EUR 31.7 m^{-2}

per module. Considering an active membrane area of 37 m^2 per module, the cost of the recycled membrane support could approach EUR 0.9 m^{-2}. In addition, the cost of the coating layer, regarding the current price in market of the employed reagents (displayed in Table S2 as Supplementary Material) and excluding CAPEX and OPEX considerations, could approach approximately EUR 2.4 m^{-2}. Under these considerations, the costs of the materials required to prepare the Pur-RE membrane could approach EUR 3.3 m^{-2}. It is worth noting that this preliminary estimation does not consider the costs associated with module opening and repacking, which should be evaluated in future studies. For comparison, the selling price of AMH-PES Ralex ® membrane is between EUR 70 and 100 m^{-2} [50]. Since transmembrane NO_3^- flux through Pur-RE membrane was lower than in the case of the reference commercial membrane, for the sake of a roughly preliminary comparison, Table 2 shows the equivalent surface of Pur-RE membrane to reach the same removal rate of AMH-PES membrane. Two different scenarios have been considered, the first one calculating the equivalent membrane surface of Pur-RE considering the transmembrane ion flux to reach nitrate concentration below the maximum allowed (50 ppm), and a second one considering the transmembrane ion flux to reach a level below the recommended nitrate concentration (25 ppm). The transmembrane fluxes were calculated based on the data reported in Section 3.2.1.

Table 2. Approximated cost of Pur-RE considering the differences between the target ion transmembrane flux $\left(J_{NO_3^-}\right)$ in respect to the reference membrane (AMH-PES). $J_{NO_3^-\,(C_0-C_{50})}$, nitrate transmembrane flux from the initial conditions to reach a concentration below the maximum allowed level (50 mg L^{-1}), $J_{NO_3^-\,(C_0-C_{25})}$, nitrate transmembrane flux from the initial conditions to reach a concentration below the maximum recommended level (25 mg L^{-1}).

Transmembrane NO_3^- Flux ($J_{NO_3^-}$)	Pur-RE	AMH-PES	Pur-RE Membrane Surface (m^2) Equivalent to Performance of 1 m^2 AMH-PES *	Pur-RE Membrane Cost (EUR) Equivalent to Performance of 1 m^2 AMH-PES **
$J_{NO_3^-\,(C_0-C_{50})}$ (g m^{-2} h^{-1}) in DD	0.82	6.34	7.73	25.2
$J_{NO_3^-\,(C_0-C_{25})}$ (g m^{-2} h^{-1}) in DD	0.18	2.70	14.81	48.3
$J_{NO_3^-\,(C_0-C_{50})}$ (g m^{-2} h^{-1}) in the IEMB	0.81	6.00	7.39	24.1
$J_{NO_3^-\,(C_0-C_{25})}$ (g m^{-2} h^{-1}) in the IEMB	0.19	2.61	13.50	44.0

* Considering the data in Section 3.2.1. ** Considering the estimated material cost for Pur-RE as 3.3 EUR m^{-2}.

Certainly, a deeper economic assessment should be conducted (e.g., the cost of module opening and repacking should be evaluated, and CAPEX and OPEX should be considered), as well as an improvement in the production process (e.g., reducing the costs associated to the coating process, while improving the ion transport performance of the resulting membranes), but even in that case, it seems evident a significant economic benefit if this concept is brought to market. In addition, the European green policies for a sustainable transition (i.e., the European Green Deal [51] and the Circular Economy Action Plan [52]) advocate giving rise to taxation on landfill disposal, encouraging the extension of the service time of goods and products and making recycling alternatives more attractive from an economic point of view.

From an environmental point of view, the recycling membrane concept presented in this study is an alternative to conventional ion-exchange membrane production. In this sense, in a previous study [49], savings of more than 175 kg CO_2 equivalent and 0.28 m^3 of freshwater (among other environmental indicators) were estimated trough the Life Cycle Assessment (LCA) of the recycling process of one end-of-life RO module into UF-like properties. Therefore, even if CO_2 equivalent emission, related to the coating process, is not yet accounted for, it is possible to envisage a positive environmental impact of the proposed recycling concept.

Overall, membrane recycling is a promising alternative to waste disposal by RO industry and it is a commitment to circular economy in membrane technology and the water sector. Furthermore, it has been demonstrated that passive mass transport processes such as DD and IEMB could be a promising technological niche for the application of such AEMs.

4. Conclusions

In the present work, the feasibility of using AEMs prepared from upcycled end-of-life RO membranes for the removal of nitrate from water using DD and IEMB systems is demonstrated. The main findings of the study are summarized as follows:

1. µ Raman spectroscopy confirmed the successful deposition of the ion-exchange resin in the coated (top) surface of the membrane. However, the analysis revealed an asymmetric distribution of the ion-exchange sites in the membrane cross-section, which might contribute to the relatively low ion-exchange capacity and water content and the relatively high electrical resistance of the membranes and, therefore, have a great influence on transport properties of the membrane.
2. In respect to nitrate removal under DD and IEMB operation, the membrane incorporating Purolite® A600/9149 ion-exchange resin (Pur-RE membrane) achieved, among the prepared membranes, the best removal yields (57% and 56% of nitrate removal within 24 h under DD and IEMB operation, respectively), probably due to its higher ion-exchange capacity and lower electrical resistance, in comparison with the Amb-RE and the Lew-RE membranes. Furthermore, nitrate was biologically eliminated in the IEMB, favouring the reuse of the receiver solution and avoiding the discharge of the pollutant into the environment.
3. 2D fluorescence spectroscopy can effectively detect alterations in the excitation emission spectra of the membranes, caused by fouling and/or chemical cleaning. In this respect, the Pur-RE membrane was found to be the most stable membrane under the tested conditions.
4. The relative low cost of the employed materials in the formulation of Pur-Re membrane anticipated an economic benefit of the presented membrane recycling concept.

Overall, this work shows the potential applicability of RO upcycled membranes as AEMs in electrochemical potential driven processes (DD and IEMB), for an efficient removal of nitrate from water at a minimum energy requirement. The developed membranes could be obtained at a lower cost than commercial ones and therefore facilitate the practical implementation of these passive transport processes. Lastly, membrane recycling represents a commitment to the transition to a circular economy.

Supplementary Materials: The following supporting information can be downloaded at https://www.mdpi.com/article/10.3390/membranes12020101/s1. Table S1: Main characteristic Raman bands (cm^{-1}), between 200–2000 cm^{-1}, for Polyvinyl chloride (PVC), the ion exchange resin (Purolite® A600/9149), Polysulfone (PSf), and Polyester (PET), identified in the membrane Pur-RE. Figure S1: µ-Raman spectra between 200–2000 cm^{-1} of polyvinyl chloride (PVC), the ion exchange resin (Purolite® A600/9149), polysulfone (PSF), and polyester (PET), identified in the membrane Pur-RE. Figure S2: Time course concentration of nitrite ([NO_2^-]) in IEMB experiments (a) feed compartment and (b) receiver compartment. Membranes: Pur-RE, Amb-RE, Lew-RE, and AMH-PES. Figure S3. Scores for PC1 and PC2 obtained from the PCA of all EEMs of the tested membranes. Side a and side b in AMH-PES refers to the membrane surfaces in contact with the feed and the receiver compartments, respectively. Figure S4: Scores for PC1 and PC2 obtained from the EEMs of the PET surface in pristine membranes and (a) membranes after the accelerated fouling test, (b) membranes after the chemical cleaning, and (c) membranes after their use in the bioreactor (i.e., after the IEMB experiments) and posterior cleaning. Figure S5: ATR-FTIR spectra of studied membranes in pristine state, after the accelerated fouling test, and after the chemical cleaning treatment cleaning, (a) Pur-RE, (b) Amb-RE, (c) Lew-RE, (d) AMH-PES. Table S2: Calculation of the price of consumable materials for the preparation of 1 m^2 of AEM using the methodology described in Section 2.2. of the manuscript. References [4,23,38,53–56] are cited in the Supplementary Materials.

Author Contributions: Conceptualization, A.L.-L., S.P., S.V. and J.G.C.; methodology, A.L.-L., S.P., C.F.G., V.O. and S.V.; validation, A.L.-L.; formal analysis, S.P. and C.F.G.; investigation, A.L.-L., S.M.

and V.O.; resources, J.M.O., S.M. and S.V.; data curation, A.L.-L., S.P. and C.F.G.; writing—original draft preparation, A.L.-L., J.M.O., S.M., S.P., C.F.G., V.O., S.V. and J.G.C.; writing—review and editing, A.L.-L., J.M.O., S.M., S.P., C.F.G., V.O., S.V. and J.G.C.; visualization, A.L.-L., J.M.O., S.M. and S.P.; supervision, J.M.O., S.M., E.G.-C., S.V. and J.G.C.; project administration, S.M.; funding acquisition, S.M., E.G.-C. and J.G.C. All authors have read and agreed to the published version of the manuscript.

Funding: This research was funded by the Spanish Ministry of Science and Innovation (MCI), the State Research Agency (AEI), the European Regional Development Fund (ERDF), grant numbers CTM2015-65348-C2-1-R (MINECO/FEDER, UE), RTI2018-096042-B-C21 (MCI/AEI/FEDER, UE), and CTM2015-74695-JIN (AEI/FEDER/UE), also by MCI, AEI, and the European Social Fund (FSE), grant number BES-2016-076244 (MCI/AEI/FSE, UE). This work was also supported by the Associate Laboratory for Green Chemistry—LAQV which is financed by national Portuguese funds from FCT/MCTES (UIDB/50006/2020).

Conflicts of Interest: The authors declare no conflict of interest.

References

1. Mayor, B. Growth patterns in mature desalination technologies and analogies with the energy field. *Desalination* **2019**, *457*, 75–84. [CrossRef]
2. Greenlee, L.F.; Lawler, D.F.; Freeman, B.D.; Marrot, B.; Moulin, P. Reverse osmosis desalination: Water sources, technology, and today's challenges. *Water Res.* **2009**, *43*, 2317–2348. [CrossRef]
3. Landaburu-Aguirre, J.; García-Pacheco, R.; Molina, S.; Rodríguez-Sáez, L.; Rabadán, J.; García-Calvo, E. Fouling prevention, preparing for re-use and membrane recycling. Towards circular economy in RO desalination. *Desalination* **2016**, *393*, 16–30. [CrossRef]
4. Lejarazu-Larrañaga, A.; Molina, S.; Ortiz, J.M.; Navarro, R.; García-Calvo, E. Circular economy in membrane technology: Using end-of-life reverse osmosis modules for preparation of recycled anion exchange membranes and validation in electrodialysis. *J. Memb. Sci.* **2020**, *593*, 117423. [CrossRef]
5. Lejarazu-Larrañaga, A.; Molina, S.; Ortiz, J.M.; Riccardelli, G.; García-Calvo, E. Influence of acid/base activation treatment in the performance of recycled electromembrane for fresh water production by electrodialysis. *Chemosphere* **2020**, *248*, 126027. [CrossRef] [PubMed]
6. Lejarazu-Larrañaga, A.; Ortiz, J.M.; Molina, S.; Zhao, Y.; García-Calvo, E. Nitrate-selective anion exchange membranes prepared using discarded reverse osmosis membranes as support. *Membranes* **2020**, *10*, 377. [CrossRef]
7. Tanaka, Y. *Ion Exchange Membranes, Fundamentals and Applications*, 1st ed.; Tanaka, Y., Ed.; Elsevier: Amsterdam, The Netherlands, 2007; ISBN 978-0-444-51982-5.
8. Zhao, B.; Zhao, H.; Dockko, S.; Ni, J. Arsenate removal from simulated groundwater with a Donnan dialyzer. *J. Hazard. Mater.* **2012**, *215–216*, 159–165. [CrossRef]
9. Asante-Sackey, D.; Rathilal, S.; Kweinor Tetteh, E.; Ezugbe, E.O.; Pillay, L.V. Donnan membrane process for the selective recovery and removal of target metal ions—A mini review. *Membranes* **2021**, *11*, 358. [CrossRef]
10. Crespo, J.G.; Velizarov, S.; Reis, M.A. Membrane bioreactors for the removal of anionic micropollutants from drinking water. *Curr. Opin. Biotechnol.* **2004**, *15*, 463–468. [CrossRef] [PubMed]
11. Bijay-Singh; Craswell, E. Fertilizers and nitrate pollution of surface and ground water: An increasingly pervasive global problem. *SN Appl. Sci.* **2021**, *3*, 518. [CrossRef]
12. Vo, T.K.Q.; Lee, J.J.; Kang, J.S.; Park, S.; Kim, H.S. Nitrogen removal by sulfur-based carriers in a membrane bioreactor (MBR). *Membranes* **2018**, *8*, 115. [CrossRef]
13. World Health Organization. *Guidelines for Drinking-Water Quality*, 4th ed.; World Health Organization: Geneva, Switzerland, 2017; ISBN 9789241549950.
14. Schoeman, J.J.; Steyn, A. Nitrate removal with reverse osmosis in a rural area in South Africa. *Desalination* **2003**, *155*, 15–26. [CrossRef]
15. Elmidaoui, A.; Elhannouni, F.; Sahli, M.A.M.; Chay, L.; Elabbassi, H. Pollution of nitrate in Moroccan ground water: Removal by electrodialysis. *Desalination* **2001**, *136*, 325–332. [CrossRef]
16. Samatya, S.; Kabay, N.; Yüksel, Ü.; Arda, M.; Yüksel, M. Removal of nitrate from aqueous solution by nitrate selective ion exchange resins. *React. Funct. Polym.* **2006**, *66*, 1206–1214. [CrossRef]
17. Aslan, S.; Cakici, H. Biological denitrification of drinking water in a slow sand filter. *J. Hazard. Mater.* **2007**, *148*, 253–258. [CrossRef]
18. Huno, S.K.M.; Rene, E.R.; Van Hullebusch, E.D.; Annachhatre, A.P. Nitrate removal from groundwater: A review of natural and engineered processes. *J. Water Supply Res. Technol. Aqua* **2018**, *67*, 885–902. [CrossRef]
19. Soares, M.I.M. Biological denitrification of groundwater. *Water Air Soil Pollut.* **2000**, *123*, 183–193. [CrossRef]
20. Velizarov, S.; Crespo, J.G.; Reis, M.A. Ion exchange membrane bioreactor for selective removal of nitrate from drinking water: Control of ion fluxes and process performance. *Biotechnol. Prog.* **2002**, *18*, 296–302. [CrossRef] [PubMed]

21. Fonseca, A.D.; Crespo, J.G.; Almeida, J.S.; Reis, M.A. Drinking water denitrification using a novel ion-exchange membrane bioreactor. *Environ. Sci. Technol.* **2000**, *34*, 1557–1562. [CrossRef]
22. Matos, C.T.; Fortunato, R.; Velizarov, S.; Reis, M.A.M.; Crespo, G.Ã. Removal of mono-valent oxyanions from water in an ion exchange membrane bioreactor: Influence of membrane permselectivity. *Water Res.* **2008**, *42*, 1785–1795. [CrossRef] [PubMed]
23. Rodríguez-Sáez, L.; Landaburu-Aguirre, J.; Molina, S.; García-Payo, M.C.; García-Calvo, E. Study of surface modification of recycled ultrafiltration membranes using statistical design of experiments. *Surf. Interfaces* **2021**, *23*, 100978. [CrossRef]
24. Campos Pozuelo, E.; Terrero Rodríguez, P.; Zarzo Martínez, D.; Molina Serrano, F.J.; Calzada Garzón, M.; García Pacheco, R.; Molina Martínez, S.; Rodríguez Sáez, L.; Rabadán, F.J.; Landaburu Aguirre, J.; et al. Transformation of Spiral Wound Polyamide Membranes after Its Industrial Lifespan. Spanish Patent PCT/EP2016/30931, 8 July 2016.
25. García-Pacheco, R.; Landaburu-Aguirre, J.; Terrero-Rodríguez, P.; Campos, E.; Molina-Serrano, F.; Rabadán, J.; Zarzo, D.; García-Calvo, E. Validation of recycled membranes for treating brackish water at pilot scale. *Desalination* **2018**, *433*, 199–208. [CrossRef]
26. García-Pacheco, R.; Landaburu-Aguirre, J.; Molina, S.; Rodríguez-Sáez, L.; Teli, S.B.; García-Calvo, E. Transformation of end-of-life RO membranes into NF and UF membranes: Evaluation of membrane performance. *J. Memb. Sci.* **2015**, *495*, 305–315. [CrossRef]
27. Molina, S.; Landaburu-Aguirre, J.; Rodríguez-Sáez, L.; García-Pacheco, R.; de la Campa, J.G.; García-Calvo, E. Effect of sodium hypochlorite exposure on polysulfone recycled UF membranes and their surface characterization. *Polym. Degrad. Stab.* **2018**, *150*, 46–56. [CrossRef]
28. García-Pacheco, R.; Landaburu-Aguirre, J.; Lejarazu-Larrañaga, A.; Rodríguez-Sáez, L.; Molina, S.; Ransome, T.; García-Calvo, E. Free chlorine exposure dose (ppm·h) and its impact on RO membranes ageing and recycling potential. *Desalination* **2019**, *457*, 133–143. [CrossRef]
29. Mega Group Companies Product Data Sheet. Ralex®Membrane AMHPES. Available online: https://www.mega.cz/files/datasheet/MEGA-RALEX-AMH-PES-en.pdf (accessed on 3 June 2021).
30. Ricardo, A.R.; Velizarov, S.; Crespo, J.G.; Reis, M.A.M. Validation of the ion-exchange membrane bioreactor concept in a plate-and-frame module configuration. *Process. Biochem.* **2012**, *47*, 1832–1838. [CrossRef]
31. Matos, C.T.; Velizarov, S.; Crespo, J.G.; Reis, M.A.M. Simultaneous removal of perchlorate and nitrate from drinking water using the ion exchange membrane bioreactor concept. *Water Res.* **2006**, *40*, 231–240. [CrossRef]
32. Pawlowski, S.; Galinha, C.F.; Crespo, J.G.; Velizarov, S. 2D fluorescence spectroscopy for monitoring ion-exchange membrane based technologies—Reverse electrodialysis (RED). *Water Res.* **2016**, *88*, 184–198. [CrossRef]
33. Velizarov, S. Transport of arsenate through anion-exchange membranes in Donnan dialysis. *J. Memb. Sci.* **2013**, *425–426*, 243–250. [CrossRef]
34. American Public Health Association; American Water Works Association; Water Environment Federation. *APHA Method 4500-NO3: Standard Methods for the Examination of Water and Wastewater*, 18th ed.; American Public Health Association: Washington DC, USA, 1992; Volume 552, ISBN 0-87553-207-1.
35. Pawlowski, S.; Galinha, C.F.; Crespo, J.G.; Velizarov, S. Prediction of reverse electrodialysis performance by inclusion of 2D fluorescence spectroscopy data into multivariate statistical models. *Sep. Purif. Technol.* **2015**, *150*, 159–169. [CrossRef]
36. Golubenko, D.; Yaroslavtsev, A. Development of surface-sulfonated graft anion-exchange membranes with monovalent ion selectivity and antifouling properties for electromembrane processes. *J. Memb. Sci.* **2020**, *612*, 118408. [CrossRef]
37. Bro, R. PARAFAC. Tutorial and applications. *Chemom. Intell. Lab. Syst.* **1997**, *38*, 149–171. [CrossRef]
38. Hummel, D.O. *Atlas of Plastics Additives. Analysis by Spectrometric Methods*; Springer: Berlin, Germany, 2002.
39. Berezina, N.P.; Kononenko, N.A.; Dyomina, O.A.; Gnusin, N.P. Characterization of ion-exchange membrane materials: Properties vs structure. *Adv. Colloid Interface Sci.* **2008**, *139*, 3–28. [CrossRef]
40. Pyrzynska, K. Preconcentration and recovery of metal ions by Donnan dialysis. *Microchim. Acta* **2006**, *153*, 117–126. [CrossRef]
41. Stenina, I.A.; Yaroslavtsev, A.B. Ionic mobility in ion-exchange membranes. *Membranes* **2021**, *11*, 198. [CrossRef]
42. Velizarov, S.; Rodrigues, C.M.; Reis, M.A.; Crespo, J.G. Mechanism of charged pollutants removal in an ion exchange membrane bioreactor: Drinking water denitrification. *Biotechnol. Bioeng.* **2000**, *71*, 245–254. [CrossRef]
43. Yurova, P.A.; Karavanova, Y.A.; Gorbunova, Y.G.; Yaroslavtsev, A.B. Transport properties of asymmetric ion-exchange membranes based on MC-40, MF-4SC, and polyaniline. *Pet. Chem.* **2014**, *54*, 551–555. [CrossRef]
44. Hosseini, S.M.; Nemati, M.; Jeddi, F.; Salehi, E.; Khodabakhshi, A.R.; Madaeni, S.S. Fabrication of mixed matrix heterogeneous cation exchange membrane modified by titanium dioxide nanoparticles: Mono/bivalent ionic transport property in desalination. *Desalination* **2015**, *359*, 167–175. [CrossRef]
45. Hosseini, S.M.; Jeddi, F.; Nemati, M.; Madaeni, S.S.; Moghadassi, A.R. Electrodialysis heterogeneous anion exchange membrane modified by PANI/MWCNT composite nanoparticles: Preparation, characterization and ionic transport property in desalination. *Desalination* **2014**, *341*, 107–114. [CrossRef]
46. Mubita, T.; Porada, S.; Aerts, P.; Wal, A. Van Der Heterogeneous anion exchange membranes with nitrate selectivity and low electrical resistance. *J. Memb. Sci.* **2020**, *607*, 118000. [CrossRef]
47. Breytus, A.; Hasson, D.; Semiat, R.; Shemer, H. Ion exchange membrane adsorption in Donnan dialysis. *Sep. Purif. Technol.* **2019**, *226*, 252–258. [CrossRef]
48. Galinha, C.F.; Carvalho, G.; Portugal, C.A.M.; Guglielmi, G.; Reis, M.A.M.; Crespo, J.G. Two-dimensional fluorescence as a fingerprinting tool for monitoring wastewater treatment systems. *J. Chem. Technol. Biotechnol.* **2011**, *86*, 985–992. [CrossRef]

49. Senán-Salinas, J.; García-Pacheco, R.; Landaburu-Aguirre, J.; García-Calvo, E. Recycling of end-of-life reverse osmosis membranes: Comparative LCA and cost-effectiveness analysis at pilot scale. *Resour. Conserv. Recycl.* **2019**, *150*, 104423. [CrossRef]
50. Melnikov, S.; Loza, S.; Sharafan, M.; Zabolotskiy, V. Electrodialysis treatment of secondary steam condensate obtained during production of ammonium nitrate. Technical and economic analysis. *Sep. Purif. Technol.* **2016**, *157*, 179–191. [CrossRef]
51. Fetting, C. *The European Green Deal*; European Commission: Brussels, Belgium, 2019.
52. European Commission. *Circular Economy Action Plan. For a cleaner and more competitive Europe*; European Commission: Brussels, Belgium, 2020. [CrossRef]
53. Galinha, C.F.; Carvalho, G.; Portugal, C.A.M.; Guglielmi, G.; Oliveira, R.; Crespo, J.G.; Reis, M.A.M. Real-time monitoring of membrane bioreactors with 2D-fluorescence data and statistically based models. *Water Sci. Technol.* **2011**, *63*, 1381–1388. [CrossRef]
54. Zaman, M.; Liu, H.; Xiao, H.; Chibante, F.; Ni, Y. Hydrophilic modification of polyester fabric by applying nanocrystalline cellulose containing surface finish. *Carbohydr. Polym.* **2013**, *91*, 560–567. [CrossRef] [PubMed]
55. Tetrahydrofuran Market Price & Analysis. Available online: https://www.echemi.com/productsInformation/pd20150901266-tetrahydrofuran.html (accessed on 7 September 2021).
56. Statista Price of Polyvinyl Chloride Worldwide from 2017 to 2020 with Estimated Figures for 2021 to 2022. Available online: https://www.statista.com/statistics/1171131/price-polyvinyl-chloride-forecast-globally/#statisticContainer (accessed on 6 September 2021).

membranes

MDPI

Article

Assessment of the Potential of Using Nanofiltration Polymeric and Ceramic Membranes to Treat Refinery Spent Caustic Effluents

Ana Isabel Rita [1,2,3], Ana Rita Nabais [3], Luisa A. Neves [3], Rosa Huertas [3,4], Maria Santos [1], Luis M. Madeira [2] and Sandra Sanches [4,*]

1 Sines Refinery, Petrogal S.A., 7520-952 Sines, Portugal; aibr1910@gmail.com (A.I.R.); maria.santos@galp.com (M.S.)
2 LEPABE-Laboratory for Process Engineering, Environment, Biotechnology and Energy, Faculty of Engineering, University of Porto, Rua Dr. Roberto Frias, 4200-465 Porto, Portugal; mmadeira@fe.up.pt
3 LAQV/REQUIMTE, Chemistry Department, Nova School of Science and Technology, Universidade NOVA de Lisboa, 2829-516 Caparica, Portugal; a.nabais@campus.fct.unl.pt (A.R.N.); lan11892@fct.unl.pt (L.A.N.); rosa.huertas@ibet.pt (R.H.)
4 IBET-Instituto de Biologia Experimental e Tecnológica, Apartado 12, 2780-157 Oeiras, Portugal
* Correspondence: sandramsanches@gmail.com

Abstract: Spent caustic effluents are very challenging due to their very hazardous nature in terms of toxicity as well as their extreme pH (approximately 12–14). Spent caustic has presented a challenge for wastewater treatment in refineries, due to its composition rich in mercaptans, sulfides and other aromatic compounds. To address such problems, membrane filtration was studied using real effluents from Sines Refinery, in Portugal. The present study attempts to assess the potential for spent caustic treatment with nanofiltration (NF) polymeric and ceramic membranes, assessing membrane life expectancy. For that, membrane aging studies in static mode were performed with the polymeric membrane before attempting NF treatment (dynamic studies). A ceramic membrane was also tested for the first time with this type of effluents, though only in dynamic mode. Although the polymeric membrane performance was very good and in accordance with previous studies, its lifespan was very reduced after 6 weeks of contact with spent caustic, compromising its use in an industrial unit. Contrarily to expectations, the ceramic membrane tested was not chemically more resistant than the polymeric one upon direct contact with spent caustic (loss of retention capacity in less than 1 h in contact with the spent caustic). The results obtained suggest that a pH of 13.9 is very aggressive, even for ceramic membranes.

Keywords: naphthenic spent caustic effluent; nanofiltration; aging; ceramic vs. polymeric membranes

1. Introduction

Worldwide, attempts have been made to treat complex wastewaters and assess the separation of recalcitrant organic compounds by membrane filtration. Wastewater compositions may vary completely, some of them being toxic to human beings and the environment. Most of these wastewaters are produced by heavy chemical industries, such as metallurgy, refineries and the paper industry.

In the specific case of refinery wastewaters, the treatment of spent caustic is presently a challenge due to its composition rich in mercaptans, sulfides, aromatic structures with ketones and alcohols, etc. [1]. The mercaptans that are extracted to the caustic solution in the kerosene prewashing step, along with naphthenic acids and other organosulfur species, constitute naphthenic spent caustic [2–8]. Due to the very high pH value of spent caustic, around 13, this wastewater becomes extremely difficult to handle [9,10].

Easy scale-up, low-temperature operation and space-efficiency of plants are among some of the advantages of the use of membranes for fluid filtration. However, fouling of the membrane and even complete destruction of its structure are common problems that make membrane selection critical [11]. Aging studies have been performed to understand membrane degradation and loss of performance with time when exposed to a certain agent under study [11–21]. Such studies usually assess membrane resistance to cleaning procedures such as oxidant agents (e.g., hypochlorite) or alkaline agents (e.g., NaOH). Most of the aging studies assess the impact of cleaning agents on membranes to evaluate the membrane integrity over time due to cleanings carried out during short periods of time with a certain frequency. In this context, several characterization techniques are usually applied to address membrane modifications (active layer, cross-section and support layer), such as scanning electron spectroscopy (SEM), energy dispersive spectroscopy (EDS) and Fourier-transform infrared spectroscopy (FT−IR) [22].

Nanofiltration (NF) has been applied to treat complex wastewaters with recalcitrant compounds [23–26]. Several membrane filtration processes have been studied to treat refinery effluents, although the studied effluents usually present milder pH values and organic compound concentrations compared to spent caustic; for example, Salahi et al. [27], Abadi et al. [28], Ishak et al. [29], Shariati et al. [30], Zhong et al. [31] and Zhu et al. [32] only refer to oily wastewaters, which usually mean end-chain wastewater with free and/or emulsified oil particles.

Polymeric membranes are the most common membranes used for NF treatment; however, ceramic membranes have also started to be considered as their molecular weight cut-off (MWCO) has been decreased, alongside their greater chemical stability. A study on chemical cleaning of an Inopor ceramic membrane, similar to the one tested in the present study, showed that the alkaline agent NaOH seems to compromise membrane performance significantly [33]. Not many studies exist for low MWCO ceramic membranes addressing membrane performance during aging studies [34] and/or membrane separation processes such as NF, especially for extreme pH effluents, like spent caustic.

Studies with other matrixes show that, for example in the removal of polycyclic aromatic hydrocarbons (PAHs) from wastewaters, size exclusion may be the primary removal mechanism for ceramic membranes while, for example, hydrophobic interactions played a minor role [35]. Zhao et al. [35] studied the impact of organic fouling on ceramic and polymeric membranes, concluding that ceramic membranes present a higher resistance to fouling and that their permeability can be restored after chemical cleaning cycles. Given the possible propensity for fouling formation in complex matrixes with a higher concentration of organics, it seems that ceramic membranes could enable an effective separation. A Koch MPF-34 polymeric membrane and a TiO_2 ceramic membrane supported on Al_2O_3 were used to treat acid-mine waters (pH 1.0) with positive results (maximum concentration factor of 4.19 for the polymeric membrane and 3.29 for the ceramic membrane to achieve 80% of permeate recovery, which was the maximum recovery tested). However, the authors refer to a need for a lower MWCO ceramic membrane [36]. From the work of López et al. [36], it seems that both ceramic and polymeric membranes tested may be chemically resistant to very low pH values, but no conclusion is made regarding very high pH values. A study performed by Dalwani et al. [37] addressed NF performance with a modified polyethersulfone (PES) composite membrane in acidic and alkaline media, observing a stable membrane performance even after prolonged exposure (up to several weeks). The authors reported that especially in strongly alkaline conditions, the membrane appeared to have wider pores; they describe the phenomenon as membrane matrix swelling caused by different interactions between the ions in the solution and the resulting membrane charge density at different pH. Santos et al. [24] attempted to treat refinery spent caustic with an NF Koch MPF-34 spiral wound configuration polymeric membrane. In their study, very high rejections were obtained for key components of spent caustic effluents (99.9% for oil and grease and 97.7% for chemical oxygen demand (COD)) at the optimal concentration factor of 3. On the other hand, this study did not include testing the membrane in the long term (i.e., assessing retention and permeability profiles with time) or assessing the main structure modifications due to high pH exposure time, making these results only valid for first-time application (fresh membrane). Therefore, more studies are necessary about the lifespan of this membrane (and others) with refinery spent caustic and will provide feedback for a potential industrial treatment.

The present study attempts to assess the potential for spent caustic treatment with NF polymeric and ceramic membranes. For that, membrane aging studies in static mode were performed with the polymeric membrane before testing its performance for spent caustic treatment (under dynamic conditions), and results were compared with the work performed by Santos et al. [24]. A ceramic membrane was also tested, though only in dynamic conditions. To the best of the authors' knowledge, previous studies did not evaluate membrane aging in filtration mode throughout long-term operations, also with the particularity of such an aggressive effluent.

2. Materials and Methods

2.1. Reagents

H_2SO_4, 95–97%, was provided by Fluka, Honeywell (Morristown, NJ, USA); NaOH pellets, reagent grade, were supplied by PanReac; silica gel (grade 923) was provided by Grace Davidson (Columbia, MD, USA); tetrachloroethylene was provided by PanReac AppliChem (Barcelona, Spain); and $NaHSO_3$ was provided by AnalaR NORMAPUR, VWR (Radnor, PA, USA).

Cotton filter papers, grade 40, with a diameter of 150 mm and a pore size of 8 μm were provided by Whatman, GE (Little Chalfont, Buckinghamshire, UK). This material was used as part of SMEWW5520 C/F—analysis of oil and grease in wastewater (cf. Section 2.6.1).

2.2. Membranes

Both polymeric and ceramic nanofiltration membranes were assessed in terms of their potential towards naphthenic spent caustic effluents. The commercial proprietary composite polymeric Koch membrane, flat-sheet type ($18' \times 18'$) SelRO MPF-34 with a MWCO of 200 Da, was selected based on its stability under acid/base conditions as well as promising results previously obtained [24]. A commercial ceramic NF membrane from Inopor (Germany) with active and support layers composed of titania (TiO_2) and alumina (Al_2O_3), respectively, was also studied. This is a single-channel tubular membrane with a MWCO of 200 Da and a filtration area of 0.011 m^2.

2.3. Experimental Setup

The assessment of NF membranes' potential to treat spent caustic effluents was conducted in a bench-scale unit, illustrated in Figure 1.

The bench-scale unit consisted of two vessels for the feed and the permeate, an Ifimoto Iberica centrifugal high-pressure pump and a manual valve, through which the pressure inside the system could be adjusted. Transmembrane pressure (TMP) was measured with two pressure transducers from Aplisens (model WW-45) placed before and after the NF module. Pretreatment of the effluent was ensured with an 80 μm prefilter, which can retain solid particles that could promote membrane fouling.

The bench-scale unit in this setup is very versatile, as it allows testing different types of membrane configurations and even MWCO ranges. The NF module in Figure 1 may represent a crossflow module with the polymeric flat sheet membrane or a tubular module with the ceramic tubular membrane. For the crossflow module, a stainless steel crossflow cell was used with a 15 cm^2 flat-sheet SelRO MPF-34 membrane (Koch, Spain) that was placed between two rectangular cross-section channels (feed and permeate) measuring 150 mm $\times$ 10 mm $\times$ 1 mm (length $\times$ wide $\times$ height) each. In the case of the tubular configuration, a stainless steel tubular single-channel housing of 500 mm $\times$ 10 mm (length $\times$ internal diameter) was used with a 110 cm^2 tubular single-channel ceramic membrane, a TiO_2 with 200 Da (Inopor, Germany).

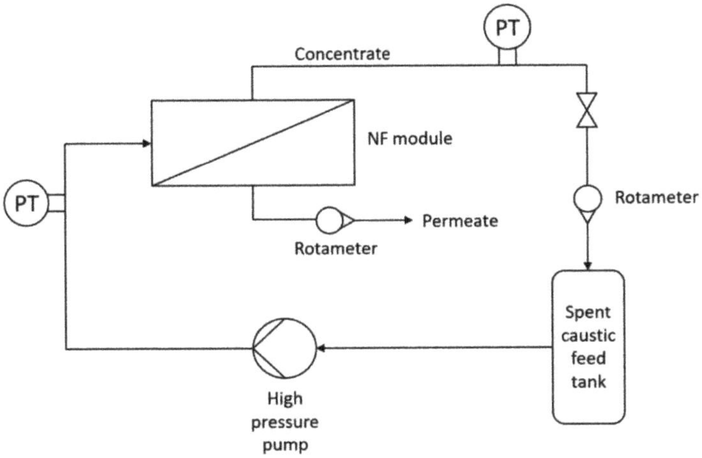

Figure 1. Schematic representation of the NF experimental setup.

2.4. Aging Experiments

Membrane aging is typically employed to address membrane degradation, where alterations can be evaluated in both the active layer and the support. Aging experiments were executed firstly in static conditions for 12 weeks in duplicate assays and secondly in dynamic conditions for 6 and 12 weeks (static aging prior to NF test).

The aging experiments in static conditions allowed studying possible alterations in membrane dimensions and contact angle values and were performed in duplicate with 4 cm^2 membrane (2 × 2 cm^2). The 4 cm^2 pieces were individually soaked in the aging solution (spent caustic effluent No. 1 (see Table 1)) inside glass flasks and were kept in the dark and under static conditions for 12 weeks. Dimensions were taken each two weeks, while the contact angle measurements were performed at the end of the 12 weeks. The contact angles of the polymeric membranes were measured to evaluate their hydrophilic/hydrophobic nature over the soaking time while the dimensions were taken in order to evaluate a possible loss of volume (cf. Section 2.6.2). To compare with the pristine membrane, another 4 cm^2 piece was rinsed with distilled (DI) water before analyses.

Table 1. Characteristics of spent caustic effluents tested.

Effluent No.	1	2
Date of collection	11 January 2019	10 August 2019
COD (mg O_2/L)	102,190	88,360
Phenolic compounds (mg/L)	1908	2742
Sulfides (mg/L)	34.9	31.6
Total O&G (mg/L)	18,970	17,837
Polar O&G (mg/L)	15,053	10,090
Nonpolar O&G (mg/L)	3917	7747
Conductivity (mS/cm)	60.7	61.2
pH (Soresen scale)	13.9	13.9

The aging experiments in dynamic conditions were performed with two pieces of polymeric membranes of 15 cm^2 each that were cut to enable a perfect fitting in the filtration unit for subsequent filtration tests. These membranes were rinsed with DI water to remove all preservatives and then characterized as described in Section 2.6.2. After that, the pieces were individually soaked in the aging solution (spent caustic effluent No. 1 (Table 1)) inside glass flasks and were kept in the dark and under static conditions for 6 or 12 weeks. After these periods, the aged membranes were taken, rinsed with DI water and placed in a $NaHSO_3$ solution (5 g/L) for preservation until nanofiltration tests were carried out. Then, the aged membranes and the pristine membrane (used as control, rinsed with DI water before NF test) were submitted to NF tests (described in Section 2.5), after which were characterized by means of scanning electron microscopy–energy-dispersive X-ray spectroscopy (SEM-EDS) and Fourier-transform infrared spectroscopy (FT−IR), in order to draw conclusions about possible structure modifications caused by changes in the retention capacity and permeability over time, as a means to simulate long-term industrial treatment processes.

2.5. Nanofiltration Experiments

The hydraulic permeability for both polymeric and ceramic membranes was determined prior to any experiment. Polymeric membranes were cut in rectangles of 15 cm^2 of filtration area and placed in the crossflow module. The ceramic membrane was directly placed in the Inopor housing. Water permeability was determined by filtering DI water at 10.0 bar, 8.0 bar and 6.0 bar for 2 h (also ensuring compaction of the polymeric membrane for NF tests). Permeate weight was measured over time using a 250 mL beaker. Temperature throughout the test was monitored with a thermometer (VWR). Temperature and transmembrane pressure (TMP) were measured every 15 min during water permeability tests. Permeability, L_P, was calculated as shown in Equation (1), where V_P stands for permeate volume, t stands for the duration of a permeate sample collection, A stands for filtration area and ΔP stands for the TMP. No correction was included for temperature.

$$L_P = \frac{V_P}{t \times A \times \Delta P} \quad (1)$$

Another important parameter in assessing membrane performance is compound rejection. The apparent rejection (R) can be calculated as shown in Equation (2), where C_P and C_F are the concentrations of a given compound in the permeate and feed, respectively.

$$R\ (\%) = 100 \times \left(1 - \frac{C_P}{C_F}\right) \quad (2)$$

After hydraulic permeabilities were determined, the membrane performance was assessed through NF tests. Nanofiltration experiments with polymeric membranes were performed with effluent No. 1 (Table 1) and conducted at 10 bar with a cross-flow velocity of 1.22 m/s (equivalent to a flow rate of 2.82 L/min), representing a Reynolds number (Re) of 2485 (almost laminar flow). Nanofiltration experiments with ceramic membranes were performed with effluent No. 2 (Table 1) and were conducted at 10 bar with a cross-flow velocity of 1.93 m/s (equivalent to 4.45 L/min), representing a Re of 15,134 (which is turbulent).

It was necessary to apply a pretreatment to the raw effluent in order to remove particulates and oily droplets that could cause fouling on the membrane. The pretreatment was also carried out in the experimental setup (Figure 1), where the NF module was replaced by the prefilter. A total volume of 100 L of naphthenic spent caustic was prefiltered before the NF tests were executed. NF tests were carried out in concentration mode, meaning that the permeate was collected in a different flask and the retentate was recirculated to the feed flask. Because the flux was so low during NF tests, it was not possible to perform flux intermediate measurements during each test. Therefore, each test lasted until enough permeate volume was collected to perform all the physicochemical analyses (around 7 h for pristine polymeric membrane tests, and approximately 1.3 h for each aged membrane). After the conclusion of the NF test, a sequence of basic and acidic washings was applied. The NF unit was rinsed first with tap water to remove the working solution, and then the unit was rinsed with DI water. After that, the basic cleaning focused on feeding the system with a sodium hydroxide solution (5 g/L) recirculated for 15 min. After rinsing the system with DI water, an acid cleaning with a citric acid solution (5 g/L) was carried out for 15 min, followed by a final rinsing with DI water. Hydraulic permeabilities were taken in each step. At the end of the cleaning procedure, the unit was fed with a solution of sodium bisulfite (5 g/L) to preserve the membrane for future use.

In the case of the pristine polymeric membrane, two more NF tests were attempted to test the robustness of the membrane, modifying the pH of the spent caustic stream feeding the NF unit. Spent caustic (effluent No. 1 in Table 1) pH was lowered to 10.0 and to 3.0 with H_2SO_4. In the case of pH 3.0, two phases were formed; the phase of interest was the polar one, so only this phase was collected to be treated by NF. In the case of ceramic membranes, the 200 Da membrane was tested only with the spent caustic at its original pH (effluent No. 2 in Table 1).

All samples throughout the study were characterized, as described in the following section, in terms of concentrations of contaminants such as COD, oil and grease (O&G) (polar and nonpolar), phenolic compounds and sulfides; pH; and conductivity in order to evaluate the retention capacity of the membranes. Before and after NF tests, the polymeric membrane was characterized by SEM-EDS and FT−IR, and the ceramic membrane was characterized by SEM-EDS (as described in Section 2.6.2). Inductively coupled plasma atomic emission spectrometry (ICP-AES) was used to determine the concentration of metal elements, namely titanium and aluminum, in spent caustic solutions (feed, permeates collected during the test, total permeate and retentate) in the NF test carried out with the 200 Da ceramic membrane (see Section 2.6.2).

2.6. Characterization Methods

2.6.1. Effluent Characterization

COD was measured using the LCK 014 kit (HACH) and a DR 3900 spectrophotometer (HACH) after the digestion of the samples in an HT 200 S digester (HACH LANGE), according to APHA guide [38]. Sulfides and phenolic compounds were determined by spectrophotometry using a DR 3900 spectrophotometer (HACH) as well as the LCK 653 kit and the LCK 346 kit from HACH, respectively. O&G (polar and nonpolar) was measured according to the Standard Method SMEWW 5520 C/F [38], after sample dilution to ensure that the concentration would fit into the calibration curve of the NICOLET 6700 FT−IR instrument. The conductivity was measured using a conductivity meter, model LF 320 (WTW); pH was measured using a pH 1100 L pH meter (VWR), and temperature was measured with an alcohol thermometer (VWR).

2.6.2. Membrane Characterization

Dimensions and Contact Angle

The thickness of the polymeric membranes was monitored with a micrometer (Elcometer 124, Manchester, UK).

The contact angles were determined by the sessile drop method, by manually depositing a drop of DI water and spent caustic on the membrane surface. Multiple replicates (9 measurements on 3 different locations of the active layer and support) were performed and the mean angle was determined. All images were acquired by CAM2008 (KSV) software, where the drop shape fitting was executed by mathematical functions.

Scanning Electron Microscopy with Energy Dispersive Spectroscopy (SEM-EDS) and Mapping

Scanning electron microscopy (SEM) with energy dispersive spectroscopy (EDS) analyses were performed on polymeric membranes, while SEM with element mapping was performed on the surface of ceramic membranes. The morphology of the support and active layers of the membranes was studied by a Hitachi S-2400 scanning electron microscope. The microscope was equipped with a silicon drift detector (SDD) to characterize the elemental composition of the surface based on the EDS spectroscopy and a digital acquisition system with Esprit 7.1 software to detect Bruker light elements. Prior to SEM-EDS and SEM mapping characterizations, membranes were coated with gold to make them electrically conductive.

Cross-sections of each membrane were processed using the software ImageJ (University of Wisconsin, Madison, WI, USA) [39,40] for the ×300 SEM image amplification obtained. The images were spatially scaled, and their total membrane surface area was calculated and then binarized after setting a certain threshold level (149 for pristine polymeric membrane and 255 for aged polymeric membranes). Pore density (number of pores divided by area of membrane), porosity (area of pores divided by membrane area), total pore area and the average Feret's diameter (the longest distance between any two points along the selection boundary) were further automatically determined by the software.

Fourier-Transform Infrared Spectroscopy (FT−IR)

FT−IR coupled with attenuated total reflectance (ATR) was used to analyze and compare the surface chemical structure of membranes on both sides (active and support layers). Bruker Spectrometer IFS 66/S instrument (USA) equipped with an H-ATR and a ZnSe crystal was used as equipment, and normalized spectra were recorded with accumulated 60 scans in the range of 4000 to 550 cm^{-1} and with 4 cm^{-1} resolution. All membranes were cleaned with DI water and dried before analysis. Three different random zones were selected for the analysis of each membrane layer to assess potential differences in the chemical structure in different membrane sections.

Inductively Coupled Plasma Atomic Emission Spectroscopy (ICP-AES)

The equipment used for ICP-AES analyses was a Horiba Jobin Yvon S.A.S., Model Ultima2 (France). This was equipped with a 40.68 Hz RF generator and a Czerny-Turner 1.00 m spectrophotometer. Data acquisition and equipment control were performed with a computer equipped with JY v5.4 software, which allows visualizing all parameters and obtained data online. Sample volume was 1 mL per metallic element.

3. Results and Discussion

3.1. *Effluents*

Spent caustic is the solution that results from the kerosene prewashing step, before the sweetening process, in the kerosene merox unit of a petroleum refinery. In the specific case of Sines Refinery, in Portugal, this extremely aggressive effluent accounts for 90% of the impact on the final refinery wastewater in terms of polar oil and grease (O&G) [41].

The characteristics of spent caustic vary over time, due to the different crudemix being processed in the refinery and the operating conditions employed. Two different spent caustic effluents were tested due to constraints related to the plant operation procedures: one to assess NF on polymeric membranes (effluent No. 1) and the second collection to assess NF on ceramic membranes (effluent No. 2). Moreover, the robustness of the polymeric membrane was tested at different feed pH. Table 1 exhibits the compositions of both effluents.

3.2. *Koch SelRO Polymeric Membrane*

3.2.1. NF Tests to Evaluate Membrane Performance

Pure water permeability is an intrinsic membrane characteristic mainly governed by the active layer pore size distribution. The hydraulic permeability for the polymeric membrane was determined to be 1.81 ± 0.28 $Lh^{-1}bar^{-1}m^{-2}$, at 22 ± 2 °C. The fabricant of the polymeric membrane, Koch, lists a hydraulic permeability from 1.5 to 2.0 $Lh^{-1}bar^{-1}m^{-2}$, which is in line with the obtained experimental value.

Membrane performance was addressed after aging for 6 and 12 weeks with spent caustic (effluent No. 1 in Table 1). The values of L_P/L_{P0} (hydraulic permeability at a given time normalized by the hydraulic permeability of the pristine membrane) obtained for 6 weeks and 12 weeks aged membranes were, respectively, 4.46 and 4.62. A considerable increase in the hydraulic permeability of the membrane after soaking in the spent caustic effluent can be observed. It is interesting that the increase in the permeability mainly occurs within the initial period of 6 weeks since similar permeabilities were attained after 6 and 12 weeks of aging (8.07 ± 0.31 $Lh^{-1}bar^{-1}m^{-2}$ vs. 8.36 ± 0.34 $Lh^{-1}bar^{-1}m^{-2}$ after 6 and 12 weeks, respectively).

Figure 2 shows the rejection of contaminants during membrane separation tests. COD removals were very high for the pristine membrane (99.6%); however, a continuous decrease is observed with the aging time (95.8% for 6 weeks and 90.0% for 12 weeks aged membranes (Figure 2A)).

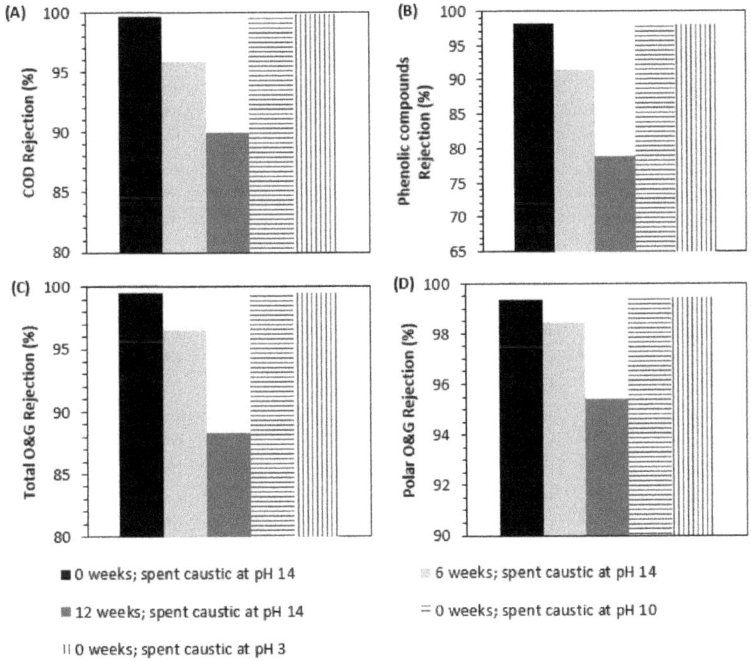

Figure 2. Apparent rejection of contaminants during membrane separation with aged membranes for 0 (pristine membrane), 6 and 12 weeks with spent caustic at pH 14 and pristine membranes with spent caustic at pH 10 and 3. (**A**) COD, (**B**) phenolic compounds, (**C**) total O&G, (**D**) polar O&G.

Considering the first 6 weeks of aging, COD removal decreased on average 0.63% per week, but in the second period of 6 weeks, COD removal decreased on average 0.97% per week. For the membranes aged for 6 and 12 weeks, phenolic compound removals reached 91.3% and 78.9%, respectively, which are much lower than the pristine membrane performance (98.1%) (Figure 2B). It should be noted that phenolic compounds are the target compound class with the lowest MW of all (180–220 Da) [42]; therefore, it may be deduced that the more significant loss of retention is for phenolic compounds. The decrease in their rejection is aligned with the decrease in membrane retentive properties and increase in membrane permeability, which may be related to swelling effects [37]. Sulfide compounds are not presented in Figure 2 since their rejection is 100% in all samples, which may be related to the natural degradation of sulfides into the atmosphere (NF tests occurred during summertime in the open air). Total O&G removal reached a maximum of 99.5% for the pristine membrane. Considering aged membranes, the performance in O&G removal decreased also (96.6% removal for 6 weeks aged membrane and 88.3% removal for 12 weeks aged membrane) (Figure 2C). For aged membranes, polar O&G removals reached 98.4% for 6 weeks and 95.4% for 12 weeks (0.15% average removal loss per week during the first 6 weeks of aging and 0.51% average removal loss per week during the second 6 weeks of aging) (Figure 2D). These results indicate a loss of rejection of polar O&G contaminants on aged membranes, especially for the 12 weeks aged membrane. This loss of retention capacity was more relevant during the second 6 weeks of aging.

As mentioned previously, the major increase in permeability occurred during the first 6 weeks of aging, while the highest decreases in the rejection capacities were observed during the second 6 weeks of aging. The small increase in permeability in the second 6 weeks of aging seems not to be able to justify the increase in the loss of retention capacity in the same period. It could be related to molecular exclusion mechanisms, as it seems that the membrane pore integrity may have been compromised during the aging period, as reported elsewhere [11–13,15,16,18–20,22].

One possible reason to change membrane pore integrity in such a way that compromises permeability and retention capacity could be the very high pH of the spent caustic. Data from FT−IR observations (described hereinbelow) suggest that the active layer is not likely to have been modified since the differences observed are related to the adsorption of organic components that are present in spent caustic. However, permeability is a property largely governed by the narrowest pore (that of the active layer), which leads to the conclusion that, to some extent, permeability data variations observed with aging are related to the possible deterioration of the active layer, although not observable. On the other hand, the support layer seems to have been considerably modified, according to the alterations in retention properties.

Although the membrane is to some extent pH-resistant, it could happen that the extreme pH conditions negatively affect the membrane structure. In order to test this, spent caustic pH was decreased from 14 to 10 and 3. The values of L_P/L_{P0} obtained for pH of 14, 10 and 3 were, respectively, 3.04, 2.37 and 2.07. Hydraulic permeability seems to suffer more the higher the pH of the solution, particularly between pH 10 and 14. Figure 2 shows the impact of pH on the retention capacity of the polymeric membrane. Retention capacity of the pristine membrane is nearly independent of the pH value, with all contaminant rejections being similar in all experiments. Therefore, it can be concluded that pH manipulation will only cause an alteration in the permeability and is not responsible for retention capacity alterations.

Santos et al. [24] attempted to treat naphthenic spent caustic with a Koch NF polymeric SelRO spiral-wound membrane module. It was suggested that the contaminants present in the spent caustic do not significantly affect the membrane performance. The results herein obtained corroborate the study of Santos et al. [24] for a pristine membrane; however, a great loss of performance was observed for aged membranes: a great loss of performance in permeability during the first 6 weeks of ageing and a great loss of performance in retention capacity in the second 6 weeks of ageing.

3.2.2. Contact Angle and Thickness

Contact angle depends highly on the surface composition and, to a lesser extent, on the surface roughness. The contact angle is a measure of the tendency for the water to wet the membrane surface (it can be measured on both sides). For the Koch SelRO polymeric membrane, the contact angle with DI water on the active and support layers was 30.62° and 36.28°, respectively, showing the hydrophilic nature of the membrane on both sides. Interestingly, the contact angle with spent caustic was approximately 0° on both sides, demonstrating the higher affinity of the membrane towards the spent caustic effluent, potentially towards polar organic compounds (with a polarity lower than the water polarity).

Figure 3 presents the thickness evolution during 12 weeks of aging in duplicate samples. In the present work, both length and width were determined as 2 cm for all sampling points throughout the 12 weeks.

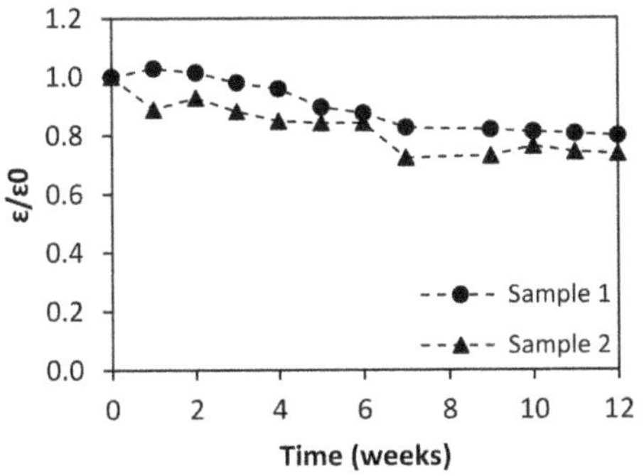

Figure 3. Thickness profile measured for the two immersed polymeric membranes along the 12 weeks of aging (ε—thickness at any time during the aging procedure; ε_0—thickness at the beginning of the aging procedure).

Figure 3 shows a consistent loss of thickness for the two studied membranes, and therefore a loss of membrane volume is observed during aging. This observation could be related to modifications to the membrane structure on the active layer, support or both. These results are consistent with permeability and retention property data obtained and discussed above, in the previous section.

3.2.3. SEM-EDS

Active and support layers as well as cross-sections of pristine and aged membranes (for 6 and 12 weeks) were analyzed by scanning electron microscopy (SEM). Figure 4 shows the results obtained.

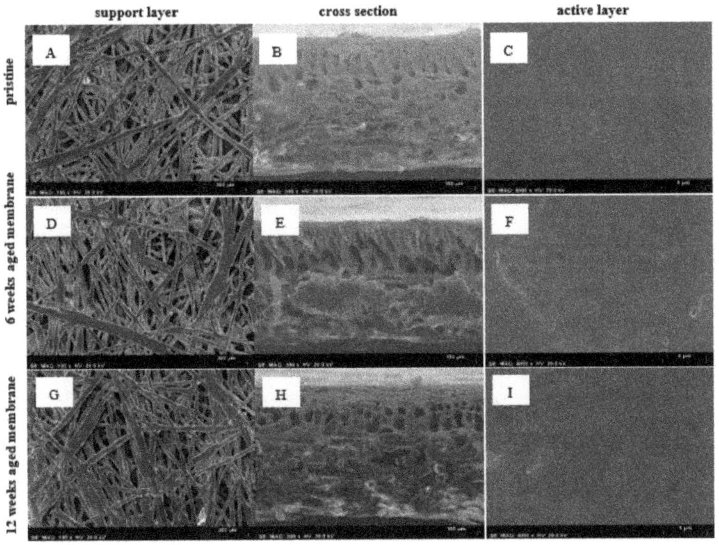

Figure 4. SEM analysis for the polymeric membrane. (**A**) Pristine membrane support layer; (**B**) pristine membrane cross-section; (**C**) pristine membrane active layer; (**D**) 6 weeks aged membrane support layer; (**E**) 6 weeks aged membrane cross-section; (**F**) 6 weeks aged membrane active layer; (**G**) 12 weeks aged membrane support layer; (**H**) 12 weeks aged membrane cross-section; (**I**) 12 weeks aged membrane active layer. Support layer magnification is ×100, cross-section magnification is ×300 and active layer magnification is ×4000.

In terms of physical modifications of the membrane structure, it appears that the intermediate structure (right after the active layer in the cross-section view) becomes slightly looser with aging (Figure 4B,E,H), although it is not completely clear. Pore coalescence may be the cause for this observation. The support layer (Figure 4A,D,G) seems unchanged. No conclusions can be drawn regarding the active layer since the structure is very compact and SEM maximum magnification does not enable comparisons among the membranes with the required level of detail. On the other hand, variations in permeability and thickness were observed with time, thus making it impossible to draw a clear conclusion of a connection between membrane structure and function (L_P, and rejection capacity) based on SEM results alone. To further study membrane structure, FT−IR profiles were collected and are discussed in the next section. Chemical surface composition (EDS) was also studied to determine elementary composition (Table 2).

Table 2. EDS analysis of active and support layers of polymeric membranes (pristine, 6 and 12 weeks aged prior to NF test).

Membrane	Active Layer			Support Layer	
	C (%)	S (%)	O (%)	C (%)	S (%)
Pristine membrane	61.6	19.9	18.5	99.0	1.0
Aged for 6 weeks	64.7	19.0	16.3	99.0	1.0
Aged for 12 weeks	67.7	16.6	15.7	99.3	0.7

It is interesting to observe a continuous increase in the carbon content with aging in the active layer while other elements (S and O) are decreasing, which could mean a possible degradation of this layer. On the support, an alteration occurs between 6 and 12 weeks of aging and is related to a loss of sulfur. Although the membrane composition is not available, SEM-EDS analysis suggests that its composition essentially includes carbon, oxygen and sulfur in the active layer and mostly carbon in the support. Especially for the active layer, which has the selectivity property, several polymers used in NF polymeric membranes contain such composition, such as polyethersulfone (PES) and polysulfone (PS) [43].

Pore density analyses were performed on the pristine membrane, as well as for the 6 and 12 weeks aged membranes, prior to NF tests. An SEM cross-section image (×300 image magnification) of each membrane was processed using the ImageJ software. For that, the total cross-section area was calculated over SEM images that were also binarized (Figure S1 in the Supplementary Material section), and the values obtained are summarized in Table 3, comprising only the pore features in the membrane support (due to the very low MWCO of the active layer, SEM could not show any pores). As such, the discussion of the impact of membrane aging may only be performed considering the main porous parameters for the membrane cross-section. However, a cross-sectional analysis also has some limitations since it only indicates possible distortion of the pores due to membrane aging.

The cross-section area obtained, calculated pore density (number of pores divided by area of membrane), porosity (area of pores divided by area of membrane), circularity (Equation (3)) and Feret's diameter (the longest distance between any two points along the selection boundary) [44] are presented in Table 3.

$$\text{Circularity} = 4\pi\frac{\text{Area}}{\text{Perimeter}^2} \tag{3}$$

From the data presented in Table 3, is clear that the pores in the support are of micro scale, even in the pristine membrane. However, the membrane support aged 6 and 12 weeks showed an increase in the porosity, number of pores and pore density. Extremely high Feret's diameter values were detected, indicating the highly irregular shape of pores, particularly for the 6 weeks aged membrane. The estimated average circularity value was around 0.8 in all measurements with estimated values obtained that varied between 0.6 and

1 (perfect circles) but with a high standard deviation (Figure S2). Therefore, a considerable variation in the average Feret's diameter values was also obtained, indicating a highly irregular shape of pores in the support layer, particularly finger-like for this part of the membrane support. Moreover, the increase in the maximum Feret's diameter obtained for the 6 weeks aged membrane indicates a possible distortion of the porous structure (clearly shown for the wider porous section area) after aging. On the other hand, the decrease in this value with the aging time indicates a possible compaction at the end of 12 weeks.

Table 3. Pore measurements for the support layer of the pristine membrane, as well as for the 6 and 12 weeks aged membranes prior to NF test based on the SEM cross-section ×300 magnification.

	Pristine	Membrane 6	Membrane 12
Porosity (%)	12.10	19.91	27.93
Number of pores	2196	2962	4039
Pore density (μm^{-2})	0.032	0.041	0.063
Minimum pore area (μm^2)	0.074	0.074	0.075
Maximum pore area (μm^2)	442.45	3821.37	714.92
Total pore area (μm^2)	8345.47	14,257.44	18,022.10
Average circularity	0.828 ± 0.252	0.878 ± 0.221	0.837 ± 0.252
Minimum circularity	0.078	0.038	0.007
Maximum circularity	1.000	1.000	1.000
Average Feret diameter (μm)	1.662 ± 3.745	1.441 ± 4.974	1.503 ± 3.570
Minimum Feret's diameter (μm)	0.384	0.384	0.386
Maximum Feret's diameter (μm)	39.761	141.438	66.347

3.2.4. FT−IR

The objective of these studies was to determine the possible structure alterations in the membranes that occurred during aging experiments and after NF tests. The tested membranes are divided as the pristine membrane without any treatment ("Pristine") and with an NF test ("Pristine + NF test"), the duplicate aged membranes after immersion for 12 weeks ("Membrane 1 + 12 weeks" and "Membrane 2 + 12 weeks") and the membranes aged for 6 and 12 weeks followed by an NF test ("Membrane 6 weeks + NF test" and "Membrane 12 weeks + NF test"). It is important to mention that both the side in contact directly with the effluent in the NF tests (active layer) and the opposite side (support layer) were tested.

The FT−IR spectra obtained for the active layers for two membranes immersed under the same conditions in the effluent were analyzed and compared with the membrane that was not immersed; data are shown in Figure 5.

The results obtained by FT−IR for the active layer of 12 weeks aged membranes after immersion (duplicate: Membrane 1 and Membrane 2) depicted in Figure 5 and amplified in Figure S3 were nearly identical but showed some differences compared to the pristine membrane. Thus, among similarities, bands associated with the C=C bond at around 3100 cm^{-1}, 871 cm^{-1} and 833 cm^{-1} (Figures 5 and S3) were identified in all spectra. The pristine and both the 12 weeks aged membranes showed bands at around 3370, ~3100, 2950–2924 and 2860 cm^{-1} (Figure S3A), characteristic of –OH, =C-H, Cl-C-H and H-C-H bonds, respectively [45]. The peak at 1142 cm^{-1} was associated with C-O st and was also present in all spectra, as were peaks in the range of 800–600 cm^{-1}, also assigned to C=CH stretching vibration [45] and occasionally C-Cl [46,47]. A peak at around 1576 cm^{-1} was also observed in all spectra, probably associated with the stretching mode of -COO st group. Peaks at 1665 cm^{-1} and 1435 cm^{-1} were linked to asymmetric and symmetric vibration of -COO- groups, respectively [48,49], and peaks at 1070 and 1031 cm^{-1} associated with S=O bonds were also present, seeming to be in accordance with the presence of those atoms in SEM-EDS results (Table 2).

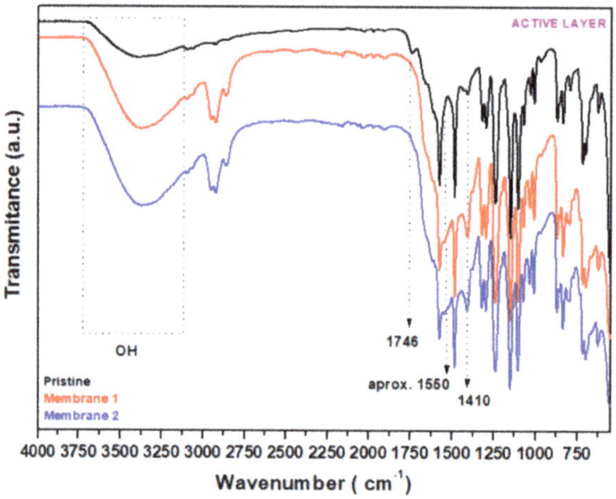

Figure 5. FT−IR spectra of the active layer of the pristine membrane and of the active layer for membranes after 12 weeks immersion in spent caustic (made in duplicate).

Among the differences observed, the pristine membrane spectrum presented a characteristic band at 1746 cm^{-1} due to the stretching vibration of the C=O group, while for the 12 weeks aged membranes this peak disappeared (Figures 5 and S3B). This fact could be attributed to the aminolysis reaction with the urethane groups with ammonia or alkanolamines, present in the effluent and giving as subproducts alcohols and amides, as previously reported [50]. Moreover, this is in line with the increase in intensity observed in the spectra for the band associated with these groups (OH bond), around 3370 cm^{-1} (Figure 5) and 1410 cm^{-1}, as well as the differences in the C=O band in the amide region (around 1690–1630 cm^{-1}) (Figure S3B). The increase in the intensity for bands at 2950 cm^{-1} and 2924 cm^{-1} in the spectra of 12 weeks aged membranes could be attributed to the presence of compounds from the effluent (Figure S3A). A broad band appearing at around 1520 cm^{-1} is associated with the presence of ammonia (NH_4^+) and salt derivates (-NH_3^+) and could be also related to the effluent. Bands at around 1550–1550 cm^{-1} associated with N–O asymmetric stretch; 1377 cm^{-1}, symmetric stretch; and 773 cm^{-1}, out-of-plane stretch modes for deformation of nitro compounds (nitrates (NO_3^-) and nitrites (NO_2^-)) [45], which are present as residues in the aged membranes, are also often found in the spectra of other oily effluents [51]. New peaks at 809 cm^{-1} and at 685 cm^{-1} are also associated with the effluent residue.

Thus, after analyzing and comparing the spectra, we can conclude that the immersion of the membranes does not have an impact on the main bands associated with the membrane initial polymeric structure (-OH, C-H, C-N, C-O and Cl-C bonds) and the changes observed in the spectra are likely due to the adsorption of components from the effluent as well as chemical reactions due to the presence of potential additives employed during membrane production.

The impact of filtration in the chemical structure of the membrane was also checked by comparing the pristine membrane before and after the NF test as well as two membranes immersed for 6 and 12 weeks prior to an NF test (Figure 6).

Spectra depicted in Figure 6, for all membranes tested by filtration, did not show significant differences apart from the new signals detected in the region of 2090–1952 cm^{-1}, which could be assigned to isothiocyanate bonds (N=C=S st) [45,52], and a new peak at 1819 cm^{-1} (C=O st) associated with acid halide or anhydride, which is an indicator of chemical modifications in the urethane structure. The intensity of the peaks at around 3413 cm^{-1}, around 2960–2854 cm^{-1} and around 1737 cm^{-1} (Figure 6) increases as the contact time with the effluent increases. Thus, similarly to the analysis shown in Figure 5, the presence of the bands at 1410 cm^{-1} and 1377 cm^{-1} is probably related to the presence of alcohols and effluent components (Figure S4).

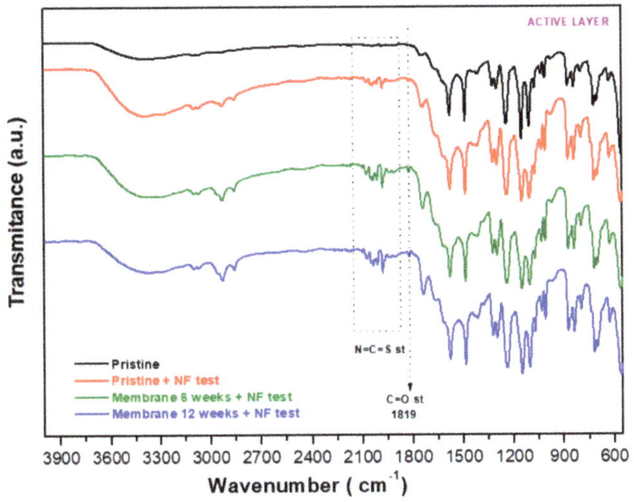

Figure 6. FT−IR spectra of the active layer for the pristine membrane before and after the NF test as well as the active layer of membranes aged for 6 and 12 weeks and after NF tests.

The support layers of the membranes were also analyzed and compared by FT−IR (Figure 7). First, the chemical structure for pristine and aged duplicate membranes (named Membrane 1 and Membrane 2), immersed in caustic effluent for 12 weeks (Figure S5), was analyzed. Similarly to the analysis conducted for the active layer (Figure 6), FT−IR characterization was conducted for the support layer in aged membranes immersed for 6 and 12 weeks in caustic effluent and subjected to NF tests (Figure 7), and the results were compared to NF-tested and untested pristine membrane support.

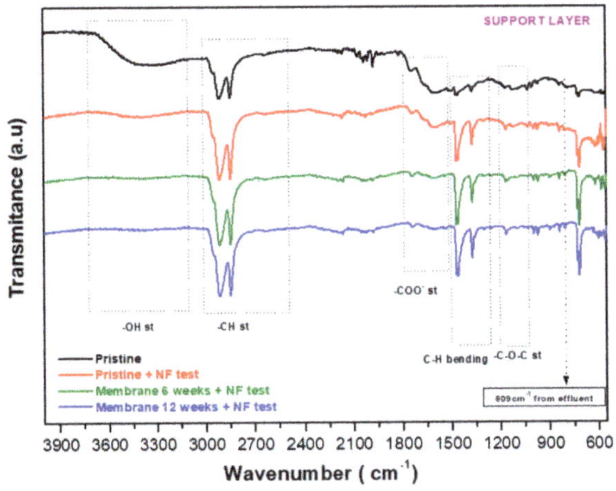

Figure 7. FT−IR spectra of the support layer in the pristine membrane before and after NF test, as well as aged FT−IR spectra of the support layers for membranes aged for 6 and 12 weeks used in NF tests.

The spectra for the support layer of membranes immersed for 12 weeks in spent caustic effluent were identical to those of the pristine membrane and showed peaks associated with C-H bonds (2985–2800 cm^{-1}, 1457 cm^{-1} and 1383 cm^{-1}) that remained in the main structure. However, an evident deacetylation with the loss of -OH (region around 3400 cm^{-1}) and COO- groups (around 1660 cm^{-1}) and an important change in the region for C-O bonds (1165–1015 cm^{-1}) groups indicates that the polymeric structure of the support layers might have been considerably damaged after exposure to the spent caustic effluent.

As expected, these changes in the chemical structure were similar to those found in submerged membranes subjected to filtration (Figure 7). The peak at 809 cm^{-1} present in both spectra (Figures S5 and 7) could be also related to the detected effluent components (Figure S4).

As a conclusion, the data discussed above suggest that the active layer is not likely to have been modified since the differences observed are related to the adsorption of organic components that are present in spent caustic, affecting its retentive properties, as shown above. On the other hand, the support layer seems to have been considerably altered, in line with the changes observed in the permeability of the membranes, porosity properties and EDS results.

3.3. *Inopor Ceramic Membrane*

Similarly to what has been shown in Section 3.2, the performance of the ceramic membrane was also evaluated.

Although compaction is not required for ceramic membranes, this membrane was allowed to compact for half an hour, with DI water. Hydraulic permeability was determined to be 5.61 ± 0.14 Lh^{-1}bar^{-1}m^{-2} at 23 ± 2 °C. The fabricant of the ceramic membranes, Inopor, lists a hydraulic permeability of 9.0 Lh^{-1}bar^{-1}m^{-2}, which is relatively higher than the obtained value, but typically ceramic membranes may present a slight deviation due to the rigid structure [32].

Tests were performed for a maximum period of 6 h. Figure 8 shows the permeability evolution for the ceramic membrane during the NF test.

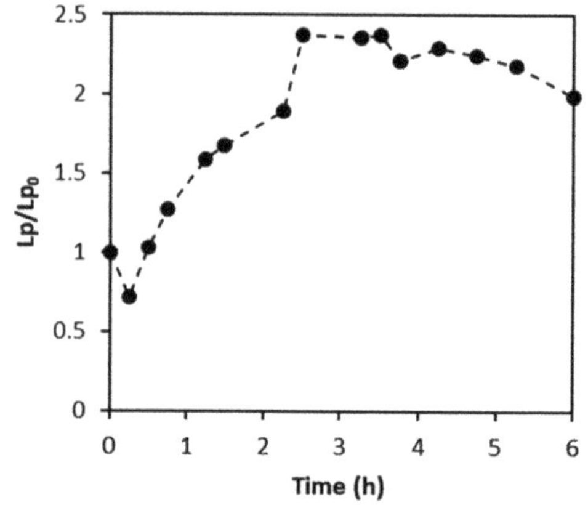

Figure 8. Relative permeability of the 200 Da ceramic membrane throughout the test with spent caustic effluent.

Figure 8 shows that permeability increased between 0.5 and 2.5 h of testing, slightly decreasing afterward for the 200 Da membrane. TiO_2 is a nontoxic material, and according to the manufacturer, it is very resistant to extreme pH conditions (1–13 between 20 to 40 °C) [53]. Spent caustic is a difficult wastewater due to extreme pH, around 14 (Table 1). Contrary to expectations, the 200 Da ceramic membrane did not resist spent caustic.

The results obtained evidence that the membranes addressed are not suitable to treat the tested spent caustic effluents. After the filtration test, the membrane was rinsed with DI water and L_P was assessed. For the 200 Da membrane, L_P was determined to be 47.72 ± 1.23 $Lh^{-1}bar^{-1}m^{-2}$, at 22 ± 2 °C. The original hydraulic permeability was determined as 5.61 ± 0.14 $Lh^{-1}bar^{-1}m^{-2}$ at 23 ± 2 °C, from where it may be observed that L_P increased more than 8 times, not being able to recover the initial permeability.

Figure 9 shows the evolution in the rejection of all contaminants (or lumped parameters) addressed (COD, phenolic compounds, sulfides, total O&G and polar O&G) throughout the last 3 h of the test (because the flux was too low before that to allow for a permeate collection).

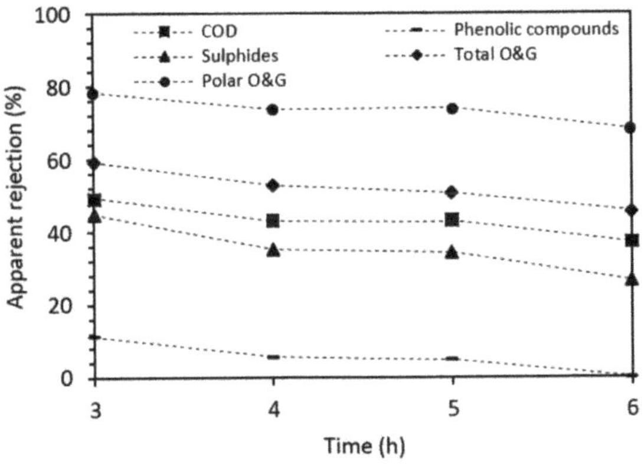

Figure 9. Apparent rejection of contaminants during separation with the 200 Da ceramic membrane.

The loss in retentive properties could be related to changes in the membrane structure. Considering the rejection capabilities of the ceramic membrane, phenolic compounds seem to be the contaminants with the lowest rejection, possibly due to their low molecular weight. The comparison of the removals obtained for total O&G and polar O&G suggests that the membrane performed better towards the removal of polar O&G. Some authors argue that NF TiO_2 ceramic membranes possess a tendency to lose rejection when the contaminants are positively charged and maintain high rejection for negatively charged compounds and low molecular weight neutral organics [54]. In a strongly alkaline medium, organic compounds with functional groups located at the end of their structures are usually present in their ionic form.

In order to further study the reasons behind the loss of rejection and to confirm if loss of membrane integrity could have occurred, an ICP analysis was performed on the feed, permeate and retentate samples (Table 4). SEM–EDS analyses were also performed on the membrane.

Table 4. ICP analysis of feed, retentate and permeate samples after/during the NF test with the 200 Da membrane.

Sample Name	[Al], mg/L	[Ti], mg/L
Feed (spent caustic)	2.06	0.02
Retentate—Total *	2.07	0.05
Permeate—Total *	1.38	0.43
Permeate at 3 h of NF test	1.45	0.50
Permeate at 4 h of NF test	1.37	0.29
Permeate at 5 h of NF test	1.62	0.27
Permeate at 6 h of NF test	1.81	0.28

* The analyses refer to the total volume of retentate/permeate collected during the 6 h of the NF test.

Table 4 shows a similar aluminum concentration in both the feed and retentate, but a slightly lower aluminum concentration in the permeate samples, therefore showing that there was no release of aluminum from the membrane structure. For the titanium element, concentration was slightly increased in the retentate compared with feed concentration, but concentration increased considerably in the permeate samples, leading to an indication of titanium being released from the membrane structure.

Cross-sections, surfaces and element mapping of the ceramic membrane were analyzed by SEM. Figure 10 shows the surfaces and cross-section of the membrane, along with the mapping of its most significant elements. Table 5 contains the EDS analysis for the active layer of the ceramic membrane after the NF tests.

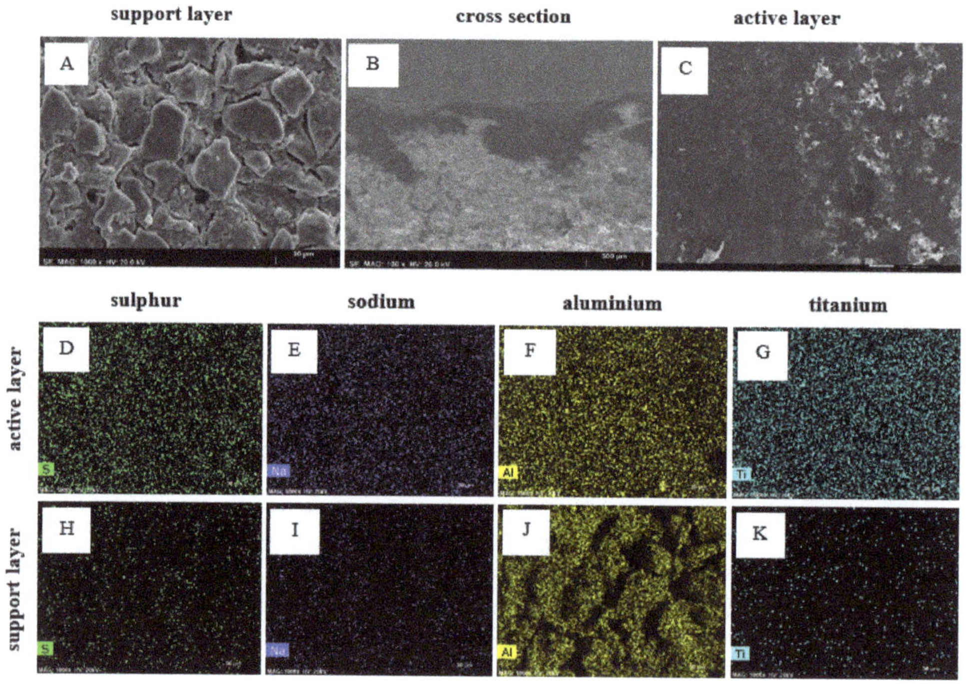

Figure 10. SEM-EDS analysis of the 200 Da ceramic membrane after the NF test as well as the mapping of the most significant elements on the active layer and support layer after NF test. (**A**) Support layer; (**B**) cross-section; (**C**) active layer; (**D**) sulfur element on the active layer; (**E**) sodium element on the active layer; (**F**) aluminum element on the active layer; (**G**) titanium element on the active layer; (**H**) sulfur element on the support layer; (**I**) sodium element on the support layer; (**J**) aluminum element on the support layer; (**K**) titanium element on the support layer. Layer magnifications are $\times 1000$ and cross-section magnification is $\times 100$.

Table 5. EDS analysis of active layer of the ceramic membrane after the NF test (200 Da).

Membrane	O (%)	C (%)	Al (%)	Ti (%)	Zr (%)	Na (%)	S (%)	Fe (%)	Si (%)
200 Da	22.1	10.2	18.0	23.8	10.8	9.9	3.2	1.6	0.4

Although SEM analysis (Figure 10) does not appear to evidence damages on the membrane structure (e.g., cracks) or differences in the distribution of the elements over the membrane, the increase in the permeability of the membrane and the loss of its retention properties is corroborated by ICP observations, which may indicate a disintegration mechanism. Possibly, because the loss of Ti detected by ICP from the membrane to the permeate is not so high and/or can be homogeneous throughout the membrane area, SEM-EDS images do not show it as ICP analysis did. Although the membrane was thoroughly rinsed with DI water after the experiment, the deposition of some salts could be observed in both the active and support layers. The distribution of some elements originally only present in spent caustic, such as sulfur and sodium, was analyzed through mapping. The distribution of the metallic elements that compose the membrane (aluminum and titanium) was also analyzed by mapping. The concentration of sulfur and sodium is considerably higher in the selective layer, which might indicate a higher retention capacity for these compounds. The aluminum distribution seems to be similar on both sides of the membrane. Likewise, titanium is also present in both layers. ICP results indicate a loss of only titanium. Nevertheless, the presence of aluminum occurs in both layers due to a deposition of compounds from the matrix on the active layer, where there is the aluminum element. It is possible that the loss of elements observed by ICP has no such expression in the Figure 10 results, presumably because this loss might be not so relevant. Due to this membrane presenting high rejections, the concentration of elements on the active layer is expected to be higher. On the other hand, the support layer pores are wider and, therefore, may accumulate some elements from the active layer, which could mean that the titanium element released by the active layer will accumulate in the support layer.

4. Conclusions

Hydraulic permeability results of the polymeric membrane showed a noteworthy decrease in performance, mainly during 6 weeks of aging, and a smaller loss of permeability in the second period of 6 weeks (final permeability was 4.6 times higher than the hydraulic permeability of the pristine sample). A significant loss in rejection was also observed, following the permeability results (between 5 and 20% less rejection in the analyzed contaminants/lumped parameters). SEM analysis showed that the cross-section of the membrane presented a different pore density, as if the structure loosens up with time in direct contact with spent caustic. Thickness was also decreased during aging. EDS analyses indicated an increase in the carbon element on the active layer, while FT−IR results suggest that the active layer did not present any structural modifications; however, the adsorption of the organic compounds naturally present in spent caustic may have been responsible for the selectivity decay. On the other hand, the support layer presented several modifications that may justify the permeability changes observed.

The hydraulic permeability of the 200 Da ceramic membrane decreased during the first 10 min of testing, but after that, the membrane started increasing in permeability regularly (and lost rejection properties) until 2.5 h of testing, contrary to expectations. EDS and SEM mapping results suggest the disintegration of the membrane, corroborated by the loss of titanium during the NF test. This membrane could be further tested in the future with spent caustic modified to lower pH values, given that the original pH of spent caustic is extremely high.

Supplementary Materials: The following supporting information can be downloaded at: https://www.mdpi.com/article/10.3390/membranes12010098/s1, Figure S1: SEM binarized cross section images, using a threshold of 149 for Pristine Membrane (up), and 255 for Membrane 6 (middle) and Membrane 12 (down); Figure S2: Representation of the circularity (left) and porous area (μm^2) (right) distribution, for all porous (counts) detected by the ImageJ software: Pristine (Up), Membrane 6 (middle) and Membrane 12 (down). (Obtained with Origin Software); Figure S3: FT-IR spectra magnifications in regions 4000–2700 cm^{-1} (A) and 2200–550 cm^{-1} (B) of the active layer of the pristine membrane and membranes after 12 weeks immersion in spent caustic (made in duplicate); Figure S4: FT-IR spectra of spent caustic samples collected in the 17 March 2018 in original form (A) and acidified form (C) and collected in the 22 March 2018 in original form (B) and acidified form (D); Figure S5: FT-IR spectra of support layer for pristine membrane and membranes (made in duplicate) after 12 weeks immersed in spent caustic effluent.

Author Contributions: Conceptualization, A.I.R., L.M.M. and S.S.; methodology, A.I.R., L.A.N., R.H., L.M.M. and S.S.; formal analysis, A.I.R., R.H., M.S., L.M.M., S.S.; investigation, A.I.R., A.R.N., L.A.N., R.H.; resources, R.H., M.S., L.M.M.; data curation, A.I.R.; writing—original draft preparation, A.I.R., R.H., L.M.M., S.S.; writing—review and editing, A.I.R., A.R.N., L.A.N., M.S., L.M.M., S.S.; supervision, M.S., L.M.M., S.S. All authors have read and agreed to the published version of the manuscript.

Funding: This research was funded by the PhD grant PD/BDE/114356/2016, co-financed by Fundação para a Ciência e Tecnologia (FCT) and Galp group, Portugal.

Institutional Review Board Statement: Not applicable.

Informed Consent Statement: Not applicable.

Acknowledgments: The present work was co-financially supported by Fundação para a Ciência e Tecnologia (FCT) within the program Human Capital Operational Programme (HCOP) (co-financed by PT2020 and European Social Funds) and national funds from MCTES and Galp group, through the grant PD/BDE/114356/2016. iNOVA4Health-UID/Multi/04462/2013, a program financially supported by Fundação para a Ciência e Tecnologia/Ministério da Educação e Ciência, through national funds and co-funded by FEDER under the PT2020 Partnership Agreement, and funding from INTERFACE Programme, through the Innovation, Technology and Circular Economy Fund (FITEC), are gratefully acknowledged. This work was supported by the Associate Laboratory for Green Chemistry (LAQV), which is financed by national funds from FCT/MCTES (UIDB/50006/2020). This work was also financially supported by Base Funding-UIDB/00511/2020 of the Laboratory for Process Engineering, Environment, Biotechnology and Energy (LEPABE) funded by national funds through FCT/MCTES (PIDDAC).

Conflicts of Interest: The authors declare no conflict of interest.

Nomenclature

A	Filtration area (m^2)
L_P	Hydraulic permeability ($Lh^{-1}bar^{-1}m^{-2}$)
L_{P0}	Hydraulic permeability for the pristine membrane ($Lh^{-1}bar^{-1}m^{-2}$)
TMP	Transmembrane pressure (bar)
Δt	Duration of nanofiltration test (h)
V	Permeate volume (L)
Acronym list	
ATR	Attenuated total reflectance
COD	Chemical oxygen demand
DI	Distilled
EDS	Energy dispersive spectroscopy
FT−IR	Fourier-transform infrared spectroscopy
ICP-AES	Inductively coupled plasma atomic emission spectroscopy
MWCO	Molecular weight cut-off
NF	Nanofiltration
O&G	Oil and grease
PAHs	Polycyclic aromatic hydrocarbons
SEM	Scanning electron spectroscopy
SDD	Silicon drift detector
SMEWW	Standard method for examination of water and wastewater
PES	Polyethersulfone
PS	Polysulfone

References

1. Rita, A.; Rodrigues, C.; Santos, M.; Sanches, S.; Madeira, L. Comparison of different strategies to treat challenging refinery spent caustic effluents. *Sep. Purif. Technol.* **2020**, *253*, 117482. [CrossRef]
2. Boczkaj, G.; Makoś, P.; Przyjazny, A. Application of dispersive liquid-liquid microextraction and gas chromatography with mass spectrometry for the determination of oxygenated volatile organic compounds in effluents from the production of petroleum bitumen. *J. Sep. Sci.* **2016**, *39*, 2604–2615. [CrossRef]

3. Makoś, P.; Przyjazny, A.; Boczkaj, G. Methods of assaying volatile oxygenated organic compounds in effluent samples by gas chromatography—A review. *J. Chromatogr. A* **2019**, *1592*, 143–160. [CrossRef] [PubMed]
4. Boczkaj, G.; Makoś, P.; Fernandes, A.; Przyjazny, A. New procedure for the control of the treatment of industrial effluents to remove volatile organosulfur compounds. *J. Sep. Sci.* **2016**, *39*, 3946–3956. [CrossRef]
5. Boczkaj, G.; Makoś, P.; Fernandes, A.; Przyjazny, A. New procedure for the examination of the degradation of volatile organonitrogen compounds during the treatment of industrial effluents. *J. Sep. Sci.* **2017**, *40*, 1301–1309. [CrossRef]
6. Makoś, P.; Fernandes, A.; Boczkaj, G. Method for the determination of carboxylic acids in industrial effluents using dispersive liquid-liquid microextraction with injection port derivatization gas chromatography–mass spectrometry. *J. Chromatogr. A* **2017**, *1517*, 26–34. [CrossRef] [PubMed]
7. Makoś, P.; Fernandes, A.; Przyjazny, A.; Boczkaj, G. Sample preparation procedure using extraction and derivatization of carboxylic acids from aqueous samples by means of deep eutectic solvents for gas chromatographic-mass spectrometric analysis. *J. Chromatogr. A* **2018**, *1555*, 10–19. [CrossRef]
8. Makoś, P.; Fernandes, A.; Boczkaj, G. Method for the simultaneous determination of monoaromatic and polycyclic aromatic hydrocarbons in industrial effluents using dispersive liquid-liquid microextraction with gas chromatography-mass spectrometry. *J. Sep. Sci.* **2018**, *41*, 2360–2367. [CrossRef] [PubMed]
9. Alnaizy, R. Economic analysis for wet oxidation process for the treatment of mixed refinery spent caustic. *Environ. Prog.* **2008**, *27*, 295–3011. [CrossRef]
10. Berne, F.; Cordonnier, J. *Industrial Water Treatment—Refining, Petrochemical and Gas Processing Techniques*; Gulf Publishing Company: Houston, TX, USA, 1995; Volume 37.
11. Benavente, L.; Coetsier, C.; Venault, A.; Chang, Y.; Causserand, C.; Bacchin, P.; Aimar, P. FTIR mapping as a simple and powerful approach to study membrane coating and fouling. *J. Membr. Sci.* **2016**, *520*, 477–489. [CrossRef]
12. Antón, E.; Álvarez, J.R.; Palacio, L.; Prádanos, P.; Hernandez, A.; Pihlajamäki, A.; Luque, S. Ageing of polyethersulfone ultrafiltration membranes under long-term exposures to alkaline and acidic cleaning solutions. *Chem. Eng. Sci.* **2015**, *134*, 178–195. [CrossRef]
13. Antón, F.; Alvarez, J.; Luque, S. Cleaning and Ageing of Ultrafiltration Membranes. *Procedia Eng.* **2012**, *44*, 689. [CrossRef]
14. Cui, L.; Feng, Z.; Goodwin, C.; Gao, W.; Liao, B. Structure and properties of membrane at different ages in drinking water treatment. *Desalination Water Treat.* **2017**, *74*, 44–52. [CrossRef]
15. Ettori, A.; Gaudichet-Maurin, E.; Schrotter, J.-C.; Aimar, P.; Causserand, C. Permeability and chemical analysis of aromatic polyamide based membranes exposed to sodium hypochlorite. *J. Membr. Sci.* **2011**, *375*, 220–230. [CrossRef]
16. Ettori, A.; Gaudichet-Maurin, E.; Aimar, P.; Causserand, C. Pilot scale study of chlorination-induced transport property changes of a seawater reverse osmosis membrane. *Desalination* **2013**, *311*, 24–30. [CrossRef]
17. Fraga, M.; Sanches, S.; Pereira, V.; Crespo, J.; Yuan, L.; Marcher, J.; de Yuso, M.M.; Rodríguez-Castellón, E.; Benavente, J. Morphological, chemical surface and filtration characterization of a new silicon carbide membrane. *J. Eur. Ceram. Soc.* **2016**, *37*, 899–905. [CrossRef]
18. El Mansour, M.; Ettori, A.; Luque, S.; Alvarez, J.; Causserand, C.; Aimar, P. Accelerated Ageing of Crosslinked Polyamide Membranes. *Procedia Eng.* **2012**, *44*, 789. [CrossRef]
19. Pellegrin, B.; Prulho, R.; Rivaton, A.; Thérias, S.; Gardette, J.; Gaudichet-Maurin, E.; Causserand, C. Hypochlorite Cleaning of Polyethersulfone/Polyvinylpyrrolidone Ultrafiltration Membranes: Impact on Performances. *Procedia Eng.* **2012**, *44*, 472–475. [CrossRef]
20. Pellegrin, B.; Prulho, R.; Rivaton, A.; Thérias, S.; Gardette, J.-L.; Gaudichet-Maurin, E.; Causserand, C. Multi-scale analysis of hypochlorite induced PES/PVP ultrafiltration membranes degradation. *J. Membr. Sci.* **2013**, *447*, 287–296. [CrossRef]
21. Pellegrin, B.; Mezzari, F.; Hanafi, Y.; Szymczyk, A.; Remigy, J.-C.; Causserand, C. Filtration performance and pore size distribution of hypochlorite aged PES/PVP ultrafiltration membranes. *J. Membr. Sci.* **2015**, *474*, 175–186. [CrossRef]
22. Causserand, C.; Pellegrin, B.; Rouch, J.-C. Effects of sodium hypochlorite exposure mode on PES/PVP ultrafiltration membrane degradation. *Water Res.* **2015**, *85*, 316–326. [CrossRef] [PubMed]
23. Alipour, Z.; Azari, A. COD removal from industrial spent caustic wastewater: A review. *J. Environ. Chem. Eng.* **2020**, *8*, 103678. [CrossRef]
24. Santos, B.; Crespo, J.G.; Santos, M.A.; Velizarov, S. Oil refinery hazardous effluents minimization by membrane filtration: An on-site pilot plant study. *J. Environ. Manag.* **2016**, *181*, 762–769. [CrossRef]
25. Sanches, S.; Penetra, A.; Rodrigues, A.; Ferreira, E.; Cardoso, V.V.; Benoliel, M.J.; Crespo, M.T.B.; Pereira, V.J.; Crespo, J.G. Nanofiltration of hormones and pesticides in different real drinking water sources. *Sep. Purif. Technol.* **2012**, *94*, 44–53. [CrossRef]
26. Sanches, S.; Penetra, A.; Granado, C.; Cardoso, V.V.; Ferreira, E.; Benoliel, M.J.; Crespo, M.T.B.; Pereira, V.J.; Crespo, J.G. Removal of pesticides and polycyclic aromatic hydrocarbons from different drinking water sources by nanofiltration. *Desalination Water Treat.* **2011**, *27*, 141–149. [CrossRef]
27. Salahi, A.; Mohammadi, T.; Rekabdar, F.; Mahdavi, H. Reverse osmosis of refinery oily wastewater effluents. *J. Environ. Heal. Sci. Eng.* **2010**, *7*, 413–422.
28. Abadi, S.R.H.; Sebzari, M.R.; Hemati, M.; Rekabdar, F.; Mohammadi, T. Ceramic membrane performance in microfiltration of oily wastewater. *Desalination* **2011**, *265*, 222–228. [CrossRef]
29. Ishak, S.; Malakahmad, A.; Isa, M.H. Refinery wastewater biological treatment: A short review. *J. Sci. Ind. Res.* **2012**, *71*, 251–256.

30. Shariati, S.R.P.; Bonakdarpour, B.; Zare, N.; Ashtiani, F.Z. The effect of hydraulic retention time on the performance and fouling characteristics of membrane sequencing batch reactors used for the treatment of synthetic petroleum refinery wastewater. *Bioresour. Technol.* **2011**, *102*, 7692–7699. [CrossRef] [PubMed]
31. Zhong, J.; Sun, X.; Wang, C. Treatment of oily wastewater produced from refinery processes using flocculation and ceramic membrane filtration. *Sep. Purif. Technol.* **2003**, *32*, 93–98. [CrossRef]
32. Zhu, Y.; Wang, D.; Jiang, L.; Jin, J. Recent progress in developing advanced membranes for emulsified oil/water separation. *NPG Asia Mater.* **2014**, *6*, e101. [CrossRef]
33. Li, J. Chemical Stability of Ceramic Membrane. Delft University of Technology: Delft, The Netherlands, 2016; Unpublished manuscript.
34. Chiu, T. Effect of ageing on the microfiltration performance of ceramic membranes. *Sep. Purif. Technol.* **2011**, *83*, 106–113. [CrossRef]
35. Zhao, Y.-Y.; Wang, X.-M.; Yang, H.-W.; Xie, Y.-F.F. Effects of organic fouling and cleaning on the retention of pharmaceutically active compounds by ceramic nanofiltration membranes. *J. Membr. Sci.* **2018**, *563*, 734–742. [CrossRef]
36. López, J.; Reig, M.; Vecino, X.; Gibert, O.; Cortina, J.L. Comparison of acid-resistant ceramic and polymeric nanofiltration membranes for acid mine waters treatment. *Chem. Eng. J.* **2019**, *382*, 122786. [CrossRef]
37. Dalwani, M.; Bargeman, G.; Hosseiny, S.S.; Boerrigter, M.; Wessling, M.; Benes, N.E. Sulfonated poly(ether ether ketone) based composite membranes for nanofiltration of acidic and alkaline media. *J. Membr. Sci.* **2011**, *381*, 81–89. [CrossRef]
38. Standard Methods for the Examination of Water and Wastewater. American Public Health Association: Washington, DC, USA, 2012. [CrossRef]
39. Abramoff, M.D.; Magalhaes, P.J.; Ram, S.J. Image Processing with ImageJ. *Biophotonics Int.* **2004**, *11*, 36–42.
40. Masselin, I.; Durand-Bourlier, L.; Laine, J.M.; Sizaret, P.Y.; Chasseray, X.; Lemordant, D. Membrane characterization using microscopic image analysis. *J. Memb. Sci.* **2001**, *186*, 85–96. [CrossRef]
41. Santos, B.; Galinha, C.F.; Crespo, J.G.; Santos, M.A.; Velizarov, S. Prediction of polar oil and grease contamination levels in refinery wastewater through multivariate statistical modeling. *Sep. Purif. Technol.* **2013**, *119*, 51–57. [CrossRef]
42. Rita, A.; Monteiro, A.; Albuquerque, R.; Santos, M.; Ribeiro, J.; Madeira, L.; Sanches, S. Unravelling the relation between processed crude oils and the composition of spent caustic effluents as well as the respective economic impact. *J. Hazard. Mater.* **2021**, *421*, 126629. [CrossRef] [PubMed]
43. Sousa, M.R.S.; Lora-García, J.; López-Pérez, M.-F.; Heran, M. Identification of Foulants on Polyethersulfone Membranes Used to Remove Colloids and Dissolved Matter from Paper Mill Treated Effluent. *Water* **2020**, *12*, 365. [CrossRef]
44. Walton, W.H. Feret's Statistical Diameter as a Measure of Particle Size. *Nature* **1948**, *162*, 329–330. [CrossRef]
45. Pretsch, E.; Clerc, T.; Seibl, J.; Simon, W. *Tables of Spectral Data for Structure Determination of Organic Compounds*, 2nd ed.; Springer-Verlag: Berlin/Heidelberg, Germany, 1989. [CrossRef]
46. Kazemnejadi, M.; Shakeri, A.; Nikookar, M.; Shademani, R.; Mohammadi, M. Selective and metal-free epoxidation of terminal alkenes by heterogeneous polydioxirane in mild conditions. *R. Soc. Open Sci.* **2018**, *5*, 171541. [CrossRef]
47. Wang, N.; Raza, A.; Si, Y.; Yu, J.; Sun, G.; Ding, B. Tortuously structured polyvinyl chloride/polyurethane fibrous membranes for high-efficiency fine particulate filtration. *J. Colloid Interface Sci.* **2013**, *398*, 240–246. [CrossRef]
48. Cheng, F.; He, J.; Yan, T.; Liu, C.; Wei, X.; Li, J.; Huang, Y. Antibacterial and hemostatic composite gauze of N,O-carboxymethyl chitosan/oxidized regenerated cellulose. *RSC Adv.* **2016**, *6*, 94429–94436. [CrossRef]
49. Khan, J.A.; Sayed, M.; Shah, N.S.; Khan, S.; Zhang, Y.; Boczkaj, G.; Khan, H.M.; Dionysiou, D.D. Synthesis of eosin modified TiO_2 film with co-exposed {001} and {101} facets for photocatalytic degradation of para-aminobenzoic acid and solar H_2 production. *Appl. Catal. B Environ.* **2019**, *265*, 118557. [CrossRef]
50. Bhuvaneswari, G.H. Degradability of Polymers. In *Recycle Polyurethane Foam;* Sabu Thomas, M.G.T., Rane, A.V., Kanny, K., Abitha, V.K., Eds.; William Andrew Publishing: Norwich, NY, USA, 2018; pp. 29–44. [CrossRef]
51. Radelyuk, I.; Tussupova, K.; Zhapargazinova, K.; Yelubay, M.; Persson, M. Pitfalls of Wastewater Treatment in Oil Refinery Enterprises in Kazakhstan—A System Approach. *Sustainability* **2019**, *11*, 1618. [CrossRef]
52. Nyquist, R.A. (Ed.) Azines, Isocyanates, Isothiocyanates, and Carbodiimides. In *Interpreting Infrared, Raman, and Nuclear Magnetic Resonance Spectra*; Academic Press: London, UK, 2001; pp. 45–63. [CrossRef]
53. Li, C.; Sun, W.; Lu, Z.; Ao, X.; Li, S. Ceramic nanocomposite membranes and membrane fouling: A review. *Water Res.* **2020**, *175*, 115674. [CrossRef] [PubMed]
54. Fujioka, T.; Khan, S.J.; McDonald, J.A.; Nghiem, L.D. Nanofiltration of trace organic chemicals: A comparison between ceramic and polymeric membranes. *Sep. Purif. Technol.* **2014**, *136*, 258–264. [CrossRef]

membranes

MDPI

Article

Low-Temperature Hydrophilic Pervaporation of Lactic Acid Esterification Reaction Media

Elena González Díaz [1], Sonia Álvarez-García [2], Susana Luque [2] and José R. Álvarez [2,*]

[1] Sulzer Chemtech Ltd., 4123 Allschwil, Switzerland; elena.gonzalez@sulzer.com
[2] Department of Chemical and Environmental Engineering, University of Oviedo, 33071 Oviedo, Spain; alvarezsonia@uniovi.es (S.Á.-G.); sluque@uniovi.es (S.L.)
* Correspondence: jras@uniovi.es; Tel.: +34-985103028

Abstract: Esterification reactions show a limited conversion due to the presence of water, which favors the opposite reaction. The removal of water from the reaction mixture increases the production of the ester. Pervaporation is an effective dehydration technique, usually applied to binary mixtures. The effect on pervaporation of a reactive multicomponent system involving water, ethanol, ethyl lactate and lactic acid with high acid concentration (13.5 wt. %) at relatively low temperatures (40–80 °C) was studied. Three hydrophilic membranes mainly fabricated for dehydration purposes from Sulzer Chemtech were used, i.e., PERVAP™ 3100, PERVAP™ 2216 and PERVAP™ 1131. The last one revealed as the most suitable for the application and it was further characterized with binary and ternary solutions. The membrane showed high affinity for the lactic acid. The acid permeation played a key role in the water/ethanol and water/ethyl lactate selectivity. Lactic acid permeates and crystalizes in the permeate side of the membrane at very low water concentration (below 2 wt. %), causing a drop in flux and membrane selectivity. Ethyl lactate is responsible of the loss of integrity of the membranes.

Keywords: pervaporation; hydrophilic membrane; temperature

1. Introduction

The production of lactic acid esters is one the most promising uses of lactic acid, since they can be used in the pharmaceutical, food, automotive and aerospace industries. Lactic esters have a great potential as green solvents, because of their high boiling points and nontoxic and biodegradable properties, replacing common organic solvents like methyl isobutyl ketone [1].

Lactic acid esterification is commonly catalyzed by organic and inorganic acids, ionic exchange resins and zeolites [2], but enzymes as lipases have also been employed [3]. Water content is a key parameter that affects the equilibrium concentration of reactants and products in the reaction medium, because it favors the ester hydrolysis and limits the yield of the ester. Moreover, in the case of enzyme-catalyzed esterification, it also affects the thermal inactivation of the enzymes.

Pervaporation (PV) is an attractive method for water removal in esterification reactions, helping the system to shift the conversion beyond the equilibrium limitation [4,5]. Hydrophilic membranes, i.e., those able to separate mainly water from an organic–water mixture, are employed. Membrane common materials are: polyvinylalcohol (PVA), polyvinilalcohol/polyacrilonitrile (PVA/PAN), polyetherimide (PEI) [6], with PVA the most frequently employed [7]. The use of nonpolymeric materials such as zeolites is also remarkable [8–10].

One of the first applications in esterification reactions was introduced in 1987 by Kita et al. [11] who found that PV was able to shift the conversion of oleic acid with ethanol beyond the reaction equilibrium value and, therefore, reduce the need for excess reactants. The first kinetic studies on the coupling of PV and esterification were published by David et al. [12,13] in the reaction of propionic acid with 1-propanol and 2-propanol using a PVA membrane. The results showed yield improvements of 20%.

Bagnell et al. [14] used catalytically active PV Nafion tubes in the reaction of acetic acid with methanol and *n*-butanol, increasing the conversions from 73% and 70% to 77% and 95%, respectively. Butyl acetate was also obtained by Zhu and Chen [15] with a catalytic crosslinked PVA active layer on a porous ceramic plate support; by Xuehei and Lefu [16] with a phosphatic PVA/PAN membrane; by Liu and Chen [17] with a PVA/porous ceramic one; by Peters et al. [18] with a catalytic zeolite coat on a ceramic hollow fiber silica support; by Qing et al. [19] with a layer of PVA and zirconium sulfate tetrahydrate on a PVA/polyethersulfone membrane; and by Zhang et al. [20] with a coat of PVA and ion-exchange resins on a dense PVA/polyethersulfone membrane.

Benedict et al. [5,21] esterified succinic and lactic acid with ethanol in the presence of solid catalysts (Amberlyst Xn-1010 and Nafion NR50) using GFT-1005 and T1-b pervaporation membranes. The esterification of lactic acid and ethanol in the presence of Amberlyst 15 was done by Ma et al. [22] with a chitosan–tetraethoxysilane organic–inorganic hybrid membrane, by Delgado et al. [23] with the hydrophilic PERVAP™ 2201, by Pereira et al. [24] with a microporous silica membrane, and by Zhang et al. [25] with three home-made composite membranes of glutaral, chitosan, gelatin, carbomer, hyaluronic acid and carbomer on PAN. Budd et al. [26] chose zeolite/polyelectrolyte multilayer membranes and *p*-toluene sulfonic acid for obtaining ethyl lactate. This acid and the PERVAP™ 2205 membrane were combined with an ionic solvent to produce bornyl acetate from acetic acid and borneol by Izák et al. [27]. Amberlyst-15 and PERVAP™ 2201 were also employed by Assabumrungrat et al. [28] in the synthesis of methyl acetate and by Sanz and Gmehling [29,30] in the reaction of acetic acid with isopropanol.

The aim of this work is to find an adequate PV membrane (initially designed for the dehydration of solvents) for the dehydration of the mixture involved in the esterification of lactic acid with excess ethanol. The chosen initial acid to alcohol molar ratio was between 1:5 and 1:10, the pH is close to 3 units and the initial amount of water is around 10 wt. %, since aqueous solutions of lactic acid (80 wt. %) and azeotropic ethanol were used, aiming to conduct a study with mixtures close to the actual industrial reaction conditions. Concentration ranges differ considerably from the study of Delgado et al. [31], in which the water content was always higher than 10 wt. %. In the present study, water concentration was reduced to 2 wt. % and below.

After a preliminary screening at 60 °C using three hydrophilic membranes, a more thorough study was performed on the most suitable membrane using different feed composition (binary, ternary, and quaternary solutions) and two other temperatures (40 °C, the optimum temperature for biocatalysis, and 80 °C, closer to the usual working temperature for dehydration by PV).

2. Materials and Methods

2.1. Materials and Analytical Methods

L(+) Lactic acid from Riedel de Häen (80 wt. % aqueous solution; Seelze, Germany), ethyl lactate (99%, Fluka, Buchs, Switzerland), and azeotropic ethanol (96%, Panreac, Barcelona, Spain) were all of analytical grade and used without further purification.

Ethyl lactate, ethanol, and water concentration in permeate were measured by gas chromatography with a thermal conductivity detector (GC-TCD Shimatzu GC-8, Kyoto, Japan), using a Supelco 30 m capillary column. Injection temperature was 200 °C and the column was kept at 190 °C.

Water concentration in feed was analyzed by a Karl Fischer coulometer (Metrohm 737 KF, Herisau, Switzerland). Lactic acid concentration in feed and permeate was analyzed by titration with NaOH 0.05 M, using phenolphthalein as indicator.

All concentrations are expressed as weight per cent.

2.2. Pervaporation Membranes

In this work, the commercial PERVAP™ 2216 and the experimental PERVAP™ 1131 and PERVAP™ 3100 membranes, manufactured by Sulzer Chemtech Ltd. (Neunkirchen, Germany) [32], were tested. These membranes were chosen after a preliminary screening of several commercial PV membranes (results not shown), most of which had severe damage within a short time when exposed to concentrated lactic acid/ethyl lactate aqueous solutions.

PERVAP™ membranes are asymmetric composite membranes for water permeation. PERVAP™ 2216 membrane is made of crosslinked PVA supported on an ultrafiltration membrane made of PAN, which is cast on a layer of non-woven porous polyester. PVA has relatively good chemical stability, film-forming ability, and high hydrophilicity, but has a poor stability in aqueous solutions. Therefore, it needs to be insolubilized by crosslinking (with maleic acid, glutaraldehyde, etc.) or otherwise be modified to get good mechanical properties and selective permeability towards water. This membrane is similar to other commercial PV membranes, such as PERVAP™ 2201D, PERVAP™ 2200, PERVAP™ 2201 or PERVAP™ 2211, differences being in the degree of crosslinking [33].

PERVAP™ 3100 is similar to PERVAP™ 2216 made with a thin top layer and an additional nonwoven supporting structure of polypropylene, which aims to provide higher fluxes while maintaining good mechanical resistance.

PERVAP™ 1131 is an experimental membrane, based on polyethylene polymerization by plasma application, which makes it more resistant towards acid aqueous solutions.

According to the manufacturer, the three membranes are designed to work between 40 and 105 °C.

2.3. Pervaporation Apparatus and Procedure

PV experiments were carried out in a commercially available test equipment from Sulzer Chemtech, a description of which can be found elsewhere [34]. The effective membrane area of the test cell was 178 cm^2 and the volume of the feed tank was approximately 2 L. In the feed side, an additional water cooler was also used when the required temperature was 40 °C. The permeate side pressure was always maintained at 11 ± 1 mbar. Two cold traps of dry ice in ethanol located in the permeate line allowed the condensation and collection of permeate samples. Prior to be used in the PV experiments, the membranes were placed in the test cell and contacted with the feed solution for at least 16 h. During this conditioning, permeate pressure was always kept at 11 mbar.

Feed was pumped to the membrane cell at a flow rate of 40 L/h ensuring a turbulent flow (Reynolds number of 3000), to achieve a similar hydrodynamic regime as in industrial modules.

Four feed mixtures were employed to assess the influence of each component and the possible coupling effects. The initial concentrations of water, lactic acid and ethyl lactate are always the same in the feed stream, while the ethanol is adjusted accordingly:

- Binary mixture: 10 wt. % water and 90 wt. % ethanol.
- Ternary mixtures: One with ester, water, and alcohol (17.7 wt. % ethyl lactate, 10 wt. % water and 72.3 wt. % ethanol) and another with acid, water, and alcohol (13.5 wt. % lactic acid, 10 wt. % water and 76.5 wt. % ethanol). There is no lactic acid + ethyl lactate + ethanol mixture because ethanol and lactic acid were used from commercial aqueous solutions, as described in Section 2.1.
- Quaternary mixture: 13.5 wt. % lactic acid, 17.7 wt. % ethyl lactate, 10 wt. % water and 58.8 wt. % ethanol.

Experiments proceeded until values below 2 wt. % of water in the feed were achieved, which corresponded to almost no permeation through the membrane. Concentration profiles for all the compounds were followed by sampling feed and permeate and analyzing them as described previously. Permeation experiments were on average week-long tests (except for the 22,016 membrane that spanned twice as long). Some tests were replicated and the errors were below 2%.

The separation factor (β) was calculated according to the ratio of concentrations between permeate (p) and feed (f) for two substances (i and j) with the following expression:

$$\beta_{pervap\ i,j} = \frac{C_{pi}}{C_{pj}}\frac{C_{fj}}{C_{fi}} \quad (1)$$

Wijmans and Baker [35] considered pervaporation as a combination of evaporation and membrane separation steps, and therefore including two contributions in the separation factor: $\beta_{pervap} = \beta_{evap} \cdot \beta_{mem}$, where β_{evap} is the separation factor resulting of the vapor-liquid equilibrium:

$$\beta_{evap\ i,j} = \frac{p'_i}{C_{fi}}\frac{C_{fj}}{p'_j} \quad (2)$$

and β_{mem} is the separation factor due to the membrane. β_{mem} can be expressed, in turn, as the ratio of the permeate fluxes (J) and partial pressures (p') of the compounds i and j in the feed.

The flux of an i component is described as a function of its permeance (P_i) times the partial pressure difference between the feed (p'_i) and the permeate (p''_i):

$$J_i = \mathrm{P}_i(p'_i - p''_i) \quad (3)$$

Partial pressures in the feed side are calculated using liquid composition, pure component vapor pressure (calculated using an extended Antoine equation, found in the Appendix A) and the activity coefficient, which was obtained from the UNIQUAC activity model (the parameters are also listed in the Appendix A) which proved to be the best model for this system, based on the multicomponent vapor-liquid equilibria data experimentally determined by Delgado et al. [36]. Activity coefficients are close to one for both lactic acid and ethanol, between 1.7 and 2 for water, and between 6 and 7 for ethyl lactate, depending on the composition. The partial pressures in the permeate can be calculated directly from the total permeate pressure times the permeate concentration.

The ratio of permeate fluxes between components i and j is

$$\frac{J_i}{J_j} = \frac{\mathrm{P}_i}{\mathrm{P}_j}\frac{(p'_i - p''_i)}{\left(p'_j - p''_j\right)} \quad (4)$$

As the ratio of permeances ($\mathrm{P}_i/\mathrm{P}_j$) is the i/j membrane selectivity (α_{ij}), and the ratio of fluxes is equal to the ratio of permeate pressures, Equation (4) leads to

$$\frac{J_i}{J_j} = \frac{p''_i}{p''_j} = \alpha_{ij}\frac{(p'_i - p''_i)}{\left(p'_j - p''_j\right)} \quad (5)$$

Finally, combining (1), (2) and (5), the resulting expression is

$$\beta_{pervap} = \beta_{evap}\,\alpha_{ij}\frac{(p'_i - p''_i)}{\left(p'_j - p''_j\right)}\frac{p'_j}{p'_i} \quad (6)$$

If permeate pressure is considerably lower than the feed pressure, it simplifies to:

$$\beta_{pervap} = \beta_{evap}\,\alpha_{ij} \quad (7)$$

Equation (7) will be used in this study. β_{pervap} is obtained experimentally, β_{evap} is calculated from vapour-liquid equilibria, and α_{ij} will be ratio between them.

3. Results

3.1. Membrane Selection

To select the most suitable membrane for the mixture representing the reaction to produce ethyl lactate, a 13.5 wt. % lactic acid and 17.7 wt. % ethyl lactate solution in ethanol (acid to alcohol molar ratio of 1:10), with 10 wt. % of water content, was pumped to the membrane cell.

Figure 1 shows the influence of the feed water concentration on the permeate water flux for the three membranes tested. PERVAP™ 2216 gave the lowest flux values, in fact more than 12 days were necessary to reduce the feed water content from 10 to 2 wt. %. However, this membrane shows the best selectivity, with water contents in the permeate above 99.5 wt. %. This is coherent with a highly cross-linked top layer (and likely thicker than the other two). The selectivity of the other two membranes was very similar, although the flux of the PERVAP™ 3100 was twice as much that of PERVAP™ 1131. However, the former had a very poor chemical resistance and had to be replaced weekly during the study. Its thinner top layer was not robust enough for this application.

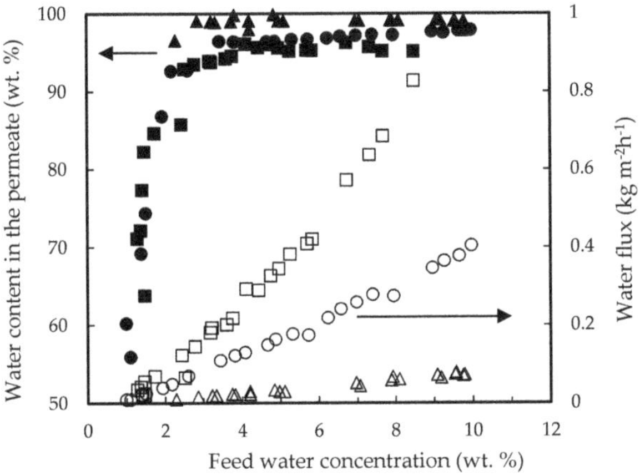

Figure 1. Water permeation fluxes (open symbols) and water concentration in permeate (filled symbols) as a function of feed water content for the three PV membranes tested. (■,□) PERVAP™ 3100, (●,○) PERVAP™ 1131 and (▲,△) PERVAP™ 2216. Initial feed content: 13.5% lactic acid, 17.7% ethyl lactate, 10% water and 58.8% ethanol. Temperature = 60 °C.

The permeation fluxes of all the components and the selectivity profiles water/ethanol and water/ethyl lactate as a function of the water content in the feed mixture are shown in Figures 2 and 3, respectively. Ethanol fluxes corresponding to the PERVAP™ 2216 were below the detection limit (under 0.005 wt. %), and thus, only two non-infinite selectivity values could be calculated. Component fluxes, as shown in Figure 2, follow the trend: water > ethanol > ethyl lactate ≈ lactic acid, in accordance with the molecular size. The molecular weights of water, ethanol, lactic acid, and ethyl lactate are 18.01, 46.07, 90.08, and 118.13 g/mol, respectively.

With respect to the membranes, the permeation flux decreased in the order PERVAP™ 3100 > PERVAP™ 1131 > PERVAP™ 2216. Moreover, two differentiated behaviors could be observed depending on the water content in the feed:

- Above 2 wt. %: The fluxes for all feed components increased linearly as the water concentration in the feed increased (Figure 2). There was an order of magnitude difference between water and ethanol fluxes, and almost another order of magnitude between ethanol and both ethyl lactate and lactic acid fluxes, respectively. Selectivity of water/ethanol and ethyl lactate/ethanol (Figure 3) was constant.

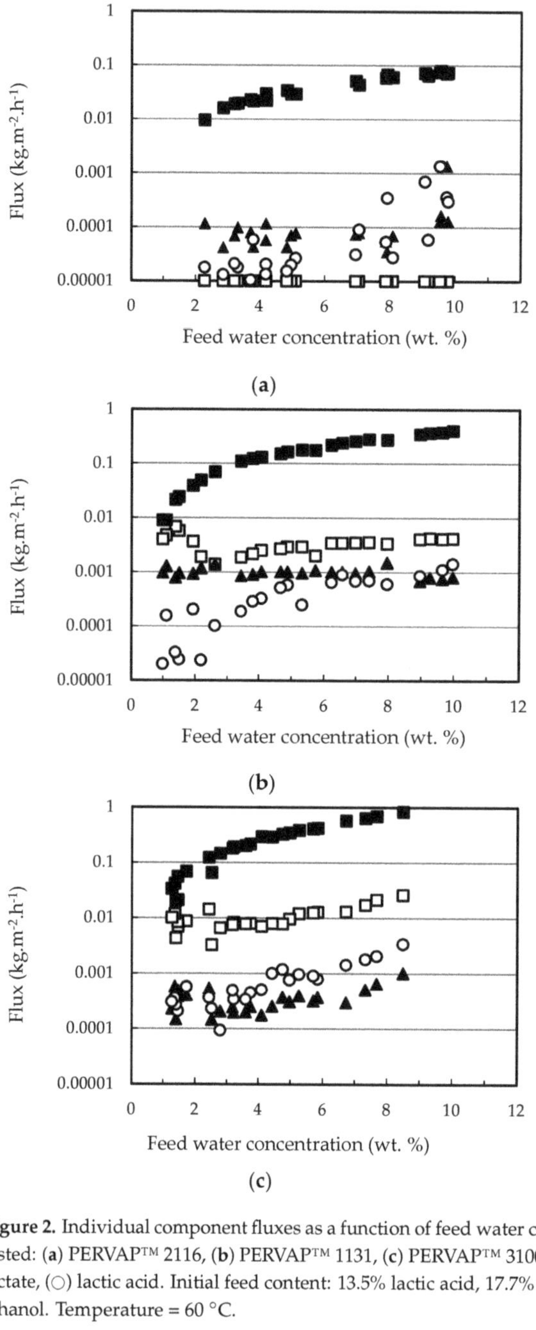

Figure 2. Individual component fluxes as a function of feed water content for the three PV membranes tested: (**a**) PERVAP™ 2116, (**b**) PERVAP™ 1131, (**c**) PERVAP™ 3100. (■) Water, (□) ethanol, (▲) ethyl lactate, (○) lactic acid. Initial feed content: 13.5% lactic acid, 17.7% ethyl lactate, 10% water and 58.8% ethanol. Temperature = 60 °C.

- Below 2 wt. %: Water concentration in the permeate dropped exponentially when water content in the feed decreased below 2 wt. %. PERVAP™ 2216 fluxes could be considered negligible, even before reaching 2 wt. % (Figures 1 and 2a). PERVAP™ 1131 showed a constant ethanol and ethyl lactate fluxes, while lactic acid flux decreased dramatically like water flux (Figure 2b). There was, however, significant scattering below 2 wt. %. Finally, PERVAP™ 3100 also showed a sharp decrease in the selectivity of water/ethanol and water/ethyl lactate as the water feed content was low.

This behavior has been previously reported in literature [37] and can be explained considering the effect of different substances in membrane swelling.

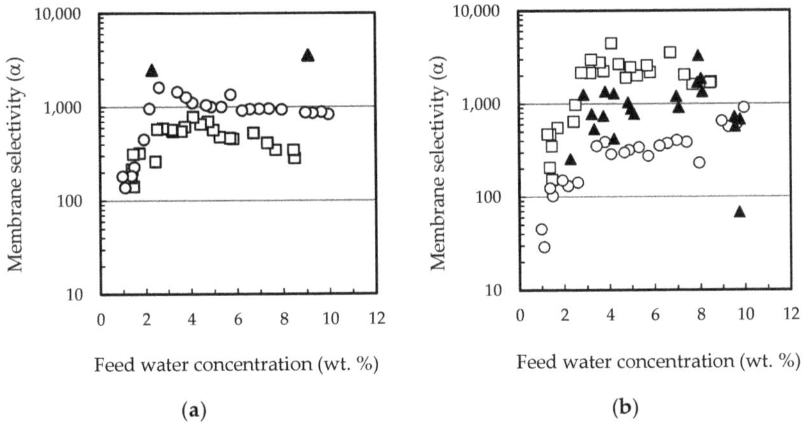

Figure 3. Membrane selectivity as a function of feed water content for the three PV membranes tested (**a**) water/ethanol, (**b**) water/ethyl lactate. (□) PERVAP™ 3100, (○) PERVAP™ 1131 (▲) PERVAP™ 2216. Initial feed content: 13.5% lactic acid, 17.7% ethyl lactate, 10% water and 58.8% ethanol. Temperature = 60 °C.

3.2. PERVAP™ 1131 Membrane Characterization

Due to the low water flux of PERVAP™ 2216, to the poor resistance of PERVAP™ 3100, PERVAP™ 1131, which was acid-resistant and gave high selectivity, was selected as the most appropriate membrane for further analysis. The performance of PERVAP™ 1131 with different feed composition and temperatures was, thus, studied more thoroughly.

In the case of feed composition, four systems were employed to assess the influence of each component and the possible coupling effects: one binary, two ternary and the quaternary reaction mixture used for membrane screening.

Figure 4 shows fluxes and water permeate contents for the four above-mentioned mixtures. The highest content of water in permeate corresponded to the quaternary mixture, but sharply dropped when feed water decreases below 2 wt. %. This was also found but less pronounced in the other mixtures. The flux of water was the same independently of the feed mixture tested.

Figure 5 shows the influence of the different substances when added to a water-ethanol mixture. Figure 5a shows that the flux of water was initially one order of magnitude higher than the flux of ethanol, that stays relatively constant. However, both fluxes became very similar as the feed water content dropped down to 2 wt. %. When ethyl lactate was added as third component there was no effect on water or ethanol fluxes with respect to the binary mixture, and the flux of ethyl acetate was two orders of magnitude lower than water, as shown in Figure 5b. In contrast, when the third component was lactic acid, the flux of ethanol decreased at the same rate as that of water, the flux of lactic acid being two to three orders of magnitude lower, as shown in Figure 5c. When the four compounds are in the mixture (Figure 5d), there was a combined effect of lactic acid and ethyl lactate in the flux of ethanol. The latter was at that point two orders of magnitude lower than the flux of water until the water feed content dropped to 2 wt. %, at which point it sharply increased to equal that of water. Lactic acid flux also decreased more significantly when the water feed content dropped down 5 wt. %. On the other hand, the flux of ethyl lactate stayed constant, as in the tertiary mixture shown in Figure 5b. The resulting water/ethanol and water/ethyl lactate membrane selectivity is shown in Figure 6 as a function of water feed composition.

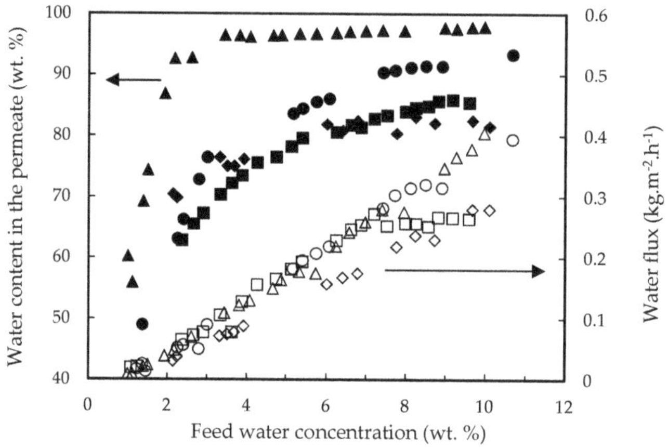

Figure 4. Water permeation fluxes (open symbols) and water concentration in permeate (closed symbols) as a function of feed water content for four mixtures: (▲,△) 13.5% lactic acid, 17.7% ethyl lactate, 10% water and 58.8% ethanol; (■,□) 17.7% ethyl lactate, 10% water and 72.3% ethanol; (◆,◇) 13.5% lactic acid, 10% water and 76.5% ethanol; (●,○) 10% water and 90% ethanol. PERVAP™ 1131 membrane at T = 60 °C.

Figure 6a shows that the water/ethanol selectivity for binary and ternary mixtures was the same for all water feed compositions, and the selectivity for the quaternary mixture was one order of magnitude higher in favor of water permeation, which indicates a strong effect of the presence at the same time of both lactic acid and ethyl lactate, reducing ethanol permeation by an order of magnitude as mentioned. When ethyl lactate was added to water-ethanol mixture, there was no change of membrane selectivity, as described from Figure 5a,b, and water and ethanol permeation rates were about the same. When lactic acid was added to water-ethanol mixture, there was an increase in water-ethanol selectivity at low feed water concentration, because both water and ethanol fluxes decreased unlike in the other cases, as shown in Figure 5c. When lactic acid and ethyl lactate were both added to the water-ethanol mixture, the water/ethanol selectivity increased tenfold because lactic acid and ethyl lactate were preferentially dissolved in the membrane, allowing the passage of water but hindering the passage of ethanol (this will be further analyzed in Section 3.3). Moreover, water-ethyl lactate selectivity was not affected (Figure 6b), as expected from the fluxes shown in Figure 5b,d.

Membranes were not fouled by any of the components of the mixture. As shown in Figure 4 water permeation fluxes were the same for any of the binary, ternary or quaternary mixtures. The same membrane was used for all experiments and no cleaning was performed between the tests.

Nevertheless, a relative long exposure of the membrane (30 days) to the ethyl lactate showed membrane deterioration (coloring and the presence of pores or defects in the surface), in fact the membrane started to dissolve, due to the combined action of acid and the ethyl lactate as solvent, losing its integrity.

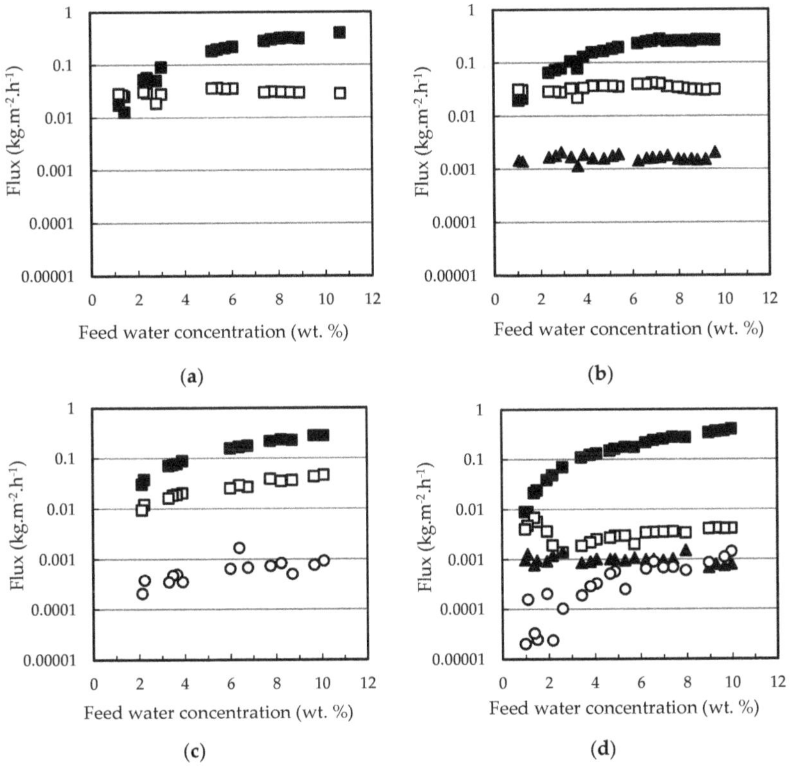

Figure 5. Individual component fluxes as a function of feed water content for mixtures of different composition: (**a**) Water + ethanol, (**b**) water + ethanol + ethyl lactate, (**c**) water + ethanol + lactic acid, (**d**) water + ethanol + ethyl lactate + lactic acid. (■) Water, (□) ethanol, (▲) ethyl lactate, (○) lactic acid. PERVAP™ 1131 membrane at T = 60 °C.

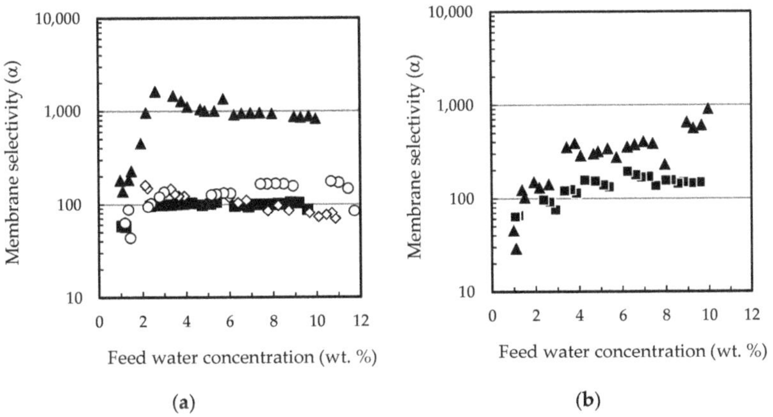

Figure 6. Selectivity as a function of feed water content for mixtures of different composition: (**a**) water/ethanol, (**b**) water/ethyl lactate. (▲) contains 13.5% lactic acid, 17.7% ethyl lactate, 10% water and 58.8% ethanol; (■) 17.7% ethyl lactate, 10% water and 72.3% ethanol; (◇) 13.5% lactic acid, 10% water and 76.5% ethanol; (○) 10% water and 90% ethanol. PERVAP™ 1131 membrane at T = 60 °C.

3.3. Degree of Swelling

To study the interaction of all the components with the membrane, the degree of membrane swelling when subjected to static contact was determined. Swelling depends not only on the structure of membrane but also on the affinity of the components towards the membrane, the interactions between components, and the way the interactions of each component with the membrane affects the interaction of the rest of components, as expected from a nonideal mixture [38–40].

The degree of swelling was calculated for all compounds according to the variation in the weight of a piece of membrane placed in a vial for 15 days at 60 °C, as follows:

$$Degree\ of\ Swelling\ (\%) = \frac{W_w - W_d}{W_d} 100 \quad (8)$$

where W_w and W_d are the weights of the membrane in wet and dry states, respectively.

Pure lactic acid and ethyl lactate are the components with more affinity for the membrane, the degree of swelling for lactic acid being twice the value for ethyl lactate (Table 1). For the other components, only water showed some relevant swelling, but five times lower than ethyl lactate. This suggests a very strong affinity of the PERVAP™ 1131 towards acids. To verify this hypothesis, several pieces of membrane were soaked with water, ethyl lactate and water respectively at 60 °C, adding 0.5 mL of concentrated sulfuric acid (to have a pH below 2 units, as with the lactic acid). The results show an increase in swelling in accordance with the suggested acid affinity.

Table 1. Degree of swelling calculated as the change of weigh of a piece of membrane incubated for 15 days at 60 °C.

Pure Components	Swelling (%)
Ethanol	1.4
Water	7.0
Ethyl lactate	33
Lactic acid	67
Acidic components	
Ethanol + 0.5 mL H_2SO_4	8.8
Water + 0.5 mL H_2SO_4	20
Ethyl lactate + 0.5 mL H_2SO_4	60
Mixtures	
10% water in ethanol	2.9
17.7% ethyl lactate + 10% water in ethanol	1.4
13.5% lactic acid + 10% water in ethanol	12
17.7% ethyl lactate + 13.5% lactic acid + 10% water in ethanol	15

Comparing the degree of swelling of the four solutions in the characterization of the PERVAP™ 1131, the highest value was obtained for the four-component solution, which is the actual case in an esterification reaction and the one which yields the highest selectivity for water. Figure 6a shows, as mentioned, the higher water/ethanol selectivity, which clearly accounts for the effect of lactic acid in competition with the ethyl lactate. In fact, it is the combined effect of the acid and the ethyl lactate responsible for that change. Moreover, during the pervaporation experiments, due to the high permeate rate of water with respect to the rest of the components, there is a surprising reduction of lactic acid concentration in the feed, while its concentration should increase, being the least permeable. The lactic acid disappearance cannot be justified by the esterification reaction since there is no match in the mass balance that accounts for ethanol or ethyl lactate variations. This, together with the high degree of interaction, indicates that the acid that disappears from the solution permeates the membrane. This was found to happen at the permeate side of the membrane, where the vacuum conditions result in reduced water and ethanol concentrations and gave rise to the formation of lactic acid crystals, as proven by microscopy (Figure 7). The reduction of water and the presence of ethyl lactate in the membrane promotes the crystallization of lactic acid in a nonaqueous environment.

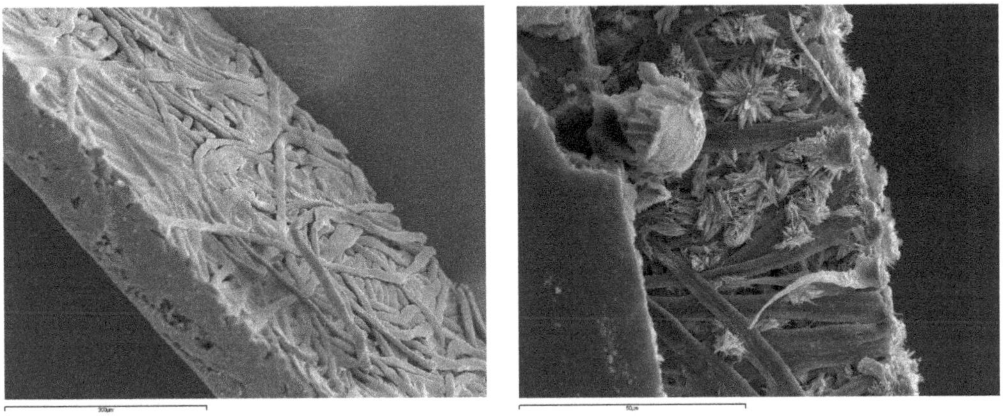

(**a**) (**b**)

Figure 7. PERVAP™ 1131 membrane SEM pictures: (**a**) new membrane, (**b**) membrane with lactic acid crystals after pervaporation.

3.4. *Temperature Effect*

Experiments were also performed at 40 and 80 °C. Figure 8 shows the water permeation fluxes together with the water content in the permeate. Data at 40 °C were scarce due to the low permeation rate. Temperature clearly increases water flux, and the water permeate content sharply decreases when the water feed content drops down 2 wt. %, as it was shown for the three membranes tested in Figure 1.

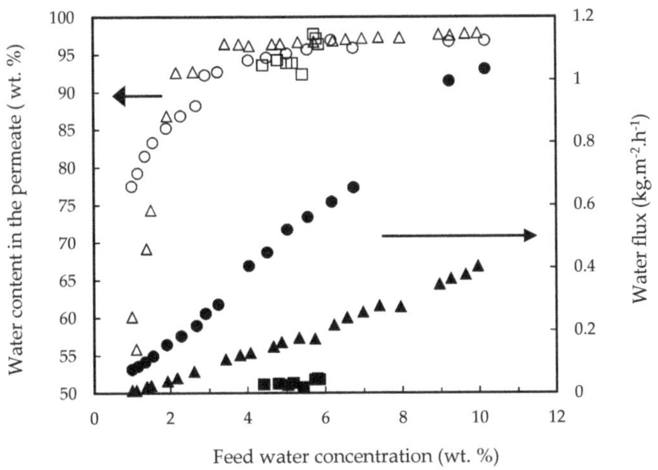

Figure 8. Water permeation fluxes (open symbols) and water concentration in permeate (closed symbols) as a function of feed water content for the PERVAP™ 1131 membrane at different temperatures: (●,○) T = 80 °C, (▲,△) T = 60 °C, (■,□) T = 40 °C. Initial feed content: 13.5% lactic acid, 17.7% ethyl lactate, 10% water and 58.8% ethanol.

The fluxes of the rest of substances in the quaternary mixture increased with temperature as shown in Figure 9.

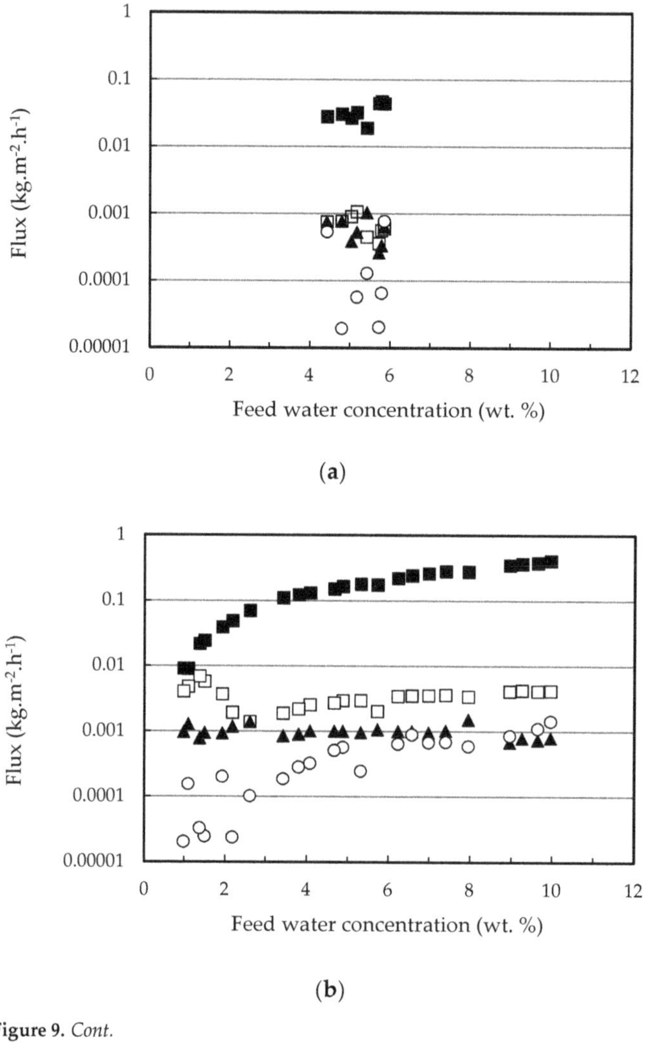

Figure 9. *Cont.*

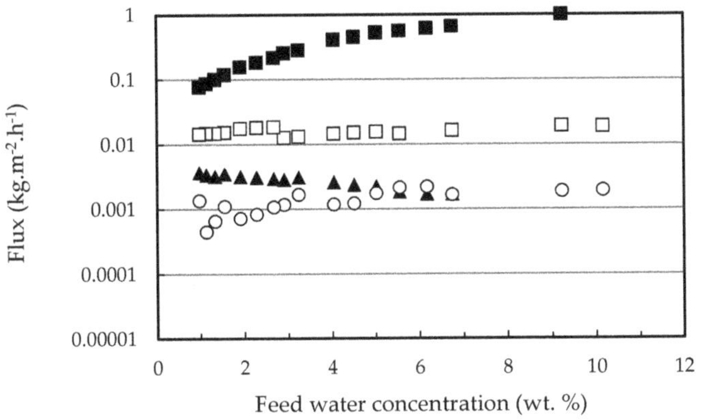

(c)

Figure 9. Individual component fluxes as a function of feed water content for the PERVAP™ 1131 membrane at: (**a**) 40 °C, (**b**) 60 °C, (**c**) 80 °C. (■) Water, (□) ethanol, (▲) ethyl lactate, (○) lactic acid. Initial feed content: 13.5% lactic acid, 17.7% ethyl lactate, 10% water and 58.8% ethanol.

Temperature did not significantly affect water/ethanol selectivity, as shown in Figure 10a, but water/ethyl lactate selectivity showed a higher increase from 40 to 60 °C than from 60 to 80 °C and decreasing as the feed water content was reduced (Figure 10b). Both effects are always in favor of the more volatile component: ethanol in the case of water/ethanol and water in the case of water/ethyl lactate. These effects were expected, since temperature weakens hydrogen bonds and increase the motion of the polymer chains, which facilitates diffusion. The lactic acid is more soluble at higher temperatures, so it crystallizes at lower water concentration, and this allows the membrane to perform better [40].

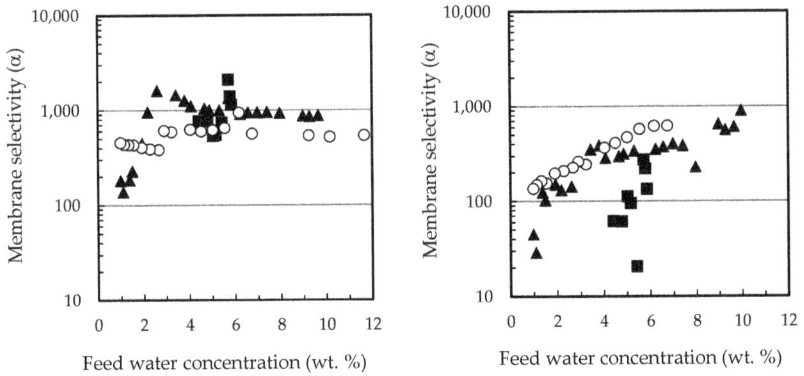

(a) (b)

Figure 10. Membrane selectivity as a function of feed water content for the PERVAP™ 1131 membrane at different temperatures: (**a**) water/ethanol, (**b**) water/ethyl lactate. (○) T = 80 °C, (▲) T = 60 °C, (■) T = 40 °C. Initial feed content: 13.5% lactic acid, 17.7% ethyl lactate, 10% water and 58.8% ethanol.

The dependence of the total permeation flux at any water feed percentage, J, with the absolute temperature is described by the formula:

$$J = J_o \, exp\left(-E_p/RT\right) \tag{9}$$

where J_o is the pre-exponential factor and E_p the activation energy. J_o changes with the feed composition, but E_p should remain constant. The calculated activation energy between 5.85 and 4.41 wt. % of water was 106.55 kJ/mol, a value in the same order of magnitude but twice the one given by Delgado et al. [41] with the PERVAP™ 2201 from 12 to 63 wt. % of water in the feed.

4. Conclusions

Pervaporation membranes PERVAP™ 2216, PERVAP™ 3100 and PERVAP™ 1131 were tested in the dehydration of a mixture of 13.5 wt. % lactic acid, 17.7 wt. % ethyl lactate, 10 wt. % water and 58.8 wt.% ethanol at 60°C. PERVAP™ 1131 was the most appropriate for the process. PERVAP™ 3100 had chemical resistance problems and poor selectivity, and the water fluxes of PERVAP™ 2216 were extremely low.

The effect of feed composition, swelling and temperature effects were studied for the PERVAP™ 1131 membrane. Binary and ternary mixtures have comparable water, ethyl lactate and lactic acid fluxes but lower water/ethanol selectivity than the quaternary one. Lactic acid together with ethyl lactate plays a key role in the water/ethanol selectivity due to the higher affinity of the membrane for them. An increase of temperature increases the permeation fluxes of all substances and increases the passage of the more volatile component in the water/ethanol and water/ethyl lactate systems. Lactic acid is the more soluble component in the membrane, but the one with the lowest vapor pressure, leading to crystallization on the permeate side of the membrane at very low water concentrations, where the nonaqueous environment enhances the crystallization.

Author Contributions: Conceptualization, S.L. and J.R.Á.; methodology, J.R.Á., S.L. and. E.G.D.; validation, E.G.D., J.R.Á. and S.Á.-G.; formal analysis, J.R.Á., S.Á.-G., S.L. and. E.G.D.; investigation, E.G.D., J.R.Á. and S.L.; resources, S.L., data curation, J.R.Á. and S.Á.-G., writing—original draft preparation, E.G.D. and J.R.Á., writing—review and editing, J.R.Á., S.Á.-G. and S.L.; supervision, J.R.Á. and S.L.; project administration, S.L.; funding acquisition, S.L. and J.R.Á. All authors have read and agreed to the published version of the manuscript.

Funding: This research was funded by the Spanish Ministry of Education and Science (MEC) for financial support (PPQ02-03118).

Institutional Review Board Statement: Not applicable.

Informed Consent Statement: Not applicable.

Data Availability Statement: Data available upon request to the corresponding author.

Acknowledgments: Sulzer Chemtech Ltd. is gratefully acknowledged for providing materials and equipment used in this study.

Conflicts of Interest: The authors declare no conflict of interest.

Appendix A

The vapor pressure of pure substances was calculated using an extended Antoine equation, where the temperature T is given in K and the vapor pressure P^S in Pa:

$$lnP^S = k_1 + \frac{k_2}{T} + k_3 lnT + k_4 T^{k_5} \quad \text{(A1)}$$

The parameters for the compounds are given in Table A1. T_{min} and T_{max} indicate the minimum and maximum temperature for which the expression is appropriate for each compound (in K, as well).

Table A1. Extended Antoine equation parameters.

Parameter	Lactic Acid	Ethanol	Water	Ethyl Lactate
k_1	225.19	73.304	73.469	78.774
k_2	−18757	−7122.3	−7258.9	−6715.3
k_3	−28.816	−7.1424	−7.3037	−9.5666
k_4	1.3×10^{-5}	2.89×10^{-6}	4.17×10^{-6}	0.014993
k_5	2	2	2	1
T_{min}	289.9	159.05	273.16	247.15
T_{max}	675	514	647.13	588

The UNIQUAC parameters needed for the activity coefficient calculations are given in Tables A2 and A3.

Table A2. UNIQUAC volume and area parameters.

Parameter	Lactic Acid	Ethanol	Water	Ethyl Lactate
r_i	3.1648	2.1055	0.92	4.4555
q_i	2.88	1.972	1.4	3.928

Table A3. UNIQUAC binary interaction parameters.

a_{ij}	Lactic Acid	Ethanol	Water	Ethyl Lactate
Lactic acid	–	−43.32	155.18	125.29
Ethanol	191.28	–	728.97	−148.67
Water	−39.61	−756.95	–	64.53
Ethyl lactate	52.64	341.77	99.8	–
b_{ij}	**Lactic Acid**	**Ethanol**	**Water**	**Ethyl Lactate**
Lactic acid	–	0	0	0
Ethanol	0	–	−2.0046	0
Water	0	2.4936	–	0
Ethyl lactate	0	0	0	–

References

1. Sanz, M.A.; Luque, S.; Berrueta, J.; Coca, J. Ésteres de ácido láctico. *Ing. Quim.* **2002**, *395*, 51–56.
2. de la Iglesia, Ó.; Mallada, R.; Menéndez, M.; Coronas, J. Continuous zeolite membrane reactor for esterification of ethanol and acetic acid. *Chem. Eng. J.* **2007**, *131*, 35–39. [CrossRef]
3. Yahya, A.R.M.; Anderson, W.A.; Moo-Young, M. Ester synthesis in lipase-catalyzed reactions. *Enzyme Microb. Technol.* **1998**, *23*, 438–450. [CrossRef]
4. Wasewar, K.; Patidar, S.; Agarwal, V.K. Esterification of lactic acid with ethanol in a pervaporation reactor: Modeling and performance study. *Desalination* **2009**, *243*, 305–313. [CrossRef]
5. Benedict, D.J.; Parulekar, S.J.; Tsai, S.-P. Pervaporation-assisted esterification of lactic and succinic acids with downstream ester recovery. *J. Membr. Sci.* **2006**, *281*, 435–445. [CrossRef]
6. Lipnizki, F.; Field, R.W.; Ten, P.-K. Pervaporation-based hybrid process: A review of process design, applications and economics. *J. Membr. Sci.* **1999**, *153*, 183–210. [CrossRef]
7. Smitha, B.; Suhanya, D.; Sridhar, S.; Ramakrishna, M. Separation of organic–organic mixtures by pervaporation—A review. *J. Membr. Sci.* **2004**, *241*, 1–21. [CrossRef]
8. Ahn, H.; Lee, H.; Lee, S.-B.; Lee, Y. Pervaporation of an aqueous ethanol solution through hydrophilic zeolite membranes. *Desalination* **2006**, *193*, 244–251. [CrossRef]
9. Li, Y.; Liu, J.; Yang, W. Formation mechanism of microwave synthesized LTA zeolite membranes. *J. Membr. Sci.* **2006**, *281*, 646–657. [CrossRef]
10. Li, X.; Kita, H.; Zhu, H.; Zhang, Z.; Tanaka, K. Synthesis of long-term acid-stable zeolite membranes and their potential application to esterification reactions. *J. Membr. Sci.* **2009**, *339*, 224–232. [CrossRef]
11. Kita, H.; Tanaka, K.; Okamoto, K.; Yamamoto, M. The esterification of oleic acid with ethanol accompanied by membrane separation. *Chem. Lett.* **1987**, *16*, 2053–2056. [CrossRef]
12. David, M.O.; Graf, R.; Nguyen, T.Q.; Neel, J. Pervaporation—Esterification coupling: Part I. Basic kinetic model. *Trans. IchemE.* **1991**, *69*, 335–340.

13. David, M.O.; Graf, R.; Nguyen, T.Q.; Neel, J. Pervaporation—Esterification coupling: Part II. Modelling of the influence of different operating parameters. *Trans. IchemE.* **1991**, *69*, 341–346.
14. Bagnell, L.; Cavell, K.; Hodges, A.M.; Mau, A.W.H.; Seen, A.J. The use of catalytically active pervaporation membranes in esterification reactions to simultaneously increase product yield, membrane permselectivity and flux. *J. Membr. Sci.* **1993**, *85*, 291–299. [CrossRef]
15. Zhu, Y.; Chen, H. Pervaporation separation and pervaporation-esterification coupling using crosslinked PVA composite catalytic membranes on porous ceramic plate. *J. Membr. Sci.* **1998**, *138*, 123–134.
16. Xuehui, L.; Lefu, W. Kinetic model for an esterification process coupled by pervaporation. *J. Membr. Sci.* **2001**, *186*, 19–24. [CrossRef]
17. Liu, Q.L.; Chen, H.F. Modeling of esterification of acetic acid with *n*-butanol in the presence of $Zr(SO_4)_2 \cdot 4H_2O$ coupled pervaporation. *J. Membr. Sci.* **2002**, *196*, 171–178. [CrossRef]
18. Peters, T.A.; Benes, N.E.; Keurentjes, J.T.F. Zeolite-coated ceramic pervaporation membranes; pervaporation−esterification coupling and reactor evaluation. *Ind. Eng. Chem. Res.* **2005**, *44*, 9490–9496. [CrossRef]
19. Qing, W.; Wu, J.; Chen, N.; Liu, L.; Deng, Y.; Zhang, W. A genuine in-situ water removal at a molecular lever by an enhanced esterification-pervaporation coupling in a catalytically active membrane reactor. *Chem. Eng. J.* **2017**, *323*, 434–443. [CrossRef]
20. Zhang, W.; Qing, W.; Chen, N.; Ren, Z.; Chen, J.; Sun, W. Enhancement of esterification conversion using novel composite catalytically active pervaporation membranes. *J. Membr. Sci.* **2014**, *451*, 285–292. [CrossRef]
21. Benedict, D.J.; Parulekar, S.J.; Tsai, S.-P. Esterification of lactic acid and ethanol with/without pervaporation. *Ind. Eng. Chem. Res.* **2003**, *42*, 2282–2291. [CrossRef]
22. Ma, J.; Zhang, M.; Lu, L.; Yin, X.; Chen, J.; Jiang, Z. Intensifying esterification reaction between lactic acid and ethanol by pervaporation dehydration using chitosan–TEOS hybrid membranes. *Chem. Eng. J.* **2009**, *155*, 800–809. [CrossRef]
23. Delgado, P.; Sanz, M.T.; Beltrán, S.; Núñez, L.A. Ethyl lactate production via esterification of lactic acid with ethanol combined with pervaporation. *Chem. Eng. J.* **2010**, *165*, 693–700. [CrossRef]
24. Pereira, C.S.M.; Silva, V.M.T.M.; Pinho, S.P.; Rodrigues, A.E. Batch and continuous studies for ethyl lactate synthesis in a pervaporation membrane reactor. *J. Membr. Sci.* **2010**, *361*, 43–55. [CrossRef]
25. Zhang, M.; Chen, L.; Jiang, Z.; Ma, J. Effects of dehydration rate on the yield of ethyl lactate in a pervaporation-assisted esterification process. *Ind. Eng. Chem. Res.* **2015**, *54*, 6669–6676. [CrossRef]
26. Budd, P.M.; Ricardo, N.M.P.S.; Jafar, J.J.; Stephenson, B.; Hughes, R. Zeolite/polyelectrolyte multilayer pervaporation membranes for enhanced reaction yield. *Ind. Eng. Chem. Res.* **2004**, *43*, 1863–1867. [CrossRef]
27. Izák, P.; Mateus, N.M.M.; Afonso, C.A.M.; Crespo, J.G. Enhanced esterification conversion in a room temperature ionic liquid by integrated water removal with pervaporation. *Sep. Purif. Technol.* **2005**, *41*, 141–145. [CrossRef]
28. Assabumrungrat, S.; Phongpatthanapanich, J.; Praserthdam, P.; Tagawa, T.; Goto, S. Theoretical study on the synthesis of methyl acetate from methanol and acetic acid in pervaporation membrane reactors: Effect of continuous-flow modes. *Chem. Eng. J.* **2003**, *95*, 57–65. [CrossRef]
29. Sanz, M.T.; Gmehling, J. Esterification of acetic acid with isopropanol coupled with pervaporation: Part I: Kinetics and pervaporation studies. *Chem. Eng. J.* **2006**, *123*, 1–8. [CrossRef]
30. Sanz, M.T.; Gmehling, J. Esterification of acetic acid with isopropanol coupled with pervaporation: Part II. Study of a pervaporation reactor. *Chem. Eng. J.* **2006**, *123*, 9–14. [CrossRef]
31. Delgado, P.; Sanz, M.T.; Beltrán, S. Pervaporation of the quaternary mixture present during the esterification of lactic acid with ethanol. *J. Membr. Sci.* **2009**, *332*, 113–120. [CrossRef]
32. Yave, W. The improved pervaporation PERVAP membranes. *Filtr. Sep.* **2017**, *54*, 14–15. [CrossRef]
33. Kujawski, J.K.; Kujawski, W.M.; Sondej, H.; Jarzynka, K.; Kujawska, A.; Bryjak, M.; Rynkowska, E.; Knozowska, K.; Kujawa, J. bDewatering of 2,2,3,3-tetrafluoropropan-1-ol by hydrophilic pervaporation with poly(vinyl alcohol) based Pervap™ membranes. *Sep. Purif. Technol.* **2017**, *174*, 520–528. [CrossRef]
34. Lipnizki, F.; Hausmanns, S. Hydrophobic Pervaporation of Binary and Ternary Solutions: Evaluation of Fluxes, Selectivities, and Coupling Effects. *Sep. Sci. Technol.* **2005**, *39*, 2235–2259. [CrossRef]
35. Wijmans, J.G.; Baker, R.W. A simple predictive treatment of the permeation process in pervaporation. *J. Membr. Sci.* **1993**, *79*, 101–113. [CrossRef]
36. Delgado, P.; Sanz, M.T.; Beltrán, S. Isobaric vapor–liquid equilibria for the quaternary reactive system: Ethanol+water+ethyl lactate+lactic acid at 101.33kPa. *Fluid Phase Equilib.* **2007**, *255*, 17–23. [CrossRef]
37. Clément, R.; Bendjama, Z.; Nguyen, Q.T.; Néel, J. Extraction of organics from aqueous solutions by pervaporation. A novel method for membrane characterization and process design in ethyl acetate separation. *J. Membr. Sci.* **1992**, *66*, 193–203. [CrossRef]
38. Kariduraganavar, M.Y.; Kulkarni, S.S.; Kittur, A.A. Pervaporation separation of water–acetic acid mixtures through poly(vinyl alcohol)-silicone based hybrid membranes. *J. Membr. Sci.* **2005**, *246*, 83–93. [CrossRef]
39. Oh, B.-K.; Lee, Y.M. Effects of functional group and operating temperature on the separation of pyridine-water mixture by pervaporation. *J. Membr. Sci.* **1996**, *113*, 183–189. [CrossRef]
40. Upadhyay, D.J.; Bhat, N.V. Pervaporation studies of gaseous plasma treated PVA membrane. *J. Membr. Sci.* **2004**, *239*, 255–263. [CrossRef]
41. Delgado, P.; Sanz, M.T.; Beltrán, S. Pervaporation study for different binary mixtures in the esterification system of lactic acid with ethanol. *Sep. Purif. Technol.* **2008**, *64*, 78–87. [CrossRef]

membranes

MDPI

Article

Feasibility of Adsorption Kinetic Models to Study Carrier-Mediated Transport of Heavy Metal Ions in Emulsion Liquid Membranes

Gerardo León [1,*], Elisa Gómez [2], Beatriz Miguel [1], Asunción María Hidalgo [2], María Gómez [2], María Dolores Murcia [2] and María Amelia Guzmán [1]

1 Departamento de Ingeniería Química y Ambiental, Universidad Politécnica de Cartagena, Paseo Alfonso XIII 52, 30206 Cartagena, Spain; beatriz.miguel@upct.es (B.M.); maguzmanmv@gmail.com (M.A.G.)
2 Departamento de Ingeniería Química, Campus de Espinardo, Universidad de Murcia, 30100 Murcia, Spain; egomez@um.es (E.G.); ahidalgo@um.es (A.M.H.); maria.gomez@um.es (M.G.); md.murcia@um.es (M.D.M.)
* Correspondence: gerardo.leon@upct.es

Abstract: Emulsion liquid membranes have been successfully used for the removal of different types of organic and inorganic pollutants by means of carrier-mediated transport mechanisms. However, the models that describe the kinetics and transport of such mechanisms are very complex due to the high number of model parameters. Starting from an analysis of the similarity between the elemental mechanisms of carrier-mediated transport in liquid membranes and of transport in adsorption processes, this paper presents an experimental analysis of the possibility of applying kinetic and mechanistic models developed for adsorption to carrier-mediated transport in emulsion liquid membranes. We study the removal of a target species, in this case, Cu(II), by emulsion liquid membranes containing membrane phase solutions of benzoylacetone (carrier agent), Span 80 (emulsifying agent) and kerosene (diluent), and hydrochloric acid as a stripping agent in the product phase. The experimental results fit the pseudo-second-order adsorption kinetic model, showing good relationships between the experimental and model parameters. Although both Cu(II) diffusion through the feed/membrane interface boundary layer and complex Cu-benzoylacetone diffusion through the membrane phase controls Cu(II) transport, it is the former step that mainly controls the transport process.

Keywords: emulsion liquid membranes; carrier-mediated transport; adsorption; kinetic; mechanism

1. Introduction

The presence of heavy metals, elements with atomic weights of between 63.5 and 200.6, and a specific gravity greater than 5.0 [1], in industrial effluents and wastewater, is a serious environmental problem since they are non-biodegradable in natural conditions, have a tendency to accumulate in living organisms, causing a variety of diseases and disorders, and have an inhibiting effect on the biodegradation of organic pollutants which may also be present in wastewater [2,3]. Therefore, their concentrations should be reduced to acceptable levels before being discharged into the environment. Many industries are responsible for the direct or indirect discharge of heavy metals into the environment: mining, hydrometallurgy, textiles, tanneries, pulp and paper, pesticides, petrochemicals, refining, batteries, fertilizers, electroplating, among others [4,5].

Over the past few years, several techniques have been described for removing heavy metal ions wastewater, among them chemical precipitation [6], cementation [7], coagulation-flocculation [8], adsorption [9–12], electrochemical processes [13,14], and photocatalytic methods [15].

Since the 1970s, when Bhattacharyya et al. [16] and Sato et al. [17] used ultrafiltration and reverse osmosis membranes, respectively, for treating metal plating wastewaters, several membrane-based processes have been used to remove heavy metal ions from wastewaters, including micellar enhanced ultrafiltration [18,19], nanofiltration [20,21], reverse osmosis [22], forward osmosis [23], electrodialysis [24] and polymer inclusion membranes [25].

Liquid membranes have also been used for the efficient removal of heavy metal ions from aqueous solutions [26–30]. A liquid membrane system consists of two miscible liquids (feed and product phases) separated by a third liquid (immiscible in both) as a membrane phase. The efficiency of metal ion removal by liquid membrane processes is significantly improved through the use of so-called carrier-mediated transport [31]. In such transport, a reagent (carrier agent) is incorporated into the membrane phase to transport the heavy metal ion from the feed to the product phase through the membrane phase. At the feed/membrane interface, the carrier and the heavy metal ion reversibly form a compound that is soluble in the membrane phase. This compound is transported through the membrane phase due to its concentration gradient and is broken at the membrane/product interface by reacting with a stripping agent; the heavy metal ion is released into the product phase and, as a consequence of its concentration gradient, the carrier diffuses to the opposite interface through the membrane phase before initiating a new separation cycle.

There are three basic types of liquid membranes: bulk, supported, and emulsion liquid membranes [32]. In emulsion liquid membranes, a primary emulsion formed by the membrane phase and the product phase is dispersed in the feed phase, the encapsulated internal droplets of the product phase and the external feed phase being separated by the membrane phase.

The elementary mechanism of heavy metal ion permeation through an emulsion liquid membrane by a carrier-mediated transport includes four basic steps (Figure 1) [33]:

Step 1. Heavy metal ion diffusion through the boundary layer of the feed phase to the feed/membrane interface.
Step 2. Heavy metal ion reaction, at the feed/membrane interface, with the carrier present in the membrane phase forming a metal-carrier complex soluble in the membrane phase.
Step 3. Complex diffusion, in the membrane phase, from the feed/membrane interface toward the membrane/product interface.
Step 4. Complex breakdown at the membrane/product interface due to the reaction between the heavy metal ion and stripping agent present in the product phase and carrier regeneration.

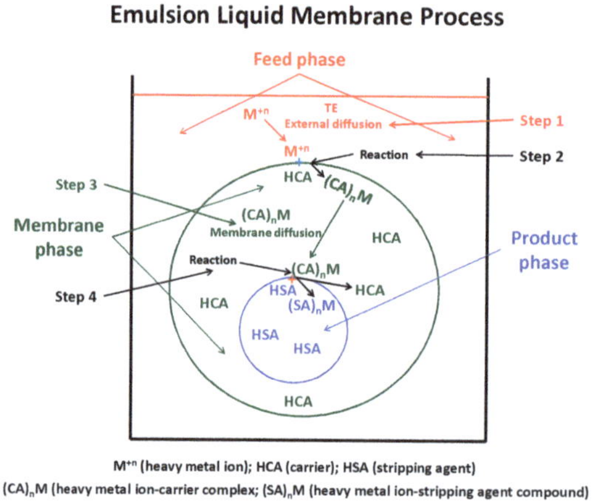

Figure 1. Mechanism of heavy metal ion permeation through an emulsion liquid membrane by carrier-mediated transport.

The models described in the literature to study the kinetics and mechanism of carrier-mediated transport in emulsion liquid membranes tend to be highly complex as they include a large number of model parameters [33–35]. It must be taken into account that from using models to fit experimental data, it is necessary to make simplifying assumptions and that all theoretical models are approximated, but some of them are useful [36]. For this reason, it may be of interest the look for simpler mathematical models to which the experimental results are adequately adjusted, allowing the explanation for the phenomenon under study.

Bearing this in mind, let us highlight the similarity that the above described carrier-mediated transport mechanism in emulsion liquid membranes has with the elemental mechanism of adsorption of an adsorbate onto an adsorbent, which includes three basic steps (Figure 2) [37,38]:

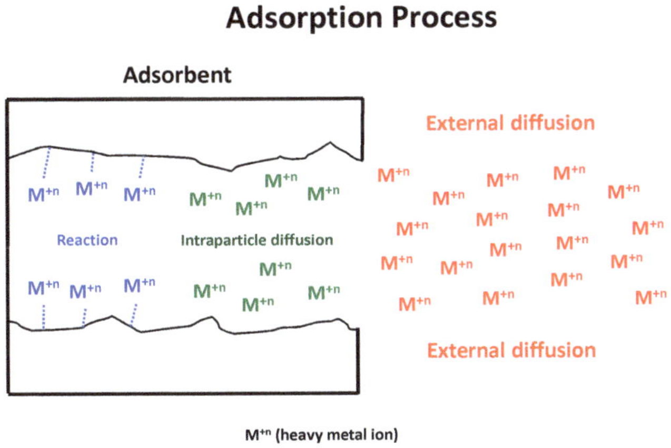

Figure 2. Mechanism of the adsorption of an adsorbate onto an adsorbent.

Step 1. External diffusion (film diffusion), which is the transport of the heavy metal ion from the bulk phase to the external surface of the adsorbent.
Step 2. Intraparticle diffusion (pore diffusion), which is the transport of the heavy metal ion from the external surface into the pores.
Step 3. Surface reaction, which is the attachment of the heavy metal ion to the internal surface of the sorbent.

As the reaction steps are much faster than those of diffusion, the transport kinetics in a carrier-mediated emulsion liquid membrane process will be governed by diffusion of the metal ion through the boundary layer of the feed phase to the feed/membrane interface or by diffusion of the metal-complex in the membrane phase. In the same way, in an adsorption process, the transport kinetics will be governed by external diffusion or by intraparticle diffusion.

This great similarity between the transport mechanisms of both processes makes the kinetic and mechanistic study of the carrier-mediated transport in emulsion liquid membrane processes by means of the kinetics and mechanistic models developed for adsorption processes an interesting option for experimental analyses, especially because no such study seems to have been carried out to date.

Accordingly, in this paper, we analyze the carrier-mediated transport of Cu^{2+} (as target species) through emulsion liquid membranes using benzoyl acetone (carrier), kerosene (diluent), and Span 80 (emulsifying agent) in the membrane phase and hydrochloric acid as stripping agent in the product phase, in different experimental conditions, and fitting the obtained results to kinetic and mechanistic models that have been described for adsorption processes.

2. Materials and Methods

2.1. Materials

Benzoylacetone (99%), kerosene, and Span 80 (sorbitan monooleate) were supplied by Sigma Aldrich; copper (II) chloride (98.5%), hydrochloric acid (37%), acetic acid (96%), and sodium acetate 3-hydrate (for analysis) were obtained from Panreac.

2.2. Methods

The internal aqueous phase consisted of an aqueous solution of hydrochloric acid ranging from 0.05 M to 0.50 M. The organic membrane phase was prepared by dissolving the appropriate amounts of the carrier (benzoylacetone, 0.1% to 2.0%) and the surfactant (Span 80, 5%) in kerosene (organic diluent) by gently mixing using a magnetic stirrer.

The water in oil primary emulsion (*w*/*o*) was prepared by mixing the aqueous product phase with the organic membrane phase, at a volume ratio of 1/1, under stirring, before emulsifying the mixture at a high stirring rate (2700 rpm) for 5 min to obtain a stable emulsion, using an OMNI MIXER homogenizer (Omni International, Kennesaw, GA, USA). This emulsion was dispersed into the external feed Cu(II) aqueous phase (0.025 M in acetate buffer, pH ranging from 4.0 to 5.5) in a cylindrical glass cell to form the secondary emulsion (*w*/*o*/*w*), at a 1/2 emulsion phase/feed phase volume ratio. The content of the glass cell was stirred (50 to 200 rpm) in order to disperse the *w*/*o* emulsion in the external phase to form the *w*/*o*/*w* double emulsions.

The external feed phase was periodically sampled and, after settling for 3 min to separate feed and emulsion phases, the Cu(II) content in the feed phase was analyzed by atomic absorption spectrometry, using a ContrAA 700 (Analytik Jena, Edinburgh Instruments, Livingston, UK) instrument at a wavelength of 324.8 nm.

The duration of the experiments was 20 min to ensure that no increase occurred in the concentration of Cu(II) in the feed phase with time as a consequence of emulsion breakage in any of the studied experimental conditions.

Experiments were conducted in duplicate at room temperature, the results showing a maximum deviation of 3%. No removal of Cu(II) from the feed phase was detected in the absence of a carrier agent in the membrane phase.

Analysis of the efficiency of an elimination process using emulsion liquid membranes is usually carried out using three parameters: removal percentage, flux, and permeability.

The percentage of Cu(II) removal from the feed phase (RP) was determined according to Equation (1), where $C_{f,0}$ and $C_{f,t}$ are the initial and time t concentrations of Cu(II) in the feed phase.

$$RP = \frac{C_{f,0} - C_{f,t}}{C_{f,t}} \times 100 \quad (1)$$

Initial apparent fluxes (J_a) and apparent permeabilities (P_a) of Cu(II) through the feed/membrane interface were obtained from the plots against the time of $C_{f,t}$ and $\ln(C_{f,t}/C_{f,0})$, respectively, during the first five minutes of the experiment, according to Equations (2) and (3), in which it was assumed that membrane area is proportional to the emulsion volume ($A = k \cdot V_{emul}$), that the stripping reaction is fast enough so that there is no accumulation of Cu(II) in the membrane phase and that the same preparation conditions of the emulsion lead to the same size uniformity of the emulsion globules [27,39].

$$J_a = -\frac{V_f \cdot dC_f}{V_{emul} \cdot dt} \quad (2)$$

$$\ln \frac{C_{f,t}}{C_{f,0}} = -\frac{V_{emul} \cdot P_a \cdot t}{V_f} \quad (3)$$

where V_f and V_{emul} are the volumes of feed and of emulsion, respectively.

Additionally, the initial Cu(II) removal rate ($V_{0,exp}$) was obtained by extrapolating to zero (intercept) the representation of the Cu(II) removal rate ($mg_{Cu(II)remov}/L_{emul} \cdot t$) against time during the first three minutes of the experiment.

In an adsorption process, the parameter adsorption capacity is defined for any time t (q_t) or at equilibrium (q_e), as the quantity of adsorbate (usually expressed in mg) retained per mass (or volume) unit of adsorbent (m/g or mg/L, respectively) [37,38].

In the case of emulsion liquid membranes, we will define this parameter as the amount of compound removed from the feed phase (expressed in milligrams) per volume unit of emulsion phase (expressed in liters).

Accordingly, the amount of Cu(II) removed from the feed phase per volume unit of emulsion phase, at any time t, q_t (mg/L), and at equilibrium, q_e (mg/L), was estimated from the relationships Equations (4) and (5), where $C_{f,e}$ is the equilibrium concentration (at 20 min) of Cu(II) in the feed phase (mg/L).

$$q_t = (C_{f,0} - C_{f,t}) \cdot \frac{V_f}{V_{emul}} \tag{4}$$

$$q_e = (C_{f,0} - C_{f,e}) \cdot \frac{V_f}{V_{emul}} \tag{5}$$

The kinetics of the carrier-mediated transport of Cu(II) through emulsion liquid membranes was analyzed using four adsorption kinetic models: Lagergren pseudo-first-order, Ho pseudo-second-order, Elovich and Avrami.

The linear form of the Lagergren pseudo-first-order model [40] is expressed by Equation (6). The values of the model constants, $q_{e,pfo}$ (mg/L), and k_{pfo} (pseudo-first-order rate constant, min^{-1}) can be obtained from the plot of ln(qe − qt) vs. time.

$$\ln\left(q_{e,exp} - q_t\right) = \ln q_{e,pfo} - k_{pfo} \cdot t \tag{6}$$

Equation (7) shows the linearized form of the Ho pseudo-second-order kinetic model [41]. The values of the model constants, $q_{e,pso}$ ($mg \cdot L^{-1}$), k_{pso} (pseudo-second-order rate constant, $L \cdot mg^{-1} \cdot min^{-1}$), and $V_{0,pso}$ (initial adsorption rate, defined as $k_{pso} \cdot (q_{e,pso})^2$, $mg \cdot L^{-1} \cdot min^{-1}$) can be obtained by plotting t/qt against time.

$$\frac{t}{q_t} = \frac{1}{k_{pso} \cdot q_{e,pso}^2} + \frac{t}{q_{e,pso}} \tag{7}$$

The Elovich kinetic model [42] is generally expressed, assuming that $\alpha \cdot \beta \cdot t > 1$, by Equation (8). The values of the model constants, α (initial adsorption rate, $mg \cdot L^{-1} \cdot min^{-1}$) and β (number of available sites for adsorption, $L \cdot mg^{-1}$), can be obtained from the plot of q_t versus lnt.

$$q_t = \frac{\ln(\alpha \cdot \beta)}{\beta} + \frac{\ln t}{\beta} \tag{8}$$

The linear form of the Avrami kinetic model [43] is usually described by Equation (9). The values of model constants K_{av} (Avrami's constant rate, min^{-1}) and n_{av} (Avrami's order model) can be obtained from the plot of $\ln[-\ln(1 - q_t)]$ versus lnt.

$$\ln[-\ln(1 - q_t)] = \ln K_{av} + n_{av} \cdot \ln t \tag{9}$$

The above described kinetic models do not provide information about the rate-controlling step of the carrier-mediated transport of Cu(II) through an emulsion liquid membrane, which is characterized, as mentioned above, by Cu(II) diffusion through the boundary layer of the feed/membrane interface (equivalent to the external diffusion in an adsorption process), by Cu-carrier complex diffusion through the membrane phase (equivalent to the intraparticle diffusion in an adsorption process), or by both. To obtain information about that rate-controlling step, Weber and Morris intraparticle diffusion and Boyd adsorption models were used.

The Weber and Morris intraparticle diffusion model [44] is commonly expressed by Equation (10). The values of the model constants k_{intp} (intraparticle diffusion rate constant,

$mg \cdot L^{-1} \cdot min^{-1/2}$) and C_i (effect of extraparticle diffusion, $mg \cdot L^{-1}$) can be obtained by plotting q_t against $t^{1/2}$.

$$q_t = k_{intp} \cdot t^{1/2} + C_i \tag{10}$$

When that plot is linear and passes through the origin, intraparticle diffusion is the rate-controlling step of the adsorption process. Otherwise, both extraparticle and intraparticle diffusion steps are involved in controlling the rate.

If the latter occurs, it is necessary to identify which of these two steps is the one that mainly controls the rate of the overall process, usually using for this purpose the Boyd kinetic model [45], which is expressed in the form:

$$Bt = -0.4977 - \ln\left(1 - \frac{q_t}{q_e}\right) \tag{11}$$

If the plot of Bt versus time is a straight line and passes through the origin, the adsorption process is mainly controlled by the intraparticle diffusion step; otherwise, it is mainly controlled by the extraparticle diffusion step.

3. Results and Discussion

3.1. Kinetics of Cu(II) Transport

The fitting of the data to these models in the different studied experimental conditions is shown in Figure 3. The reliability of the fit was determined based on the values of the determination coefficient (R^2).

Data of the carrier-mediated transport of Cu(II) through emulsion liquid membranes, in the different experimental conditions studied, are best fitted by the pseudo-second-order kinetic model, which showed determination coefficient values higher than 0.99 (0.9903–0.9975) in all cases, while pseudo-first-order (0.9478–0.9984), Elovich (0.9816–0.9957) and Avrami (0.9812–0.9985) models provided several determination coefficients lower than 0.99.

The pseudo-second-order was originally proposed for the removal of divalent metal cations from water using zeolites, assuming adsorption/reaction mechanisms [46]. The model assumes that the uptake rate is second order with respect to the available sites [37] and describes adsorption processes in which the chemical bonding between adsorbates and functional groups on the surface of adsorbents is responsible for the adsorption capacity of the adsorbent [45,47]. This is what occurs in the process of Cu(II) removal by carrier-mediated transport in emulsion liquid membranes, which includes a stripping reaction, that takes place at the membrane/product interface to release the metal ion into the permeate phase. In this reaction, each Cu(II) ion is striped in the product phase by a reaction with two HCl molecules.

In addition, it is known that the pseudo-second-order kinetic adsorption model closely fits the experimental results when the initial concentration of the species to be removed is low, the experiment time range included in the fit is broad, and there are many active sites to react with the species to be removed [38]. These three aspects occurred in our study of Cu(II) removal by emulsion liquid membranes using carrier-mediated transport.

Therefore, the pseudo-second-order model is the one that best describes all the possible steps included in the global transport mechanism, with a k value that is a result of a complex interplay between different controlling mechanisms, although no information is provided concerning the mass transfer mechanism.

From an analysis of the results obtained from fitting the experimental data to the pseudo-second-order kinetic model, several conclusions can be drawn that support the feasibility of the use of adsorption kinetic models to study the carrier-mediated transport of heavy metal ions through emulsion liquid membranes.

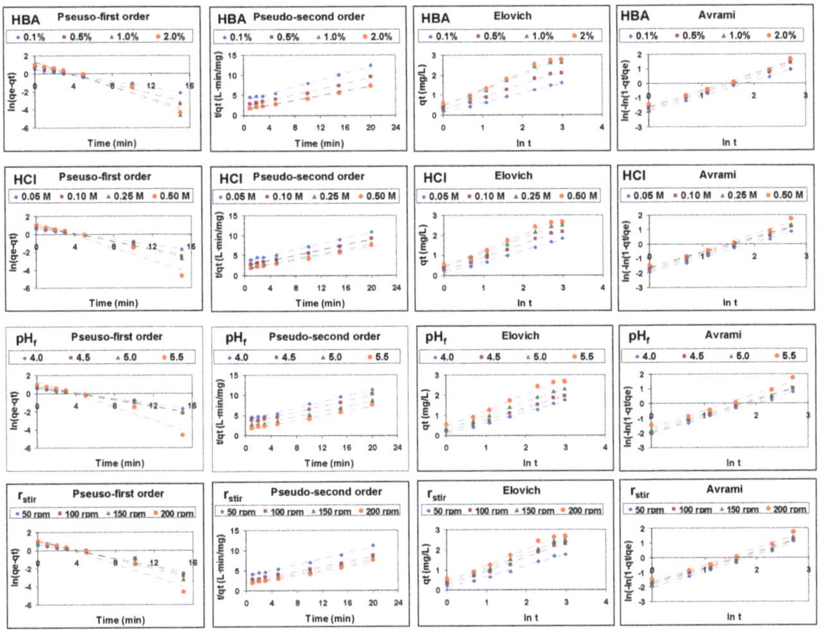

Figure 3. Fitting of the data obtained in the different experimental conditions to the studied kinetic models.

First, as can be seen from Figure 4, there is a good relationship between the values of the theoretical parameters of the pseudo-second-order model ($q_{e,pso}$ and $V_{0,pso}$) and the corresponding experimental values of these parameters (with coefficients of determination that range between 0.983 and 0.997)

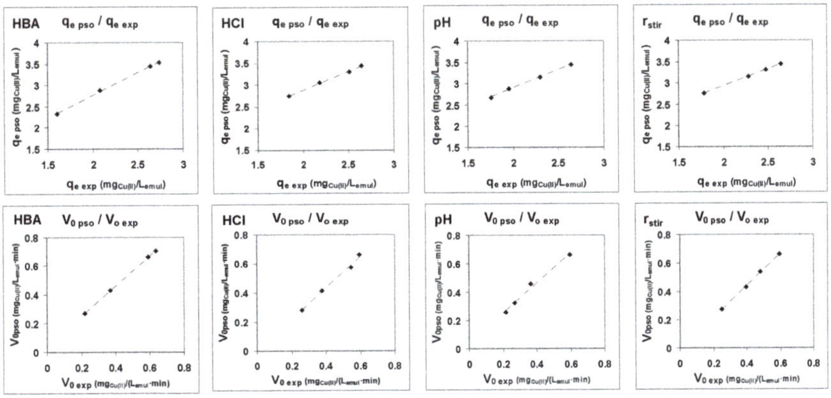

Figure 4. Relationship between experimental and pseudo-second-order model values of parameters q_e and V_0.

Second, the variation of the values of these theoretical parameters of the pseudo-second-order model with the values of the different experimental parameters studied follows a trend that is practically similar to that which follows the experimentally calculated removal percentage of Cu(II) from the feed phase (Figure 5). As the removal percentage is usually used as a parameter to analyze the influence of the different experimental conditions on the efficiency of an emulsion liquid membrane process, any of the three parameters obtained from fitting of the experimental results to the pseudo-second-order kinetic model can also be used to analyze this influence.

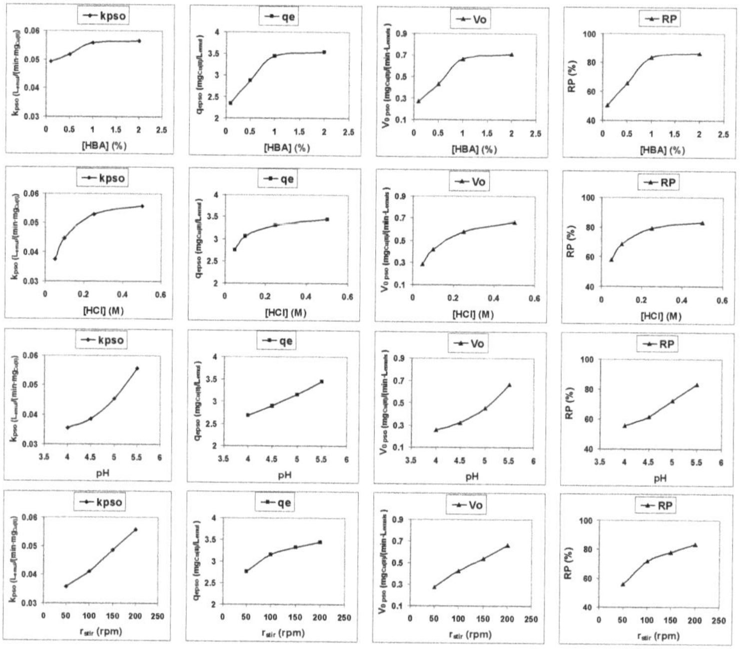

Figure 5. Variation in the values of pseudo-second-order model parameter (k_{pso}, $q_{e,pso}$, $V_{0,pso}$) and those of the removal percentage (RP) with the different experimental conditions studied.

Accordingly, using both the elimination percentage and any of the three parameters of the model (k_{pso}, $q_{e,pso}$ or $V_{0,pso}$), it can be observed that the efficiency of the process increases as the HBA concentration in the membrane phase increases from 0.1% to 1%, above which it remains constant; the efficiency increases as the HCl concentration in the product phase increases from 0.05 M to 0.50 M, the efficiency increases as the pH of the feed phase increases from 4.0 to 5.5 and the efficiency increases as the stirring rate of the secondary emulsion increases from 50 rpm to 200 rpm.

The increase in benzoylacetone concentration in the membrane phase from 0.1% to 1.0% favors the removal of Cu(II) from the feed phase due to the greater number of Cu(II) ions that can form the complex Cu-benzoylacetone, which increases its concentration gradient through the membrane phase, favoring its diffusion to the membrane/product interface. At higher benzoylacetone concentrations, this effect is partially neutralized by the increase in the emulsion viscosity, which leads to an increase in the emulsion globules size and consequently to a decrease in the mass transfer surface area, leading to removal efficiencies that remain constant.

The increase in both the concentration of HCl in the product phase and the pH in the feed phase increase the gradient of the proton concentration between the product and the feed phases, which is the driving force of the Cu(II) transport process, leading to an increase in the Cu(II) removal efficiency.

The increase in the stirring rate leads to the increase in the number of smaller emulsion globules, leading to an increase in the mass transfer surface area and consequently to an increase in Cu(II) removal efficiency.

Thirdly, there is a good linear relationship between the parameters of the model and the values of the initial apparent fluxes and permeabilities calculated from the experimental results, as can be seen in Figure 6.

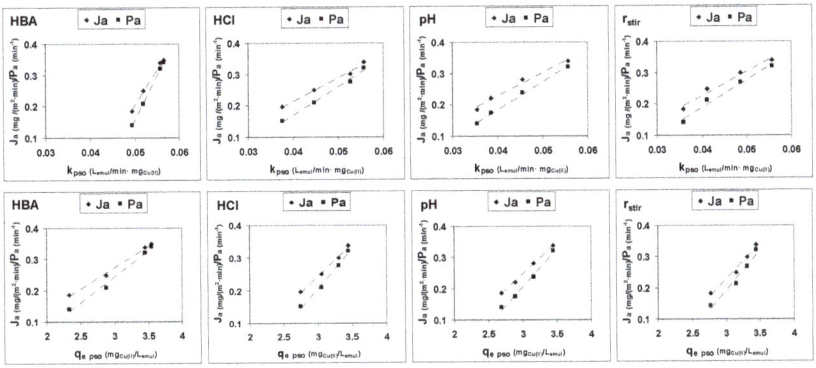

Figure 6. Relationship between the values of pseudo-second-order model parameters (k_{pso} and $q_{e,pso}$) and the experimental initial apparent fluxes (Ja) and permeabilities (Pa).

3.2. Mechanism of Cu(II) Transport

The carrier-mediated transport of Cu(II) through emulsion liquid membranes was analyzed by the Weber-Morris and Boyd mechanistic models to obtain information about the rate-controlling step of the carrier-mediated transport of Cu(II) through an emulsion liquid membrane (Figure 7).

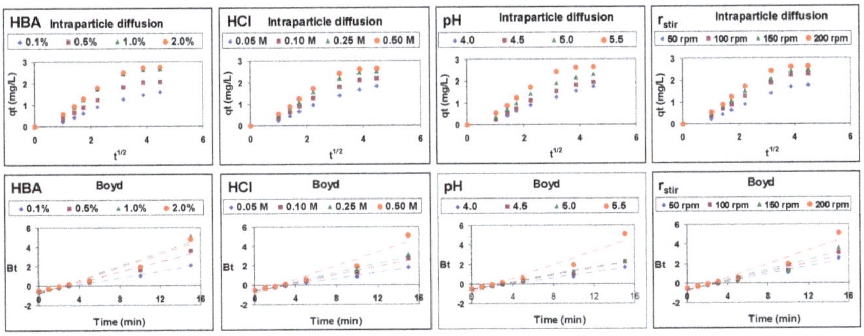

Figure 7. Analysis of the experimental data of carrier-mediated transport of Cu(II) through emulsion liquid membranes by intraparticle diffusion and Boyd models.

Intraparticle diffusion model representations are not linear over the entire time range. The existence of a clear multilinearity means that Cu-benzoylacetone complex diffusion in the membrane phase (from the feed/membrane interface toward the membrane/product interface) is not the only rate-controlling step, but rather that more than one step is involved. That is, the dual nature of the intraparticle model graphs confirms that both Cu(II) diffusion through the boundary layer of the feed/membrane interface and Cu-benzoylacetone complex diffusion through the membrane phase control the Cu(II) transport from the feed phase to the product phase.

As Boyd model representations are not linear and they do not pass through the origin, it can be deduced that Cu(II) diffusion through the boundary layer of the feed phase to the feed/membrane interface is the step that mainly controls the rate of the carrier-mediated transport process of Cu(II) through an emulsion liquid membrane.

4. Conclusions

We have studied the possibility of applying kinetic and mechanistic models developed for the adsorption process to the carrier-mediated transport in emulsion liquid membranes by analyzing Cu(II) removal from aqueous solutions (pH 4.0–5.5) by emulsion liquid membranes containing solutions of benzoylacetone (carrier agent, 0.5–2.0%) and Span 80 (emulsifying agent, 5%) in kerosene, as membrane phase, and hydrochloric acid (0.05 M–0.50 M) as stripping agent in the product phase. The membrane phase/product phase volume ratio used was 1/1, the feed phase/emulsion phase volume ratio was 2/1, and the stirring rate ranged from 50 to 200 rpm. The experimental results were analyzed by four kinetic adsorption models, the best fit being obtained with the pseudo-second-order kinetic model ($R^2 > 0.99$). A good relationship between the experimental parameters and those of the model was observed, and it was verified that variations in these model parameters with the different experimental conditions follow the same trend as that of the removal percentage calculated experimentally, the parameter normally used to analyze the efficiency of the liquid membrane process. Likewise, a linear relationship was observed between the model parameters and the initial apparent fluxes and permeabilities determined experimentally. The study of the transport mechanism suggests that although both Cu(II) diffusion through the feed/membrane interface boundary layer and complex Cu-benzoylacetone diffusion through the membrane phase control Cu(II) transport, it is the former step that mainly controls the transport process.

Author Contributions: For Conceptualization, G.L., E.G. and B.M.; methodology, G.L., E.G., B.M., A.M.H., M.G., M.D.M. and M.A.G.; formal analysis, G.L., E.G. and B.M.; investigation, A.M.H., M.G., M.D.M. and M.A.G.; data curation, A.M.H., M.G., M.D.M. and M.A.G.; writing—original draft preparation, G.L., E.G. and B.M.; writing—review and editing, G.L., E.G. and B.M. All authors have read and agreed to the published version of the manuscript.

Funding: This research received no external funding.

Institutional Review Board Statement: Exclude this statement. The study did not require ethical approval.

Informed Consent Statement: Not applicable.

Data Availability Statement: Please exclude this statement. The study did not report any data.

Conflicts of Interest: The authors declare no conflict of interest.

References

1. Srivastava, N.K.; Majumder, C.B. Novel biofiltration methods for the treatment of heavy metals from industrial wastewater. *J. Hazard. Mater.* **2008**, *151*, 1–8. [CrossRef]
2. Han, R.; Zhou, B.; Huang, Y.; Lu, X.; Li, S.; Li, N. Bibliometric overview of research trends on heavy metal health risks and impacts in 1989–2018. *J. Clean. Prod.* **2020**, *276*, 123249. [CrossRef]
3. Zaynab, M.; Al-Yahyai, R.; Ameen, A.; Sharif, Y.; Ali, L.; Fatima, M.; Khan, K.A.; Li, S. Health and environmental effects of heavy metals. *J. King Saud Univ. Sci.* **2022**, *34*, 101653. [CrossRef]
4. Qasem, N.A.A.; Mohammed, R.H.; Lawal, D.U. Removal of heavy metal ions from wastewater: A comprehensive and critical review. *NPJ Clean Water* **2021**, *4*, 36. [CrossRef]
5. Kumar, M.; Seth, A.; Singh, A.K.; Rajput, M.S.; Sikandar, M. Remediation strategies for heavy metals contaminated ecosystem: A review. *Environ. Sustain. Indic.* **2021**, *12*, 100155. [CrossRef]
6. Chen, Q.; Yao, Y.; Li, X.; Lu, J.; Zhou, J.; Huang, Z. Comparison of heavy metal removals from aqueous solutions by chemical precipitation and characteristics of precipitates. *J. Water Process. Eng.* **2018**, *26*, 289–300. [CrossRef]
7. Ahmeda, M.S.; Zewaila, T.M.; El-Ashtoukhya, E.S.Z.; Faraga, H.A.; El Azabb, I.H.; Albatatid, F.A.; Al-Hossainye, A.F.; Zorombag, M.S.; Abdel-Azizg, M.H. Enhancement of heavy metals recovery from aqueous solutions by cementation on a rotating cylinder using a stationary wiper. *J. Ind. Eng. Chem.* **2021**, *97*, 460–465. [CrossRef]
8. Shrestha, R.; Ban, S.; Devkota, S.; Sharma, S.; Joshi, R.; Tiwari, A.P.; Kim, H.Y.; Joshi, M.K. Technological trends in heavy metals removal from industrial wastewater: A review. *J. Environ. Chem. Eng.* **2021**, *9*, 105688. [CrossRef]
9. Burakov, A.E.; Galunin, E.V.; Burakova, I.V.; Kucherova, A.E.; Agarwal, S.; Tkachev, A.G.; Gupta, V.K. Adsorption of heavy metals on conventional and nanostructured materials for wastewater treatment purposes: A review. *Ecotoxicol. Environ. Saf.* **2018**, *148*, 702–712. [CrossRef] [PubMed]

10. Chai, W.S.; Cheun, J.Y.; Kumar, P.S.; Mubashir, M.; Majeed, Z.; Banat, F.; Ho, S.H.; Show, P.L. A review on conventional and novel materials towards heavy metal adsorption in wastewater treatment application. *J. Clean. Prod.* **2021**, *296*, 126589. [CrossRef]
11. Bilal, M.; Ihsanullah, I.; Younas, M.; Shah, M.U.H. Recent advances in applications of low-cost adsorbents for the removal of heavy metals from water: A critical review. *Sep. Purif. Technol.* **2022**, *278*, 119510. [CrossRef]
12. Qiu, B.; Tao, X.; Wang, H.; Li, W.; Ding, X.; Chu, H. Biochar as a low-cost adsorbent for aqueous heavy metal removal: A review. *J. Anal. Appl. Pyrolysis* **2021**, *155*, 105081. [CrossRef]
13. Tran, T.K.; Chiu, K.F.; Lin, C.Y.; Leu, H.J. Electrochemical treatment of wastewater: Selectivity of the heavy metals removal process. *Int. J. Hydrogen Energy* **2017**, *42*, 27741–27748. [CrossRef]
14. Ya, V.; Martin, N.; Chou, Y.H.; Chen, Y.M.; Choo, K.H.; Chen, S.S.; Li, C.W. Electrochemical treatment for simultaneous removal of heavy metals and organics from surface finishing wastewater using sacrificial iron anode. *J. Taiwan Inst. Chem. Eng.* **2018**, *83*, 107–114. [CrossRef]
15. Gao, X.; Meng, X. Photocatalysis for Heavy Metal Treatment: A Review. *Processes* **2021**, *9*, 1729. [CrossRef]
16. Bhattacharyya, D.; Moffitt, M.; Grieves, R.B. Charged Membrane Ultrafiltration of Toxic Metal Oxyanions and Cations from Single and Multisalt Aqueous Solutions. *Sep. Sci. Technol.* **1978**, *13*, 449–463. [CrossRef]
17. Sato, T.; Imaizumi, M.; Kato, S.; Taniguchi, Y. RO Applications in wastewater reclamation for reuse. *Desalination* **1977**, *23*, 65–76. [CrossRef]
18. Juang, R.S.; Xu, Y.Y.; Chen, C.L. Separation and removal of metal ions from dilute solutions using micellar-enhanced ultrafiltration. *J. Membr. Sci.* **2003**, *218*, 257–267. [CrossRef]
19. Yaqub, M.; Lee, S.H. Heavy metals removal from aqueous solution through micellar enhanced ultrafiltration: A review. *Environ. Eng. Res.* **2019**, *24*, 363–375. [CrossRef]
20. Al-Rashdi, B.A.M.; Johnson, D.J.; Hilal, N. Removal of heavy metal ions by nanofiltration. *Desalination* **2013**, *315*, 2–17. [CrossRef]
21. Yang, X.; Liu, Y.; Hu, S.; Yu, F.; He, Z.; Zeng, G.; Feng, Z.; Sengupta, A. Construction of Fe_3O_4@MXene composite nanofiltration membrane for heavy metal ions removal from wastewater. *Polym. Adv. Technol.* **2021**, *32*, 1000–1010. [CrossRef]
22. Ozaki, H.; Sharma, K.; Saktaywin, W. Performance of an ultra-low-pressure reverse osmosis membrane (ULPROM) for separating heavy metal: Effects of interference parameters. *Desalination* **2002**, *144*, 287–294. [CrossRef]
23. Vital, B.; Bartacek, J.; Ortega-Bravo, J.C.; Jeison, D. Treatment of acid mine drainage by forward osmosis: Heavy metal rejection and reverse flux of draw solution constituents. *Chem. Eng. J.* **2018**, *332*, 85–91. [CrossRef]
24. Mohammadi, T.; Moheb, A.; Sadrzadeh, M.; Razmi, A. Modeling of metal ion removal from wastewater by electrodialysis. *Sep. Purif. Technol.* **2005**, *41*, 73–82. [CrossRef]
25. Keskin, B.; Zeytuncu-Gökoğlu, B.; Koyuncu, I. Polymer inclusion membrane applications for transport of metal ions: A critical review. *Chemosphere* **2021**, *279*, 130604. [CrossRef] [PubMed]
26. Ahmad, A.L.; Kusumastutib, A.; Derek, C.J.C.; Ooi, B.S. Emulsion liquid membrane for heavy metal removal: An overview on emulsion stabilization and destabilization. *Chem. Eng. J.* **2011**, *171*, 870–882. [CrossRef]
27. León, L.; León, G.; Senent, J.; Pérez-Sirvent, C. Optimization of Copper Removal from Aqueous Solutions Using Emulsion Liquid Membranes with Benzoylacetone as a Carrier. *Metals* **2017**, *7*, 19. [CrossRef]
28. Amini, M.; Rahbar-Kelishami, A.; Alipour, M.; Vahidi, O. Supported Liquid Membrane in Metal Ion Separation: An Overview. *J. Membr. Sci. Res.* **2018**, *4*, 121–135.
29. León, G.; Hidalgo, A.M.; Miguel, B.; Guzmán, M.A. Pertraction of Co(II) through Novel Ultrasound Prepared Supported Liquid Membranes Containing D2EHPA. Optimization and Transport Parameters. *Membranes* **2020**, *10*, 436. [CrossRef]
30. Rzelewska-Piekut, M.; Regel-Rosocka, M. Liquid membranes for separation of metal ions from wastewaters. *Phys. Sci. Rev.* **2021**, 20210049. [CrossRef]
31. León, G. Facilitated transport. In *Encyclopedia of Membranes*, 1st ed.; Drioli, E., Giorno, L., Eds.; Springer: Berlin/Heidelberg, Germany, 2016; pp. 763–764.
32. Kislik, V.S. Introduction, General Description, Definitions and Classification. Overview. In *Liquid Membranes. Principles and Applications in Chemical Separation and Wastewater Treatment*, 1st ed.; Kislik, V.S., Ed.; Elsevier: Amsterdam, The Netherlands, 2010; pp. 1–15.
33. Ma, G.; Jiang, Y.; Kun, S.C. A general mass transfer model for liquid surfactant membrane. *Chem. Eng. Sci.* **1997**, *52*, 433–441.
34. Ho, W.S.; Hatton, T.A.; Lightfoot, E.N.; Li, N.N. Liquid surfactant membranes: A diffusion controlled model. *AIChE J.* **1982**, *28*, 662–671. [CrossRef]
35. Teramoto, M.; Sakai, T.; Yamagawa, K.; Ohsuga, M.; Miyake, Y. Extraction of phenol and cresol by liquid surfactant membrane. *Sep. Sci. Technol.* **1983**, *18*, 397–406. [CrossRef]
36. Box, G.E.P. Science and statistics. *J. Am. Stat. Assoc.* **1976**, *71*, 791–799. [CrossRef]
37. Tan, K.L.; Hameed, B.H. Insight into the adsorption kinetics models for the removal of contaminants from aqueous solutions. *J. Taiwan Inst. Chem. Eng.* **2017**, *74*, 25–48. [CrossRef]
38. Wang, J.; Guo, X. Adsorption kinetic models: Physical meanings, applications, and solving methods. *J. Hazard. Mater.* **2020**, *390*, 122156. [CrossRef]
39. Kulkarni, P.S.; Mahajani, V.V. Application of liquid emulsion membrane (LEM) process for enrichment of Molybdenum from aqueous solutions. *J. Membr. Sci.* **2002**, *201*, 123–135. [CrossRef]
40. Lagergren, S. About the theory of so-called adsorption of soluble substances. *K. Sven. Vetensk. Handl.* **1898**, *24*, 1–39.

41. Ho, Y.S. Review of second-order model for adsorption systems. *J. Hazard. Mater.* **2006**, *136*, 681–689. [CrossRef] [PubMed]
42. Elovich, S.Y.; Larinov, O.G. Theory of adsorption from solutions of non-electrolytes on solid (I) equation adsorption from solutions and the analysis of its simplest form, (II) verification of the equation of adsorption isotherm from solutions. *Izvestyja Akad. Nauk SSSR Otd. Khimicheskikh Nauk* **1962**, *2*, 209–216.
43. Avrami, M. Kinetics of phase change. I General theory. *J. Chem. Phys.* **1940**, *7*, 1103–1112. [CrossRef]
44. Weber, W.J.; Morris, J.C. Kinetics of adsorption on carbon from solution. *J. Sanit. Eng. Div.* **1963**, *89*, 31–59. [CrossRef]
45. Boyd, G.E.; Adamson, A.W.; Myers, L.S. The exchange adsorption of ions from aqueous solutions by organic zeolites. II. Kinetics. *J. Am. Chem. Soc.* **1947**, *69*, 2836–2848. [CrossRef] [PubMed]
46. Coleman, N.T.; McClung, A.C.; Moore, D.P. Formation constants for Cu(II)-peat complexes. *Science* **1956**, *123*, 330–331. [CrossRef] [PubMed]
47. Ho, Y.S.; McKay, G. The kinetics of sorption of divalent metal ions onto sphagnum moss peat. *Water Res.* **2000**, *34*, 735–742. [CrossRef]

membranes

MDPI

Article

Separation Iron(III)-Manganese(II) via Supported Liquid Membrane Technology in the Treatment of Spent Alkaline Batteries

Francisco J. Alguacil and Félix A. Lopez *

Centro Nacional de Investigaciones Metalurgicas (CENIM), Consejo Superior de Investigaciones Científicas (CSIC), Avda. Gregorio del Amo 8, 28040 Madrid, Spain; fjalgua@cenim.csic.es
* Correspondence: f.lopez@csic.es

Abstract: In this paper, the transport of iron(III) from iron(III)-manganese(II)-hydrochloric acid mixed solutions, coming from the treatment of spent alkaline batteries through a flat-sheet supported liquid membrane, is investigated (the carrier phase being of Cyanex 923 (commercially available phosphine oxide extractant) dissolved in Solvesso 100 (commercially available diluent)). Iron(III) transport is studied as a function of hydrodynamic conditions, the concentration of manganese and HCl in the feed phase, and the carrier concentration in the membrane phase. A transport model is derived that describes the transport mechanism, consisting of diffusion through a feed aqueous diffusion layer, a fast interfacial chemical reaction, and diffusion of the iron(III) species-Cyanex 923 complex across the membrane phase. The membrane diffusional resistance (Δ_m) and feed diffusional resistance (Δ_f) are calculated from the model, and their values are 145 s/cm and 361 s/cm, respectively. It is apparent that the transport of iron(III) is mainly controlled by diffusion through the aqueous feed boundary layer, this being the thickness of this layer calculated as 2.9×10^{-3} cm. Since manganese(II) is not transported through the membrane phase, the present system allows the purification of these manganese-bearing solutions.

Keywords: membrane transport; Cyanex 923; iron(III); manganese(II); separation

1. Introduction

Recycling is an issue of increasing necessity in the 21st century which is caused, among other factors, by the decrease and depletion of certain raw materials, and more stringent environmental pressures about the discharge of contaminants.

One example of materials to be recycled are alkaline batteries, which also contain valuable metals that can be recovered and reuse (allowing one to gain a further profit from the recycling process). CENIM-CSIC developed investigations about the use of ammonium carbonate processing in the recycle of zinc-bearing materials such as electric arc furnace (EAF) flue dusts and Waelz oxides [1–3]. Since zinc is an important component of alkaline batteries, this type of processing was applied on the recycle of these spent alkaline batteries [4,5]. In the process, the mass yielded after breaking of the battery case is subjected to a leaching process in ammoniacal ammonium carbonate solutions. Zinc is recovered from the leachate, whereas the solid residue contains iron and manganese (VII) compounds. This residue is treated via leaching in HCl medium, yielding a solution containing Fe(III) and Mn(II), since in the operational conditions, manganese(VII) oxidizes chloride ions and, thus, it is reduced to Mn(II). From the solution, Fe(III) and Mn(II) can be separated using liquid-liquid extraction technology. However, the use of liquid membranes, specifically supported liquid membranes technologies, can be also of interest due to the economical and operational advantages that these technologies possess.

In fact, supported liquid membrane technologies have been proposed for the treatment of strategic and toxic metals. Some of their uses have recently been reviewed [6,7]

and specifically published in the case of heavy metals [8–10], germanium [11], rare earth metals [12,13], scandium [14], europium [15], and tungsten [16].

Before scaling up the flat-sheet supported liquid membrane to a more dynamic membrane methodology, such as hollow fiber modules, a model of the liquid membrane system is convenient in order to design this more efficient process.

The present investigation presents a kinetic modelling of the active transport of Fe(III), from Fe(III)-Mn(II)-HCl solutions, using Cyanex 923-Solvesso 100 solutions immobilized on a microporous hydrophobic support. These aqueous solutions originated in the processing of spent alkaline batteries. Diffusional resistances due to the membrane and the feed phase are calculated from the model, and the influence of hydrodynamic and chemical conditions are established in order to yield an efficient flat-sheet liquid membrane system.

2. Materials and Methods

2.1. Reagents and Solutions

The extractant used in the investigation was Cyanex 923 (Solvay, Brussels, Belgium, commercially available phosphine oxide), which was dissolved in Solvesso 100 (Exxon Chem Iberia, Barcelona, Spain, aromatic diluent) in order to obtain an adequate range of extractant concentrations to the transport experiments. Also, the composition of the aqueous feed solutions was scaled down to the metal concentrations range convenient to this type of membrane experimentation. All of the chemicals used in the experimentation were of AR grade.

The solid support used in the present work was Millipore Durapore GVHP4700 (Celanese Plastics, Irving, TX, USA) (microporous polyvinylidene fluoride film) of 75% porosity, 1.67 tortuosity, 12.5×10^{-3} cm thickness, and 0.22 μm effective pore size.

2.2. Methods

2.2.1. Liquid-Liquid Extraction Experiments

Iron (III) liquid-liquid extraction experiments were performed by mixing, in thermostatted separatory funnels, equal volumes of organic solutions of Cyanex 923 (from 1.9 $\times 10^{-2}$ to 0.13 M) in Solvesso 100 with 0.01 g/L iron(III) in 4 M HCl solutions for 10 min at 20 °C. The careful optimization of experimental setups indicates that equilibrium reaches within 5 min of contact of both the phases. After the quick phase disengagement (less than 30 s), iron(III) was analyzed in the aqueous solutions by atomic absorption spectrometry, reproducibility within ±3%, and the metal concentration in the equilibrated organic solutions was estimated by the corresponding mass balance.

2.2.2. SLM Experiments

Transport experiments were carried out in a two compartments cell which consisted of a feed solution half-cell (200 cm^3) separated from the receiving solution half-cell (200 cm^3) by the solid support with an effective membrane area of 11.3 cm^2. The feed and the receiving solutions were mechanically stirred at 20 °C to avoid concentration polarization conditions at the support interfaces and in the bulk of both solutions.

The supported liquid membrane was prepared by impregnation of the solid support with the corresponding carrier solution by immersion for 24 h. Then, it was left to drip for 20 s before being placed in the cell.

Metals transport were determined by monitoring Fe(III) and Mn(II) concentrations in the feed and receiving phases as a function of time and by atomic absorption spectrometry. Metal concentrations in the solutions were found to be reproducible within ±3%. The iron overall mass transfer coefficient (K_{Fe}) was computed using the next equation:

$$\ln \frac{[Fe]_{f,t}}{[Fe]_{f,0}} = -\frac{AK_{Fe}}{V} t \quad (1)$$

where A was the support area, V is the volume of the feed solution, $[Fe]_{f,t}$ and $[Fe]_{f,0}$ were the iron concentrations in the feed solution at an elapsed time and time zero, respectively, and t was the elapsed time.

The percentage of iron transported to the receiving solution was calculated by:

$$\%T = \frac{[Fe]_{r,t}}{[Fe]_{f,0} - [Fe]_{f,t}} 100 \quad (2)$$

where $[Fe]_{r,t}$ was the iron concentration in the receiving solution at an elapsed time. Manganese(II) transport was evaluated as above, and under the present experimental conditions, basically manganese(II) was not transported ($K_{Mn} < 1 \times 10^{-6}$ cm/s) across the membrane.

3. Results

3.1. Iron(III) Liquid-Liquid Extraction Equilibrium

The extraction of iron(III) by the phosphine oxide was based on a solvation reaction [17], represented by the next equilibrium [18,19]:

$$nL_{org} + FeCl^{-}_{4_{aq}} + H^{+}_{aq} \Leftrightarrow HFeCl_4 \cdot nL_{org} \quad (3)$$

where L represented the extractant molecule, n was an stoichiometric factor (the number of extractant molecules involved in the extraction process), and the subscripts $_{aq}$ and $_{org}$ referred to the equilibrated aqueous and organic phases, respectively.

The extraction equilibrium can be described by the next equation:

$$K_{ext} = \frac{[HFeCl_4L_n]_{org}}{[FeCl_4^-]_{aq}[H^+]_{aq}[L]^n_{org}} \quad (4)$$

In order to validate Equation (3) and calculate both the extraction constant and the stoichiometric factor n values, the experimental data were treated by a tailored computer program which minimized the U function, defined as:

$$U = \Sigma(\log D_{cal} - \log D_{exp})^2 \quad (5)$$

D_{exp} and D_{cal} being the experimental distribution coefficients and the corresponding values calculated by the program. The experimental distribution coefficient D_{exp} was calculated as:

$$D_{exp} = \frac{[Fe]_{org}}{[Fe]_{aq}} \quad (6)$$

In the above equation, $[Fe]_{org}$ and $[Fe]_{aq}$ are the iron concentrations in the equilibrated organic and aqueous phases, respectively.

The results derived from the experimental data and treated by the program indicated that the extraction of iron (III) by Cyanex 923 corresponded to the reaction showed in Equation (3), with log K_{HCl} 4.86 and U 0.093. The stoichiometric factor n was calculated as 4.

3.2. Iron(III) Transport through the Supported Liquid Membrane

Figure 1 showed a probable transport scheme for Fe(III) with Cyanex 923 (represented as L) dissolved in Solvesso 100 through a supported liquid membrane. The driving force for iron(III) transport was the difference in acidity between the feed and receiving phases, and thus iron(III) transport was coupled to the acid co-transport from the feed to the receiving phase.

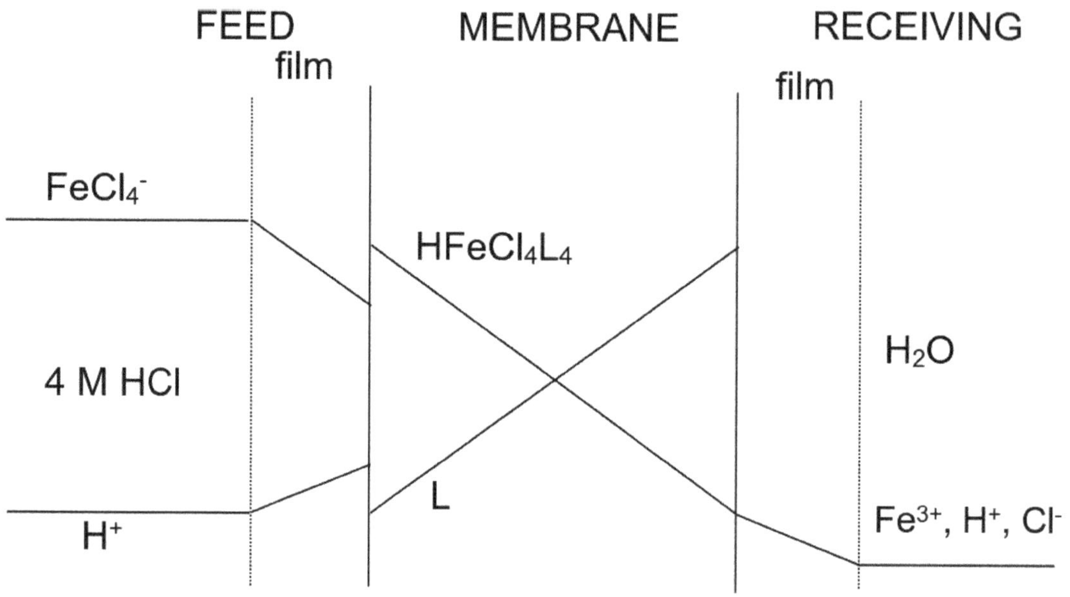

Figure 1. Concentration profile of the species through the supported liquid membrane.

3.2.1. Influence of the Stirring Speed Applied on the Feed Phase on Iron(III) Transport

In order to yield effective iron(III) transport through the supported liquid membrane, it was of importance to establish the influence of the hydrodynamic conditions, firstly, and the influence of the stirring speed, applied to the feed phase on the overall mass transfer coefficient, was investigated. The transport of iron(III) across the supported liquid membrane was dominated by diffusional resistances which can be of two types: (i) one associated to the feed phase boundary layer, and (ii) other associated to the membrane support. It is relatively usual that the magnitude of the first competed with the value of the support resistance [20].

In the present work, stirring of the feed solution was carried out from 500 to 1500 min^{-1} (Table 1).

Table 1. Influence of the stirring speed on transport of Fe(III) as a function of the overall mass transfer coefficient (K_{Fe}).

Stirring Speed, min^{-1}	K_{Fe} 10^3, cm/s	% Fe Recovery [a]
500	0.7	99
1000	1.1	99
1250	1.4	99
1500	1.5	99

Feed phase: 0.01 g/L Fe(III) and 0.17 g/L Mn(II) in 2 M HCl. Feed phase stirring speed: variable. Membrane phase: 10% *v/v* Cyanex 923 in Solvesso 100 supported on GVHP4700. Receiving phase: distilled water. Receiving phase stirring speed: 500 min^{-1}. Temperature: 20 °C. [a] Recovery in the receiving phase after 3 h.

It can be seen that the overall mass transfer value increased with the increase of the stirring speed up to 1250 min^{-1} and then remained constant. Thus, at 1250 min^{-1}, the thickness of the aqueous feed boundary layer reached a minimum, and iron(III) transport maximized. Also, from the results showed in this Table 1, it can be observed that iron(III) recovery in the receiving phase was practically quantitative; the transport of Mn(II) can be considered as negligible (see Section 2.2.2). Thus, in the first instance, the present system (carrier phase and support) can be used to separate Fe(III) from Mn(II). It is also worth noting here that the pH of the receiving phase became more acidic (from pH 5 to pH 1.5–2) as the time was elapsed. This result can be attributed to the co-transport of HCl from the feed to the receiving phase (see Section 3.2).

3.2.2. Influence of the Stirring Speed Applied on the Receiving Phase on Iron(III) Transport

The influence of the stirring speed on iron(III) transport, applied to the receiving phase, was also investigated using the same experimental conditions showed in Table 1, albeit using a stirring speed in the feed phase of 1250 min^{-1} and varying the stirring speed in the receiving phase from 500 to 1000 min^{-1} The results showed that this variation had no influence on iron(III) transport or in the recovery of the element in this phase. This behavior was attributable to that if the stirrer in the receiving phase half-cell was very close to the membrane support, the thickness of the boundary layer was considered to be minimized, and the resistance in this side can be neglected [21].

3.2.3. Influence of the HCl Concentration in the Feed Phase on Iron(III) Transport

The influence of the variation of the acid concentration in the feed phase on iron(III) transport was investigated using the same experimental conditions showed in Table 1, but with acid concentrations ranging from 1 to 8 M. In these series of experiments, the stirring speeds applied on the feed and receiving phases were of 1250 and 500 min^{-1}, respectively. The results from these series of experiments were summarized in Table 2. It can be seen that iron(III) transport increased when the HCl concentration increased in the 1–4 M range, and then decreased at HCl concentrations of 6 and 8 M. This decrease can be explained due to the competitive transport of HCl across the supported liquid membrane impregnated with Cyanex 923 dissolved in Solvesso 100, it was shown [22] that Cyanex 923 extracted mineral acids from aqueous solutions. In the case of manganese(II), negligible transport of this element was detected along the experiments.

Table 2. Influence of HCl concentration on iron(III) transport.

[HCl], M	$K_{Fe} \cdot 10^{-3}$, cm/s	% Fe Recovery [a]
1	0.08	99
1.5	0.56	99
2	1.4	99
4	3.4	72
6	3.1	68
8	2.3	65

[a] Recovery in the receiving phase after 3 h.

3.2.4. Influence of Cyanex 923 Concentration on Iron(III) Transport

Previously, it was detected that the solid support impregnated only with Solvesso 100 did not transport neither iron(III) or manganese(II). Thus, the presence of the extractant or carrier was essential to achieve metal transport from the feed to the receiving solution. This influence was investigated using a feed phase of 0.01 g/L Fe(III) and 0.17 g/L Mn(II) in 4 M HCl medium, whereas the receiving phase was of distilled water; in these series of experiments, the extractant concentration was varied from 1.3 to 40% *v/v* (0.03–1 M) dissolved in Solvesso 100. The results of these experiments are summarized in Table 3.

Table 3. Influence of Cyanex 923 concentration in the carrier phase on iron(III) transport.

[Cyanex 923], % *v/v*	$K_{Fe} \cdot 10^{-3}$, cm/s	% Fe Recovery [a]
1.3	1.0	90
2.5	2.3	99
5	2.5	98
10	3.4	72
20	2.9	71
30	2.3	72
40	2.1	72

[a] Recovery in the receiving phase after 3 h.

Results showed in the above Table 3 indicated that there was an increase of iron(III) permeation with the increase from 1.3 to 10% *v/v* Cyanex 923 concentration in the carrier solution, and then a decrease in iron(III) transport resulted when higher Cyanex 923 concentrations (20–40% *v/v*) were used. This decrease, at a first instance, can be attributed to the increase of the carrier phase viscosity, which resulted in lower values of the overall mass transfer coefficient. In the case of manganese(II), negligible transport was observed, even using the most concentrated Cyanex 923 solution. At 10% *v/v* Cyanex 923 in Solvesso 100 concentration, a maximum in iron(III) transport was attained, this maximum or limiting mass transfer coefficient (K_{lim}) was explained by the assumption that in this experimental condition, the diffusion in the membrane (Δ_m) was negligible in comparison with feed diffusion (Δ_f), and the transport process was entirely controlled by the diffusion in the film of the feed solution. Thus:

$$K_{lim} = \frac{1}{\Delta_f} = \frac{D_f}{d_f} \tag{7}$$

In the above equation, $D_f = 10^{-5}$ cm^2/s represented the average diffusion coefficient in the aqueous feed phase, and d_f to the thickness of the feed phase boundary layer. Accordingly, in the present system, d_f was calculated as 2.9×10^{-3} cm. This value represents the minimum thickness of the boundary layer under the present experimental conditions.

3.2.5. Influence of Manganese(II) Concentration in the Feed Phase on Iron(III) Transport

The possible variation of manganese(II) concentration in the feed phase on iron(III) transport was also investigated. The carrier phase was of 10% *v/v* Cyanex 923 in Solvesso 100 impreganting Durapore GVHP4700 support, whereas the receiving solution was water. The aqueous feed phase contained 0.01 g/L Fe(III) in 4 M HCl medium and varying Mn(II) concentrations from nil to 0.17 g/L. The results indicated that the presence of Mn(II) in the solution had not detrimental effect on iron(III) transport ($K_{Fe} = 3.2 \pm 0.2 \times 10^{-3}$ cm/s), thus, crowding or population effect [23], due to the presence of Mn(II) in the solution, was negligible on iron(III) permeation under these experimental conditions.

3.2.6. Estimation of Diffusional Parameters and Evaluation of Mass Transfer Resistances

The permeation rate of iron(III) was estimated by the rates associated to the diffusion of Fe(III)-bearing species through the feed phase diffusion layer and the diffusion of the Fe(III)-Cyanex 923 species through the membrane. Fick's first diffusion law was applied to the diffusion layer in the feed and the membrane sides [24]. The fluxes in each of the above phases can be represented by:

$$J_f = \Delta_f^{-1}\left([Fe]_{f,0} - [Fe]_{f,i}\right) \tag{8}$$

$$J_m = \Delta_m^{-1}\left([HFeCl_4L_4]_{f,i} - [HFeCl_4L_4]_{r,i}\right) \tag{9}$$

where $[Fe]_{f,0}$ was the initial iron concentration in the feed phase, $[Fe]_{f,i}$ was the iron concentration at the feed phase/membrane interface, and $[HFeCl_4L_4^-]_{f,i}$ and $[HFeCl_4L_4]_{r,I}$ were the concentrations of the species at the feed phase/membrane interface and the membrane/receiving phase interface, respectively.

Usually, the concentration of the metal-organic ligand species ($HFeCl_4L_4$ in the present investigation) in the membrane phase at the receiving phase side was negligible when compared with the corresponding one at the feed phase side. Thus, Equation (9) was rewritten as:

$$J_m = \Delta_m^{-1}[HFeCl_4L_4]_{f,i} \quad (10)$$

It can be considered that the chemical reaction (Equation (3)) was fast if compared to the diffusion rate, thus, local equilibrium at the feed phase-membrane interface was attained and concentrations in this interface was related by Equation (4). At steady state, $J = J_f = J_m$, being J expressed as:

$$J = K_{Fe}[Fe]_{f,0} \quad (11)$$

By combination of Equations (4), (8), (10) and (11), the following expression was derived:

$$K_{Fe} = \frac{K_{ext}[L]_m^4[H^+]_f}{\Delta_m + \Delta_f\left(K_{ext}[L]_m^4[H^+]_f\right)} \quad (12)$$

where Δ_f and Δ_m were the transport resistances due to diffusion by the feed phase boundary layer and the membrane, respectively.

This last equation combined in one expression the equilibrium and diffusional parameters involved in the iron(III) transport though a flat-sheet supported liquid membrane containing Cyanex 923 in Solvesso 100 as carrier phase.

The resistances to the mass transfer were evaluated by the use of Equation (12), from this equation the next expression can be derived:

$$\frac{1}{K_{Fe}} = \Delta_f + \frac{\Delta_m}{K_{ext}[L]_m^4[H^+]_f} \quad (13)$$

A plot of $1/K_{Fe}$ versus $1/K_{ext}[L]_m{}^4[H^+]_f$, for various extractant concentrations in Solvesso 100 and an aqueous feed solution of 4 M HCl, allowed to determine Δ_m (slope) and Δ_f (ordinate). From the above plot ($r^2 = 0.978$), it was determined that $\Delta_m = 145$ s/cm and $\Delta_f = 361$ s/cm. These values are the transport resistances due to diffusion through the membrane pores and the feed boundary layer, respectively.

The diffusion coefficient in the membrane phase is defined as

$$D_m = \frac{d_m}{\Delta_m} \quad (14)$$

was calculated as 8.6×10^{-5} cm^2/s. In the above equation, d_m represented the membrane thickness. The diffusion coefficient of the phosphine oxide-iron(III) species in the bulk organic phase [25]:

$$D_{b,m} = D_m\frac{\tau^2}{\varepsilon} \quad (15)$$

was estimated as 3.2×10^{-4} cm^2/s. This value was greater than the value of the diffusion coefficient in the membrane phase (Equation (14)), this was attributable to the diffusional resistance caused by the membrane thickness, separating the feed and the receiving phases, In Equation (15), τ is the membrane tortuosity (1.67) and ε is the support porosity (75).

Assuming that the extractant concentration in the membrane phase was constant, an apparent diffusion coefficient for iron (III)-Cyanex 923 dissolved in Solvesso 100 species was estimated as:

$$D_m^a = \frac{Jd_m}{[L]_m} \quad (16)$$

where J was calculated accordingly to Equation (11) and an initial iron(III) concentration in the feed phase of 0.01 g/L; considering an extractant concentration of 10% *v/v* Cyanex 923 in Solvesso 100, this apparent coefficient van be calculated as 3×10^{-8} cm^2/s.

Since the overall mass transfer resistance was the result of the sum of the different resistances participating in the permeation process, Equation (13) can be rewritten as:

$$R = R_f + R_m \quad (17)$$

where R_f and R_m were the resistances in the feed and the membrane phases, respectively. Table 4 showed the contribution of these resistances to the iron(III) transport.

Table 4. Contribution of mass transfer resistances to iron(III) transport.

Experimental Condition	[a] R, s/cm	[b] R_f, s/cm	$\%R_f^0$	$\%R_m^0$
1–10% *v/v* Cyanex 923	1000–294	361	36–100	64–0
10–40% *v/v* Cyanex 923	294–476	361	100–76	0–24
1–4 M HCl	12,195–294	361	3–100	97–0
4–8 M HCl	294–435	361	100–83	0–17
0–0.17 g/L Mn(II)	294	361	100	0

[a] Experimental values. [b] Model value.

Furthermore, the fractional resistance of each step ($\%R_f^0$ and $\%R_m^0$) to the transport process was calculated under various experimental conditions. As it can be seen from the results presented in Section 4, in almost all of the conditions, diffusion by the aqueous feed boundary layer was the controlling step for the iron(III) transport process.

4. Conclusions

The use of Cyanex 923, dissolved in Solvesso 100, coupled to a solid supported liquid membrane (flat-sheet mode) technology, allowed the selective separation of Fe(III) over Mn(II) from solutions derived in the treatment of spent alkaline batteries. A kinetic model for Fe(III) transport is developed and finding that Fe(III) transport is mainly controlled by diffusion of protons and $FeCl_4^-$ species across the feed aqueous layer. This contributes, under certain experimental conditions, to the diffusion by the membrane of the neutral complex formed by the $HFeCl_4$ species and Cyanex 923.

Author Contributions: Conceptualization, F.J.A. and F.A.L.; methodology, F.J.A.; validation, F.J.A. and F.A.L.; formal analysis, F.J.A.; investigation, F.J.A. and F.A.L.; resources, F.A.L.; writing—original draft preparation, F.J.A.; writing—review and editing, F.J.A. and F.A.L. All authors have read and agreed to the published version of the manuscript.

Funding: This research received no external funding.

Institutional Review Board Statement: Not applicable.

Informed Consent Statement: Not applicable.

Data Availability Statement: All the data supporting the findings of this study are available within the article.

Acknowledgments: We acknowledge support towards the publication fee by the CSIC Open Access Publication Support Initiative through its Unit of Information Resources for Research (URICI).

Conflicts of Interest: The authors declare no conflict of interest.

References

1. Caravaca, C.; Cobo, A.; Alguacil, F.J. Considerations about the recycling of EAF flue dusts as source for the recovery of valuable metals by hydrometallurgical processes. *Resour. Conserv. Recycl.* **1994**, *10*, 35–41. [CrossRef]
2. Alguacil, F.J.; de Goicoechea, N.; Dañobeitia, I.; Cobo, A.; Caravaca, C.; García, F. Procedimiento para la Obtención de un Óxido de Cinc de Alta Pureza Mediante Lixiviación de Óxidos Waelz con Disoluciones de Carbonato Amónico. Spanish Patent 9,500,605, 28 March 1995.
3. Ruiz, O.; Clemente, C.; Alonso, M.; Alguacil, F.J. Recycling of an electric arc furnace flue dust to obtain high grade ZnO. *J. Hazard. Mat.* **2007**, *141*, 33–36. [CrossRef] [PubMed]
4. Cebriano, T.; García-Díaz, I.; López Fernández, A.; Fernández, P.; López, F.A. Synthesis and characterization of ZnO micro- and nanostructures grown from recovered ZnO from spent alkaline batteries. *J. Environ. Chem. Eng.* **2017**, *5*, 2903–2911. [CrossRef]
5. López, F.A.; Cebriano, T.; García-Díaz, I.; Fernández, P.; Rodríguez, O.; López Fernández, A. Synthesis and microstructural properties of zinc oxide nanoparticles prepared by selective leaching of zinc from spent alkaline batteries using ammoniacal ammonium carbonate. *J. Clean. Prod.* **2017**, *148*, 795–803. [CrossRef]
6. Rzelewska-Piekut, M.; Regel-Rosocka, M. Liquid membranes for separation of metal ions from wastewaters. *Phys. Sci. Rev.* **2021**, in press. [CrossRef]
7. Zante, G.; Boltoeva, M.; Masmoudi, A.; Barillon, R.; Trébouet, D. Supported ionic liquid and polymer inclusion membranes for metal separation. *Sep. Purif. Rev.* **2021**, in press. [CrossRef]
8. Yesil, H.; Molaey, R.; Calli, B.; Tugtas, A.E. Removal and recovery of heavy metals from sewage sludge via three-stage integrated process. *Chemosphere* **2021**, *280*, 130650. [CrossRef] [PubMed]
9. Othmen, K.; Ncib, S.; Barhoumi, A.; Dammak, L.; Bouguerra, W. Recovery of nickel ions by supported liquid membrane (Slm) using d2ehpa as carrier. *Desalin. Water Treat.* **2021**, *225*, 422–429. [CrossRef]
10. Lee, L.Y.; Morad, N.; Ismail, N.; Rafatullah, M. Selective separation of cadmium (II), copper (II) and nickel (II) ions from electroplating wastewater using dual flat sheet supported liquid membrane. *Desalin. Water Treat.* **2021**, *224*, 291–301. [CrossRef]
11. Patel, M.; Karamalidis, A.K. Germanium: A review of its US demand, uses, resources, chemistry, and separation technologies. *Sep. Purif. Technol.* **2021**, *275*, 118981. [CrossRef]
12. Patil, A.B.; Struis, R.P.W.J.; Testino, A.; Ludwig, C. Extraction of Rare Earth Metals: The New Thermodynamic Considerations Toward Process Hydrometallurgy. *Miner. Met. Mater. Ser.* **2021**, 187–194. [CrossRef]
13. Mahanty, B.; Karak, A.; Mohapatra, P.K.; Egberink, R.J.M.; Valsala, T.P.; Sathe, D.B.; Bhatt, R.B.; Huskens, J.; Verboom, W. Carrier mediated transport of actinides using hexa-n-hexylnitrilotriacetamide (HHNTA). *Chem. Eng. Process. Process Intensif.* **2021**, *161*, 108323. [CrossRef]
14. Rout, P.C.; Sarangi, K. A systematic study on extraction and separation of scandium using phosphinic acid by both solvent extraction and hollow fibre membrane. *Miner. Process. Extr. Metall. Trans. Inst. Min. Metall.* **2021**, in press. [CrossRef]
15. Kumar, R.; Kandwal, P. Mathematical Modeling of facilitated transport of Eu (III) ion by CMPO in modified diluents as extractant. *J. Phys. Conf. Ser.* **2021**, *1849*, 01200. [CrossRef]
16. Ali, N.; Shah, S.; Khan, A.; Khan, S.B.; Kamal, T.; Asiri, A.M. Selective separation of tungsten from the model and industrial effluents through supported liquid membrane. *Chem. Pap.* **2021**, *75*, 553–563. [CrossRef]
17. Regel-Rosocka, M.; Alguacil, F.J. Recent trends in metals extraction. *Rev. Metal.* **2013**, *49*, 292–316. [CrossRef]
18. Alguacil, F.J.; Martínez, S. Permeation of iron (III) by an immobilised liquid membrane using Cyanex 923 as mobile carrier. *J. Membr. Sci.* **2000**, *176*, 249–255. [CrossRef]
19. Alguacil, F.J.; Alonso, M. Iron (III) transport using a supported liquid membrane containing Cyanex 921. *Hydrometallurgy* **2000**, *58*, 81–88. [CrossRef]
20. Bohrer, M.P. Diffusional boundary layer resistance for membrane transport. *Ind. Eng. Chem. Fundam.* **1983**, *22*, 72–78. [CrossRef]
21. Pavon, S.; Fortuny, A.; Coll, M.T.; Bertaub, M.; Sastre, A.M. Permeability dependencies on the carrier concentration and membrane viscosity for Y (III) and Eu (III) transport by using liquid membranes. *Sep. Purif. Technol.* **2020**, *239*, 116573. [CrossRef]
22. Alguacil, F.J.; López, F.A. The extraction of mineral acids by the phosphine oxide Cyanex 923. *Hydrometallurgy* **1996**, *42*, 245–255. [CrossRef]
23. De Gyves, J.; De San Miguel, E.R. Metal ion separations by supported liquid membrane. *Ind. Eng. Chem. Res.* **1999**, *38*, 2182–2202. [CrossRef]
24. Sastre, A.M.; Alguacil, F.J.; Alonso, M.; Lopez, F.; Lopez-Delgado, A. On cadmium (II) membrane-based extraction using Cyanex 923 as carrier. *Solvent Extr. Ion Exch.* **2008**, *26*, 192–207. [CrossRef]
25. Huang, T.C.; Juang, R.S. Rate and mechanism of divalent metal transport through supported liquid membrane containing di (2-ethylhexyl) phosphoric acid as a mobile carrier. *J. Chem. Technol. Biotechnol.* **1988**, *42*, 3–17. [CrossRef]

membranes

MDPI

Article

Reduction of Cost and Environmental Impact in the Treatment of Textile Wastewater Using a Combined MBBR-MBR System

Xuefei Yang [1] and Víctor López-Grimau [2,*]

[1] Institute of Textile Research and Industrial Cooperation of Terrassa (INTEXTER), Universitat Politècnica de Catalunya, Carrer de Colom 15, 08222 Terrasa, Spain; xuefei.yang@upc.edu
[2] Department of Project and Construction Engineering, Universitat Politècnica de Catalunya, Carrer de Colom 11, 08222 Terrassa, Spain
* Correspondence: victor.lopez-grimau@upc.edu; Tel.: +34-937-397-316

Abstract: A hybrid Moving Bed Biofilm Reactor—Membrane Bioreactor (MBBR-MBR) was developed for the treatment of wastewater from a Spanish textile company. Compared with conventional activated sludge (CAS) treatment, the feasibility of this hybrid system to reduce economic and environmental impact on an industrial scale was conducted. The results showed that, technically, the removal efficiency of COD, TSS and color reached 93%, 99% and 85%, respectively. The newly dyed fabrics performed with the treated wastewater were qualified under the standards of the textile industry. Economically, the values of Capital Expenditure (CAPEX) calculated for the hybrid MBBR-MBR system are profitable because of the reduction in Operational Expenditure (OPEX) when compared with CAS treatment, due to the lower effluent discharge tax thanks to the higher quality of the effluent and the decolorizing agent saved. The result of Net Present Value (NPV) and the Internal Rate of Return (IRR) of 18% suggested that MBBR-MBR is financially applicable for implantation into the industrial scale. The MBBR-MBR treatment also showed lower environmental impacts than the CAS process in the life cycle assessment (LCA) study, especially in the category of climate change, thanks to the avoidance of using extra decolorizing agent, a synthetic product based on a triamine.

Keywords: textile wastewater; moving bed biofilm reactor; membrane bioreactor; economic feasibility; life cycle assessment; water reuse; decolorizing agent

1. Introduction

The textile industry processes a wide variety of fibers: naturals (cotton, wool, etc.), artificial (viscose, acetate, etc.) or synthetic (polyester, acrylic, etc.). Each textile fiber needs a specific processing technology and corresponding sizing agents, dyes, and auxiliaries [1]. The dyeing and finishing processes of the textile industry involve the generation of large volumes of wastewater. The composition of textile wastewater is very complex, and it is often characterized by variable pH, high concentrations of Chemical Oxygen Demand (COD), high turbidity, problems of color and limited biodegradability due to the dyes remaining in the wastewater [2]. The chemical stability and low biodegradability of compounds like dyes in the wastewater especially cause significant environmental concerns [3,4]. In response to such a complex type of wastewater, researchers and industries have developed various treatment processes, such as physico-chemical (coagulation–flocculation, adsorption and filtration), biological technologies and combined treatment processes [5–7]. Compared with physico-chemical methods, biological processes are more environmentally friendly because of the complete degradation of contaminants without producing secondary pollutants [8]. In contrast, conventional biological processes by conventional activated sludge (CAS) are not able to properly eliminate the color of textile effluents and they require the application a tertiary treatment for color removal in order to accomplish current regulations [4]. In Spain, the discharge of industrial wastewater to public sanitation systems must comply with the limits established by the local and regional authorities, which manage the treatment facilities [9]. In addition, companies must pay a discharge tax to cover the costs of public sanitation infrastructures. This discharge tax depends on the wastewater volume and the pollutant load [10]. To lower the discharge tax, industries have to increase the efficiency of their wastewater treatment systems, in order to obtain effluents of higher quality that can be reused in the production process.

Among different advanced biological treatments, Membrane Bioreactor (MBR), as a promising process combining biological treatment and membrane filtration, has been increasingly applied for industrial wastewater treatment, including the textile sector [11]. The MBR process has shown several advantages over CAS treatment, such as small footprint, stable effluent quality, high tolerance to high concentrations of organic matters, and lower sludge production [12,13]. Due to the benefits of MBR over CAS reflected in better effluent quality, no additional chemical products being needed and less sludge production, the MBR process has been proved in Life Cycle Assessment (LCA) studies to be a more eco-friendly option, especially in environmental impact related to global warming potential, abiotic depletion and acidification [14–16]. In previous techo-economic research, MBR showed a higher cost due to the large energy consumption, but the fact that an MBR plant can reduce the Hydraulic Retention Time (HRT) compensates for the higher power expense [17,18]. However, even though MBR can efficiently treat wastewater with a higher organic load than that of the CAS process, the concentration of biomass is limited in practical applications to avoid affecting the oxygen transfer coefficient [19].

Another biological treatment that has been attracting more and more attention in textile wastewater treatment in recent years is the Moving Bed Biofilm Reactor (MBBR), thanks to its ability to withstand a much higher biomass concentration [20–22]. Due to the large number of biofilms fixed on the carriers, the biomass concentration of MBBR can be higher than that of CAS [23]. Previous studies have shown that MBBR can effectively remove COD, but its ability to remove color is limited because of its incompleteness of sludge decantation [24,25]. In practical applications, it is essential to add coagulant to obtain a well-clarified effluent and, in many cases, extra decolorizing agent needs to be used to improve the removal of color [26]. From the economic and environmental perspective, these additional products generate extra cost and result in environmental impacts. For example, most of the decolorizing agents used are quaternary ammonium salt [27], which has a high impact on the toxicity category [14].

A hybrid MBBR-MBR system will be able to improve the sludge decantation of the MBBR system by way of the MBR membrane filtration. Moreover, since a part of the biofilm is formed on the MBBR carriers, the oxygen transfer coefficient will not be influenced, allowing energy saving for aeration. Two previous studies have been published on the performance of biofilm-membrane filtration with complex processes, including anaerobic and aerobic tanks, in the wastewater treatment of the textile industry, showing that this treatment is viable for treating textile wastewater [28,29]. With the effective treatment of MBBR-MBR, the effluent does not need to add a coagulant or decolorizing agent, notably reducing the environmental impact. It should be noted that the previous studies used separate tanks for MBBR and MBR. It could be an attractive option to keep the MBBR and MBR process within one reactor to maximize the advantages of both systems. To the best of our knowledge, no studies have been performed on the reuse of treated textile wastewater by MBBR-MBR. Economically, water reuse in the textile industry will allow particularly important savings, not only for the water cost as a consumer but also for the environmental tax.

In this study, a hybrid system, MBBR-MBR, was studied in the treatment of wastewater from a local textile industry. The study centered on the treating efficiency of eliminating organic compounds, suspended solids and color. Subsequently, new dyeing processes were performed with the treated water to assess the viability of water recovery. The objectives of this study are to evaluate the economic and environmental feasibility of the implementation of the hybrid MBBR-MBR on an industrial scale according to the results of its treating efficiency on which, to our knowledge, no such research has been done. The feasibility analysis of the MBBR-MBR system is based on the comparison with the results of the CAS system of the textile industry that provided us wastewater for this study.

Capital Expenditures (CAPEX) and Operational Expenditures (OPEX) of the MBBR-MBR system are calculated. OPEX are compared with those calculated for the current CAS system. The consumption of electricity and decolorizing agent are considered, as well as the discharge taxes derived from the different efficiency of the two systems. The Net Present Value (NPV) and the Internal Rate of Return (IRR) of MBBR-MBR are also calculated, taking into account that the savings derived from water reuse that entails a reduction in acquisition costs and wastewater discharge.

From the environmental point of view, an LCA study is carried out, comparing the environmental impacts of the combined MBBR-MBR system with respect to the CAS system. These impacts come from the electrical consumption of the two treatments and the consumption of decolorizing agent that is essential in CAS, while the MBBR-MBR system avoids the use of this synthetic product based on a triamine.

2. Materials and Methods

2.1. Caracterization of Textile Wastewater

The wastewater for the input of the MBBR-MBR reactor comes from a Catalan textile finishing industry, Acabats del Bages, S.A. (Monistrol de Montserrat, Spain). The wastewater was obtained from the outlet of its homogenization tank. Table 1 presents the average values of the main pollutant parameters of the wastewater.

Table 1. Characteristics of the wastewater.

Parameter	Unit	Value
pH	–	8.44 ± 0.54
conductivity	mS/cm	5.15 ± 0.47
Chemical Oxygen Demand (COD)	mg/L	1996 ± 440
5 days Biological Oxygen Demand (BOD_5)	mg/L	403 ± 88
Total Suspended Solids (TSS)	mg/L	940 ± 121
Total Kjeldahl Nitrogen (TKN)	mg/L	55 ± 21
Total Phosphorus (TP)	mg/L	10 ± 2
Color	mg Pt-Co/L	700 ± 234

2.2. Reactor Description

An MBBR-MBR pilot plant was designed and built to determine the efficiency of the combination of moving bed biofilm and membranes technologies on the treatment of textile wastewater. The diagram of the hybrid MBBR-MBR reactor is shown in Figure 1. The effective volume of the reactor is 110 L. The reactor consists of two parts connected at the bottom: MBBR tank (79 L) and MBR tank (31 L). The components of the reactor are presented in Figure 1.

The carriers used in the study are BIOFILL C-2 plastic carriers (BIO-FIL, Barcelona, Spain). The filling ratio of the carriers in the MBBR tank is 25 vol.%. The detailed information is shown in Table 2.

The overall operation of the treatment was 222 days. The ultrafiltration membrane was installed in the MBR tank after the growth of biofilm on the carriers of MBBR was stabled. The flow rate was raised gradually to maintain a steady status of the sludge. At this point, the flow rate was fixed at 4.5 L/h. The period of filtration and backwashing was set at 15 min and 30 s, respectively. The membrane used was a MOTIMO BT01 hollow fiber flat plat membrane (MOTIMO Membrane Technology Co., Ltd. (Tianjin, China)). Characteristics of the UF membrane are included in Table 3.

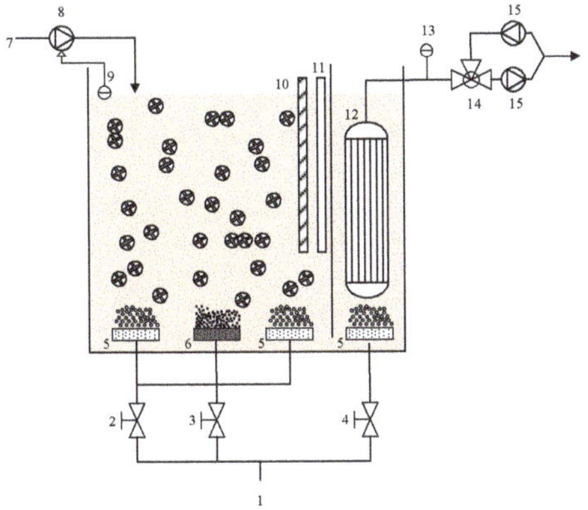

1. Compressed air
2. Needle valve
3. Valve controlled by DO (Labview)
4. Time controlled valve (Labview)
5. Thick bubble air diffuser (for mixing carriers and for air scouring the membrane)
6. Fine bubble diffuser (for oxygen diffusion)
7. Wastewater
8. Feed pump
9. Level sensor (controls the feed pump)
10. DO sensor
11. Thermostat
12. Membrane
13. Membrane pressure sensor
14. Three-way valve
15. Permeate and backwash pump

Figure 1. Diagram of the MBBR-MBR pilot plant.

Table 2. Carriers characteristics of the MBBR system.

Material	Polypropylene (PP)/Polyethylene (PE)
Specific Surface (m^2/m^3)	590
Free volume	90%
Carrier diameter (mm)	25
Weight per piece (g)	2.1
Carrier density (kg/m^3)	<1

Table 3. UF Membrane characteristics of the MBR system.

Module	UF BT01 MOTIMO
Configuration	hollow fiber flat plat membrane
Membrane surface (m^2)	1
Material	polyvinylidene difluoride (PVDF)
Pore size (μm)	0.03
Maximum TMP (kPa)	80
Operating TMP (kPa)	10–60

2.3. *New Dyeing Processes Reusing the Treated Water*

The wastewater treated by the MBBR-MBR process was reused in new dyeing processes in a laboratory dyeing machine Ti-Color (Prato, Italy). Dyeing tests were performed at the following conditions: 10 g of cotton fabric was mixed at a liquor ratio of 1/10 with the dyestuff at a dye concentration 3% o.w.f (on the weight of fiber). Three commercial reactive dyes: Yellow Procion HEXL, Crimson Procion HEXL and Navy Procion HEXL, were supplied by Dystar Inc. 60 g/L of NaCl were included as a dyeing electrolyte. The dyeing process started at 50 °C for 15 min, then the temperature was raised to 80 °C at a gradient of 1.4 °C/min. After 30 min at 80 °C, 20 g/L of Na_2CO_3 was added as alkali. Finally, the dyeing lasted for 60 min more. All the experiments were run in triplicate.

After the dyeing process, a washing process was performed to eliminate the unfixed dyes from the fabric. This process consists of nine successive washes performed at a liquor ratio of 1/10.

2.4. Analytical Methods

During this study, the treating efficiency of the combined MBBR-MBR system was studied by monitoring the removal of COD, TSS and Color, according to the Standard Methods 23rd edition [30]. TKN, TP and conductivity were also determined, as these parameters are also included in the calculation of the discharge tax.

The performance of the water reusing study was determined following the Standard UNE-EN ISO 105-J03 by comparing color differences between the dyed fabrics made with treated water and the reference made with softened tap water [31]. The calculation of the total color differences ($DE_{CMC(2:1)}$) used Equation (1), considering differences in three parameters: lightness (DL_{CMC}), chroma (DC_{CMC}), and Hue (DH_{CMC}).

$$DE_{CMC(2:1)} = [(DL_{CMC})^2 + (DC_{CMC})^2 + (DH_{CMC})^2]^{1/2} \quad (1)$$

A spectrophotometer, MINOLTA CM 3600d (Osaka, Japan), was used for the determination of ($DE_{CMC(2:1)}$) following the Standard illuminant D65/10°. In general, the acceptance limit for color differences in the textile industry is one unit ($DE_{\ CMC\ (2:1)} \leq 1$). This criterion is widely used in dyeing quality control to compare the color differences between two fabric samples [32].

2.5. Economical Analysis

The economic analysis of capital expenditures (CAPEX) and operational expenditures (OPEX) for the MBBR-MBR system is determined by considering individual cost contributions to the treating process. The results are compared with the values of CAPEX and OPEX of the current system applied by the textile company to treat its wastewater. The factory has a conventional biological activated sludge system (CAS). A CAPEX of zero is assigned to the CAS system.

Additionally, the financial feasibility analysis was conducted by examining the Net Present Value (NPV) and the Internal Rate of Return (IRR). The calculation of NPV and IRR is taken into consideration of the investment payback period as 15 years. NPV is the summation of the present value of the net income obtained by folding the income and cost flow back to the starting point of the period. NPV is calculated using the following Equation (2):

$$NPV = \sum_{t=1}^{n} \frac{s}{(1+i)^t}, \quad (2)$$

in which "s" represents the cash flow, "i" is the interest rate and "t" is the life cycle of investment project [33].

IRR is the interest rate when the cumulative NPV is zero. This IRR means the rate of the largest currency devaluation that the project can withstand. It is also calculated using Equation (2).

2.6. Environmental Impact Analysis

In order to evaluate the sustainability and the environmental impact of the hybrid MBBR-MBR system, compared to the existing CAS process, a life cycle assessment (LCA) is performed according to the standard ISO 14040 [34]. Simapro is used with the database Ecoinvent 3.1. The methodologies for the calculation of environmental impact are the ReCiPe, midpoint and endpoint approaches, and the Hierarchist perspective. The functional unit (FU) is set for "1 m^3 of treated effluent". The information used in the LCA analysis is taken from the experimental results of the MBBR-MBR pilot plant. In the case of CAS treatment, the system inputs and outputs data are provided by the company.

3. Results and Discussion

3.1. Efficiency of the MBBR-MBR

The treating performance of the hybrid system was discussed in our previous technical study [35]. Throughout the experiment, the wastewater inlet had an average concentration of COD of 2000 mg/L. The MBBR-MBR treatment achieved an average removal of 93% of COD, with an HRT of 1 day. The removal rate maintained stable although the COD values of influent fluctuated greatly, thanks to the strong resistance of MBBR to shock organic loading [36]. The hybrid MBBR-MBR treatment achieved halving the HRT (2 days) of the actual CAS treatment of the textile industry. Other studies of CAS treatment set a similar HRT within 2 days to achieve the optimum removal efficiency of the organic load in the textile water [37,38]. The decrease of HRT allows significant space-saving or an improvement in the treating capacity. The average membrane pressure for the filtration was 15 kPa, which is much lower than the maximum pressure (80 kPa) that the membrane can withstand, showing that the filtration and backwash was stable, and no membrane fouling was observed during the operation, thanks to the combination with MBBR.

Color of the influent varied between 300 and 1000 mg Pt-co/L during the experiment. The average color removal efficiency was increased to 85% when the treatment was stable. This level of discoloration makes it possible to comply with emission restrictions, which stipulate that no color will appear in samples diluted by 1/30 [9], which allowed the saving on the addition of the decolorizing agent that generally results in significant environmental impact.

TSS removal rate was up to 99%. The MBR part of the hybrid system can produce highly clarified water without the need to add the coagulation products normally required by conventional MBBR. Generally, the use of coagulation agents generates high environmental impact [38].

In the previous study of CAS treating the same wastewater, the average removal rate of COD, color and TSS was 83%, 55% and 66%, respectively [39]. The comparison of the treating efficiency of the hybrid MBBR-MBR system and the CAS system is shown in Table 4. The color removal of 55% of the CAS system was insufficient to comply with current legislation and decolorizing agent must be added, while no decolorizing agent was needed for MBBR-MBR treatment. Compared to the previous study of CAS, MBBR-MBBR obtained a better effluent quality with halved HRT. The elimination of nitrogen and phosphorus was completed in both systems because, generally, textile wastewater does not contain a high concentration of nutrients and, after the biological treatment, the concentrations of TN and TP were very low. These experimental results of the present study and the previous study are used to calculate the economic costs and environmental impacts of the LCA study.

Table 4. Comparison of the treating efficiency of the hybrid MBBR-MBR system and CAS system.

Removal Efficiency	MBBR-MBR	CAS
COD	93%	83%
TSS	99%	66%
Color	85%	55%
TN	100%	100%
TP	100%	100%

3.2. *Reuse of the Treated Water*

Treated water obtained after the MBBR-MBR process was reused in a new dyeing process. Normally, the evaporation and the adsorption of water into the fiber causes 30% of the water loss during textile production. The quality of the dyes with 100% reused water (ideal reuse) was analyzed to evaluate the possibility of achieving a fully circular dyeing process; however, it would only represent 70% of the input water [40]. Our previous work studied the reuse of the treated water by the hybrid MBBR-MBR, applying tone changes indicating the color differences and the reuse of salt [35]. A new dyeing process was performed with softened tap water as the reference. The color differences of dyed fabrics with treated water and reference fabrics are shown in Table 5. As shown in the table, the $DE_{CMC(2:1)}$ value of Crimson Procion HEXL and Navy Procion HEXL were below 1 within the acceptable range. $DE_{CMC(2:1)}$ value of Yellow Procion HEXL was 1.04, which is on the acceptance limit.

Table 5. Chromatic coordinates and color differences between fabrics dyed with the treated effluent and the reference dyeing.

100% Effluent Reused	DL_{CMC}	DC_{CMC}	DH_{CMC}	$DE_{CMC(2:1)}$
Yellow Procion HEXL	0.34	−0.38	0.90	1.04
Crimson Procion HEXL	−0.29	−0.39	−0.37	0.61
Navy Procion HEXL	0.38	0.16	−0.24	0.48

These low color differences are due to the high quality of the water treated by the MBBR-MBR System, which conveniently removed organic material and color. The residual organic matter makes it difficult to fix the new colorant in the reuse processes, and the presence of residual colorants imply changes in hue [14].

3.3. *Economic Analysis of the Hybrid System*

The local textile industry from where the wastewater was taken produces 222,700 m^3 of wastewater annually. A Conventional Activated Sludge (CAS) system is the current wastewater treatment of the industry, and the daily treatment flow is 920 m^3/d with HRT of 2 days.

3.3.1. Capital Expenditures (CAPEX)

The CAPEX of the CAS system was taken as the reference (0 €) in the economic analysis. The CAPEX of the MBBR-MBR treatment was added to the reference directly.

For the MBBR part, the expenditure on the carriers with the filling ratio of 25 vol.% (96,250 €) has been taken as the CAPEX estimation according to the suppliers' information. For the MBR part, the membrane and the installation fees (366,153 €) have been considered for the CAPEX estimation according to a previous study about the cost of a small MBR [41]. In total, the CAPEX of the hybrid system is 462,403 €.

3.3.2. Operational Expenditures (OPEX)

Energy Consumption, information about the decolorizing agent, and the environmental tax generated due to wastewater discharge and sludge production were collected to calculate the operational expenditures (OPEX) of the MBBR-MBR system.

Moreover, the membrane replacement accounted for 2.4% of the energy cost, the maintenance and renovation accounted for 19.5% of the energy expenditure [42] and the average lifetime of the UF membrane was considered to be 10 years. MBBR-MBR can resist a higher organic load with more extended sludge retention time (SRT) than the CAS system, which produced less sludge after the treatment and consequently reduces the frequency of sludge disposal [43]. Throughout the treatment of MBBR-MBR, sludge concentration did not outdo the tolerance limit of the membrane. The production of sludge is estimated based on the increase rate of biomass concentration and the tolerance limit of the membrane.

The OPEX values of a CAS plant in our previous study [39] are listed for the comparison with MBBR-MBR. The detailed OPEX calculation of the existing CAS plant and the MBBR-MBR plant is demonstrated in Tables 6 and 7, respectively.

Table 6. CAS operational cost for treating 1 m^3 wastewater.

Concept						Total Price €/m³	Reference
(a) Consumption	**Unit**	**Amount**	**Unit**	**Unit price**	**Convert to €/m³**	**0.55**	
Electricity	kWh/m³	0.96	€/kWh	0.187	0.17952		[44]
Decolorizing agent	kg/m³	0.2	€/kg	1.85	0.37		[45]
(b) Environmental tax	**Unit**	**Amount**	**Unit**	**Unit price**		**0.86**	
Sludge generation	kg/m³	0.83	€/kg	0.158	0.013114		[46]
Wastewater discharge							[10]
OM [1]	kg/m³	0.23	€/kg	1.0023	0.230529		
TSS	kg/m³	0.32	€/kg	0.5011	0.160352		
N	kg/m³	0.008	€/kg	0.761	0.006088		
P	kg/m³	0.003	€/kg	1.5222	0.0045666		
Conductivity	S/cm	0.00598	€/Sm³/cm	8.0198	0.0479584		
Summation					0.449494		
ST [2] = 1.5 × SUM					0.67424101		
GT [3]					0.163		
Total price						**1.41**	

[1] OM: organic matter (OM = 2/3COD); [2] ST: specific tax; [3] GT: general tax.

Table 7. MBBR-MBR operational cost for treating 1 m^3 wastewater.

Concept						Total Price €/m³	Reference
(a) Consumption	**Unit**	**Amount**	**Unit**	**Unit price**	**Convert to €/m³**	**0.21**	
Electricity	kWh/m³	1.12	€/kWh	0.187	0.20944		[43]
Decolorizing agent	kg/m³	0	€/kg	1.85	0		[45]
(b) Environmental tax	**Unit**	**Amount**	**Unit**	**Unit price**		**0.35**	
Sludge generation	kg/m³	0.023	€/kg	0.158	0.003634		[46]
Wastewater discharge							[10]
OM	kg/m³	0.11	€/kg	1.0023	0.110253		
TSS	kg/m³	0.006	€/kg	0.5011	0.003006		
N	kg/m³	0.007	€/kg	0.761	0.005327		
P	kg/m³	0.001	€/kg	1.5222	0.001522		
Conductivity	S/cm	0.00482	€/Sm³/cm	8.0198	0.038655		
Summation					0.123742		
ST = 1.5 × SUM					0.185613		
GT					0.163		
(c) Membrane replacement						**0.01**	
(d) Maintenance and repair						**0.04**	
Total price						**0.61**	

Regarding the consumption section, MBBR-MBR had higher electricity consumption (0.21 €/m^3) because it required more electricity to operate and to maintain the membrane filtration. However, CAS operation cost more in the consumption section (0.55 €/m^3) due to the use of the decolorizing agent. The decolorizing agent was not required for MBBR-MBR because it achieved the color removal requirement.

In regard to environmental tax, thanks to the great performance of MBBR-MBR treatment on organic matter, TSS removal and sludge generation, it had a lower expense (0.35 €/m^3) than the expense of CAS (0.86 €/m^3).

Adding the membrane replacement, maintenance and repair cost, the total OPEX of MBBR-MBR was 0.61 €/m^3, while CAS is more than twice as expensive to operate with the total OPEX of 1.41 €/m^3.

Concerning the operation cost of the hybrid MBBR-MBR system, which is much lower than CAS, the initial investment CAPEX is profitable for an industrial plant of wastewater treatment. To confirm that, the economic feasibility is studied in the next section.

3.3.3. Evaluation of the Economic Feasibility (NPV and IRR)

The CAPEX and OPEX of the MBBR-MBR system have been commented on in the above sections. The values of expenditures refer to year zero and have been re-adjusted at a rate of 1.4% yearly in the following years. This rate is taken from the average value of

Spain's inflation target in the next five years [47]. Furthermore, due to wastewater reuse and the pollutant reduction in the wastewater, discharge revenues and some costs being avoided could be achieved. By MBBR-MBR treatment, it was demonstrated that water recovery could reach approximately 70%. The water recovery will allow savings in water consumption cost and discharge tax. The textile industry with CAS treatment pays 0.56 € for each m^3 of water used [48], and has also been paying 0.86 € for each m^3 of wastewater discharged. Considering the daily treatment flow of 920 m^3/d, it was deduced that 644 m^3 of water could be recovered, and 644 m^3 of water was not discharged daily. Therefore, the avoided cost of water consumption is 360.64 € daily and the avoided cost of water discharge is 553.84 € daily.

With all these cost data, the assessment of cash flow for 15 years is presented in Table 8.

Table 8. Cash flow (€) assessment for membrane filtration alternative.

Year	Revenues		Total Revenues	Expenditures		Total Expenditures	Net Cash Flow
	Water Recovery	Reduction in Discharge		CAPEX	OPEX		
1	0	0	0	462,403	0	462,403	−462,403
2	87,275	134,029	221,304	0	135,847	135,847	85,457
3	88,497	135,905	224,402	0	137,749	137,749	86,653
4	89,736	137,808	227,544	0	139,677	139,677	87,867
5	90,992	139,737	230,730	0	141,633	141,633	89,097
6	92,266	141,694	233,960	0	143,616	143,616	90,344
7	93,558	143,677	237,235	0	145,626	145,626	91,609
8	94,868	145,689	240,556	0	147,665	147,665	92,891
9	96,196	147,729	243,924	0	149,732	149,732	94,192
10	97,542	149,797	247,339	0	151,829	151,829	95,511
11	98,908	151,894	250,802	0	153,954	153,954	96,848
12	100,293	154,020	254,313	0	156,110	156,110	98,204
13	101,697	156,177	257,874	0	158,295	158,295	99,578
14	103,121	158,363	261,484	0	160,511	160,511	100,972
15	104,564	160,580	265,145	0	162,758	162,758	102,386

In terms of considering the NPV for the alternatives, the discount rate was set to 10%. Additionally, as described in the cash flow calculation, the economic life of the MBBR-MBR system was taken as 15 years. By using Equation (2), NPV was calculated. As can be observed, the amortization of the investments will occur in 2 years, where the NPV will be in the order of 85,457 € and for 15 years of operation, the NPV value of the MBBR-MBR system is 193,990 €.

The internal rate of return (IRR) is an evaluation method for investment that finds out the potential rate of return of an asset. When IRR is discounted, the project's NPV will be zero. It can also be understood as the expected rate of return of a project [49]. IRR is calculated assuming the value of NPV to be zero using the same Equation (2). When a project has a high IRR value, then it can be concluded that the project has high financial feasibility. However, if the IRR value is lower than the discount rate, the application of the system would be unappealing. The IRR value calculated for the MBBR-MBR system is 18%, which is higher than the discount rate assumed (10%), denoting economic feasibility.

Both NPV value and IRR suggest that the MBBR-MBR system is financially applicable for the implantation into industrial scale.

3.4. LCA Analysis

3.4.1. Inventory Results

The inventory results of the MBBR-MBR treatment are shown in Table 9. All information is related to the functional unit (1 m^3 treated water).

Table 9. Inventory analysis of the MBBR-MBR treatment.

Processes Included in LCA	MBBR Input	MBBR Output	Unit/FU	Ecoinvent Unit Process
COD	2	0.13	kg	
TSS	0.94	0.01	kg	
N	0.055	0.003	kg	
P	0.010	0.001	kg	
Color	700	105	g Pt-co	
Conductivity	6.46	5.42	mS/cm	
Wastewater	1	1	m^3	
Sludge		0.021	kg	
Decolorizing agent	0		kg	diethylenetriaminepentaacetic acid (DTPA), at plant/RER(Europe) Unit
Electricity	1.12		kWh	Electricity, medium voltage, production ES, at grid/ES(Spain) Unit

3.4.2. Environmental Impact Evaluation

The environmental impact generated by MBBR-MBR treatment conforming to the LCA results is analyzed with an endpoint approach, and the results of MBBR-MBBR treatment are compared with the CAS treatment of our previous study, using the same criteria with reference to their total environmental impact [39].

The results of the environmental impact evaluation are shown in milipoints (mPt) and the differences between categories could be compared. The environmental impact of the MBBR-MBR process was compared with the environmental impact of CAS. The comparison of environmental impacts is demonstrated in Table 10.

Table 10. Comparison of environmental impacts of CAS vs. MBBR-MBR: endpoint analysis.

Treatment	Inputs		Human Health (mPt)	Ecosystems (mPt)	Resources (mPt)
CAS	Electricity (kWh/m^3)	0.96	22.8	1.9	31.8
	Decolorizing agent(kg/m^3)	0.2	34.4	3.4	81.2
	TOTAL		57.2	5.3	113.0
MBBR-MBR	Electricity (kWh/m^3)	1.12	26.6	2.2	37.1
	Decolorizing agent(kg/m^3)	0	0	0	0
	TOTAL		26.6	2.2	37.1

During the MBBR-MBR operation, as can be observed, no decolorizing agent was added because the treatment achieved the color removal requirement. The consumption of electricity throughout the operation accounted for the total environmental impact. The results demonstrate that less impact was generated on the Ecosystem; at the same time, the major impacts occurred on Resources and Human Health.

As shown, the MBBR-MBR treatment had a lower impact on all categories, comparing with CAS treatment. Although, according to Tables 6 and 7, MBBR-MBR had a slightly higher energy consumption of 1.12 kWh/m^3 compared with 0.96 kWh/m^3 of CAS, the endpoint results have demonstrated that avoiding the use of decolorizing agent fully compensated for the environmental impact due to higher energy consumption.

The environmental impacts generated from the MBBR-MBR and CAS treatment are compared in Figure 2, on the specific categories related to Human Health, Ecosystem, and Resources.

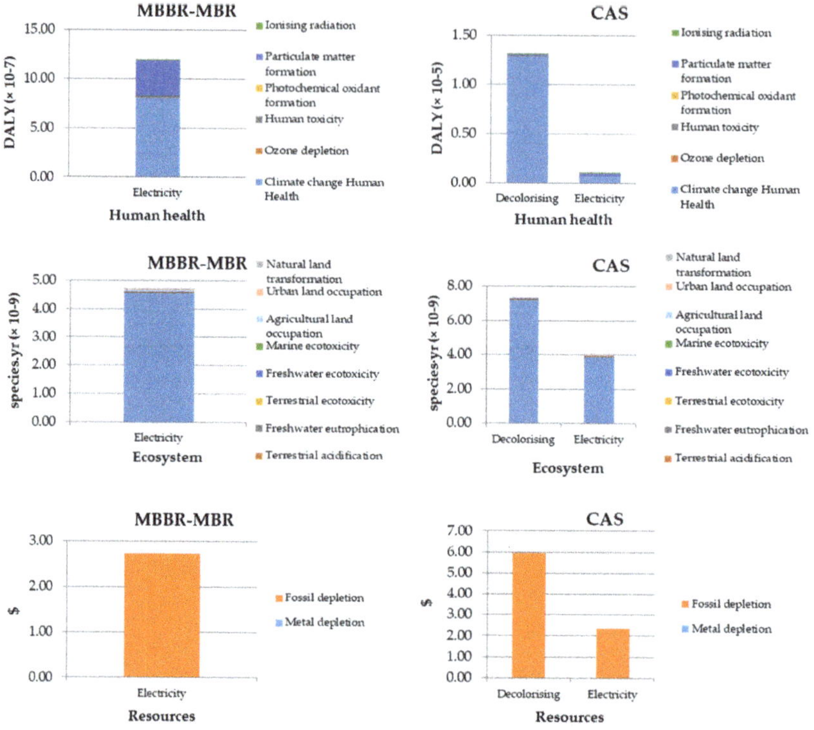

Figure 2. Comparative analysis of the effect of electricity and decolorizing agent consumption on the impacted categories of MBBR-MBR and CAS.

Regarding MBBR-MBR treatment, the main factors affecting Human Health due to electricity consumption are Climate Change, Human Health and Particulate Matter formation categories. Meanwhile, the major impact on the ecosystem is attributed to climate change, while agricultural land, terrestrial ecotoxicity, natural land transformation, urban land occupation, and terrestrial acidification had a lower impact on the Ecosystem category. This is due to the composition of the Spanish electricity mix, which in 2020 had a 45.5% electricity generation from renewable sources [50]. In addition, the main impact on Resources came from the Fossil depletion category, while the Metal depletion category had almost no impact.

At the same time, similar results can be observed with CAS treatment; the main environmental impact generated by energy and decolorizing agent consumption related to Human Health and Ecosystem is the Climate change ecosystem. It should be noted that, although the hybrid MBBR-MBR system had higher energy consumption due to the filtration process, no major differences were produced in the related environmental impacts of MBBR-MBR and CAS treatment. In addition, the use of a decolorizing agent in CAS treatment caused many more impacts on all three categories, especially on Human Health. The reason that the consumption of decolorizing agent has a great impact on the category of Climate Change is that the decolorizing agent is DTPA, a triamine. This synthetic product is produced from the alkylation of ammonia. Ammonia production in Europe has an emission factor of 2104 t CO_2/t NH_3. Ammonia is synthesized from hydrogen produced by reforming natural gas [51].

While the endpoint methods are useful for decision-making because they can compare results in points, then midpoint analysis can help identify issues of specific environmental concern [52]. The results of the midpoint assessment of MBBR-MBR were also compared with the previous CAS study, shown in Table 11. MBBR-MBR demonstrated more environmental advantages since its impacts were the lower in all the categories, especially in Climate Change, Human Health, Marine eutrophication, Terrestrial ecotoxicity, Freshwater ecotoxicity and Marine ecotoxicity, thanks to the high quality of the effluent treated by MBBR-MBR and the avoidance of using extra decolorizing agent.

Table 11. Comparison of CAS and MBBR-MBR: midpoint analysis.

Impact Category	Unit	CAS	MBBR-MBR	Impact Reduction of MBBR-MBR
Climate change Human Health	kg CO_2-eq	1.29	0.08	94%
Ozone depletion	kg CFC-11 eq	1.39×10^{-7}	2.89×10^{-8}	79%
Human toxicity	kg 1.4-DB eq	0.12	0.03	79%
Photochemical oxidant formation	kg NMVOC	3.89×10^{-3}	2.30×10^{-3}	41%
Particulate matter formation	kg PM10 eq	1.95×10^{-3}	1.40×10^{-3}	28%
Ionizing radiation	kg U235 eq	0.16	0.11	33%
Terrestrial acidification	kg SO_2-eq	6.46×10^{-3}	4.83×10^{-3}	25%
Freshwater eutrophication	kg P-eq	7.84×10^{-5}	2.79×10^{-5}	64%
Marine eutrophication	kg N-eq	2.24×10^{-3}	1.76×10^{-4}	92%
Terrestrial ecotoxicity	kg 1.4-DB eq	2.96×10^{-4}	6.69×10^{-5}	76%
Freshwater ecotoxicity	kg 1.4-DB eq	7.40×10^{-3}	1.02×10^{-4}	99%
Marine ecotoxicity	kg 1.4-DB eq	1.04×10^{-3}	2.14×10^{-4}	79%
Agricultural land occupation	m^2 year	5.51×10^{-3}	5.22×10^{-3}	5%
Urban land occupation	m^2 year	2.16×10^{-3}	1.79×10^{-3}	17%
Natural land transformation	m^2 year	1.25×10^{-5}	1.10×10^{-5}	12%
Water depletion	m^3	1.12×10^{-2}	3.33×10^{-3}	70%
Metal depletion	kg 1Fe eq	2.02×10^{-3}	8.57×10^{-4}	58%
Fossil depletion	kg oil eq	5.17×10^{-7}	1.70×10^{-7}	67%

4. Conclusions

The experimental study of a hybrid MBBR-MBR showed the efficient removal of COD (93%), color (85%) and TSS (99%) with 1 day of HRT. The HRT reduction of 50% for the application of industrial scale is very attractive, resulting in space and energy saving. Additionally, 100% of the water treated by the MBBR-MBR system achieved new dyeings, representing 70% of the textile industry's water consumption in the dyeing process.

The value of CAPEX calculated for the hybrid MBBR-MBR system is 462,403 € and the increase in CAPEX is profitable due to the reduction in OPEX, which results in lower taxes and savings in decolorizing agent with regard to a small increase in electricity consumption. Additionally, the NPV and IRR study shows that water reuse after the treatment played an important role leading to the cost saving of water consumption and discharge tax. The 18% of IRR calculated demonstrated that MBBR-MBR has great economic feasibility in industrial-scale textile wastewater treatment.

Regarding the LCA study, because of the high efficiency in color removal, the hybrid MBBR-MBR system did not require the addition of decolorizing agent, which is a tri-amine that has a significant impact on the environment. Therefore, the results suggested that MBBR-MBR treatment generated a much lower environmental impact than CAS treatment.

Author Contributions: Investigation, X.Y.; Methodology, X.Y. and V.L.-G.; Project administration, V.L.-G.; Supervision, V.L.-G.; Writing—original draft, X.Y.; Writing—review & editing, V.L.-G. All authors have read and agreed to the published version of the manuscript.

Funding: This study is co-funded by ACCIÓ (Generalitat de Catalunya, Spain) within the REGIREU Project (COMRDI16-1-0062). The preliminary results of this study were presented at the II International Congress on Water and Sustainability, 24–26 March 2021, Terrassa, Barcelona, Spain. This work was selected by the scientific committee to assume the article processing charges (APC) of publishing it in an open access journal.

Institutional Review Board Statement: Not applicable.

Informed Consent Statement: Not applicable.

Data Availability Statement: The data presented in this study are available on request from the corresponding author. The data is not publicly available because it is licensed by the Ecoinvent database or because it is owned by the textile company Acabats del Bages (Monistrol de Montserrat, Spain).

Conflicts of Interest: The authors declare no conflict of interest.

References

1. Xu, H.; Yang, B.; Liu, Y.; Li, F.; Shen, C.; Ma, C.; Tian, Q.; Song, X.; Sand, W. Recent advances in anaerobic biological processes for textile printing and dyeing wastewater treatment: A mini-review. *World J. Microbiol. Biotechnol.* **2018**, *34*, 165. [CrossRef]
2. López-Grimau, V.; Vilaseca, M.; Gutiérrez-Bouzán, C. Comparison of different wastewater treatments for colour removal of reactive dye baths. *Desalin. Water Treat.* **2016**, *57*, 2685–2692. [CrossRef]
3. Umar, K.; Ibrahim, M.N.M.; Ahmad, A.; Rafatullah, M. Synthesis of Mn-doped TiO_2 by novel route and photocatalytic mineralization/intermediate studies of organic pollutants. *Res. Chem. Intermed.* **2019**, *45*, 2927–2945. [CrossRef]
4. Ahmad, M.; Rehman, W.; Khan, M.M.; Qureshi, M.T.; Gul, A.; Haq, S.; Ullah, R.; Rab, A.; Menaa, F. Phytogenic fabrication of ZnO and gold decorated ZnO nanoparticles for photocatalytic degradation of Rhodamine B. *J. Environ. Chem. Eng.* **2021**, *9*, 104725. [CrossRef]
5. Cardoso, J.C.; Bessegato, G.G.; Boldrin Zanoni, M.V. Efficiency comparison of ozonation, photolysis, photocatalysis and photoelectrocatalysis methods in real textile wastewater decolorization. *Water Res.* **2016**, *98*, 39–46. [CrossRef] [PubMed]
6. Cao, X.L.; Yan, Y.N.; Zhou, F.Y.; Sun, S.P. Tailoring nanofiltration membranes for effective removing dye intermediates in complex dye-wastewater. *J. Memb. Sci.* **2020**, *595*, 117476. [CrossRef]
7. Tianzhi, W.; Weijie, W.; Hongying, H.; Khu, S.T. Effect of coagulation on bio-treatment of textile wastewater: Quantitative evaluation and application. *J. Clean. Prod.* **2021**, *312*, 127798. [CrossRef]
8. Sahinkaya, E.; Yurtsever, A.; Çınar, Ö. Treatment of textile industry wastewater using dynamic membrane bioreactor: Impact of intermittent aeration on process performance. *Sep. Purif. Technol.* **2017**, *174*, 445–454. [CrossRef]
9. Decret 130/2003, Approving the Regulation of Public Sanitation Services. Available online: http://aca-web.gencat.cat/aca/documents/ca/legislacio/decrets/decret_130_2003.pdf (accessed on 26 October 2021).
10. Cànon de L'aigua. Available online: http://aca.gencat.cat/web/.content/10_ACA/K_Canon_i_altres_tributs/canon_de_laigua/20_Industrial/GT_DUCA_tipus_vigents_industrials.pdf (accessed on 26 October 2021). (In Catalan)
11. Sari Erkan, H.; Bakaraki Turan, N.; Önkal Engin, G. Membrane Bioreactors for Wastewater Treatment. In *Comprehensive Analytical Chemistry*; Chormey, D.S., Bakırdere, S., Turan, N.B., Engin, G.Ö., Eds.; Elsevier B.V.: Amsterdam, The Netherlands, 2018; Volume 81, pp. 151–200, ISBN 9780444640642. [CrossRef]
12. Jegatheesan, V.; Pramanik, B.K.; Chen, J.; Navaratna, D.; Chang, C.Y.; Shu, L. Treatment of textile wastewater with membrane bioreactor: A critical review. *Bioresour. Technol.* **2016**, *204*, 202–212. [CrossRef]
13. Galiano, F.; Friha, I.; Deowan, S.A.; Hoinkis, J.; Xiaoyun, Y.; Johnson, D.; Mancuso, R.; Hilal, N.; Gabriele, B.; Sayadi, S.; et al. Novel low-fouling membranes from lab to pilot application in textile wastewater treatment. *J. Colloid Interface Sci.* **2018**, *515*, 208–220. [CrossRef]
14. Álvarez, M.D.; Buscio, V.; López-Grimau, V.; Gutiérrez-Bouzán, C. LCA study of a new electrochemical and ultraviolet (EC-UV) combined system to decolourise and reuse textile saline effluents: Environmental evaluation and proposal to improve the production process. *Chem. Eng. J.* **2020**, *392*, 123696. [CrossRef]
15. Pirani, S.; Natarajan, L.; Abbas, Z.; Arafat, H. Life cycle assessment of membrane bioreactor versus CAS wastewater treatment: Masdar City and beyond. In Proceedings of the Sixth Jordan International Chemical Engineering Conference, Amman, Jordan, 12–14 March 2012.
16. Pretel, R.; Robles, A.; Ruano, M.V.; Seco, A.; Ferrer, J. Economic and environmental sustainability of submerged anaerobic MBR-based (AnMBR-based) technology as compared to aerobic-based technologies for moderate-/high-loaded urban wastewater treatment. *J. Environ. Manag.* **2016**, *166*, 45–54. [CrossRef] [PubMed]
17. Bertanza, G.; Canato, M.; Laera, G.; Vaccari, M.; Svanström, M.; Heimersson, S. A comparison between two full-scale MBR and CAS municipal wastewater treatment plants: Techno-economic-environmental assessment. *Environ. Sci. Pollut. Res.* **2017**, *24*, 17383–17393. [CrossRef]
18. Roccaro, P.; Vagliasindi, F.G.A. Techno-economic feasibility of membrane bioreactor (MBR). In *Advances in Science, Technology and Innovation*; Springer Nature: Berlin/Heidelberg, Germany, 2020; pp. 269–270. [CrossRef]
19. Yang, X.; Crespi, M.; López-Grimau, V. A review on the present situation of wastewater treatment in textile industry with membrane bioreactor and moving bed biofilm reactor. *Desalin. Water Treat.* **2018**, *103*, 315–322. [CrossRef]
20. Liu, Y.; Wang, N.; Wei, Y.; Dang, K.; Li, M.; Li, Y.; Li, Q.; Mu, R. Pilot study on the upgrading configuration of UASB-MBBR with two carriers: Treatment effect, sludge reduction and functional microbial identification. *Process Biochem.* **2020**, *99*, 211–221. [CrossRef]
21. Gzar, H.A.; Al-Rekabi, W.S.; Shuhaieb, Z.K. Applicaion of Moving Bed Biofilm Reactor (MBBR) for Treatment of Industrial Wastewater: A mini Review. *J. Phys. Conf. Ser.* **2021**, *1973*, 012024. [CrossRef]

22. Kozak, M.; Cirik, K.; Basak, S. Treatment of textile wastewater using combined anaerobic moving bed biofilm reactor and powdered activated carbon-aerobic membrane reactor. *J. Environ. Chem. Eng.* **2021**, *9*, 105596. [CrossRef]
23. Di Trapani, D.; Mannina, G.; Torregrossa, M.; Viviani, G. Comparison between hybrid moving bed biofilm reactor and activated sludge system: A pilot plant experiment. *Water Sci. Technol.* **2010**, *61*, 891–902. [CrossRef]
24. Park, H.O.; Oh, S.; Bade, R.; Shin, W.S. Application of A2O moving-bed biofilm reactors for textile dyeing wastewater treatment. *Korean J. Chem. Eng.* **2010**, *27*, 893–899. [CrossRef]
25. Pratiwi, R.; Notodarmojo, S.; Helmy, Q. Decolourization of remazol black-5 textile dyes using moving bed bio-film reactor. In Proceedings of the IOP Conference Series: Earth and Environmental Science, Banda Aceh, Indonesia, 26–27 September 2018; Institute of Physics Publishing: Jakarta, Indonesia, 2018; Volume 106, p. 12089. [CrossRef]
26. Shin, D.H.; Shin, W.S.; Kim, Y.H.; Han, M.H.; Choi, S.J. Application of a combined process of moving-bed biofilm reactor (MBBR) and chemical coagulation for dyeing wastewater treatment. *Water Sci. Technol.* **2006**, *54*, 181–189. [CrossRef]
27. Ke, P.; Zeng, D.; Xu, K.; Cui, J.; Li, X.; Wang, G. Preparation of Quaternary Ammonium Salt-Modified Chitosan Microspheres and Their Application in Dyeing Wastewater Treatment. *ACS Omega* **2020**, *5*, 24700–24707. [CrossRef]
28. Dong, B.; Chen, H.; Yang, Y.; He, Q.; Dai, X. Treatment of printing and dyeing wastewater using MBBR followed by membrane separation process. *Desalin. Water Treat.* **2014**, *52*, 4562–4567. [CrossRef]
29. Grilli, S.; Piscitelli, D.; Mattioli, D.; Casu, S.; Spagni, A. Textile wastewater treatment in a bench-scale anaerobic-biofilm anoxic-aerobic membrane bioreactor combined with nanofiltration. *J. Environ. Sci. Health Part A* **2011**, *46*, 1512–1518. [CrossRef]
30. *Standard Methods for the Examination of Wastewater*, 23rd ed.; APHA-AWWA-WEF: Washington, DC, USA, 2017. Available online: https://cmc.marmot.org/Record/.b57522091 (accessed on 21 October 2021).
31. AENOR. Textiles—Tests for Colour Fastness—Part J03: Calculation of Colour Differences (ISO 105-J03:2009). 2009. Available online: https://www.une.org/encuentra-tu-norma/busca-tu-norma/norma?c=N0044887 (accessed on 21 October 2021).
32. Sala, M.; López-Grimau, V.; Gutiérrez-Bouzán, C. Photoassisted electrochemical treatment of azo and phtalocyanine reactive dyes in the presence of surfactants. *Materials* **2016**, *9*, 211. [CrossRef] [PubMed]
33. Andrade, L.H.; Mendes, F.D.S.; Espindola, J.C.; Amaral, M.C.S. Reuse of dairy wastewater treated by membrane bioreactor and nanofiltration: Technical and economic feasibility. *Braz. J. Chem. Eng.* **2015**, *32*, 735–747. [CrossRef]
34. *ISO 14040: Environmental Management–Life Cycle Assessment—Principles and Framework*; International Organization for Standardization: Geneva, Switzerland, 2006.
35. Yang, X.; López-Grimau, V.; Vilaseca, M.; Crespi, M.; Ribera-Pi, J.; Calderer, M.; Martínez-Lladó, X. Reuse of textile wastewater treated by moving bed biofilm reactor coupled with membrane bioreactor. *Color. Technol.* **2021**, *137*, 484–492. [CrossRef]
36. Dong, Z.; Lu, M.; Huang, W.; Xu, X. Treatment of oilfield wastewater in moving bed biofilm reactors using a novel suspended ceramic biocarrier. *J. Hazard. Mater.* **2011**, *196*, 123–130. [CrossRef]
37. Lourenço, N.D.; Franca, R.D.G.; Moreira, M.A.; Gil, F.N.; Viegas, C.A.; Pinheiro, H.M. Comparing aerobic granular sludge and flocculent sequencing batch reactor technologies for textile wastewater treatment. *Biochem. Eng. J.* **2015**, *104*, 57–63. [CrossRef]
38. Yang, X. Study of a Hybrid System: Moving Bed Biofilm Reactor-Membrane Bioreactor (MBBR-MBR) in the Treatment and Reuse of Textile Industrial Effluents. Ph.D. Thesis, Universitat Politècnica de Catalunya, Barcelona, Spain, July 2021.
39. Mohamed-Zine, M.-B.; Hamouche, A.; Krim, L. The study of potable water treatment process in Algeria (boudouaou station) -by the application of life cycle assessment (LCA). *J. Environ. Health Sci. Eng.* **2013**, *11*, 11–37. [CrossRef] [PubMed]
40. Yang, X.; López-Grimau, V.; Vilaseca, M.; Crespi, M. Treatment of textilewaste water by CAS, MBR, and MBBR: A comparative study from technical, economic, and environmental perspectives. *Water* **2020**, *12*, 1306. [CrossRef]
41. Lo, C.H.; McAdam, E.; Judd, S. The cost of a small membrane bioreactor. *Water Sci. Technol.* **2015**, *72*, 1739–1746. [CrossRef]
42. Iglesias, R.; Simón, P.; Moragas, L.; Arce, A.; Rodriguez-Roda, I. Cost comparison of full-scale water reclamation technologies with an emphasis on membrane bioreactors. *Water Sci. Technol.* **2017**, *75*, 2562–2570. [CrossRef] [PubMed]
43. Radjenović, J.; Matošić, M.; Mijatović, I.; Petrović, M.; Barceló, D. Membrane Bioreactor (MBR) as an Advanced Wastewater Treatment Technology. In *Emerging Contaminants from Industrial and Municipal Waste*; Springer: Berlin/Heidelberg, Germany, 2008; pp. 37–101. [CrossRef]
44. Endesa—Luz. Available online: https://www.endesatarifasluzygas.com/luz/?sem=sem-endesa-google-tl-75&adgroupid=77440481986&gclid=CjwKCAiAy9jyBRA6EiwAeclQhG_kGGL-xchrpwXH50-OM42n6Yx1KUOLsskbYiOKVgbKUWkNwAa99BoC-zkQAvD_BwE (accessed on 21 October 2021).
45. Supplier: TIDE 2000 S.L. (Montornès del Vallès, B.) Discoloration Agent. Available online: http://www.tide2000.net/productos/coagulantes-y-decolorantes/ (accessed on 21 October 2021).
46. Ley 2/2014, de 27 de Enero, de Medidas Fiscales, Administrativas, Financieras y del Sector Público. Diario Oficial de la Generalidad de Cataluña 2014, 61561–61567. Available online: https://www.boe.es/eli/es-ct/l/2014/01/27/2 (accessed on 21 October 2021).
47. Spain—Inflation Rate 2025 | Statista. Available online: https://www.statista.com/statistics/271077/inflation-rate-in-spain/ (accessed on 21 October 2021).
48. Agència Catalana de L'aigua Informe Anual del Preu de L'aigua a Catalunya 2020; Barcelona, 2021. Available online: http://aca.gencat.cat/web/.content/10_ACA/J_Publicacions/05-estudis-preus-i-tarifes/18_observatori_preus_2020_ca.pdf (accessed on 26 October 2021).

49. Kumar, A.K.; Sharma, S.; Dixit, G.; Shah, E.; Patel, A. Techno-economic analysis of microalgae production with simultaneous dairy effluent treatment using a pilot-scale High Volume V-shape pond system. *Renew. Energy* **2020**, *145*, 1620–1632. [CrossRef]
50. REE. El Sistema Eléctrico Español 2020. 2021. Available online: https://www.ree.es/es/datos/publicaciones/informe-anual-sistema/informe-del-sistema-electrico-espanol-2020 (accessed on 21 October 2021). (In Spanish)
51. IPCC. Guidelines for National Greenhouse Gas Inventories. Volume 3 Industrial Processes and Product Use. Chapter 3: Chemical Industry Emissions. 2006. Available online: https://www.ipcc-nggip.iges.or.jp/public/2006gl/vol3.html (accessed on 21 October 2021).
52. Van Hoof, G.; Vieira, M.; Gausman, M.; Weisbrod, A. Indicator selection in life cycle assessment to enable decision making: Issues and solutions. *Int. J. Life Cycle Assess.* **2013**, *18*, 1568–1580. [CrossRef]

membranes

MDPI

Article

Novel Cellulose Acetate-Based Monophasic Hybrid Membranes for Improved Blood Purification Devices: Characterization under Dynamic Conditions

Adriana Janeca [1], Flávia S. C. Rodrigues [1,2], Maria Clara Gonçalves [1,3] and Mónica Faria [1,2,*]

1 Departamento de Engenharia Química, Instituto Superior Técnico, Universidade de Lisboa, Av. Rovisco Pais, 1049-001 Lisboa, Portugal; adriana.janeca@tecnico.ulisboa.pt (A.J.); flavia.rodrigues@tecnico.ulisboa.pt (F.S.C.R.); clara.goncalves@tecnico.ulisboa.pt (M.C.G.)

2 CeFEMA, Center of Physics and Engineering of Advanced Materials, Av. Rovisco Pais, 1049-001 Lisboa, Portugal

3 CQE, Centro de Química Estrutural, Av. Rovisco Pais, 1049-001 Lisboa, Portugal

* Correspondence: monica.faria@tecnico.ulisboa.pt; Tel.: +351-21-841-79-46

Abstract: A novel cellulose acetate-based monophasic hybrid skinned amine-functionalized CA-SiO_2-$(CH_2)_3NH_2$ membrane was synthesized using an innovative method which combines the phase inversion and sol-gel techniques. Morphological characterization was performed by scanning electron microscopy (SEM), and the chemical composition was analyzed by Fourier transform infrared spectroscopy in attenuated total reflection mode (ATR-FTIR). The characterization of the monophasic hybrid CA-SiO_2-$(CH_2)_3NH_2$ membrane in terms of permeation properties was carried out in an in-house-built single hemodialysis membrane module (SHDMM) under dynamic conditions. Permeation experiments were performed to determine the hydraulic permeability (Lp), molecular weight cut-off (MWCO) and the rejection coefficients to urea, creatinine, uric acid, and albumin. SEM confirmed the existence of a very thin (<1 μm) top dense layer and a much thicker bottom porous surface, and ATR-FTIR showed the main bands belonging to the CA-based membranes. Permeation studies revealed that the Lp and MWCO of the CA-SiO_2-$(CH_2)_3NH_2$ membrane were 66.61 $kg \cdot h^{-1} \cdot m^{-2} \cdot bar^{-1}$ and 24.5 kDa, respectively, and that the Lp was 1.8 times higher compared to a pure CA membrane. Furthermore, the CA-SiO_2-$(CH_2)_3NH_2$ membrane fully permeated urea, creatinine, and uric acid while completely retaining albumin. Long-term filtration studies of albumin solutions indicated that fouling does not occur at the surface of the CA-SiO_2-$(CH_2)_3NH_2$ membrane.

Keywords: monophasic hybrid membrane; sol-gel; phase inversion; blood purification; ultrafiltration; crossflow filtration; convection; hemodialysis

1. Introduction

Chronic kidney disease, a growing public health concern affecting 11–13% of the global population [1], is defined by an irreversible worsening of renal function which can lead to end-stage renal disease (ESRD). Currently, about 3 million ESRD patients receive renal replacement therapies to survive, and the number is likely to reach 10 million by 2030 [1]. The progressive loss of kidney function is linked to the retention of metabolic waste products, also known as retention solutes or uremic toxins (UTs), which are normally excreted by healthy kidneys. UTs are classified based on their molecular weight (MW), removal pattern during hemodialysis (HD), and relative affinity for protein binding, and are divided into three major groups: (1) small water-soluble compounds (MW < 500 Da); (2) middle molecules (MW $\geq$ 500 Da); and (3) protein-bound uremic toxins (PBUTs; MW < 500 Da when free, >66.5 kDa when bound to albumin (MW 66.5 kDa)) [2].

Hemodialysis is the renal replacement therapy most widely used to purify the blood of ESRD patients. The hemodialyzer is the key component of the extracorporeal blood circulation circuit, as it is responsible for removing accumulated UTs and water while simultaneously retaining vital blood components, such as proteins, through a bundle of semi-permeable membranes [3]. The main mass separation mechanisms used to remove water and UTs from the blood of ESRD patients through semi-permeable membranes are diffusion, convection, and adsorption.

Although different classification schemes for HD membranes have been proposed [4], they are traditionally classified according to water flux—a term also referred to as water permeability. The clinical parameter used to characterize the water permeability of a dialyzer is the ultrafiltration coefficient, K_{UF}, expressed in mL/h/mmHg units. The water permeability of a hemodialyzer is found by in vitro ultrafiltration experiments, in which bovine blood is ultrafiltered at varying transmembrane pressures (TMPs). At low TMPs, the ultrafiltration rate (Q_{UF}) varies linearly with TMP and then reaches a plateau at relatively high TMP values [5]. The slope of the linear portion of the Q_{UF} versus the TMP curve defines the dialyzer K_{UF}. The membrane pore size is the membrane property with the largest impact on water permeability, whereas the ultrafiltrate flux (J_{UF}) is approximately proportional to the fourth power of the mean membrane pore radius [6]. As such, small changes in pore size have an exceptionally large effect on water permeability. K_{UF} is the most-used parameter for classification purposes; according to the United States Food and Drug Administration, a value of 12 mL/h/mmHg differentiates low-permeability and high-permeability dialyzers [7].

Traditionally, dialysis membranes have been broadly classified on the basis of their composition as cellulosic or synthetic [8]. The earliest hemodialysis membranes consisted of regenerated cellulose, which rendered small pore sizes—enabling the clearance of molecules smaller than 5000 Daltons into the dialysate, and were associated with an inflammatory response [8]. In the early 1970s, in an attempt to enhance the hemocompatibility of regenerated cellulose membranes, researchers turned to modified cellulose materials, such as cellulose acetate (CA) and cellulose triacetate, where a large percentage of the hydroxyl groups is replaced with an acetate radical, eliminating the active surface sites for complement protein interaction. This modification also led to the improvement of solute mass transfer [9]. This decade also witnessed the development of synthetic membranes, such as polyacrylonitrile (PAN) and polysulfone (PS) [6]. These latter membranes differed fundamentally from cellulosic materials not only in their polymeric composition, but also in a number of other features including larger pore size, decreased thickness, and hydrophobicity.

Due to their hydrophobicity, synthetic HD membranes must be rendered hydrophilic before they can be used in HD, thus adding an extra step to the preparation process. Furthermore, their final hemocompatibility and filtration performance depends on the hydrophilization processes they are subjected to [10]. In terms of pore size, they are generally larger in synthetic membranes than in cellulosic membranes [8], allowing for a faster extraction of water and potential removal of middle-sized and protein-bound retention solutes [11].

The evolution of biomaterials and improved membrane synthesis technology during the 1990s resulted in new cellulosic and synthetic membranes with specific characteristics and refined properties. The development of integral asymmetric CA membranes [12] by means of the wet-phase inversion technique paved the way to the synthesis of integrally skinned asymmetric membranes with a multitude of structures covering a wide range of membrane processes, from ultrafiltration (UF) [13] to reverse osmosis [14]. CA membranes were used to develop low-flux hemodialyzers [15], high-flux hemodialyzers [16] and even low-flux hemodialyzers with efficient β_2-microglobulin (β2m) removal [17]—a marker for middle-sized molecules.

Due to the higher water permeability and efficiency in the removal of UTs of larger molecular weight, the majority of HD membranes on the market today are made from synthetic polymers [18]. Nevertheless, the highly porous structure of synthetic membranes has disadvantages: studies focused on the dialysate/ultrafiltrate removed from patients subjected to high-flux HD identified the convective removal of useful and even vital compounds found in blood [11,18], such as albumin and vitamin D-transporting protein [19]. At a much smaller scale, melt-spun CA membranes are being successfully used in HD [20,21], with studies reporting their low toxicity and reduced complement activation [22].

Cellulose is the most abundant and available biorenewable polymer on Earth's crust. Furthermore, the long chain biopolymeric carbohydrate molecules primarily composed of monosaccharide units (which compose cellulosic materials) can be easily functionalized, making it a versatile material with immense potential in biomedical applications. Nevertheless, CA membranes exhibit some drawbacks—namely, limited chemical resistance, low mechanical strength, low shelf life, and small pore size [23]. To overcome these limitations, our research group has focused on the development of novel monophasic hybrid CA-based membranes, which combine the high mechanical and thermal stability of the inorganic material with the flexibility, ductility, and processability of CA [24–28]. In a recent study [27], a modified version of the phase inversion technique [13] coupled with the sol-gel method [29,30] was used to synthesize novel monophasic hybrid skinned amine-functionalized CA-SiO_2-$(CH_2)_3NH_2$ membranes.

It is envisioned that the incorporation of NH_2 groups into cellulosic membranes may translate into preferential interactions with proteins present in the blood—especially albumin (that may or may not be bound to PBUTs). The monophasic hybrid membranes are fabricated under acidic conditions, giving rise to the protonation of NH_2 groups. Therefore, when the hybrid membranes are in close contact with albumin, a strong electrostatic interaction between the positively charged $-NH_3^+$ group (from the hybrid membrane) and the $-COOH$ or $-NH$ (from albumin) occurs. This electrostatic interaction may allow the displacement of PBUTs from albumin.

The mechanical properties of the CA-SiO_2-$(CH_2)_3NH_2$ membranes were enhanced and permeation studies revealed that the introduction of propyl-amine groups increased the hydraulic permeability by a factor of three when compared to pure CA membranes [27].

The aim of this study is to evaluate the potential of monophasic hybrid CA-SiO_2-$(CH_2)_3NH_2$ membranes for application in blood purification devices such as HD, where the preferential permeation of uremic toxins and the retention of vital blood components such as albumin are a mandate. For this, a monophasic hybrid CA-SiO_2-$(CH_2)_3NH_2$ membrane containing 95 wt% CA and 5 wt% SiO_2 + SiO_2-$(CH_2)_3NH_2$ was synthesized and characterized in an in-house-built single hemodialysis membrane module (SHDMM) under dynamic conditions. For comparison purposes, a pure CA membrane was also synthesized and characterized. The surface and cross-section morphologies of the membranes were characterized by scanning electron microscopy (SEM) and the chemical composition was analyzed by Fourier transform infrared spectroscopy in attenuated total reflection mode (ATR-FTIR). Permeation experiments were carried out to evaluate the membranes' performance in terms of hydraulic permeability (Lp), molecular weight cut-off (MWCO), and rejection coefficients to a group of small water-soluble compounds and bovine serum albumin (BSA).

2. Materials and Methods

2.1. Materials

Membranes were synthesized with cellulose acetate from Sigma-Aldrich (Steinheim, Germany) (CA; ($C_6H_7O_2(OH)_3$, ~30,000 g/mol, reagent grade $\geq$ 97%, esterification degree ~40%), tetraethyl orthosilicate (TEOS; $Si(OC_2H_5)_4$, 208.33 g/mol, reagent grade 98%), purchased from Alfa Aesar (Karlsruhe, Germany), 3-(triethoxysilyl)–propylamine (APTES; $C_9H_{23}NO_3Si$, 221.37 g/mol, reagent grade $\geq$ 98%) purchased from Sigma-Aldrich (Steinheim, Germany), formamide (CH_3NO, 45.02 g/mol, $\geq$99.5%) purchased from Carlo Erba (Val-de-Reuil, France), acetone (C_3H_6O, 58.08 g/mol, $\geq$99.6%) purchased from Labsolve

(Zedelgem, Belgium), and nitric acid (HNO_3, 63.01 g/mol, 1.39 g/mL at 20 °C, 65% *v*/*v*) purchased from LabSolve (Zedelgem, Belgium).

Membrane drying was performed with isopropanol (≥99.8%) from Honeywell (Seelze, Germany) and n-hexane (≥95%) from Carlo Erba (Val-de-Reuil, France).

Permeation experiments were carried out with urea (MW 60.06 g/mol) purchased from Merck (Darmstadt, Germany), creatinine (MW 113.12 g/mol) purchased from Sigma-Aldrich (Steinheim, Germany), uric acid (MW 168.11 g/mol) purchased from Alfa Aesar (Kandel, Germany) and bovine serum albumin (BSA; MW 66.5 g/mol) purchased from Sigma-Aldrich (Steinheim, Germany).

MWCO was studied using polyethylene glycol (PEG) 400 (MW 400 g/mol) purchased from Sigma-Aldrich (Steinheim, Germany), PEG 3000 (MW 3000 g/mol), PEG 6000 (MW 6000 g/mol), PEG 10,000 (MW 10,000 g/mol) purchased from Merck (Hohenbrunn, Germany), PEG 20,000 (MW 20,000 g/mol) purchased from Sigma-Aldrich (Steinheim, Germany) and PEG 35,000 (MW 35,000 g/mol) purchased from Merck (Hohenbrunn, Germany).

To dissolve BSA, phosphate buffer saline (PBS) was prepared using sodium chloride (NaCl, 58.44 g/mol, ≥99.5%) purchased from Merck (Darmstadt, Germany), potassium chloride (KCl, 74.56 g/mol, ≥99.5%) purchased from Panreac (Barcelona, Spain), potassium dihydrogen phosphate (KH_2PO_4, 136.09 g/mol, ≥99.5%) purchased from Merck (Darmstadt, Germany) and disodium hydrogen phosphate dihydrate ($Na_2HPO_4 \cdot 2H_2O$, 177.99 g/mol, ≥99.5%) purchased from Merck (Darmstadt, Germany).

The quantification of BSA was carried out according to the Bradford protein assay. Bradford reagent was prepared using Coomassie Brilliant Blue G-250 ($C_{47}H_{48}N_3NaO_7S_2$, 854.04 g/mol) purchased from Panreac (Barcelona, Spain), ethanol (C_2H_6O, 46.07 g/mol, 96% *v/v*) purchased from Manuel Vieira & C^a (Torres Novas, Portugal) and phosphoric acid (H_3PO_4, 98.00 g/mol, 1.71 g/mL at 20 °C, reagent grade 85%) purchased from Honeywell (Seelze, Germany).

All chemicals used in the synthesis, drying and characterization of the monophasic membranes were used without further purification.

2.2. Membrane Fabrication

The monophasic hybrid integrally skinned asymmetric CA-SiO_2-$(CH_2)_3NH_2$ membrane was prepared by coupling the phase inversion method with the sol-gel technique.

The casting solution of the monophasic hybrid CA-SiO_2-$(CH_2)_3NH_2$ membrane was prepared in two steps. First, 16.4 g of cellulose acetate, 29.0 g of formamide and 51.1 g of acetone were mixed in a reaction vessel to allow the complete dissolution of cellulose acetate. After 5 h of mixing, 2.4 g of TEOS, 0.6 g of APTES (to uphold amine in situ functionalization) and 12 drops of nitric acid were added to the mixture, promoting hydrolysis and hetero-condensation during the casting solution homogenization step. The final solution was placed on an agitation plate for another 19 h, resulting in a total homogenization time of 24 h for CA.

A reference casting solution of the pure CA membrane was prepared by mixing 17.0 g of cellulose acetate, 30.0 g of formamide and 53.0 g of acetone for 24 h.

The pure CA and CA-SiO_2-$(CH_2)_3NH_2$ casting solutions were cast on glass plates with a 250 μm Gardner knife at room temperature and a solvent evaporation time of 30 s, after which they were quenched into a gelation bath containing water at a temperature of 4 ± 1 °C, rendering the pure CA and monophasic hybrid CA-SiO_2-$(CH_2)_3NH_2$ membranes. After a residence time of approximately 3 h in the coagulation bath, the CA and CA-SiO_2-$(CH_2)_3NH_2$ membranes were detached from the glass plate and stored separately in deionized water at 4 ± 1 °C.

The sol-gel reactions which occur during the hybrid membranes' casting solution step have been previously described [27]. Briefly, after the sol-gel hydrolysis reactions, the silanol groups from the inorganic (TEOS) and hybrid (APTES) monomers hetero-condensate with the C–OH groups from the CA polymer, forming a new Si–O–C covalent bond. The sol-gel process was developed under acid catalysis. The new hybrid casting solution is then cast to form a monophasic hybrid membrane, based on carbon, silica, and amine chemical species, originating complex carbon-silica networks.

2.3. Membrane Characterization

2.3.1. Scanning Electron Microscopy (SEM)

Prior to being imaged by scanning electron microscopy (SEM), the CA and CA-SiO_2-$(CH_2)_3NH_2$ membranes were submitted to the drying process described by Lui et al. [31]. Samples of both membranes were then fractured in liquid nitrogen, mounted on a stub and sputter-coated with gold. Micrographs of the top dense surface (magnification: 1000×), cross-sections (magnification: 700×) and the porous bottom layer (magnification: 4000×) were obtained. The active layer and total thickness of the membranes were calculated using ImageJ software (version 1.53 k, from National Institutes of Health). For each membrane, three randomly selected zones from the images depicting the cross section of the CA and CA-SiO_2-$(CH_2)_3NH_2$ membranes were measured, and the mean thickness and standard deviation of the entire membrane and active layer were calculated. In order to determine the average pore size of the porous bottom layer, the images were binarized using a threshold adjustment and subsequently analyzed via the Measure tool of the ImageJ software.

2.3.2. Attenuated Total Reflection-Fourier Transform Infrared Spectroscopy (ATR-FTIR)

Attenuated total reflection-Fourier transform infrared spectroscopy (ATR-FTIR) was used to analyze the chemical composition of the active layer surfaces of dried CA and CA-SiO_2-$(CH_2)_3NH_2$ membranes. Infrared spectra of the samples with the active layer facing upwards were obtained with a PerkinElmer Frontier FT-IR spectrometer (940 Winter Street, Waltham, MA 02451, USA), using a Pike Miracle Single Reflection ATR sampling accessory from Pike Technologies, with a Ge crystal (Graseby Specac, Smyrna; sampling depth: 0.2–1.1 μm at 4000–650 cm^{-1}). Each spectrum was obtained by averaging 256 scans with a resolution of 4 cm^{-1}. The infrared spectra were reported as transmittance versus wavenumber.

2.4. Experimental Setup

2.4.1. Single Hemodialysis Membrane Module (SHDMM)

A single hemodialysis membrane module was custom-made by micromachining of acrylic plates. Figure 1 shows the schematic representation and dimensions of the five units which compose the SHDMM. Unit II defines the feed flow chamber, while the purpose of the perforated piece, Unit III, is to support the membrane. Unit IV defines the chamber in which the permeate is collected and Units I and V seal the top feed flow and bottom permeate collecting chambers, respectively. The membrane is placed with its active layer facing upwards on top of Unit III, with a filter paper beneath it to prevent damage to the membrane when all units are sealed together by using stainless-steel screws. Because of the large thickness of Unit III (1 cm), on top of which the membrane is placed, the solutions in the feed flow channel and the permeate compartment do not interact closely enough to establish a concentration gradient. Therefore, the solute diffusion between the compartments is negligible and it is assumed that fluid and solute removal occur exclusively by convection. The feed flow and permeate collecting chambers were designed to be slit-like microchannels, where the channel height (2B) is much smaller than the other two dimensions: width (W) and length (L). As such, the dimensions of the microchannels are 2B = 0.3 cm, W = 3.0 cm, and H = 25.0 cm (2B ≪ W and L). The effective membrane surface area is defined by the area of the membrane clamped between the O-ring which seals Units I and II and is 105 cm^2.

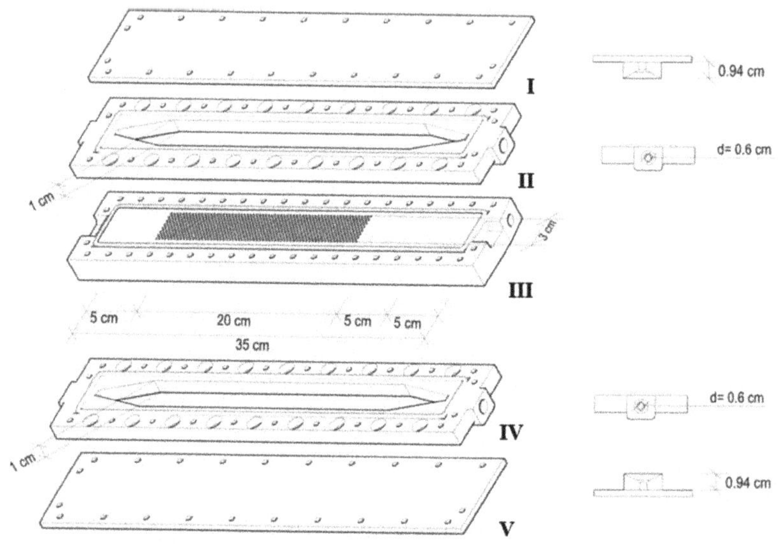

Figure 1. Schematic representation of the single hemodialysis membrane module. Unit I seals the feed flow chamber, represented by Unit II, while Unit V seals the permeate collecting chamber, represented by Unit IV. Unit III is the supporting surface for the membrane to be tested.

2.4.2. Feed Circulation Circuit

Figure 2 represents the circuit which simulates the extracorporeal blood circulation circuit found in HD machines. The feed reservoir, a 500 mL glass Kitasato, is placed in a thermostatically controlled water bath (model CORIO C, from JULABO, Seelbach, Germany) to maintain the feed solution at the average normal body temperature, 37 °C. The feed solution is pumped from the reservoir and through the circuit by a peristaltic pump similar to the ones used in extracorporeal circulation such as dialysis (model ECOLINE VC-360, from ISMATEC, Wertheim, Germany), and its pulsation effect is attenuated by a damper, allowing the feed fluid to enter the SHDMM at a constant pressure. The ultrafiltrate is collected in the permeate collecting chamber, and the fraction of the feed solution which has not been permeated is recirculated back to the feed reservoir. A three-way valve placed midway between the outlet and the reservoir enables the collection of feed samples. Throughout the entire circuit, polyvinyl chloride (PVC) tubing (Tygon Saint-Gobain, La Defense, Courbevoie, France) with an inner diameter of 3 mm and a Shore hardness of 55 A was used to ensure that the shedding of particles from the interior of the tubing—a phenomenon known as spallation [32,33]—is low.

Four medical grade pressure sensors (Deltran® Utah Medical Products, Inc., Midvale, UT, USA)—P1, P2, P3, and P4—register the pressure at crucial points of the circuit. P1 and P2 register the pressure at the inlet and outlet of the feed flow chamber, respectively, whereas P3 and P4 register the pressure at the inlet and outlet of the permeate collecting chamber, respectively. The pressure sensors are connected to a data acquisition card (National Instruments cDAQ-9172, USA) which records the pressure variations throughout time via a LabView module (National Instruments, 9237, Austin, TX, USA). The cross to the left of the sensor represented by P3 indicates that the tube is clamped, and therefore the ultrafiltrate is collected by the tube placed after P4.

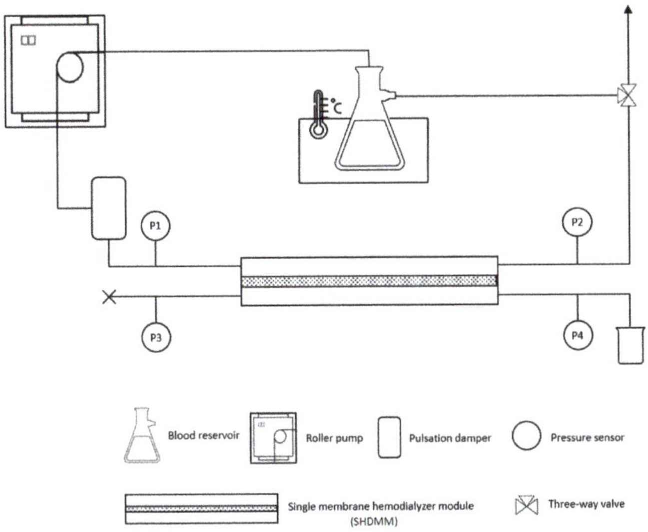

Figure 2. Schematic representation of the experimental setup used for permeation experiments under dynamic conditions.

2.4.3. Characterization of the SHDMM: Pressure Profile, Pressure Drop, Transmembrane Pressure, Microchannel Height, Shear Rate and Shear Stress at the Wall

To characterize the SHDMM, a permeation experiment was performed to obtain a typical pressure profile for the flow of deionized (DI) water at different feed volumetric flow rates (Q_F) through the SHDMM: 49 mL/min, 66 mL/min, 82 mL/min, 99 mL/min, 115 mL/min, 132 mL/min, and 148 mL/min. Figure 3 shows the pressure profile of the system registered continuously throughout the experiment. Each plateau represents the pressure registered for each pressure sensor resulting from the Q_F imposed by the peristaltic pump.

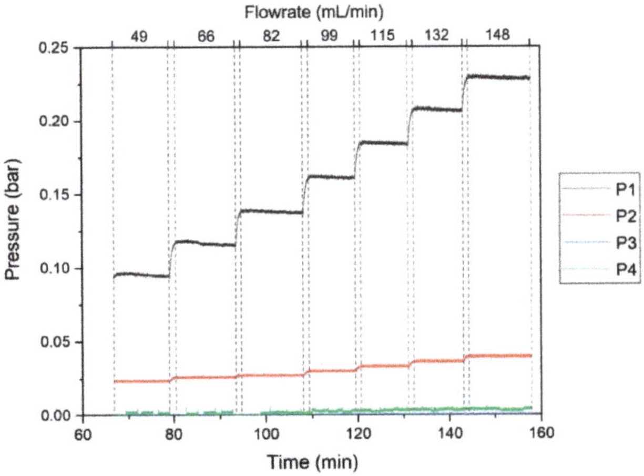

Figure 3. Pressure profile of the circuit during the circulation of DI water through the SHDMM circuit at different values of Q_F, which are indicated delimited by the dashed lines. The pressure data is color-coded according to the different sensors: black line—P1, red line—P2, green line—P3, and dark blue line—P4.

The pressure drop (ΔP) across the feed flow channel is defined by:

$$\Delta P = P_1 - P_2 \quad (1)$$

The transmembrane pressure is the hydrostatic pressure difference between the feed flow chamber and permeate collecting compartment, acting as the driving force for solute removal, and has unique values along the length of the membrane [5]. Assuming a linear variation of fluid pressure with axial distance along the channel, TMP is defined by:

$$\text{TMP} = \frac{P_1 + P_2}{2} - \frac{P_3 + P_4}{2} \quad (2)$$

As mentioned in Section 2.4.1, the feed flow microchannel has a slit-like geometry where the channel height (2B = 0.3 cm) is much smaller than the other two dimensions: width (W = 3.0 cm) and length (L = 25.0 cm). When the membrane is placed inside the SHDMM, the height of the feed flow microchannel is no longer 0.3 cm and must be accurately calculated.

The half-height of the microchannel (B) is obtained by an equation analogous to the Hagen–Poiseuille law for circular tubes and describes the fully-developed laminar flow of a Newtonian fluid in a narrow slit [34]:

$$B = \sqrt[3]{\frac{3}{2}\frac{\mu L}{W}\frac{Q_F}{\Delta P}} \quad (3)$$

where μ is the viscosity of the fluid, L is the length of the microchannel, Q_F is the feed flow rate, W is the width of the microchannel and ΔP is the pressure drop across the microchannel.

The shear stress exerted at the flow boundaries, or walls of the microchannel, τ, can be calculated by balancing the shear force at the wall against the pressure gradient for a slit channel [34]:

$$\tau = \frac{3\mu Q_F}{2B^2 W} \quad (4)$$

Shear rates, γ, at the wall are found by dividing the shear stress by the viscosity:

$$\gamma = \frac{\tau}{\mu} = \frac{3\mu Q_F}{2B^2 W} \quad (5)$$

Shear rates and shear stress in the device were varied from 4000 to 11,800 s^{-1}, and from 3 to 8 Pa, respectively.

2.5. Permeation Experiments under Dynamic Conditions

2.5.1. Water Permeability

Permeation experiments were performed in the circuit containing the SHDMM to characterize the monophasic hybrid CA-SiO_2-$(CH_2)_3NH_2$ and pure CA membranes in terms of pure water permeability.

Ultrafiltration flux, J, is defined as the amount of permeate produced per unit area of membrane surface per unit time:

$$J = \frac{V}{A \times t} \quad (6)$$

where V is the volume of permeate, A is the membrane permeation area, and t is the measuring time.

The water permeability, L_P, is obtained by the slope of the straight line obtained when J is represented as a function of TMP:

$$L_P = \frac{J}{\text{TMP}} \quad (7)$$

The range of the TMP used was 0.0591, 0.0708, 0.0819, 0.0943, 0.107, 0.120 and 0.132 bar (45, 53, 61, 71, 80, 90 and 100 mm Hg), corresponding to Q_F values of 49, 66, 82, 99, 115, 132, and 148 mL/min, respectively.

The feed temperature was kept at 37 °C, equivalent to the normal body temperature of a healthy adult, for all the permeation experiments. For comparison purposes with other membranes on the market, the water permeability found at 37 °C for the CA-SiO_2-$(CH_2)_3NH_2$ and pure CA membranes was standardized for a temperature of 25 °C by Equations (8)–(11) [35,36]:

$$\ln \mu = -6.78 + \frac{1983}{T} \; (r = 0.99923) \quad (8)$$

Equation (8) can be written in terms of mass flux and assumes the following form, when solved for J_m:

$$J_m = \rho \frac{L_P}{\mu} \Delta P \quad (9)$$

As neither L_P nor ΔP are temperature-dependent, the previous expression can be rewritten as:

$$\frac{J_m(T)\,\mu(T)}{\rho(T)} = L_P \Delta P \quad (10)$$

Ultimately, the relationship between two fluxes, $J_{m,1}$ and $J_{m,2}$, at different temperatures, T_1 and T_2, respectively, is given by:

$$\frac{J_{m,1}\mu_1}{\rho_1} = \frac{J_{m,2}\mu_2}{\rho_2} \quad (11)$$

2.5.2. Molecular Weight Cut-Off

The molecular weight cut-off (MWCO) of a membrane is defined as the molecular weight at which 90% of the solute is rejected by the membrane [37]. To determine the MWCO of each membrane, a retention curve was constructed by measuring the rejection coefficient, f, of various polyethylene glycols (PEGs) with increasing molecular weights (MWs) of 400 Da, 3000 Da, 6000 Da, 10,000 Da, 20,000 Da, and 35,000 Da, respectively. The rejection coefficient, R, is defined by:

$$R = \frac{C_F - C_P}{C_F} \quad (12)$$

where C_P corresponds to the concentration of the permeate at t = 60 min and C_F corresponds to the initial feed solution concentration. The initial concentration of the feed solutions of each PEG was 485 ± 76 mg/L and the permeation studies were performed for 1 h at a Q_F of 100 ± 5 mL/min and a TMP of 0.1 bar (76 ± 1 mmHg). Solute concentrations were determined in the permeate, initial feed, and final feed samples using a total organic carbon analyzer (TOC-VCPH/CPN, Shimadzu, Japan).

For the CA-SiO_2-$(CH_2)_3NH_2$ and pure CA membranes, two plots were made on the same graph: (1) R as a function of the solute MW with the intersection of the horizontal line at R = 90%, and (2) a plot of the linearization of R, $\log\left(\frac{R}{1-R}\right)$, as a function of solute MW with the intersection of the horizontal (dashed) line at $\log\left(\frac{R}{1-R}\right) = 0.95$.

2.5.3. Rejection Coefficients to Low-Molecular Weight Water-Soluble Uremic Toxins

The rejection coefficients, R, of the CA-SiO_2-$(CH_2)_3NH_2$ and pure CA membranes to markers for low-molecular weight water-soluble UTs: urea, creatinine, and uric acid, were determined by Equation (12). The initial feed solutions of each solute were prepared according to the highest reported concentrations of urea [38], creatinine [38], and uric acid [39]: 4.6 g/L, 240 mg/L, and 83 mg/L, respectively. The permeation studies were performed for 1.5 h at a QF of 100 ± 5 mL/min and TMP of 0.06 bar (44 ± 1 mmHg). Solute concentrations were determined in the permeate and feed samples using a UV-Vis spectrophotometer (UV-1700 PharmaSpec, Shimadzu, Japan), according to each toxin's wavelength of maximum absorbance, λmax: 200, 230, and 293 nm for urea [40], creatinine [41], and uric acid [42], respectively.

2.5.4. Bovine Serum Albumin Filtration

The rejection coefficients, R, of the CA-SiO_2-$(CH_2)_3NH_2$ and pure CA membranes to bovine serum albumin (BSA) were determined by Equation (8). The initial concentration of the feed solutions was 884 ± 4 mg/L, and the permeation studies were performed for 465 min at a Q_F of 100 ± 5 mL/min and TMP of 0.07 bar (52 ± 5 mmHg). Solute concentrations were determined in the permeate, initial feed, and final feed samples through the Bradford assay using a UV-Vis spectrophotometer (UV-1700 PharmaSpec, Shimadzu, Japan). Specifically, the absorbance of feed samples was measured at 595 nm, following the Bradford assay [43], whereas the absorbance of permeate samples was simultaneously measured at 590 nm and 450 nm according to a modified Bradford assay [44], allowing the determination of lower BSA concentrations.

3. Results and Discussion

3.1. Membrane Chacracterization

3.1.1. Scanning Electron Microscopy (SEM)

Figure 4 shows the SEM micrographs obtained for the active layer surface, cross-section, and bottom porous surface of the dried CA-SiO_2-$(CH_2)_3NH_2$ and CA membranes. The active layer of the membranes (Figure 4A,B) is responsible for the membrane selectivity, and both membranes exhibited completely dense active layers with no visible pores at a magnification of 1000×. The cross-section images (Figure 4C,D) confirmed the presence of a very thin skin layer of <1 μm for both membranes, outlining a much thicker, porous substructure. The overall thickness of the membranes was obtained from the cross-section images, yielding 103 ± 1 μm and 53 ± 1 μm for the CA-SiO_2-$(CH_2)_3NH_2$ and CA membranes, respectively. These results suggest that the introduction of silanol groups from the inorganic (TEOS) and hybrid (APTES) monomers led to membranes with increased thickness when compared to pristine CA membranes. This is in accordance with what has been observed in previous studies [25,27]. The porous structure (Figure 4E,F) confers mechanical strength to the monophasic hybrid membranes, while offering little or no resistance to the permeation of water and solutes. The average pore diameter in the porous surface for the monophasic hybrid and pristine CA membranes was 113 nm and 84 nm, respectively.

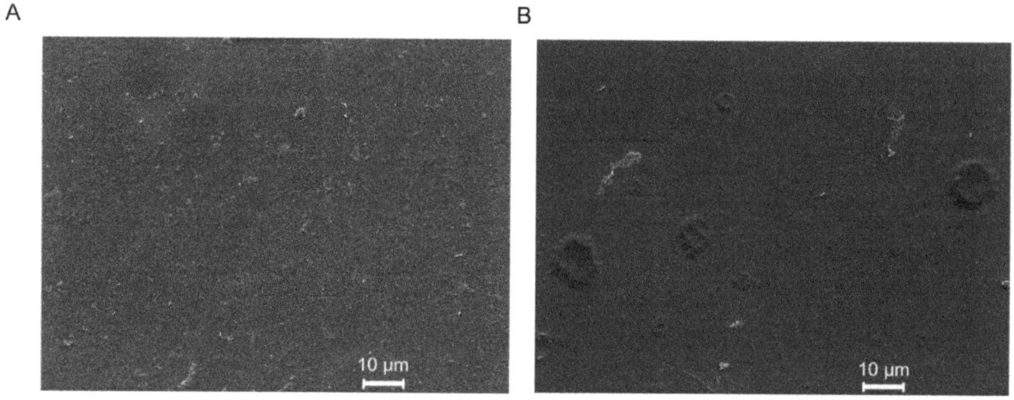

Figure 4. *Cont.*

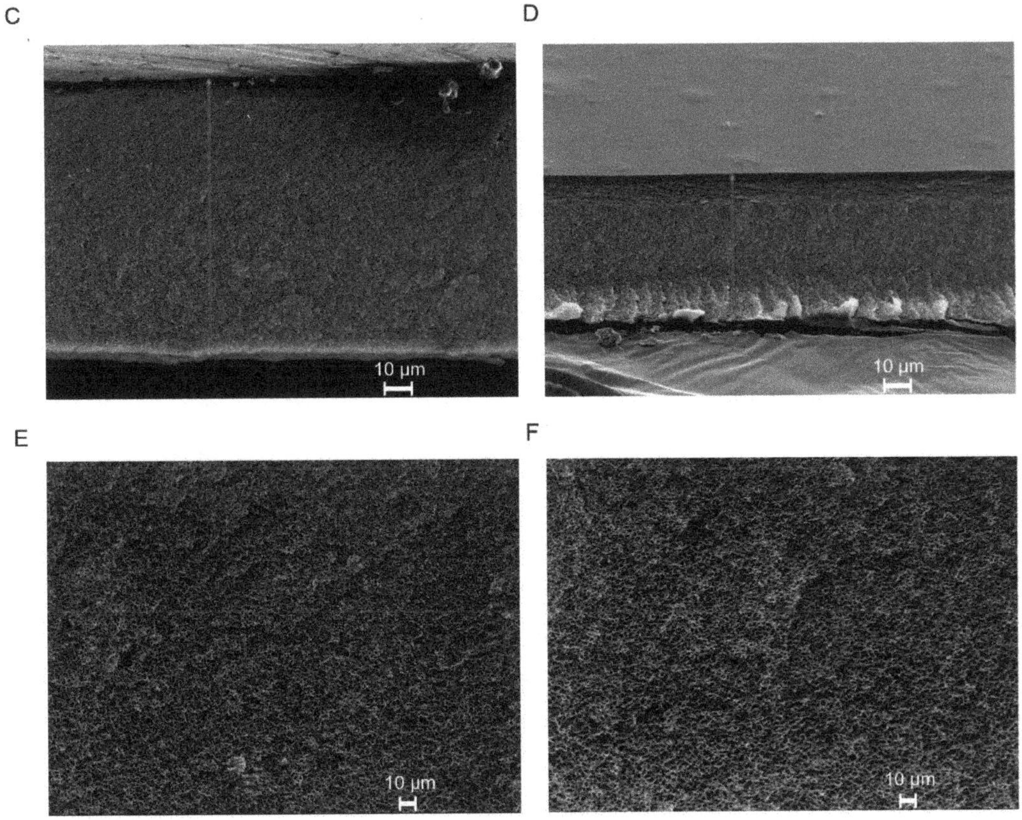

Figure 4. SEM images of the CA-SiO_2-$(CH_2)_3NH_2$ (**A**,**C**,**E**) and CA (**B**,**D**,**F**) membranes. (**A**,**B**): top active layer (1000×). (**C**,**D**): cross-section (700×) where the red lines indicate the membranes' thickness. (**E**,**F**): bottom porous surface (4000×).

3.1.2. Attenuated Total Reflection-Fourier Transform Infrared Spectroscopy (ATR-FTIR)

The ATR-FTIR spectra of the CA-SiO_2-$(CH_2)_3NH_2$ and CA membranes' active layer are compared in Figure 5.

The broad band centered at ~3360 cm^{-1} was assigned to the OH stretching mode, ν(OH), and contained contributions from the hydroxyl groups of non-esterified cellulose, as well as from adsorbed water and hydrolyzed silica precursors (TEOS and APTES). Moreover, the bending fundamental, δ(OH), for the adsorbed water was located at ~1637 cm^{-1}, which has been previously described by other authors [45].

Similar to a previous work, [46], the strong carbonyl stretching mode, ν(C=O), appeared at 1735 cm^{-1}, while the antisymmetric, ν_{as}(C–O–C), and symmetric, ν_s(C–O–C), stretching modes of the ester appeared as strong bands at 1238 cm^{-1} and 1049 cm^{-1}, respectively.

There were no distinguishable CH stretching bands of cellulose acetate, which were possibly hindered by the broad band assigned to adsorbed water. However, in the CH deformation region, there was a medium band, typical of the methyl umbrella mode of acetate groups, at 1371 cm^{-1}.

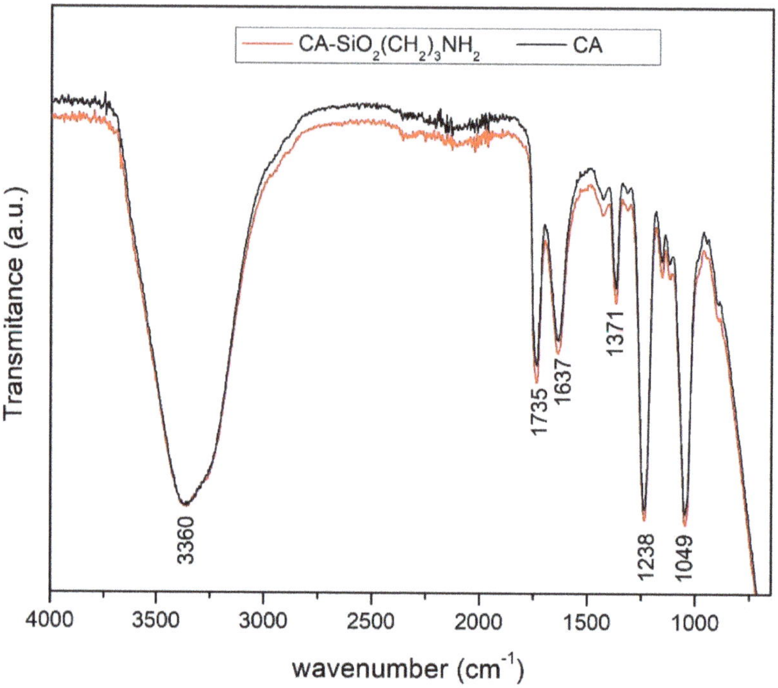

Figure 5. Wide-range ATR-FTIR spectra (4000–650 cm^{-1}) of the CA-SiO_2-$(CH_2)_3NH_2$ and CA membranes.

Amine functionalization of CA-SiO_2-$(CH_2)_3NH_2$ membranes with the same composition has been previously proven [27], despite there being no evident band of the NH bond in Figure 5. In fact, primary aliphatic amine stretching mode was assigned in the region of 3330–3340 cm^{-1}, for which the band may have overlapped with the broad band which was assigned to ν(OH) [47].

Two important bands attributed to SiO_2, namely ν(Si–O–Si) and ν(Si–O–C), occured in the regions of 1055–1165 cm^{-1} and 1115–1175 cm^{-1}, respectively. Similar to previous studies [24], and due to the low silica content in the CA-SiO_2-$(CH_2)_3NH_2$ membrane, these bands were overlaid by those attributed to the ester stretching modes.

3.2. Characterization of the SHDMM: Pressure Profile, Pressure Drop, Transmembrane Pressure, Microchannel Height, Shear Rate and Shear Stress at the Wall

Figure 6 shows the values of ΔP and TMP obtained when DI water was circulated through the SHDMM containing the CA-SiO_2-$(CH_2)_3NH_2$ membrane at different values of Q_F. The values of ΔP were 0.072, 0.091, 0.11, 0.13, 0.15, 0.17, 0.19 bar (54, 68, 83, 99, 114, 128, and 142 mmHg) for Q_F values of 49, 66, 82, 99, 115, 132, and 148 mL/min, respectively. The values of TMP were 0.059, 0.071, 0.082, 0.094, 0.11, 0.12, and 0.13 bar (45, 53, 61, 71, 80, 90, and 100 mmHg) for QF values of 49, 66, 82, 99, 115, 132, and 148 mL/min, respectively.

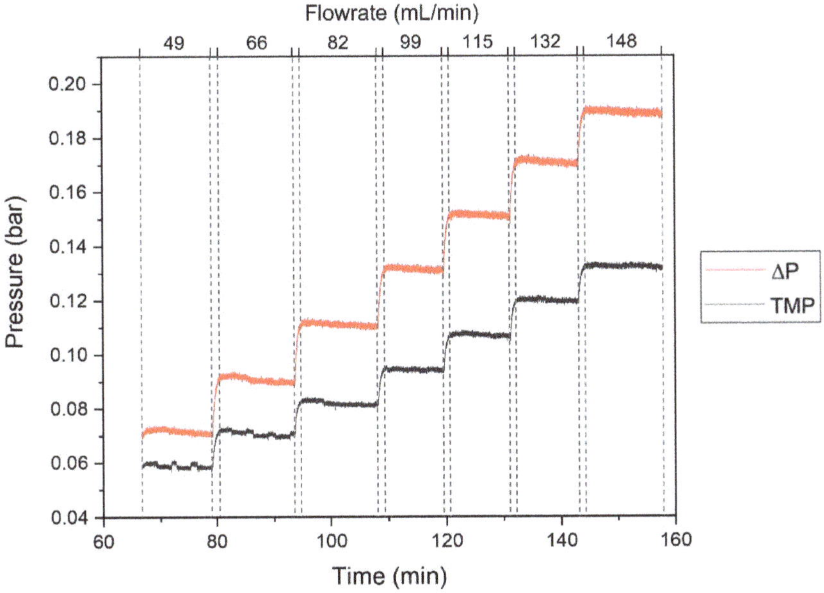

Figure 6. ΔP (red line) and TMP (black line) values obtained for the flow of DI water through the SHDMM circuit at different values of Q_F, which are delimited by the dashed lines.

Typical values of Q_F under which commercial hemodialysis equipment operates range from 250 to 400 mL/min [48,49] and TMP should not exceed 0.4 bar (300 mmHg), according to the European Renal Best Practice (ERBP) [50]—with the latter being in the same range as the operating conditions of the SHDMM circuit. The Q_F is, however, below the typical values of hemodialysis equipment in order to avoid high values of sheer rate, as described below.

Considering the viscosity of water at 37 °C, μ_{water} = 0.6913 mPa.s [51], the height of the feed microchannel, calculated by Equation (3), was 205 ± 3 μm and 255 ± 10 μm when the CA-SiO_2-$(CH_2)_3NH_2$ and CA membranes were placed inside the SHDMM, respectively. The difference in channel height of the SHDMM feed channel when it encased the CA membrane was expected, as the total thickness of this membrane was approximately 50 μm lower than the thickness of the CA-SiO_2-$(CH_2)_3NH_2$ membrane.

The shear rate for each of the Q_F values was calculated by Equation (5) and was found to be 3887, 5235, 6504, 7852, 9122, 10,470, and 11,739 s^{-1} for Q_F values of 49, 66, 82, 99, 115, 132, and 148 mL/min, respectively. It is known that high fluid shear triggers the activation of platelets and their subsequent aggregation [52]. An in vitro study by Holme et al. showed that platelet activation and aggregation was observed at shear rates of 10,500 s^{-1} and above [53]. Therefore, it is safe to use all the tested feed flow rates up until 115 mL/min.

3.3. Permeation Experiments under Dynamic Conditions

3.3.1. Water Permeability

Figure 7 shows the results of ultrafiltration flux, J, as a function of TMP obtained for the CA-SiO_2-$(CH_2)_3NH_2$ and pure CA membranes at 37 °C. The water permeability, L_P, of the CA-SiO_2-$(CH_2)_3NH_2$ and pure CA membranes was 66.61 $kg{\cdot}h^{-1}{\cdot}m^{-2}{\cdot}bar^{-1}$ and 37.09 $kg{\cdot}h^{-1}{\cdot}m^{-2}{\cdot}bar^{-1}$, respectively. For the sake of standardization, the ultrafiltration fluxes of both membranes were corrected to 25 °C through Equations (8)–(11) and revealed L_P values of 50.77 $kg{\cdot}h^{-1}{\cdot}m^{-2}{\cdot}bar^{-1}$ and 28.27 $kg{\cdot}h^{-1}{\cdot}m^{-2}{\cdot}bar^{-1}$ for the CA-SiO_2-$(CH_2)_3NH_2$ and CA membranes, respectively. The introduction of silica followed by the presence of amine groups increased the hydraulic permeability of the monophasic hybrid membrane by a factor of 1.8 when compared to the pure CA membrane.

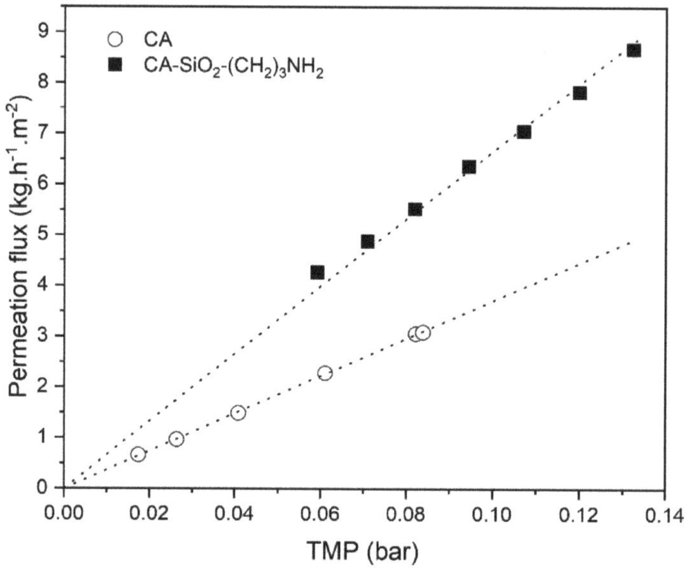

Figure 7. Ultrafiltration flux of pure water at 37 °C as a function of the transmembrane pressure for $CA\text{-}SiO_2\text{-}(CH_2)_3NH_2$ (■) and pure CA (○) membranes. The dotted lines represent the linear fit, which enables the determination of each membrane Lp.

Current hemodialysis equipment should deliver ultrafiltration rates between 10–13 mL/(h·kg) [54], under an operating TMP of between 0.133 and 0.200 bar (100 and 150 mmHg). Hence, for a 70 kg adult, the expected ultrafiltration rate in a clinical scenario should not be lower than 700 mL/h. Considering the estimated value of L_P for the $CA\text{-}SiO_2\text{-}(CH_2)_3NH_2$ membrane, 89.40 $mL\cdot h^{-1}\cdot m^{-2}\cdot mmHg^{-1}$, and for an average TMP of 0.167 bar (125 mmHg) in order to achieve the 700 mL/h threshold, a total membrane surface area of 0.06 m^2 would be enough. This is a promising result, as the estimated surface area is well below the effective permeation area of typical hemodialyzers, which ranges from 0.8 to 2.2 m^2 [55].

3.3.2. Molecular Weight Cut-Off

The MWCO was determined by two methods which are shown in Figure 8. The upper part of the curve resembled a plateau, hindering an accurate determination of the value corresponding to the intersection of the rejection coefficient curve with the horizontal rejection line (R = 90%). Hence, to overcome this ambiguity, the higher range of rejection coefficients was linearized, namely those corresponding to PEG 35,000, PEG 20,000, and PEG 10,000—enabling a precise determination of the intersection point.

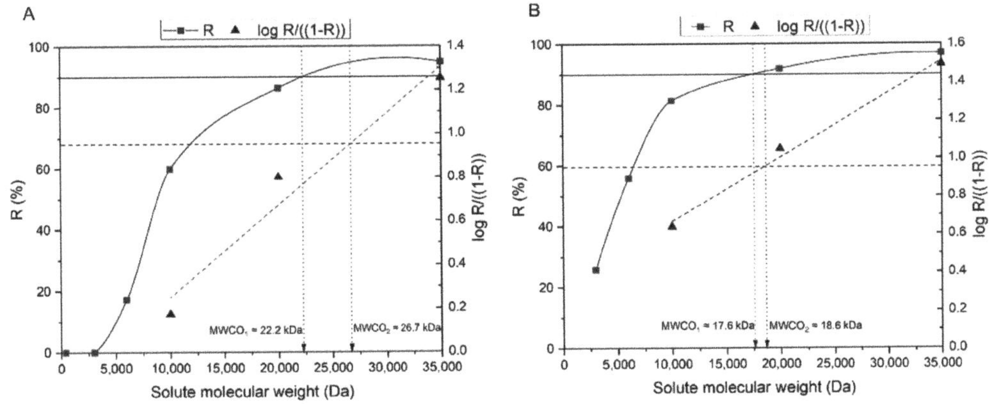

Figure 8. Rejection coefficient profiles of the CA-SiO_2-$(CH_2)_3NH_2$ (**A**) and CA (**B**) membranes to solutes with increasing molecular weights: 400, 3000, 6000, 10,000, 20,000, and 35,000 Da. The horizontal full lines are set for a rejection of 90%, i.e., $\log\left(\frac{R}{1-R}\right) = 0.95$. The dashed rejection lines represent the linearization of the rejection coefficients obtained for 3 different molecular weight solutes: 10,000, 20,000, and 35,000 Da. The dotted dropdown lines indicate the estimated MWCO according to the method employed.

The depicted values of MWCO corresponded to the intersection point of the plotted rejection coefficient curve ($MWCO_1$) and subsequent linearization ($MWCO_2$) with the respective horizontal rejection line. Hence, the MWCO was estimated to range between 22.2 and 26.7 kDa for the CA-SiO_2-$(CH_2)_3NH_2$ membrane, and between 17.6 and 18.6 kDa for the CA membrane. With the understanding that both membranes reject solutes with MWs greater than 20 kDa, it was predicted that vital blood components such as albumin, platelets, and blood cells would be rejected by both membranes. Furthermore, it is envisioned that molecules belonging to two different classes of uremic toxins proposed by the EUTox—small water-soluble compounds and middle molecules—can be removed, as they are able to cross the membrane. Prototypes to small water-soluble compounds and middle molecules include urea and β_2-microglobulin, with a respective MW of 60 Da and 11,818 Da [2]. The third group of uremic toxins, PBUTs, remains a challenge: whilst the free fraction of these molecules (MW < 500 Da) is able to cross the membrane and successfully be removed from the blood circulation, the rejection of the bound fraction will be hindered by the large MW associated with the complexes they form with albumin (>60 kDa).

3.3.3. Rejection Coefficients to Low-Molecular Weight Water-Soluble Uremic Toxins

Permeation experiments to prototype low-molecular weight water-soluble UTs (urea, creatinine, and uric acid) were performed with consideration to the highest reported concentrations found in uremic populations. Feed and permeate samples were collected at times of 0, 15, 30, 45, 60, 75, 90, and 105 min, with the respective concentrations represented in Figure 9.

The CA-SiO_2-$(CH_2)_3NH_2$ and CA membranes exhibited a similar behavior in terms of permeation of urea, creatinine, and uric acid. The initial concentrations of the feed solutions were defined according to the highest reported concentrations found in ESRD patients for the three uremic toxins [38,39], and these were measured after they had been prepared and before they were fed into the SHDMM system (t = 0 min). It should be noted that, while the highest reported uremic concentration for uric acid was 83 mg/L, the solubility limit of this toxin in water, at 20 °C, is 60 mg/L.

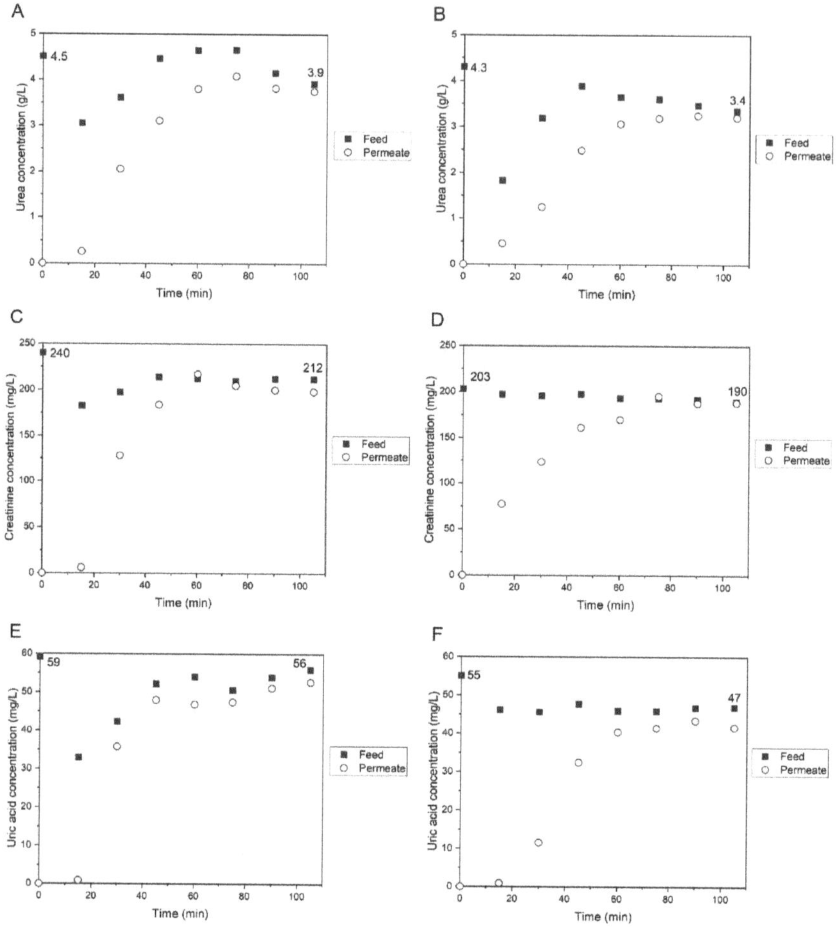

Figure 9. Concentration profiles of urea, creatinine, and uric acid for CA-SiO_2-$(CH_2)_3NH_2$ (**A**,**C**,**E**) and CA (**B**,**D**,**F**) membranes, regarding feed (■) and permeate (○) samples, for a total experiment time of 90 min. The initial feed concentrations aim to represent those of uremic populations.

Prior to each experiment, the SHDMM system was primed with water so that there are no air bubbles present in the system, as well as to prevent the CA and CA-SiO_2-$(CH_2)_3NH_2$ membranes from becoming dry. The priming volume of the system was approximately 130 mL and, when the initial feed solution containing each of the three toxins was circulated into the SHDMM system, a dilution occurs. This explains the decrease in the initial concentration of the primary feed solution at t = 0 min (measured after it was prepared and before it was fed into the SHDMM system) and the moment it was collected after 15 min of circulation in the system (t = 15 min).

In general, for both the CA and CA-SiO_2-$(CH_2)_3NH_2$ membranes, it was observed that, between 15 and approximately 50 min, the concentration of the solutes in the feed compartment increased, and this could be explained by the decrease of water in the system, which was constantly being removed by convection from the feed compartment through the membrane and into the permeate channel. Between ~50 and 105 min, the concentration of the feed solution tended to stabilize at values close to the ones of the initial feed solution. To make this clear, the values of the initial and final concentrations of the feed solution are shown in each graph of Figure 9.

Regarding the concentration of the permeate solution, at the beginning of the experiment (t = 0), the collecting permeate chamber was filled with water, and as expected, the solute concentration in the permeate chamber was 0. The concentrations of the permeate solution increased considerably between 0 and approximately 50 min and, towards the end of the experiment (t > 80 min), they approached concentration values similar to those of the feed solution (at the corresponding time). This behaviour clearly demonstrates that the membranes are permeable to the three low-molecular weight water-soluble uremic toxins evaluated—creatinine, uric acid, and urea.

The rejection coefficients (Equation (8)) were calculated after 90 min of permeation. The rejection coefficient of the CA-SiO_2-$(CH_2)_3NH_2$ and CA membranes towards uric acid was 6% and 11%, respectively, while both membranes had a rejection factor of 4% towards urea. The rejection coefficient of the CA-SiO_2-$(CH_2)_3NH_2$ and CA membranes towards creatinine was 7% and 1%, respectively. These results are in agreement with what was discussed before in terms of MWCO, given that urea (MW 60 Da), creatinine (113 Da), and uric acid (168 Da) have much lower MWs than the CA-SiO_2-$(CH_2)_3NH_2$ and CA membranes' MWCO.

3.3.4. Bovine Serum Albumin Filtration

Figure 10 shows the concentration profiles of BSA in the feed and permeate channels of the SHDMM equipped with the CA-SiO_2-$(CH_2)_3NH_2$ (Figure 10A) and CA (Figure 10B) throughout the recirculation experiments lasting 465 min. The aim of this experiment was to determine the rejection coefficients of the CA-SiO_2-$(CH_2)_3NH_2$ and CA membranes towards BSA and to assess the potential fouling of the membrane, mostly attributed to irreversible protein deposition and pore blockage.

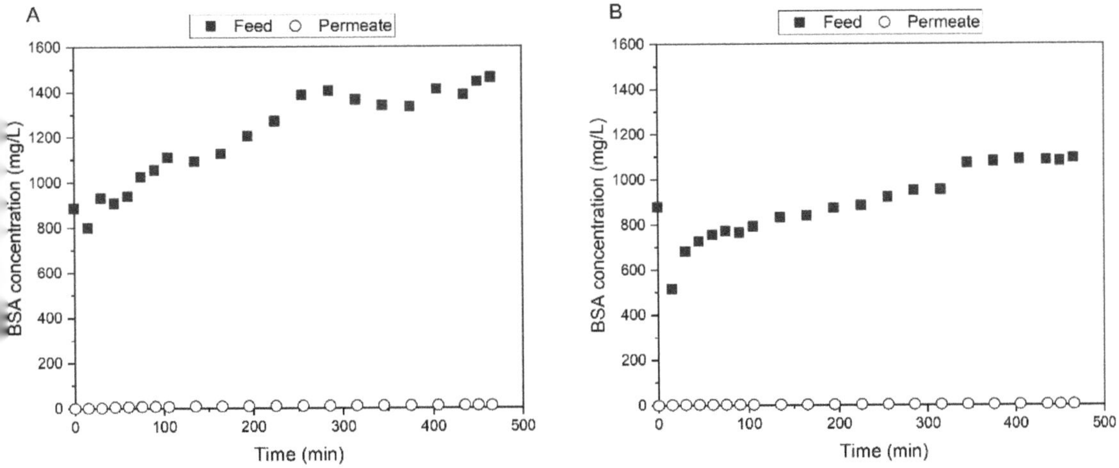

Figure 10. BSA concentration profiles for CA-SiO_2-$(CH_2)_3NH_2$ (**A**) and CA (**B**) membranes, regarding feed (■) and permeate (○) samples, for a total experiment time of 465 min.

As was described for the UT permeation experiments (Section 3.3.3), prior to initiating the filtration of the BSA solution with initial concentrations of 884 ± 4 mg/L, the SHDMM system was primed with water. The total priming volume for the permeation experiment with the CA membrane was approximately 130 mL and for the experiment with the CA-SiO_2-$(CH_2)_3NH_2$ membrane, it was reduced to 50 mL. This explains the decrease in concentration of the initial feed solutions of BSA between 0 and 15 min. From 15 to 465 min, the concentration of BSA in the feed solution of both permeation experiments through the CA and CA-SiO_2-$(CH_2)_3NH_2$ membranes increased steadily. This gives a clear indication that both membranes fully rejected BSA.

At the beginning of each experiment (t = 0 min), the permeate chamber was filled with water and therefore the permeate concentration was 0 mg/L. The permeation experiments were carried out for over 7 h and, throughout the entire experiment, the highest concentration of BSA detected in the permeate was 12 mg/L.

Furthermore, there seems to be no clear evidence of fouling events, as both the TMP and the ΔP remained approximately constant throughout the experiment, at 0.076 bar (57 $\pm$ 7 mmHg) and 0.12 bar (87 $\pm$ 3 mmHg) for the CA-SiO_2-$(CH_2)_3NH_2$ membrane, and 0.063 bar (47 $\pm$ 3 mmHg) and 0.077 bar (58 $\pm$ 2 mmHg) for the CA membrane, respectively. The ultrafiltration rate did not change significantly for any of the two membranes throughout the long-term filtration experiment, adding more evidence to support the absence of membrane fouling. At the end of the experiment, the heights of the feed microchannel of the SHDMM were 215 μm and 246 μm, when the SHDMM encased the CA-SiO_2-$(CH_2)_3NH_2$ and CA membranes, respectively. These values are very similar to those found using pure water at a Q_F of 100 mL/min (Section 3.2): 205 μm and 255 μm, for the SHDMM with the CA-SiO_2-$(CH_2)_3NH_2$ and CA membranes, respectively. This indicates that the protein deposition and adhesion is negligible, as there is no evident narrowing of the microchannel height even after 465 min of BSA filtration.

The rejection coefficient towards BSA was calculated with the BSA concentrations measured at the end of the experiment (after 465 min of filtration) and was 99% for both the CA-SiO_2-$(CH_2)_3NH_2$ and CA membranes. These results were expected given that BSA has a MW of 66.5 kDa which is similar to human serum albumin (HSA) [56], and the MWCO of the CA-SiO_2-$(CH_2)_3NH_2$ and CA membranes is approximately 27 kDa and 19 kDa, respectively.

4. Conclusions

A novel integrally skinned cellulose acetate-based monophasic hybrid amine-functionalized membrane, CA-SiO_2-$(CH_2)_3NH_2$, was synthesized by an innovative method which combines the phase inversion and sol-gel techniques.

SEM micrographs of the CA-SiO_2-$(CH_2)_3NH_2$ membrane cross-sections confirmed the existence of a thin dense active layer and a much thicker, porous layer. SEM analysis revealed that the total thickness of the CA-SiO_2-$(CH_2)_3NH_2$ membrane—103 μm—was approximately two times higher than the thickness of the CA membrane—54 μm.

Permeation studies showed enhanced mass transfer properties for the monophasic hybrid membrane when compared to the pure CA membrane, as evidenced by the increase of hydraulic permeability, measured at 37°C, from 37.09 $kg \cdot h^{-1} \cdot m^{-2} \cdot bar^{-1}$ for the CA membrane, to 66.61 $kg \cdot h^{-1} \cdot m^{-2} \cdot bar^{-1}$ for the CA-SiO_2-$(CH_2)_3NH_2$ membrane. Furthermore, the MWCO also increased from 18.1 kDa for the CA membrane, to 24.5 kDa for the CA-SiO_2-$(CH_2)_3NH_2$ membrane. Both membranes assured the permeation of water-soluble toxins such as urea, creatinine, and uric acid. Long-term continuous studies of BSA filtration revealed that both membranes completely reject albumin and that there is no evidence of irreversible fouling or albumin leakage in the SHDMM.

Author Contributions: Funding acquisition: M.F.; investigation: A.J. and F.S.C.R.; methodology: A.J., F.S.C.R., M.C.G. and M.F.; supervision: M.C.G. and M.F.; writing—original draft: A.J. and M.F.; writing—review and editing: M.F. All authors have read and agreed to the published version of the manuscript.

Funding: This research received no external funding.

Institutional Review Board Statement: Not applicable.

Informed Consent Statement: Not applicable.

Data Availability Statement: Not applicable.

Acknowledgments: Mónica Faria would like to thank the Fundação para a Ciência e a Tecnologia (FCT, Portugal) for a contract in the scope of programmatic funding UIDP/04540/2020. Flávia Rodrigues acknowledges the PhD research grant UI/BD/150949/2021 from FCT. M. Clara Gonçalves would also like to thank the Foundation for Science and Technology (FCT) through projects UIDB/00100/2020 and UIDP/00100/2020. All authors have consented to the acknowledgement and would like to thank Marta Veloso for providing linguistic proofreading of the article.

Conflicts of Interest: The authors declare no conflict of interest.

eferences

Hill, N.R.; Fatoba, S.T.; Oke, J.L.; Hirst, J.; O'Callaghan, C.A.; Lasserson, D.; Hobbs, R. Global Prevalence of Chronic Kidney Disease—A Systematic Review and Meta-Analysis. *PLoS ONE* **2016**, *11*, e0158765. [CrossRef]

Faria, M.; de Pinho, M.N. Challenges of reducing protein-bound uremic toxin levels in chronic kidney disease and end stage renal disease. *Transl. Res.* **2021**, *229*, 115–134. [CrossRef]

Peinemann, K.V.; Nunes, S.P. (Eds.) *Membranes for Life Sciences*; Wiley-VCH: Weinheim, Germany, 2008.

Bonomini, V.; Berland, Y. (Eds.) *Dialysis Membranes: Structure and Predictions. International Workshop, Marseille, June 1994*; Contrib Nephrol: Basel, Karger, 1995; Volume 113, pp. 25–31. [CrossRef]

Drukker, W.; Parsons, F.M.; Maher, J.F. *Replacement of Renal Function by Dialysis*; Springer: Dordrecht, The Netherlands, 2004. [CrossRef]

D'Amico, G.; Colasanti, G. (Eds.) *Nephrology and Dialysis Updated. 18th Course on Advances in Nephrology and Dialysis, Milano, December 1986*; Contrib Nephrol: Basel, Karger, 1988; Volume 61, pp. 1–17. [CrossRef]

Guidance for the Content of Premarket Notifications for Hemodialysis Delivery Systems. Available online: https://www.fda.gov/media/116852/download (accessed on 27 October 2021).

Clark, W.R.; Hamburger, R.J.; Lysaght, M.J. Effect of membrane composition and structure on solute removal and biocompatibility in hemodialysis. *Kidney Int.* **1999**, *56*, 2005–2015. [CrossRef]

Lipps, B.J.; Stewart, R.D.; Perkins, H.A.; Holmes, G.W.; McLain, E.A.; Rolfs, M.R.; Oja, P.D. The Hollow Fiber Artificial Kidney. *Trans. Am. Soc. Artif. Organs* **1967**, *13*, 200–207.

0. Ward, R.A.; Klein, E.; Harding, G.B.; Murchison, K.E. Response of complement and neutrophils to hydrophilized synthetic membranes. *ASAIO Trans.* **1988**, *34*, 334–337.

1. Pedrini, L.A.; Krisp, C.; Gmerek, A.; Wolters, D.A. Patterns of Proteins Removed with High-Flux Membranes on High-Volume Hemodiafiltration Detected with a MultiDimensional LC-MS/MS Strategy. *Blood Purif.* **2014**, *38*, 115–126. [CrossRef]

2. Loeb, S.; Sourirajan, S. Sea Water Demineralization by Means of an Osmotic Membrane, in: Saline Water Conversion—II. *Am. Chem. Soc.* **1963**, *38*, 117–132. [CrossRef]

3. Kunst, B.; Sourirajan, S. An approach to the development of cellulose acetate ultrafiltration membranes. *J. Appl. Polym. Sci.* **1974**, *18*, 3423–3434. [CrossRef]

4. McCray, S.B.; Vilker, V.; Nobe, K. Reverse osmosis cellulose acetate membranes II. Dependence of transport properties on acetyl content. *J. Membr. Sci.* **1991**, *59*, 317–330. [CrossRef]

5. Ingram, A.J.; Parbtani, A.; Churchill, D.N. Effects of two low-flux cellulose acetate dialysers on plasma lipids and lipoproteins-a cross-over trial. *Nephrol. Dial. Transplant.* **1998**, *13*, 1452–1457. [CrossRef]

6. Hoenich, N.A.; Woffindin, C.; Matthews, J.N.S.; Goldfinch, M.E.; Turnbull, J. Clinical comparison of high-flux cellulose acetate and synthetic membranes. *Nephrol. Dial. Transplant.* **1994**, *9*, 60–66.

7. Hoenich, N.A.; Woffindin, C.; Cox, P.J.; Goldfinch, M.; Roberts, S.J. Clinical characterization of Dicea a new cellulose membrane for haemodialysis. *Clin. Nephrol.* **1997**, *48*, 253–259.

8. Weissinger, E.M.; Kaiser, T.; Meert, N.; De Smet, R.; Walden, M.; Mischak, H.; Vanholder, R.C.; for the European Uremic Toxin Work Group. Proteomics: A novel tool to unravel the patho-physiology of uraemia. *Nephrol. Dial. Transplant.* **2004**, *19*, 3068–3077. [CrossRef] [PubMed]

9. Speeckaert, M.M.; Glorieux, G.L.; Vanholder, R.; Van Biesen, W.; Taes, Y.E.; Clément, F.; Wehlou, C.; Delanghe, J.R. Vitamin D Binding Protein and the Need for Vitamin D in Hemodialysis Patients. *J. Ren. Nutr.* **2008**, *18*, 400–407. [CrossRef]

0. Elsen, R.; Silva, L.K. Advantages of meltspun membrane technology compared to solution spinning. *Contrib. Nephrol.* **1993**, *103*, 125–138. [CrossRef]

1. Edgar, K.J.; Buchanan, C.M.; Debenham, J.S.; Rundquist, P.A.; Seiler, B.D.; Shelton, M.C.; Tindall, D. Advances in cellulose ester performance and application. *Prog. Polym. Sci.* **2001**, *26*, 1605–1688. [CrossRef]

2. Gastaldello, K.; Melot, C.; Kahn, R.-J.; Vanherweghem, J.-L.; Vincent, J.-L.; Tielemans, C. Comparison of cellulose diacetate and polysulfone membranes in the outcome of acute renal failure. A prospective randomized study. *Nephrol. Dial. Transplant.* **2000**, *15*, 224–230. [CrossRef]

3. Vos, K.D.; Hatcher, A.P.; Merten, U. Lifetime of Cellulose Acetate Reverse Osmosis Membranes. *Ind. Eng. Chem. Prod. Res. Dev.* **1966**, *5*, 211–218. [CrossRef]

4. Mendes, G.; Faria, M.; Carvalho, A.; Gonçalves, M.C.; de Pinho, M.N. Structure of water in hybrid cellulose acetate-silica ultrafiltration membranes and permeation properties. *Carbohydr. Polym.* **2018**, *189*, 342–351. [CrossRef]

25. Faria, M.; Moreira, C.; Eusébio, T.; Brogueira, P.; De Pinho, M.N. Hybrid flat sheet cellulose acetate/silicon dioxide ultrafiltration membranes for uremic blood purification. *Cellulose* **2020**, *27*, 3847–3869. [CrossRef]
26. Peixoto, I.; Faria, M.; Gonçalves, M. Synthesis and Characterization of Novel Integral Asymmetric Monophasic Cellulose-Acetate/Silica/Titania and Cellulose–Acetate/Titania Membranes. *Membranes* **2020**, *10*, 195. [CrossRef] [PubMed]
27. Andrade, M.C.; Pereira, J.C.; de Almeida, N.; Marques, P.; Faria, M.; Gonçalves, M.C. Improving hydraulic permeability, mechanical properties, and chemical functionality of cellulose acetate-based membranes by co-polymerization with tetraethyl orthosilicate and 3-(aminopropyl)triethoxysilane. *Carbohydr. Polym.* **2021**, *261*, 117813. [CrossRef] [PubMed]
28. De Pascale, M.; Faria, M.; Boi, C.; Semiao, V.; de Pinho, M.N. The effect of ultrafiltration transmembrane permeation on the flow field in a surrogate system of an artificial kidney. *Exp. Results* **2021**, *2*, 16. [CrossRef]
29. Brinker, J.; Scherer, G. *The Physics and Chemistry of Sol-Gel Processing*; Gulf Professional Publishing: Oxford, UK, 1990.
30. Gonçalves, M.C. Sol-gel Silica Nanoparticles in Medicine: A Natural Choice. Design, Synthesis and Products. *Molecules* **2018**, *23*, 2021. [CrossRef]
31. Lui, A.; Talbot, F.D.F.; Fouda, A.; Matsuura, T.; Sourirajan, S. Studies on the solvent exchange technique for making dry cellulose acetate membranes for the separation of gaseous mixtures. *J. Appl. Polym. Sci.* **1988**, *36*, 1809–1820. [CrossRef]
32. Faria, M.; Liu, Y.; Leonard, E.F. Particle Spallation in a Microfluidic Blood Processing Device: The Problem of Using Peristaltic Pumps and Silicon-based Microfilters. *Int. J. Artif. Organs* **2017**, *40*, 589–593. [CrossRef]
33. Liu, Y.; Faria, M.; Leonard, E. Spallation of Small Particles From Peristaltic Pump Tube Segments. *Artif. Organs* **2016**, *41*, 672–677. [CrossRef] [PubMed]
34. Bird, R.B.; Stewart, W.E.; Lightfoot, E.N. *Transport Phenomena*, 2nd ed.; John Wiley&Sons Inc.: Chichester, UK, 2002.
35. Rosa, M.J. Separação Seletiva de Compostos Orgânicos de Correntes Aquosas por Ultrafiltração e Nanofiltração. Ph.D. Thesis, Instituto Superior Técnico, University of Lisbon, Lisbon, Portugal, 1995.
36. Gallant, R.W.; Railey, J.M. *Physical Properties of Hydrocarbons*, 2nd ed.; Gulf Publishing Company: Oxford, UK, 1970.
37. Mulder, M. *Basic Principles of Membrane Technology*; Springer: Dordrecht, The Netherlands, 1991. [CrossRef]
38. Vanholder, R.; De Smet, R.; Glorieux, G.; Argilés, A.; Baurmeister, U.; Brunet, P.; Clark, W.; Cohen, G.; De Deyn, P.P.; Deppisch, R.; et al. Review on uremic toxins: Classification, concentration, and interindividual variability. *Kidney Int.* **2003**, *63*, 1934–1943. [CrossRef] [PubMed]
39. Duranton, F.; Cohen, G.; De Smet, R.; Rodriguez, M.; Jankowski, J.; Vanholder, R.; Argiles, A.; on behalf of the European Uremic Toxin Work Group. Normal and Pathologic Concentrations of Uremic Toxins. *J. Am. Soc. Nephrol.* **2012**, *23*, 1258–1270. [CrossRef]
40. Filutowicz, Z.; Łukaszewski, K.; Pieszyński, K. Remarks on spectra-photometric monitoring of urea in dialysate. *J. Med. Inform. Technol.* **2004**, *8*, MM105–MM110.
41. ter Beek, O.; van Gelder, M.; Lokhorst, C.; Hazenbrink, D.; Lentferink, B.; Gerritsen, K.; Stamatialis, D. In vitro study of dual layer mixed matrix hollow fiber membranes for outside-in filtration of human blood plasma. *Acta Biomater.* **2021**, *123*, 244–253. [CrossRef]
42. Montaseri, H.; Khajehsharifi, H.; Yousefinejad, S. UV Determination of epinephrine, uric acid, and acetaminophen in pharmaceutical formulations and some human body fluids using multivariate calibration. *Química Nova* **2014**, *37*, 1404–1409. [CrossRef]
43. Bradford, M.M. A rapid and sensitive method for the quantitation of microgram quantities of protein utilizing the principle of protein-dye binding. *Anal. Biochem.* **1976**, *72*, 248–254. [CrossRef]
44. Zor, T.; Selinger, Z. Linearization of the Bradford Protein Assay Increases Its Sensitivity: Theoretical and Experimental Studies. *Anal. Biochem.* **1996**, *236*, 302–308. [CrossRef]
45. Murphy, D.; de Pinho, M.N. An ATR-FTIR study of water in cellulose acetate membranes prepared by phase inversion. *J. Membr. Sci.* **1995**, *106*, 245–257. [CrossRef]
46. Figueiredo, A.S.; Garcia, A.R.; Minhalma, M.; Ilharco, L.; de Pinho, M. The Ultrafiltration Performance of Cellulose Acetate Asymmetric Membranes: A New Perspective on the Correlation with the Infrared Spectra. *J. Membr. Sci. Res.* **2020**, *6*, 70–80. [CrossRef]
47. Nyquist, R. (Ed.) Chapter 8—Aliphatic Amines. In *Interpreting Infrared, Raman, and Nuclear Magnetic Resonance Spectra*; Academic Press: San Diego, CA, USA, 2001; pp. 143–148. [CrossRef]
48. Albalate, M.; Pérez-García, R.; De Sequera, P.; Corchete, E.; Alcázar, R.; Ortega, M.; Puerta, M. Is it useful to increase dialysate flow rate to improve the delivered Kt? *BMC Nephrol.* **2015**, *16*, 20. [CrossRef] [PubMed]
49. Chang, K.Y.; Kim, S.-H.; Kim, Y.O.; Jin, D.C.; Song, H.C.; Choi, E.J.; Kim, Y.K.; Kim, Y.-S.; Kang, S.-W.; Kim, N.-H.; et al. The impact of blood flow rate during hemodialysis on all-cause mortality. *Korean J. Intern. Med.* **2016**, *31*, 1131–1139. [CrossRef] [PubMed]
50. Tattersall, J.; Martin-Malo, A.; Pedrini, L.; Basci, A.; Canaud, B.; Fouque, D.; Haage, P.; Konner, K.; Kooman, J.; Pizzarelli, F.; et al. EBPG guideline on dialysis strategies. *Nephrol. Dial. Transplant.* **2007**, *22*, ii5–ii21. [CrossRef] [PubMed]
51. Haynes, W.M. *CRC Handbook of Chemistry and Physics*, 97th ed.; CRC Press: Boca Raton, FL, USA, 2017.
52. Kroll, M.; Hellums, J.; McIntire, L.; Schafer, A.; Moake, J. Platelets and shear stress. *Blood* **1996**, *88*, 1525–1541. [CrossRef]
53. Holme, P.A.; Ørvim, U.; Hamers, M.J.A.G.; Solum, N.O.; Brosstad, F.R.; Barstad, R.M.; Sakariassen, K.S. Shear-Induced Platelet Activation and Platelet Microparticle Formation at Blood Flow Conditions as in Arteries With a Severe Stenosis. *Arterioscler. Thromb. Vasc. Biol.* **1997**, *17*, 646–653. [CrossRef]

Flythe, J.; Kimmel, S.E.; Brunelli, S.M. Rapid fluid removal during dialysis is associated with cardiovascular morbidity and mortality. *Kidney Int.* **2011**, *79*, 250–257. [CrossRef]

Magee, C.C.; Tucker, J.K.; Singh, A.K. (Eds.) *Core Concepts in Dialysis and Continuous Therapies;* Springer: Boston, MA, USA, 2016. [CrossRef]

Belinskaia, D.; Voronina, P.; Batalova, A.; Goncharov, N. Serum Albumin. *Encyclopedia* **2020**, *1*, 65–75. [CrossRef]

membranes

MDPI

Review

Overview of Membrane Science and Technology in Portugal

Liliana C. Tomé [1], Diogo M. F. Santos [2], Svetlozar Velizarov [1], Isabel M. Coelhoso [1,*], Adélio Mendes [3], João G. Crespo [1] and Maria Norberta de Pinho [2]

[1] LAQV-REQUIMTE, Department of Chemistry, NOVA School of Science and Technology, FCT NOVA, Universidade NOVA de Lisboa, 2829-516 Caparica, Portugal; liliana.tome@fct.unl.pt (L.C.T.); s.velizarov@fct.unl.pt (S.V.); jgc@fct.unl.pt (J.G.C.)

[2] Center of Physics and Engineering of Advanced Materials (CeFEMA), Laboratory for Physics of Materials and Emerging Technologies (LaPMET), Chemical Engineering Department, Instituto Superior Técnico, Universidade de Lisboa, 1049-001 Lisbon, Portugal; diogosantos@tecnico.ulisboa.pt (D.M.F.S.); marianpinho@tecnico.ulisboa.pt (M.N.d.P.)

[3] LEPABE—Laboratory for Process Engineering, Environment, Biotechnology and Energy, Faculty of Engineering, University of Porto, Rua Dr. Roberto Frias, 4200-465 Porto, Portugal; mendes@fe.up.pt

* Correspondence: imrc@fct.unl.pt

Abstract: Membrane research in Portugal is aligned with global concerns and expectations for sustainable social development, thus progressively focusing on the use of natural resources and renewable energy. This review begins by addressing the pioneer work on membrane science and technology in Portugal by the research groups of *Instituto Superior Técnico—Universidade de Lisboa* (IST), *NOVA School of Science and Technology—Universidade Nova de Lisboa* (FCT NOVA) and *Faculdade de Engenharia—Universidade do Porto* (FEUP) aiming to provide an historical perspective on the topic. Then, an overview of the trends and challenges in membrane processes and materials, mostly in the last five years, involving Portuguese researchers, is presented as a contribution to a more sustainable water–energy–material–food nexus.

Keywords: membrane processes; membrane synthesis and characterisation; integrated (hybrid) solutions; sustainable approaches; circular economy; water–energy–material–food nexus; Portuguese membrane research

1. Introduction

Over the last few decades, considerable effort has been devoted to developing better membranes and extending their range of applications to different areas. Membrane processes already have an established role in gas separation and water treatment, and their applications in the food, pharmaceutical, and health areas have been continuously increasing.

In the last few years, membrane processes proved to have a key role in biorefinery and bioenergy production processes, namely for process intensification and the recovery and purification of valuable products. Membranes are also a crucial component of electrochemical energy conversion devices, including fuel cells and electrolysers.

Moreover, the growing environmental concerns have drawn attention to the use of fossil-based polymers and toxic solvents for membrane fabrication. Therefore, the development of new membranes, using polymers from renewable sources and more sustainable fabrication methods, is being pursued.

In Portugal, active membrane-related research is now being performed in a variety of domains, ranging from the production of high-quality drinking water to agriculture and irrigation use, the recovery of valuable compounds from natural sources and industrial effluents, and membrane reactors to energy generation and storage, as well as the integration of membranes and membrane technology in carbon capture and utilisation strategies.

This review begins by addressing the pioneer work on membrane science and technology in Portugal by the research groups of *Instituto Superior Técnico—Universidade de*

Lisboa (IST), *NOVA School of Science and Technology—Universidade Nova de Lisboa* (FCT NOVA) and *Faculdade de Engenharia—Universidade do Porto* (FEUP) aiming to provide an historical perspective on the topic. Then, an overview of the trends and challenges in membrane processes and materials, mostly in the last five years, involving Portuguese researchers, is presented as a contribution to a more sustainable water–energy–material–food nexus.

2. Historical Perspective on Membrane Science and Technology

2.1. *Instituto Superior Técnico—Universidade de Lisboa (IST)*

In early times, synthetic membranes were often associated with filtration and separation unit operations and, on a smaller scale, with a myriad of applications, such as battery separators, biosensors, controlled release systems, medical devices, etc. The huge developments in the 1960s and 1970s at the level of polymeric membranes and membrane modules allowed them to be implemented on an industrial scale of a vast number of membrane processes that required the mechanisms underlying the performance of this enormous diversity of membrane materials, structures and processes to be approached systematically. This was the driving force that moved Peter Bungay (National Institutes of Health), Harry Lonsdale (Founder of Journal of Membrane Science) and Maria Norberta de Pinho (IST) to organise (June 26–July 8) a NATO Advanced Study Institute (ASI) on "Synthetic Membranes: Science, Engineering and Applications" in Portugal in 1983. The book edited by the Institute organisers [1], in addition to numerous citations in Google Scholar, earned the following comment in Polymer News: "an outstanding contribution to the educational aspects of membrane science. In fact, if there are senior-level courses on membrane science, this is, the 'textbook' to use". This NATO–ASI congregated and interconnected the major membrane groups of Europe and the U.S.A. and was the precursor of the "European Society for Membrane Science and Technology".

By that time, the Membrane group of the Chemical Engineering Department of Instituto Superior Técnico (IST), under the guidance of Maria Norberta de Pinho, progressed into two main lines of research: (1) Synthesis and Characterisation of Polymeric Membranes; (2) Design and Optimisation of Membrane Processes in Water, Food, Pulp and Paper, Cork and Chemical Industries.

From the first line of research, it is of particular interest to mention the synthesis of (1.1) Bi-Soft Segment Polyurethane Membranes for Pervaporation [2], which were further investigated (a project sponsored by the Portuguese Foundation for Science and Technology (FCT), 2003–2006) in terms of different soft segments towards the development of hemocompatible membranes for extracorporeal blood oxygenation [3], being the object of European (WO 2008/041126), U.S.A. (US2010/0111761) and Singapore (GST registration No: 200004489C) patents; (1.2) Integral Asymmetric Cellulose Acetate/Silica Hybrid Membranes for the selective removal of blood toxins in the Artificial Kidney [4]; and (1.3) Nanocomposite Membranes of Cellulose Acetate/Silver Nanoparticles and Cellulose Acetate/Silver Ion Exchanged Zeolites with bactericide activity for biofouling control [5].

In the second line of research, the design of membrane hybrid processes was carried out in the framework of European and national projects with industrial partners such as those coordinated by IST (scientific coordination of Maria Norberta de Pinho): "Treatment and Fractionation of Bleach Plant Effluents by Membrane Based Processes" (BRITE-EURAM(DGXII-EU)(1991–1994)) [6,7]; "Integrated Process for Purification/Recovery of Ammonium Containing Water in Coke Plants" (EC(DG XVII)/ESSC Coal Research (1997–2000)) [8,9]; "Valorization & Treatment of the Cork Industry Wastewaters" (JNICT/DGA(1996–1999)) [10]; "Wine Must Concentration and Rectification by Nanofiltration and Electrodialysis" (FCT/POCTI(2005–2008)) [11].

The assessment of tannic-ultrafiltration (UF) membranes' interactions and membrane fouling characteristics have led to the development of integrated processes of flocculation/flotation/UF and ozonation/UF for tannins recovery from cork processing wastewaters [10,12,13]. Furthermore, nanofiltration (NF) concentrates were used as vegetal tanning agents in the leather industry [14]. UF and NF have also been investigated for the tertiary treatment of leather industry effluents [15–17]. In the pulp and paper industry, pioneering work was carried out on water recovery by the NF of bleach plant effluents and the development of hybrid processes of NF/ED and flotation/UF [6,7,18–22].

The prediction of the membrane processes performance through the modelling of the selective mass transfer was carried out in the membranes (surface forces/pore flow model) and at the membrane/fluid interfaces (quantification of concentration polarisation through mass transfer coefficients) for pressure-driven and pervaporation processes [23–25]. Flow management and mass transfer in spiral-wound modules with recourse to computational fluid dynamics (CFD) were given special attention by the team of Viriato Semião, Vitor Geraldes and Maria Norberta de Pinho [26]. The membrane characterisation by spectroscopic (attenuated total reflection-Fourier-transform infrared spectroscopy (ATR-FTIR), X-ray photoelectron spectroscopy (XPS)) and microscopic (scanning electron microscopy (SEM), transmission electron microscopy (TEM), atomic force microscopy (AFM)) techniques provided data that were correlated to the membrane selective permeation properties [27,28].

From 2004 to 2009, the membrane groups of IST and Nova School of Science and Technology—*Universidade NOVA de Lisboa* (FCT NOVA) were members of the European Network of Excellence (Sixth Framework Programme): "Expanding Membrane Macroscale Applications by Exploring Nanoscale Material Properties", coordinated by Gilbert Rios at Institut Européen des Membranes. Divulgation activities were carried out, such as the "Membranes in Medicine" workshop.

The present research at IST addresses: (1) The role of water dynamics in the membrane pores (investigated through deuterium and proton nuclear magnetic resonance (NMR)) on the selective permeation properties; (2) UF/NF membranes with bactericide activity and biofouling control in drinking water production; (3) Water recovery and sub-products valorisation in cork and food industries, namely polyphenols and polysaccharides in the wine industry; (4) Case-specific mass transfer correlations; (5) Minimisation of concentration polarisation and fouling prevention; (6) Microfluidics and membrane microdevices for the artificial kidney and artificial lung.

2.2. NOVA School of Science and Technology—Universidade Nova de Lisboa (FCT NOVA)

Membrane research at FCT NOVA started with the PhD work of João Crespo, which was focused on the study of different bioreactors for the simultaneous production of organic acids and vitamin B12 [29]. The first contact with membranes was related to the development of membrane bioreactors in a period (1980s) where this technology was in its initial stages. We look back with a smile when we remember that João complained to his supervisor that nobody was working, or interested, on that subject. Supervisors are sometimes right! The subject of membrane bioreactors has been a hot subject in the group over the years, namely through exploring other mechanisms of transport, as in the ion-exchange membrane (bio)reactor concept, which we proposed for the first time together with Maria Reis [30], and further developed with Svetlozar Velizarov [31–33], and the development of membrane biphasic enzymatic reactors in collaboration with Elias Klein and Carlos Afonso [34,35].

In the first years of the group (early 1990s), the collaboration with Karl Böddeker (GKSS, Geesthacht)—with whom we collaborated for many years working on organophilic pervaporation (Ph.D. project of Thomas Schäfer) [36,37]—and later with John Howell (University of Bath) [38] on the concept of critical/sustainable flux for the fractionation of proteins, was extremely inspiring and fueled our research for the coming years.

The membrane group at FCT NOVA grew significantly over the years, hopefully with a few relevant contributions for the community, namely on the understanding of transport mechanisms in membrane contactors (including a pioneer work on the use of ionic liquids in membrane contactors, which started with the PhD project of Isabel Coelhoso [39–41] and later with Raquel Fortunato [42,43] and Luísa Neves [44,45]) and, more recently, new approaches to membrane contactors for protein crystallisation [46] and production of nanoemulsions [47]. We have a particular interest in the development of techniques for monitoring membrane processes, using online and real-time approaches (namely by Mass Spectrometry, started by Thomas Schäfer [48] and continued by Carla Brazinha [49,50]), and using fluorescence techniques, namely through natural fluorescence (initial work with Gundula Wolf [51,52] and Carla Portugal [53,54], continued by Cláudia Galinha [55]) or using specific molecular probes. Soon, we concluded that machine learning approaches are essential for dealing with complex monitoring signals and large data sets (see the work of Gundula Wolf and Cláudia Galinha [56,57]). More recently, we became interested in the development of membranes with a designed surface topography (work of Pavel Izak [58] and later of Sylwin Pawlowski [59]), aiming to improve mass and heat transfer at the membrane surface (where it matters!), and the use of natural (bio)polymers for membrane manufacturing.

The membrane group of NOVA is now hosted in two different locations: at the research centre, Associated Laboratory for Green Chemistry (LAQV-REQUIMTE, the largest research centre in the country, which is focused on research and education in Sustainable Chemistry and Processes) at the faculty campus in Caparica; and at the private not-for-profit research centre iBET, in Oeiras, where the lab directly coordinated by Vanessa Pereira works on the integration of (photocatalytic)membranes (with a relevant contribution on membranes development by Rosa Huertas) with advanced oxidation processes, aiming the elimination of resilient organic micropollutants from water/wastewater streams [60,61].

The research group's involvement in education programmes should be emphasised, namely the European Master programme EM3E, later EM3E-4SW, coordinated by Andre Ayral, and the Doctoral programme, Eudime, coordinated by Enrico Drioli.

2.3. *Faculdade de Engenharia—Universidade do Porto (FEUP)*

Since the end of the 1990s, Adélio Mendes' group at the University of Porto has been developing research on several topics of membrane separation technology. Since then, this group contributions to the membrane-related field have been based on four pillars: development of materials for the fabrication of membranes, such as carbon molecular sieve membranes [62–67], palladium and ceramic membranes [68]; study of membrane processes, such as membrane reactors for propyne hydrogenation [69] and methanol steam reforming [70–72], direct methanol fuel cells (DMFCs) [73,74] and proton exchange membrane (PEM) fuel cells [75]; fundamental studies, namely the development of phenomenological models and simulators for membrane separation processes [76,77], membrane reactors [78–80], PEM fuel cells [81,82], redox flow batteries [83,84]; and technological projects, namely the development of membrane processes for recovering cutting oil and extraction of beer flavours for introducing in de-alcoholised beer for partially recovering the original flavour [85–88].

The first membrane-related article of the group at the University of Porto was published in 2002 [89] and concerned the removal of volatile organic compounds, such as acetone, ethyl acetate and ethanol from the air, using GKSS (Geesthacht, Germany) polyetherimide (PEI) membranes coated with polydimethylsiloxane (PDMS). Regarding the preparation and characterisation of carbon molecular sieve membranes, the first work was published in 2004 by Lagorsse et al. [62]. Brandão et al. [90] studied the mass transport in composite PDMS membranes loaded with palladium nanoparticles. Several studies on the use of hollow fibre membranes in membrane contactors with amino acid salt solutions were also published for CO_2 removal in anaesthetic circuits [91–93]. In 2004, Janknecht et al. [85] studied the use of micro and ultrafiltration membranes for the removal of industrial cutting oil from oil emulsions.

The first Portuguese contribution to the field of direct methanol fuel cells membranes was reported by Silva et al. in 2005 [74]. Further work was developed in collaboration with Suzana Nunes from GKSS, who prepared sulfonated poly (ether ether ketone) (sPEEK) composite membranes with zirconium oxide for DMFC. Composite sPEEK membranes for DMFC with different additives were studied and compared with commercial Nafion® membranes (DuPont, Wilmington, DE, USA) [94,95]. Brandão et al. [96] modified these

Nafion® membranes with Pd composite nanoclusters to reducemethanol crossover in DMFC. Concerning the development of PEMs for fuel cells, Boaventura et al. [75] prepared and characterised proton-conducting sulfonated polytriazole membranes. Studies on the utilisation, modification, activation and ageing of PEM fuel cells have also been developed [97–102].

Mendes' research group's future perspectives of CMS membranes separation technology are focused on developing a strategy to deliver this technology as quickly as possible for widespread use. A strategy to control the pore size for the different separations with industrial relief, such as the enrichment of oxygen or nitrogen from the atmospheric air, the separation of hydrogen from different streams, the separation of carbon dioxide from natural gas or biogas, among others, is currently being developed. Moreover, the use of supported hollow fibre CMS membranes is being investigated. In the future, pervaporation will also be a very relevant research topic in this technology.

Driven by the will of curbing down the notorious impact of burning fossil fuels on the global climate, currently, in Portugal, there are two spin-off companies from the University of Porto addressing membrane processes: SEA+TECH, an environmental, energy and engineering consulting company, which aims to develop energy-efficient and sustainable systems for the production, treatment, and recycling of water, for commercial or industrial applications; and CarboPora—a company targeting the development, production and commercialisation of membrane gas separation processes, namely oxygen-enriched air, CO_2 removal from flue gas steams, methane separation/purification from biogas and hydrogen separation/purification from natural gas mixtures.

3. Activities on Membrane Science and Technology

This section describes the most recent activities on membrane processes and materials developed by Portuguese researchers in water treatment, food and health, energy and gas separation.

3.1. Water

Due to its extreme importance for the societies and ecosystems, and being at the core of their sustainable development, it is not surprising that membrane-assisted related studies for water treatment and reuse are a priority segment of the activities performed by Portuguese researchers. This is reflected by the number of studies published in the area of water treatment during the period 2017 to 2021 (13 articles related to drinking water and 41 articles related to various wastewater treatment applications by RO, NF and/or UF (Scopus)). Thus, a clear trend favouring wastewater-oriented applications is documented. This trend can be associated obviously with ever-increasingly stringent legislation criteria for wastewater discharge into the environment and with the quest for circular economy requiring water recycling and reuse. In this regard, NF (with 24 published articles) allowing a better trade-off between size-based and charge-based exclusion separation mechanisms, has attracted more attention.

3.1.1. Drinking Water

Membrane processes (especially reverse osmosis (RO)) have proven to provide high-quality drinking water worldwide, using various water sources such as groundwater, surface water, and seawater, the latter especially in countries with limited in-land water bodies. In mainland Portugal, the water surface and groundwater sources are generally sufficient, and the drinking water-producing plants have well-developed infrastructure, allowing them to satisfy the required water demands. However, on the island of Porto Santo, an island of the Madeira Archipelago, the natural groundwater, especially during summer, become insufficient and therefore seawater desalination is applied. The plant includes four RO units with a total capacity of 6800 m^3/d, operating at 50–65 bar [103]. In the framework of the operation of the Porto Santo plant, Geraldes et al. [104] carried out a techno-economic analysis of such medium-size RO plants for sea water desalination.

Although it has a much smaller capacity of less than 1000 m^3/d, RO-treated seawater has been used for irrigation and filling swimming pools in some hotels in Alvor (Algarve), thus saving the available drinking water resources for human consumption.

UF has proven to be an efficient pretreatment process for surface water before its use as a drinking water source. In Portugal, a SUEZ UF plant installed by the company Suez S.A. (formerly Suez *Environnement*) in Santa Eufémia (based north of Porto) treats water (with a capacity of 24,000 m^3/d) from the nearby Ave River. It is operated by Vimágua, the municipal authority for water management in this region. This plant is one of the largest installations of Suez ZeeWeed 1500 modules in Europe, providing high-quality drinking water to roughly 80,000 people in the Guimarães and Vizela regions.

Since membrane processes provide several advantages, such as modularity, flexibility, ability to construct small size plants when compared to thermal desalination processes, both experimental and modelling studies have been performed by Portuguese researchers. Significant pioneering contributions include the preparation, by-phase inversion, and characterisation of dense and integrally skinned cellulose acetate (CA) and cellulose acetate butyrate membranes [28], and the development of an integrated model based on the finite volume formulation to numerically simulate the fluid flow in spiral-wound RO modules [26]. Afonso et al. [105] demonstrated that RO operated at 20–30 bar can highly reduce the content of organic and inorganic compounds (rejections >98.5%) present in raw waters brackish groundwater samples collected from the Zarqa basin, Jordan, at the cost of less than 0.30 EUR/m^3.

Sanches et al. [106] studied the feasibility of NF removing pesticides and hormones from spiked distinct real water sources (surface water and groundwater) supplied by the drinking water company *Empresa Portuguesa das Águas Livres, S.A.* that treats and provides drinking water to approximately 2.8 million people in the region of Lisbon, and the effect of the water ionic composition on the rejection of target micropollutants. It was found that the NF efficiency in removing the selected compounds was not considerably affected by the conditions (static and or dynamic) of their pre-adsorption on the membrane.

The current trend in water-related membrane research is to develop integrated (hybrid) processes, which have already become preferred choices in several water treatment or usage applications. Exploring possible synergisms, benefiting from distinct process mechanisms, often leads to the development of novel, more efficient, and sustainable technologies to be applied in various domains, ranging from the production of clean water for human consumption to aquaculture and irrigation [107,108].

An integrated process (NF followed by direct or indirect UV photolysis) developed by Pereira et al. [60] has proven to be efficient for removing hormones with endocrine disruption capabilities from a real surface water matrix, thus guaranteeing the production of water of extremely high chemical quality.

A two-stage treatment to produce drinking water involving ion exchange and NF has been investigated [109]. In the first stage, the anion-exchange resin was used. A mathematical model that considers axial dispersion for the liquid phase predicted the breakthrough curves of Cl^- and $SO_4{}^{2-}$. The selectivity of separating Cl^-, Na^+, and $SO_4{}^{2-}$ ions by NF was assessed in the second stage.

Costa et al. [110–112] systematically investigated natural organic matter (NOM) removal and surface water treatment by UF and NF for the production of high-quality drinking water. More recently, the research group headed by Maria João Rosa at *Laboratório Nacional de Engenharia Civil* (LNEC) investigated powdered activated carbon (PAC)/coagulation/ceramic microfiltration (PAC/Alum/MF) treatment (Figure 1) on a pilot scale for treating low turbidity and low NOM surface waters spiked with various organic microcontaminants [113]. This treatment proved to be a reliable option for effectively removing turbidity (to below 0.03 NTU) while maintaining low residual levels of aluminium, which is an important advantage over the PAC conventional application.

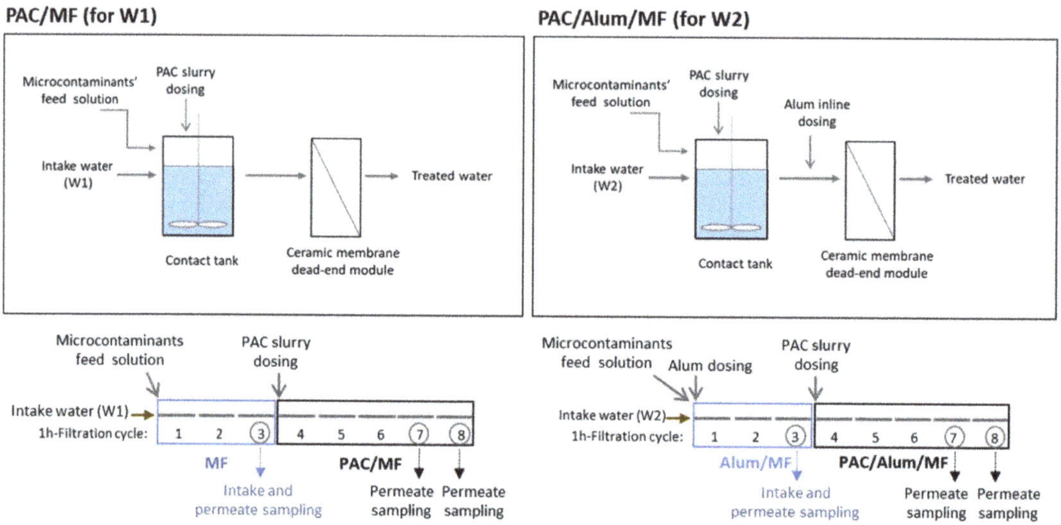

Figure 1. Scheme of powdered activated carbon/(coagulation)/microfiltration process, i.e., PAC/MF for intake water W1 and PAC/Alum/MF for intake water W2 (top) and of the respective spiking trials' procedure (bottom) (8 × 1 h filtration cycles, with 3 cycles without PAC and 5 cycles with PAC dosing). Reprinted with permission from Campinas et al. [113]. Copyright 2021, MDPI.

A novel integrated drinking water treatment process referred to as ion-exchange membrane bioreactor (IEMB) has been introduced and comprehensively studied for the removal of various toxic inorganic anions (e.g., nitrate, perchlorate, bromate, arsenate, etc.) and heavy metals such as ionic mercury. The process concept is based on Donnan dialysis (DD) principles and consists of integrating Donnan dialytic transport of the target pollutant with its simultaneous precipitation, or selective biotransformation by suitable microbial culture to harmless metabolic products, in a separate receiver compartment [114]. The process operation was successfully validated with groundwater from the region of *Ponte de Sor*, Portugal, which is naturally polluted with arsenic. Its performance was recently optimised using the response surface methodology (RSM) approach, allowing the required operating conditions to be selected, which guarantee that the arsenic concentration in the treated water is below the maximum contamination level (MCL) allowed [115].

3.1.2. Wastewater

Industrial Effluents

Regarded as a strategic pillar of the Portuguese economy, the agro-industrial sector (comprising primary agricultural production and the food and drinks industries) is extremely important and strongly contributes to the country's rising productivity and competitiveness at the macroeconomic level [116]. In this regard, it is not surprising that the valorisation of the residues and by-products, and purification before discharge or re-use of the various aqueous effluents generated by this sector has gained considerable attention. Therefore, in this overview, they will be discussed in more detail. For wastewater originating from other industries, only two recent examples of studies performed by Portuguese researchers will be briefly presented.

The role of membrane processing in the agro-industry is discussed in detail in [117,118], and a few representative recent examples are briefly presented.

As an olive oil-producing country, Portugal generates a considerable amount of olive mill wastewater (OMW), which requires purification before discharge into natural water bodies and possible use for irrigation or re-use as process water.

Fraga et al. [119] reported extremely high degrees of total suspended solids and oil and grease removal when olive mill wastewaters were filtered using new silicon carbide tubular membranes. These membranes were used at constant permeate flux to treat real OMW at the pilot scale. Recently, the same researchers designed a new submerged membrane reactor with photocatalytic membranes prepared using the sol-gel technique that can be easily scaled up [120]. The proposed reactor is a promising approach to purify wastewaters of different origins. Combining inorganic membrane technologies with other novel methods such as advanced oxidation processes (e.g., using engineered nanomaterials) can be considered among the best options for dealing with highly polluted effluents [121].

In this regard, solvent cast and electrospun photocatalytic and antimicrobial membranes of poly(vinylidene fluoride-co-hexafluoropropylene) (PVDF-HFP) were prepared with different concentrations of TiO_2 and Ag–TiO_2 to produce a multifunctional material [122]. The recyclability of the membranes has also been demonstrated.

UF, either directly [123] or in combination with biological oxidation of the organics, present has been used for partial purification of OMW. In the latter case, a jet-loop aerobic bioreactor coupled with a UF membrane with a cut-off of 300 kDa is referred to as JACTO. MBR was designed, scaled up to 1000 L and tested in Portugal and Jordan [124]. To fulfil the legislation for irrigation in Jordan or for discharging to municipal wastewater treatment plants in Portugal, the treated OMW (UF permeate) in Portugal, an additional NF(RO) treatment should be applied to decrease the TOC values below the maximum allowed limits.

Using RO with a BW30 membrane, Antónia Nunes et al. obtained an acceptable TOC level in the RO permeate [125]. Still, its pH of 5.6 should be adjusted (increased) to comply with the acceptable range of 6.5–8.4 to be used as irrigation water.

Cork processing wastewater is another environmental problem due to its high organic matter content, such as sugars and non-biodegradable compounds such as polyphenols, namely tannins.

UF pre-treatment using cellulose acetate ultrafiltration membranes prepared by phase inversion was performed in concentration and diafiltration modes [126]. Still, to be used for irrigation, NF (RO) treatment of the UF permeates would be required.

NF and/or RO performances have been recently investigated for other types of agro-industrial wastewaters, such as NF for winery [127] and UF/RO for dairy wastewater [128]. In the latter study, the authors suggest using the RO permeates for cleaning purposes.

Portuguese researchers have also investigated the performance of membrane processes applied for purification and/or reuse of wastewater originating from other industries. Only two very recent examples of studies performed with real industrial effluents will be referred to in the following.

Having to constantly rely on access to great amounts of water, the oil and gas industry has sought opportunities to reclaim water using sustainable and efficient management procedures. An on-site pilot plant treatment study on real refinery stripped sour water using RO was conducted to remove the primary pollutant blocking the possibility for water reuse in the refinery cooling tower, phenols [129]. A technical and energy demand analysis was performed, in which 90% of the initial wastewater was recovered without loss of permeate quality. It was found that installing an energy recovery device with the RO plant can decrease the operating costs from EUR 1.37 to EUR 0.41 per m^3 of treated water.

A pre-filtered real tannery effluent (collected from a plant in Algeria) was subjected to NF with (NF270 or NF90 membranes) membranes or RO (BW30 and SW30) membranes to compare their performance [130]. It was found that the RO permeate obtained with the SW30 membrane conforms to local Algerian regulations regarding being discharged directly into the natural environment (in this case, Reghaia Lake) or into urban sewers linked to wastewater biological treatment stations.

Sewage (Domestic Wastewater)

These water matrices contain liquid effluents from households and businesses, as well as pretreated or not industrial wastewater effluents, which makes their direct membrane-assisted purification extremely challenging. Therefore, membrane processes are generally combined with other treatment processes such as adsorption, precipitation, coagulation, oxidation, etc. Portuguese researchers have been actively developing and designing such integrated treatment technologies.

At LNEC, Viegas et al. [131] investigated the long-term performance of hybrid adsorption with Powdered Activated Carbon (PAC)—coagulation (with $FeCl_3$)–ceramic membrane filtration (HCMF) process operated with a sand-filtered secondary effluent at pilot scale in two test sites, namely Frielas and Beirolas wastewater treatment plants, which are two of the biggest ones in Lisbon metropolitan area, with average daily flowrates of 54,500 m^3/day and 70,000 m^3/day, respectively. A detailed cost analysis of the process, as a function of the plant flow rate, was performed pointing to total production costs of 0.21 EUR/m^3 for a 50,000 m^3/day plant and 20 years' membrane lifetime, with a higher share of OPEX (0.17 EUR/m^3), mainly associated with equipment/membranes replacement and reagents costs than with CAPEX (0.04 EUR/m^3). The membrane lifespan was shown to be a key factor in the costs.

An integrated hybrid system coupling microfiltration (MF) with UVA photolysis (UVA-MF) or TiO_2-P25 photocatalysis (UVA/TiO_2-MF) was investigated for the removal of oxytetracycline (OTC) from a real secondary effluent taken from a municipal wastewater treatment plant [132]. A significant reduction in the permeate flux and OTC removal degree was documented due to membrane fouling. Therefore, the authors pointed out that further research should focus on the possible addition of oxidants (H_2O_2, persulfate, or ozone) to improve the process performance.

NF was applied to pre-treated (filtered first with 1.2 μm filters and then with 0.45 μm polyamide membrane) domestic wastewater collected after the biological treatment and prior to the disinfection step and spiked with target antibiotics and viral genomes (Figure 2) [133]. Since high rejections were obtained, the authors consider NF a promising tertiary treatment option. In another study, Cristovão et al. [134] demonstrated the effectiveness of NF to remove several widely used anticancer drugs from hospital and wastewater effluents.

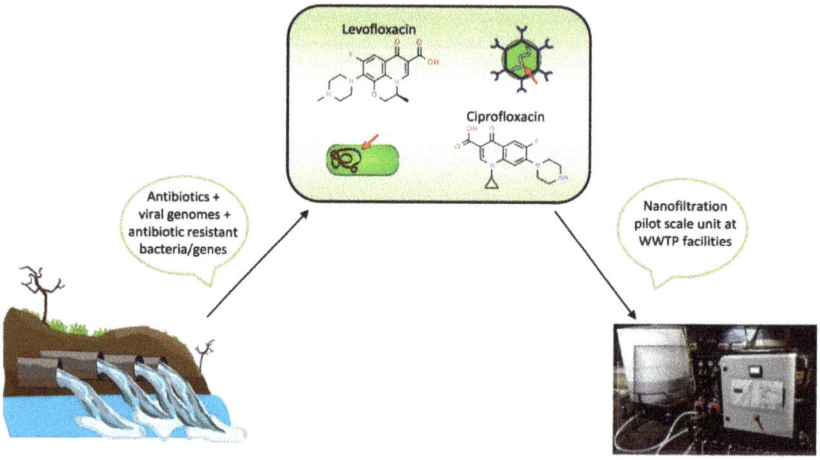

Figure 2. On-site removal of antibiotics, antibiotic resistance genes and viral genomes from a domestic wastewater effluent by a nanofiltration pilot unit. Reprinted with permission from Cristóvão et al. [133]. Copyright, MDPI.

A combination of advanced oxidation technologies and direct contact membrane distillation (DCMD) was investigated to treat municipal wastewater collected from the secondary clarifier of a municipal wastewater treatment plant located in Northern Portugal, which receives both domestic and industrial effluents [135]. It was found that the dissolved organic carbon content in the wastewater strongly affects the permeate fluxes achieved by DCMD. In this regard, the authors concluded that a combination of ozonation and DCMD is more appropriate for wastewaters with low dissolved organic carbon DOC contents (<15 mg/L).

3.2. *Membrane Applications in Food and Health*

3.2.1. Membrane Processes in Biorefinery

The biorefinery approach has enhanced the valorisation of agro-wastes reducing the environmental impact and creating added value products using sustainable practices in line with the principles of the circular economy. Membrane processes play a key role in bioactive compounds fractionation. They are an excellent alternative to conventional methodologies due to their low energy and no-additives requirements, mild conditions of temperature and pressure and easy scale-up. Pressure-driven membrane processes (microfiltration (MF), ultrafiltration (UF), nanofiltration (NF), and reverse osmosis (RO)) were largely explored by Portuguese researchers for the separation, purification, and concentration of bioactive compounds from several agri-food by-products, including grape pomaces [136,137], corn fibre and Eucalyptus bark [138,139].

Among wastes, grape pomaces are a valuable source of polyphenolic compounds (flavonoids, tannins, and benzoic acids derivatives), which can be further used for innovative formulations in the pharmaceutical and food industries. Arboleda et al. [136] evaluated the feasibility of phenolic compounds recovery from red grape pomace through a combination of ultrasound-assisted enzymatic extraction, followed by a membrane separation process. Different cellulose acetate membranes were prepared and tested. The CA400-22 membrane presented high retention coefficients towards phenolic compounds and proanthocyanidins, 73% and 92%, respectively, and exhibited low retention values for glucose and fructose (rejection values of 19.5% and 12.5%, respectively). The results indicate that combining an ultrasound-assisted enzymatic extraction with tailor-made cellulose acetate membranes can be a sustainable approach for recovering valuable fractions from red grape pomace.

Syed et al. [137] proposed an effective production of a grape pomace extract enriched in bioactive monomeric flavan-3-ols by an extraction process using 40 wt.% ethanol/water followed by dia-nanofiltration processing with a Duramem 900 membrane. The optimised fractionated extract was further concentrated by RO, allowing for the re-use of the hydroethanolic washing solution.

Giacobbo et al. [140,141] studied the recovery of wine polysaccharides and polyphenols through the fractionation of wine lees by pressure-driven membrane processes.

Corn fibre, a co-product of the starch industry, is rich in compounds with high added value, such as ferulic acid and arabinoxylans, released during alkaline extraction. Ferulic acid presents several benefits, namely antioxidant, anti-inflammatory, antiatherogenic and antiproliferative properties, making this compound widely used in the food, cosmetics, and pharmaceutical industries. Moreover, it can be used as a precursor of bio-vanillin production. Ultrafiltration was tested for separating ferulic acid from arabinoxylans due to the high molecular mass difference between the two compounds. In the work of Valério et al. [138], a Nadir UP 150 membrane was found to be an adequate choice, allowing for efficient separation of ferulic acid from arabinoxylans, with a null rejection of ferulic acid and high rejection of arabinoxylans 98%.

Polyphenols, extracted from lignocellulosic side-streams from the pulp and paper industry, have not received much attention concerning their fractionation and purification. Among the lignocellulosic materials, the bark is one of the most promising for high-added-value chemicals production due to its chemical composition and availability.

Restolho et al. [142] studied the recovery of sugars and lignosulphonates from eucalyptus spent sulphite liquor by pressure-driven membrane processes. Pinto et al. [139] investigated the processing of ethanol: water extract of *Eucalyptus globulus* bark containing bioactive polyphenols by dia-ultrafiltration (DF) and adsorption/desorption process. Preconcentration followed by DF with a 30 kDa membrane provided an increment of 21% and 64% of total phenolic compounds (TPC) and proanthocyanidins (Pac), respectively, relative to the initial extract composition. The resulting retentate was further purified by fixed-bed adsorption/desorption using a polymeric adsorbent (SP700). In the desorption step with ethanol: water 95:5% *v*/*v* about 45% of the carbohydrates were eliminated and an enriched fraction in polyphenols was obtained. Moreover, the integrated process also led to the recovery of low molecular weight phenolic compounds at the permeate stream during membrane processing and the recovery of galacturonic acid-rich carbohydrates coming from the washing step of the fixed-bed process.

3.2.2. Valorisation of Food Industry Wastewaters and Microalgae

The fishing industry is a pillar of the economy for many countries worldwide. Fish processing by-products and underused catch containing a large quantity of protein are generally converted into low-value products such as animal feed and fertilisers. However, they can be a source of value-added compounds such as peptides and amino acids, capable of modulating physiological processes and, therefore, may have a role in the prevention and control of diseases. The work of Ghalamara et al. [143] demonstrated the feasibility of using (pressure-driven) membrane technologies to purify small biopeptides from codfish blood and sardine cooking wastewaters, the most processed species in Portugal. The best-performing membrane was a UP010, with a minimum rejection of 2% of the small peptides. Results of antioxidant activity suggest that codfish blood is a potential source of natural antioxidant peptides, which may be used as a functional food additive to reduce oxidative stress-associated diseases and inhibit oxidation in foods. Furthermore, the purified peptides also exhibited an interesting antimicrobial activity against *Escherichia coli*.

Dunaliella salina is well-known for its high content in carotenoids and glycerol. Nevertheless, *D. salina* also has a high content in polar lipids, suitable for nutraceutical applications. Monte et al. [144] proposed a sustainable process to produce distinct fractions of carotenoids, glycerol, polar lipids, and proteins, which may improve the microalgae industry's revenues. This work integrates extraction with non-hazardous solvents and organic solvent nanofiltration to obtain added-value products and glycerol. High overall recoveries were obtained for carotenoids (85%), glycerol (86%), polar lipids (94%), and proteins (95%).

3.2.3. Applications of Membranes in Health

Membrane processes have been used for protein crystallisation [46,145–147], and membrane contactors have been used for CO_2 removal from anaesthesia gases [148–150] and as blood oxygenators [151–153].

Crystallisation kinetics can be enhanced by playing with membranes' morphology, topography, and chemistry, allowing for tailoring the supersaturation rate and inducing nucleation at a lower degree of supersaturation. Furthermore, a controlled diffusion of compounds during post-crystallisation treatments leads to higher crystal stability avoiding crystal handling. The use of small volumes, laminar flow, and low-cost fabrication processes of microdevices are extremely important; however, few studies are currently available regarding the use of microtechnology for protein crystallisation. Polino et al. [146] used a Nafion® membrane (Sigma-Aldrich, 1100EW, St. Louis, MI, USA) integrated with a polydimethylsiloxane (PDMS) microdevice for protein crystallisation of Hen Egg White Lysozyme (HEWL) purchased from Sigma-Aldrich and subsequent in situ crystal derivatisation based on efficient control of the diffusion rate of the derivatising agent (Hg^{2+}), assured by the ion-exchange membrane (Nafion®117) (Sigma-Aldrich, 1100EW, St. Louis, MI, USA) integrated into the microdevice. These conditions were key to guaranteeing crystal structural integrity along the derivatisation process, proving to be a more efficient alternative to the traditional crystal soaking methodologies (Figure 3).

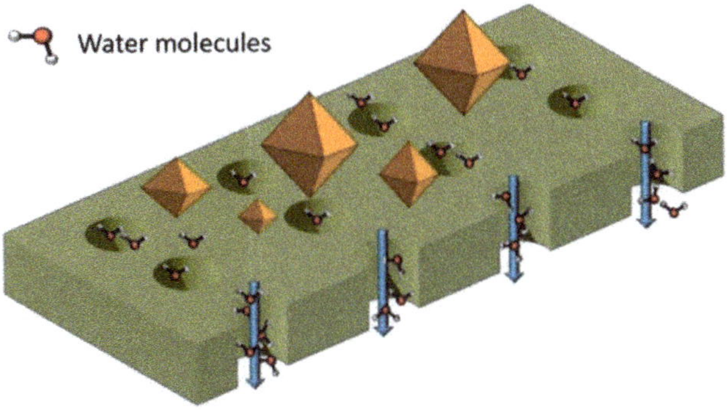

Figure 3. Membrane-assisted crystallisation principle. Protein and stripping solution are on different sides of the membrane. The volatile solvent is removed from the protein solution in the vapour phase; supersaturation is reached, and crystals form. Reprinted with permission from Polino et al. [145]. Copyright 2019, American Chemical Society.

The noble gas xenon possesses many characteristics of an ideal anaesthetic agent. However, its application is limited in daily clinical routine due to its elevated price. Martins et al. [148] propose using a gas-ionic liquid membrane contactor system, using a cholinium lysinate ionic liquid solution as CO_2 absorbent, for the removal of carbon dioxide in anaesthesia closed-circuits, allowing the recycling of the anaesthetic agents. The ability of the proposed system to recover xenon drastically reduces the cost, which can be a step forward for the routine use of this gas in surgeries.

Due to their high hemocompatibility and gas permeation capacity, bi-soft segment polyurethane/polycaprolactone (PU/PCL) polymers are promising materials for use in membrane blood oxygenators. Eusébio et al. [151] synthesised both nonporous symmetric and integral asymmetric PU/PCL membranes. The permeation properties of the atmospheric gases N_2, O_2, and CO_2 through these membranes were experimentally determined using a new custom-built gas permeation apparatus.

Oxygen mass transfer in a membrane blood oxygenator (MBO) surrogate system has been addressed by Faria et al. [152]. It consists of a slit for water circulation as a surrogate blood flow channel and a constant pressure oxygen chamber separated by an integral asymmetric hemocompatible polyurethane-based membrane. The experimental fluxes are in very good agreement with the predictions based on unidimensional axial convection and unidimensional transversal diffusion. This custom-made benchmark system will allow the flow and oxygen mass transfer to be optimised, in order to design a novel flat-sheet MBO.

Faria et al. [153] synthesised by an innovative method, combining phase inversion with the acid-catalysed sol-gel process, monophasic hybrid cellulose acetate (CA) membranes with tailored hemocompatible surfaces and permeation properties that assure the kidney metabolic functions of preferential permeation of uremic toxins and the retention of vital components of the blood such as albumin (Figure 4). Permeation studies show that the integration of silica into CA membranes results in high flux UF membranes with enhanced mass transfer properties on the metabolic functions of the kidney. The hemolysis assay shows that all the hybrid $CASiO_2$ membranes are non-hemolytic and that, even for an extended static blood/membrane surface contact time of 15 min, the thrombosis degree exhibited by all the membranes is lower than 30% of the value observed for the positive control. Shear stresses induced by UF permeation have also been analysed in a surrogate system of an artificial kidney [154].

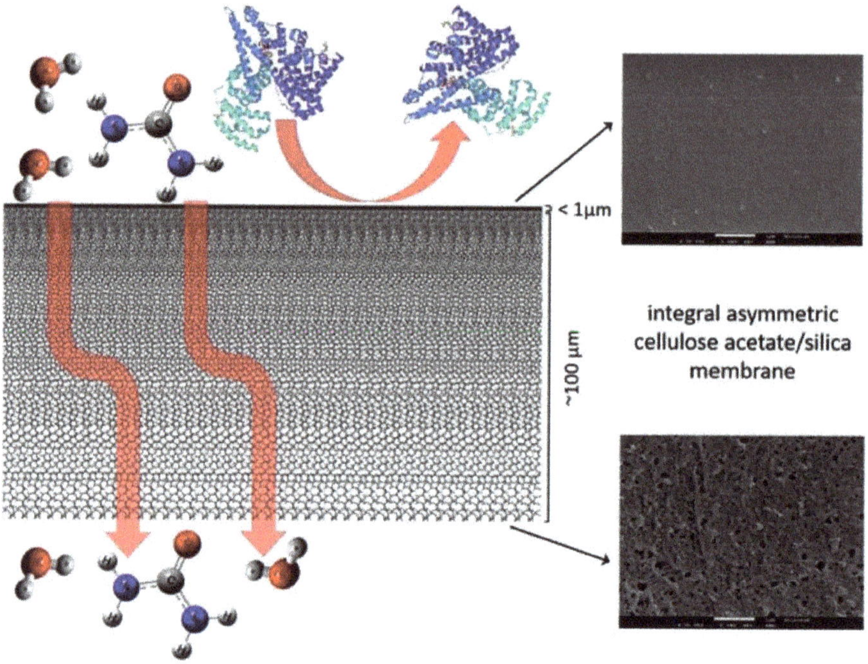

Figure 4. Monophasic hybrid cellulose acetate/silica ($CASiO_2$) integrally skinned membranes with tailored hemocompatible surfaces. Reprinted with permission from Faria et al. [153] Copyright 2020, Springer Nature.

3.3. Membranes for Energy Applications

3.3.1. Membranes for Fuel Cells

The membrane is a crucial component for most electrochemical energy conversion and storage devices, including batteries, electrolysers, and fuel cells. The latter are truly sustainable and clean energy technologies, where hydrogen and oxygen combine to generate electricity, water, and heat. Proton-exchange membranes (PEMs) remain a critical material and key challenge to fuel cell science and technology development. Further advances are urgently required, as PEM fuel cells demand increases with the fuel cell electric vehicles market growth.

In this regard, Teixeira et al. [155] doped Nafion membranes to enhance proton conductivity in PEM fuel cells. The dopants were derived from aryl mono- or bisphosphonic acid and incorporated using impregnation or casting methods. Electrochemical impedance spectroscopy (EIS) was used to assess the proton conductivity in these membranes and evaluate the influence of structure and preparation method on proton transport. The membranes prepared by casting showed higher proton conductivities than commercial Nafion®N-115 (FuelCell Store, College Station, TX, USA) at all temperatures and relative humidity conditions, with [1,4-phenylenebis(hydroxymethanetriyl)]tetrakis(phosphonic acid) offering the best proton conductivity (87 mS cm^{-1}) [155]. In another study, Teixeira et al. [156] prepared different indazole- and benzotriazolebisphosphonic acids for incorporation into Nafion®N-115 membranes (FuelCell Store, College Station, TX, USA) (up to 5 wt.% loading). Nafion-azolebisphosphonate membranes with 1 wt.% loading presented enhanced proton conductivity compared to benchmark Nafion®N-115, with the membrane doped with [hydroxy(1H-indazol-3-yl)methanediyl]bis(phosphonic acid) exhibiting a proton conductivity of 98 mS cm^{-1}. In both studies, the results suggest that Grotthuss and vehicular mechanisms are involved in the proton conduction across the membrane. The good proton conductivity of these doped membranes led Teixeira et al. [157] to evaluate their potential application in PEM fuel cells. Membrane-electrode assemblies (MEAs) containing the membrane doped with [1,4-phenylenebis(hydroxymethanetriyl)]tetrakis(phosphonic acid) presented better performance than Nafion®N-115 at all studied temperatures, with a maximum power density of ~383 mW cm^{-2} at 70 °C.

Natural polymers, such as cellulose, the most abundant biopolymer, and in particular its nanoscale forms (e.g., cellulose nanocrystals (CNCs), cellulose nanofibrils (CNFs), bacterial nanocellulose (BNC)) have been proposed as appropriate materials to engineer the membrane of PEM fuel cells [158]. Vilela et al. [159] focused on the potential of BNC as a substrate for the design of bio-based ion-exchange membranes, as they present an excellent combination of conductive and mechanical properties. In this context, they fabricated polyelectrolyte nanocomposite membranes based on poly(bis phosphate) [P(bisMEP)] and BNC via the in situ free radical polymerisation of bis[2-(methacryloyloxy)ethyl] phosphate (bisMEP) inside the BNC three-dimensional network. Vilela et al. also produced other BNC-based nanocomposites with poly(methacryloyloxyethyl phosphate) [160] and even fully bio-based ion-exchange membranes by blending BNC either with lignosulfonates [161] or algae sulphated polysaccharide (i.e., fucoidan) [162], together with tannic acid as a natural crosslinker.

In addition, Vilela et al. [163] prepared polyelectrolyte nanocomposites of BNC and poly(4-styrene sulfonic acid) as eco-friendly PEMs for microbial fuel cells (MFCs). The application of the nanocomposite membrane in a single-chamber lab-scale MFC with a pure culture of *Shewanella frigidimarina* yielded a maximum power density of 2.42 mW m^{-2}, open-circuit voltage of 0.436 V, and internal resistance of 15.1 kΩ, which was better than that obtained with a commercial Nafion®117 membrane. In another work, Vilela et al. [164] developed bio-based anion-exchange membranes (AEMs), suitable for application in alkaline fuel cells. The nanocomposite membranes were based on a tetraalkylammonium-derived polymeric ionic liquid, i.e., poly(methacroylcholine chloride), and BNC.

Recently, Ferreira et al. [165] used computational fluid dynamics (CFD) to simulate the membrane chemical degradation in PEM fuel cells. They found that degradation accelerated when voltage, temperature and pressure increased and when reactants' humidity and membrane thickness decreased.

Direct liquid fuel cells (DLFCs) are receiving increasing attention due to their advantages over conventional hydrogen fuel cells. These include higher energy density, easier transportation and storage, simplified structure, and the possibility of using a small fuel cartridge for instant recharging. Alcohols (e.g., methanol, ethanol) and sodium borohydride are the most common organic and inorganic fuels used in DLFCs.

Electrolyte membranes that restrict the methanol crossover while retaining proton conductivity are essential to improving the electrochemical selectivity in direct methanol fuel cells (DMFCs). Rambabu et al. [166] reviewed the literature on carbon nanomaterial-based PEMs published during the last decade. Based on the reported proton conductivity and methanol permeability of nanocomposite membranes containing carbon nanotubes, graphene oxide and fullerene as additives, Rambabu et al. [166] critically assessed the impact of each type of filler on those properties.

Ben Jadi et al. [167] prepared Nafion-polypyrrole membranes by a straightforward deposition method and assessed their transport properties for DMFC application. The effect of the polypyrrole coating on transport properties (e.g., water uptake, ion-exchange capacity, proton conductivity) was investigated. They found that methanol diffusion through the membrane was reduced by at least 20-fold compared to commercial Nafion®112. Ben Jadi et al. [168] also synthesised Nafion®112/polyaniline composite membranes for use in DMFCs, which led to a decrease in the methanol crossover and an increase in the proton conductivity compared to the unmodified Nafion®112.

Oliveira and Mendes [169] prepared eco-friendly polymer electrolyte membranes based on a blend of sulphonated poly(vinyl alcohol)/chitosan stabilised by nylon 6,6, using sulfosuccinic acid both as a crosslinker and sulfonating agent. The membranes presented good proton conductivity (ca. 10^{-2} S cm^{-1}) and lower H_2 and methanol permeability than benchmark Nafion®115.

Direct borohydride fuel cells (DBFCs) are a subtype of DLFCs and an excellent alternative to alcohol-fed fuel cells, as their operation does not emit greenhouse gases. DBFCs operate with sodium borohydride as the fuel and oxygen or hydrogen peroxide as oxidants. They offer advantages over other DLFCs regarding higher energy density and theoretical cell voltage [170]. To reach the initial stages of DBFCs commercialisation, it is still necessary to develop more efficient and lower-cost membranes.

Gouda et al. [171] produced composite membranes via polymer crosslinking and casting techniques. A ternary crosslinked polymer was prepared from low-cost and readily available polymers containing poly(vinyl alcohol) (PVA), poly(ethylene oxide) and poly(vinyl pyrrolidone), followed by incorporation of sulphonated graphene oxide (SGO) as a doping agent into the polymer matrix. The prepared membranes presented half the borohydride permeability of Nafion®117. In another work, Gouda et al. [172] doped the same PVA-based ternary blend with (PO_4–TiO_2) and (SO_4–TiO_2) nanotubes. The DBFCs operating with both types of membranes attained power densities close to that recorded with a Nafion®117 membrane, but at a significantly lower membrane cost [171,172]. Those results pave the way for a green, low-cost and straightforward approach to developing composite membranes for practical DBFCs.

3.3.2. Membranes for Batteries

Batteries are everywhere, from flashlights to remote controls, from mobile phones to laptops. Specifically, lithium-ion batteries (LIBs) are seeing an increasing demand due to the boom of battery electric vehicles. This requires the development of advanced materials to improve LIBs' energy/power density, conversion efficiency and safety, as well as decrease costs, extend their lifetime, and reduce their environmental impact.

The membrane separator plays a key role in LIB performance, separating the electrodes and controlling the movement of the lithium ions. Due to their suitable properties, polymer composites and blends have been frequently used as LIB separators. Nunes-Pereira et al. [173] prepared porous poly(vinylidene fluoride-co-trifluoroethylene) (PVDF-TrFE) based composite membranes filled with clays (montmorillonite), zeolites (Y zeolite), ceramics (barium titanate) and carbonaceous (multiwalled-carbon nanotubes) fillers. The results demonstrated that the filler type deeply affected the membrane separator performance in LIBs. The PVDF-TrFE membrane with MMT filler presented the best performance among all the evaluated fillers.

Costa et al. [174] prepared silica/PVDF porous composite membranes for LIB separators using a non-solvent induced phase separation process with different air exposure times before immersion in a water coagulation bath and filler content of 20 wt.%. Mesoporous silica spheres (SS) of ~400 nm average diameter were synthesised using the sol-gel method and dispersed into the polymer matrix. It was demonstrated that the morphology, degree of porosity, uptake value and electrical properties of the composite membranes were influenced by the exposure time to air and the presence of SS. The membranes were assembled in Li/C-$LiFePO_4$ half-cells, and the best cycling performance was obtained for the composite membrane after 1 min exposure to air, with an ionic conductivity of 0.9 mS cm^{-1}.

Cellulose has emerged as a sustainable alternative to traditional polyolefin-based LIB separators because of its good mechanical properties and inherent hydrophilic character. Gonçalves et al. [175] fabricated high specific surface area mesoporous cellulose nanocrystal membranes with different pore morphology as novel three-dimensional porous separators. These membranes displayed ionic conductivities of 2.7 mS cm^{-1}, wide electrochemical stability, and good interfacial compatibility with the lithium electrode, delivering a specific capacity of 122 $mAhg^{-1}$ and 85 $mAhg^{-1}$ at a C/2 and 2C rate, respectively.

Barbosa et al. [176] produced electrospun PVDF fibre membranes doped with different ionic liquids sharing the same anion, bis(trifluoromethylsulfonyl)imide [TFSI]. Their potential as membrane separators for LIBs was evaluated. The membrane doped with 15% 1-ethyl-3-methylimidazolium([C_2mim])[TFSI]) showed the best performance, with an ionic conductivity of 2.8 mS cm^{-1} at room temperature and good stability in the electrochemical window 0.0–6.0 V. When evaluated in C-$LiFePO_4$ half-cells, it exhibited a maximum discharge capacity of 119 $mAhg^{-1}$ at C-rate, with a coulombic efficiency close to 100%.

Rechargeable zinc (Zn)-air batteries are seen as attractive candidates for large-scale electricity storage owing to their high volumetric energy density, safety, economic feasibility, and environmental friendliness. In Zn-air batteries, the membrane allows the transport of OH^- ions between the air electrode and the Zn electrode while providing a physical barrier that prevents short-circuiting. Zn-air battery performance is greatly affected by the physicochemical properties of the employed membrane. However, the development of appropriate membranes has received insufficient attention. Tsehaye et al. [177] critically discussed the state-of-the-art membranes for Zn-air batteries. They presented membranes synthesis and modification strategies to minimise the crossover of zincate ions and the formation/growth of Zn dendrites. The work provides guidance for researchers and industries for selecting adequate membranes to commercialise rechargeable Zn-air batteries.

The proliferation of portable electronic devices has increased the amount of e-waste generated after their use. One of the most hazardous components in e-waste is batteries due to the heavy metals and toxic chemicals that they contain. Fuel cells and redox flow batteries are sustainable alternatives to LIBs for powering portable applications. Although they provide comparable energy and power density, they face challenges related to their dependence on PEMs based on non-environmentally friendly, high-priced perfluorosulfonic acid copolymers, which require energy-intense manufacturing and recycling procedures. Alday et al. [178] synthesised eco-friendly and sustainable biopolymer electrolyte membranes (BioPEMs) from biopolymers such as chitosan, cellulose, and starch. These BioPEMs present advantages in performance, sustainability, and cost. They also show good chemical and mechanical stability, and ionic conductivities close to that of Nafion membranes. Redox flow batteries using cellulose–chitosan-based BioPEMs delivered open-circuit voltages above 0.75 V and power density up to 2.5 mW cm^{-2}. These BioPEMs show great promise in improving biodegradable batteries in sectors requiring transient electrical energy, such as environmental monitoring, agriculture, or packaging.

Minimising the redox species crossover through the membrane is a key challenge, as it leads to capacity loss and lower coulombic efficiency. This issue needs to be readily quantified before commercial application of the non-aqueous redox flow battery (NAqRFB). Thus, Mushtaq et al. [179] developed in situ crossover diagnostics to assess the membrane efficacy in NAqRFB. The successful validation of this technique enables its use as a reference testing method to measure crossover and help select the best performing membrane for a redox flow battery. As a membrane diagnosis tool, this technique has the advantage of in situ determination of crossover without the need for disassembling the devices.

3.3.3. Reverse Electrodialysis for "Blue Energy" Generation

Reverse electrodialysis (RED) is a membrane-based technology that allows salinity gradient energy to be harvested. Fouling is an undesirable operation constraint in the RED process since it decreases obtainable net power density due to increasing stack electric resistance and pressure drop. Therefore, early fouling detection is one of the main challenges for successful RED technology implementation. S. Pawlowski et al. [180] used two-dimensional (2D) fluorescence spectroscopy as a tool for fouling monitoring in RED. Fluorescence excitation-emission matrices (EEMs) of ion-exchange membrane surfaces and natural aqueous streams were acquired during one month of a RED stack operation. The fouling evolvement on the ion-exchange membrane surfaces was followed by 2D fluorescence spectroscopy and quantified using principal components analysis (PCA). The authors found that the AEM surface in contact with river water was significantly affected due to fouling by humic compounds crossing the membrane from the lower salinity (river water) to the higher salinity (seawater) stream. The combination of 2D fluorescence spectroscopy and PCA demonstrated a high potential for studying fouling development and membrane cleaning efficiency in ion-exchange membrane processes.

Still, a major hindrance to the implementation of RED is the high price of ion-exchange membranes. S. Pawlowski et al. [59] proposed using profiled membranes to overcome the shadow effect of non-conductive spacers, which reduce the membrane area available for counter-ion transport and increase the stack electric resistance. Profiled membranes can be a promising alternative as the reliefs formed on their surface keep the membranes separated and provide channels for the solutions to flow. S. Pawlowski et al. [59] also used CFD to simulate fluid behaviour in channels formed by various profiled membranes. The highest net power density values were obtained for the corrugation shape and arrangement in the form of chevrons due to the increase in the available membrane area and an excellent balance between enhancement of mass transfer and the increase in the pressure drop in the channel. When properly designed, corrugated membranes offer better performance than conductive spacers. The proposed membrane corrugation design can be extended to other electromembrane processes, such as electrodialysis and Donnan dialysis, where high ionic mass transport rates are desirable at as low as possible energy costs [59].

The performance of AEMs in RED can be hampered by the presence of multivalent ions and fouling phenomena, thus leading to reduced net power density. Merino-Garcia et al. [181] proposed a monolayer surface modification procedure to functionalise Ralex-AEMs with poly(acrylic) acid (PAA) to (i) render a monovalent permselectivity, and (ii) minimise organic fouling. Membrane surface modification was carried out by putting heterogeneous AEMs in contact with a PAA-based aqueous solution for 24 h (Figure 5). The membranes' permselectivity and fouling behaviour in the presence of humic acid was evaluated through mass transport experiments using model NaCl-containing solutions. The use of PAA-modified AEMs resulted in a significantly enhanced monovalent permselectivity (sulphate rejection improved by >35%) and membrane hydrophilicity (contact angle decreased by >15%) in comparison to the unmodified Ralex-AEMs, without compromising the membrane electro-resistance after modification. The work demonstrates the proposed membrane modification procedure's feasibility and provides a path for improving RED efficiency.

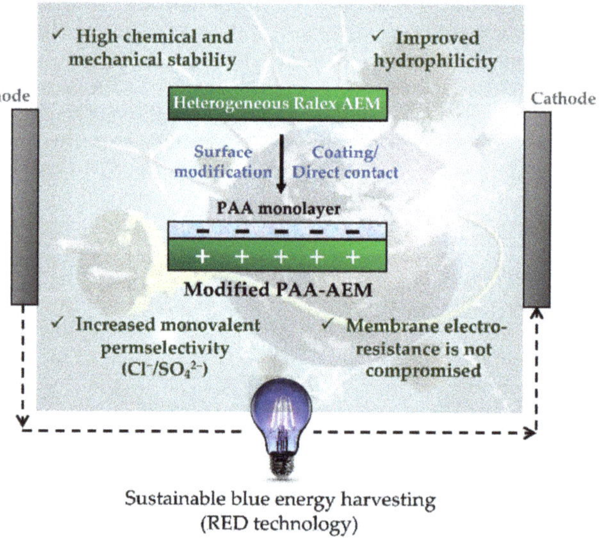

Figure 5. Scheme of the procedure used to functionalise Ralex-AEMs with poly(acrylic) acid and the advantages of the PAA modified AEMs for sustainable blue energy harvesting [181].

3.3.4. Membranes for Electrolytic Hydrogen Production

Water splitting has been proposed as a promising approach to producing clean hydrogen. The two half-reactions of water splitting, the hydrogen evolution reaction (HER) and oxygen evolution reaction (OER), occur kinetically fast in solutions with completely different pH values. Enabling HER and OER to occur under kinetically favourable conditions while using exclusively low-cost, earth-abundant electrocatalysts is highly desirable, but remains a challenge.

J. Xu et al. [182] demonstrated that using a bipolar membrane (BPM), it is possible to carry out HER in a strongly acidic solution and OER in a strongly alkaline solution, with bifunctional self-supported cobalt–nickel phosphide nanowire electrodes catalysing both reactions. Such asymmetric acid/alkaline water electrolysis could deliver a current density of 10 $mAcm^{-2}$ with ca. 100% Faradaic efficiency at 1.567 V. Moreover, using an "irregular" BPM with unintentional crossover, the voltage needing to afford 10 $mAcm^{-2}$ was reduced to 0.847 V due to the assistance of electrochemical neutralisation between acid and alkaline. J. Xu et al. [182] showed that BPM-based asymmetric water electrolysis could be accomplished in a circulated single-cell electrolyser delivering 10 $mAcm^{-2}$ at 1.550 V and splitting water very stably for at least 25 h. Water electrolysis would be enabled by a solar panel operating at 0.908 V (@13 $mAcm^{-2}$) using an "irregular" BPM. The results demonstrated that hydrogen production by BPM-based asymmetric water electrolysis is a promising alternative to conventional PEM and AEM water electrolysis.

A more integrated way to produce electrolytic hydrogen is through the chloralkali process, a major electrochemical technology based on the chloralkali industry. Hydrogen, chlorine, and sodium hydroxide are produced in the process. These compounds are used to manufacture a variety of intermediates in the chemical, pharmaceutical and crop protection industries. As the chloralkali industry is highly energy-intensive, electricity consumption represents a significant fraction (>90%) of the total production cost. The energy consumption depends on the electrolyser design, the type of membrane used, brine quality, and operating conditions. Thus, choosing the best membrane to install in a chloralkali electrolyser is a highly complex and relevant decision that should balance operating and fixed costs. Presently, the chloralkali membrane process is dominated by three membrane suppliers, and few studies are reported in the literature. This motivated A.C. Dias to carry out her PhD work [183] under the supervision of A. Mendes and to obtain valuable data for CUF—Químicos Industriais, S.A. (CUF-QI). This major Portuguese chloralkali company changed its name in 2018 to Bondalti. The work developed an experimental setup for determining the permselectivity of cation-exchange membranes and an electrochemical membrane reactor for studying the membrane cell process. It was demonstrated that significant energy savings could be achieved by choosing an adequate membrane and timely membrane replacement. The cost per ton of chlorine was calculated for three electrolysers, each using a set of a different type of membrane. It was concluded that the higher strength membrane had the lowest average total cost, with savings of around 1.5% compared to the second-best membrane. These data allowed a benchmarking methodology to be developed [183].

The same group used EIS to evaluate in situ the ohmic resistances of electrolytes and membrane in chloralkali membrane cells [184]. That allowed the effect of different operating parameters (current density, feed flow rate, brine and caustic concentrations, temperature) on the overall performance (power consumption) of a laboratory-scale chloralkali membrane cell to be assessed. It was found that cell temperature and brine concentration were the most critical variables on the cell voltage. The gas bubble evolution greatly affected the ohmic resistance, especially at the anode side, where the gas void fraction was about 10% [184].

Franco et al. [185] carried out a systematic performance history analysis of a chloralkali membrane electrolyser under industrial operating conditions. The performance was analysed as a function of its service time for about 8 years, recording the cell potential versus current density. The ohmic overpotential contribution, associated with the ion-exchange membranes, showed an average relative error smaller than 3%. The activation overpotential, related to the electrodes' performance, displayed an average relative error of 6%. This approach enabled assessing the performance of industrial chloralkali membrane electrolysers (Figure 6) for adequate maintenance scheduling, leading to significant operational and economic improvements in the chloralkali process [185].

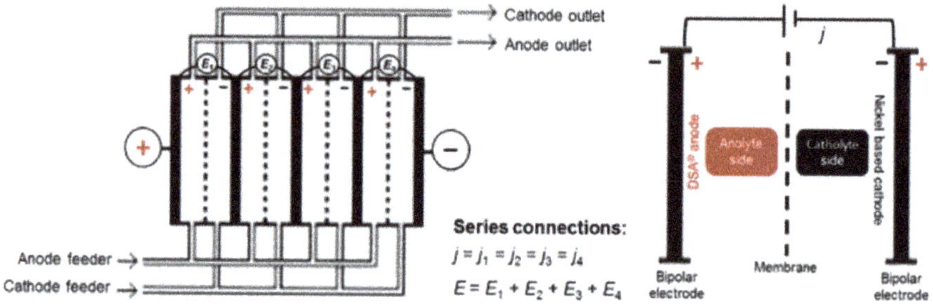

Figure 6. Scheme of a bipolar electrolyser where the stack consists of four cells connected in series: the potential between the terminals, *E*, is the sum of the individual cell potentials and the same current density, *j*, flows through the cells [185].

3.4. Membranes for Gas Separation and Pervaporation

3.4.1. Membranes for Gas Separation

Due to their close connection to global warming and unpredictable climate changes, anthropogenic greenhouse gas emissions are one of our era's most critical environmental challenges. Specifically, carbon dioxide (CO_2) is of particular concern as it currently accounts for the greatest portion of the warming associated with human activities. There are several means to curb the CO_2 emissions to the atmosphere, including CO_2 recovery from power plant flue streams (CO_2/N_2), CO_2 separation from biohydrogen and fuel gas (syngas) in hydrogen production (CO_2/H_2), and CO_2 removal from natural gas (natural gas sweetening) and biogas (CO_2/CH_4). However, the efficient separation of CO_2 from light gases is the technical and economic key. Membrane technology can offer engineering and economic advantages compared to traditional separation methods, but the major drawback in current membrane operation concerns the limited selectivity of the materials. Hence, Portuguese researchers have intensively explored the development of advanced membrane materials for CO_2 separation/capture.

Among the several membrane materials studied, a quite recent approach makes use of ionic liquids [186]. Remarkable CO_2 solubility and tunability are essential features of this unique platform. The potential of ionic liquid materials to improve membrane-based CO_2 separation processes has been demonstrated, either through the preparation of supported ionic liquid membranes (SILMs) [45,187–198], the blending of ionic liquids with poly(ionic liquid)s [199–208] or, more recently, through the development of mixed matrix membranes [209–214].

In a SILM, the selected ionic liquid is immobilised into the pores of an inert solid membrane supported by capillary forces. Inspired by the possibility of developing stimuli-responsive membranes with improved CO_2 separation, Crespo's group paid particular attention to magnetic ionic liquids containing metal-based anions with different magnetic susceptibilities in their structures [187]. The resulting SILMs exhibited CO_2 permeabilities in the range of 147–259 Barrer, while CO_2/N_2 selectivity values between 21 and 41 were obtained. The same authors also focused on the permeability modulation of these SILMs when applying an external magnetic field (0–2 Tesla) [188]. The results showed that gas permeability increases as a function of the magnetic field depending on the metal anion, while CO_2/N_2 selectivity remained constant.

Aiming at full sustainability of CO_2 separation processes, Tomé et al. [190] successfully prepared SILMs based on environmentally friendly cholinium carboxylate ionic liquids, which exhibited good CO_2/CH_4 selectivity (between 21 and 31) and CO_2/N_2 selectivity (from 39 to 50) at 20 °C under dry conditions. However, their CO_2 permeability was significantly lower (2–18 Barrer) than that of other ionic liquid families. The major bottleneck of using cholinium carboxylates under dry conditions is their high viscosities, which lead to long absorption equilibrium times. With the intent of increasing the flexibility in tailoring the permeability and selectivity of SILMs for CO_2 separation, Tomé et al. [189,191] explored the use of binary ionic liquid mixtures. The prepared SILMs revealed CO_2/N_2 separation performances on the respective Robeson upper bound, or even surpassed it, indicating their potential to separate CO_2 in low-pressure post-combustion processes. The authors concluded that using ionic liquid mixtures combining the most selective anions with the least viscous anions is a promising strategy for designing advanced liquid phases for CO_2 separation membranes. In this line, Gouveia et al. [193] proposed equimolar ionic liquid mixtures containing tricyanomethane and different amino acid-based to prepare SILMs for facilitated CO_2 separation from flue gas streams. The best results were found at low trans-membrane pressure differentials (2.5 kPa), where the mixtures with L-alaninate and taurinate anions showed the highest CO_2 permeability (765 Barrer) and CO_2/N_2 selectivity (69), respectively.

An innovative concept for removing CO_2 from flue gas streams using SILMs was also proposed by Neves et al. [45]. This approach consisted of using an ionic liquid phase comprising a carbonic anhydrase (CAn) enzyme, which catalyses the conversion of CO_2 to bicarbonate, enhancing the driving force for CO_2 transport. Using this type of thermostable enzyme allowed us to obtain SILMs with high permeability towards CO_2 (734 Barrer) at high temperature (up to 100 °C) and a reasonable transport selectivity towards CO_2 against N_2 (36) [195]. The use of CAn enzyme has recently been tested in supported liquid membranes using deep eutectic solvents (DES). For instance, a membrane containing CAn and DES composed of choline chloride and urea revealed high selectivity for CO_2/CH_4, setting above the respective Robeson upper bound [197]. In addition, a membrane prepared with CAn and a DES composed of choline chloride and levulinic acid showed a consistent behaviour up to 30 repeated cycles and stable operation at transmembrane pressure differences in the range of 0.5 to 1.25 bar [198]. These findings demonstrate the potential of DES-supported membranes for CO_2 separation, particularly in the case of biogas.

Another strategy for designing membranes for CO_2 separation is the fabrication of poly (ionic liquid) (PIL)–ionic liquid (IL) membranes. This strategy combines the intrinsic ionic liquid properties with the macromolecular structure of PILs, taking advantage of their ability to strongly interact through electrostatic interactions. Tomé et al. [199] investigated PIL/IL membranes based on PILs having different polycation functionalities (i.e., imidazolium, pyridinium, pyrrolidinium, ammonium, and cholinium) and selected the pyrrolidinium-based PILs to conduct different studies since their synthetic route is less complicated than those of the other PIL polycations. The results showed that the CO_2 permeability and selectivity of PIL–IL membranes bearing pyrrolidinium polymer matrices could be adjusted by varying the IL content [200], using random copolymers with anion mixtures [201] or PIL and IL components with different anions [202], and even by changing the polymer molecular weight [203]. The best performances were obtained for PIL/IL membranes containing cyano-functionalised anions, particularly the one with 60 wt.% of $[C_2mim][C(CN)_3]$ IL, which surpassed the Robeson 2008 upper bound for CO_2/N_2 separation (Figure 7a), with CO_2 permeability of 440 Barrer and CO_2/N_2 selectivity of 64 [204].

Pyrrolidinium-based PIL/IL membranes have also been successfully tested for CO_2/H_2 separation in hydrogen purification [205–207]. Despite the competition effect between gases in mixed gas experiments, only small differences were obtained between mixed and ideal CO_2/H_2 selectivities. The pyrrolidinium-based PIL/IL membranes showed mixed CO_2/H_2

separation performances above the upper bound even at the highest temperature (80 °C) and feed pressure (4 bar) tested. Once more, the PIL $C(CN)_3$–60 [C_2mim][$C(CN)_3$] membrane showed the best mixed gas performance, with CO_2 permeability of 325 Barrer and CO_2/H_2 selectivity of 11.4 [207]. Furthermore, Gouveia et al. [208] have recently explored the creation of microchannels (around 55–60 μm deep) on the surface of PIL/IL membranes by laser ablation using femtosecond laser radiation (Figure 7b). Considerable improvements in CO_2, N_2, and H_2 permeabilities were achieved for irradiated PIL/IL membranes, while their CO_2/N_2 and CO_2/H_2 selectivities were almost unchanged.

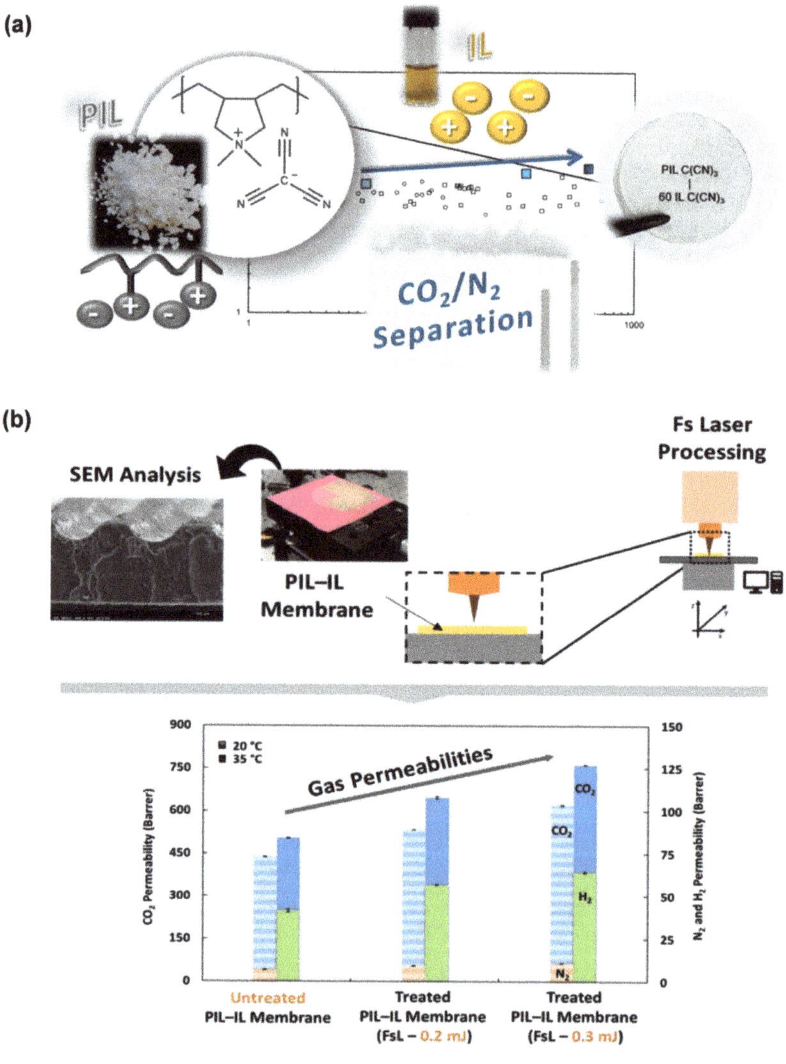

Figure 7. (**a**) CO_2/N_2 separation performance of PIL/IL membranes containing cyano-functionalised anions. Reprinted with permission from Tomé et al. [204] Copyright 2015, Elsevier. (**b**) Schematic illustration of the creation of microchannels on the surface of PIL/IL membranes by laser ablation using femtosecond laser radiation (top) and respective effect on permeability towards CO_2, N_2, and H_2 (bottom). Reprinted with permission from Gouveia et al. [208]. Copyright 2022, Elsevier.

Mixed matrix membranes combining metal-organic frameworks (MOFs) [209–212] or graphene [213,214] have also been investigated. As an example, Nabais et al. [209] incorporated different MOFs (MIL-53(Al), $Cu_3(BTC)_2$, and ZIF-8) into a pyrrolidinium-based PIL/IL (60/40 wt.%). These multi-component (PIL/IL/MOF) membranes surpassed the upper bound limit for CO_2/H_2 separation, with the one loaded with 30 wt.% of ZIF-8 showing CO_2 permeability of 97, while that of MIL-53(Al) presenting CO_2/H_2 selectivity of 13. Nevertheless, using the PIL/IL/MOF-5 system for CO_2/CH_4 separation revealed performance limitations, possibly attributed to poor chemical/physical compatibility between the PIL/IL matrix and MOF-5 particles [210].

Furthermore, inorganic porous materials have been explored for different kinds of gas mixtures [215–219]. In particular, carbon molecular sieve (CMS) membranes are strong candidates for energy-efficient gas separation applications, due to their known higher separation performances, stability to corrosive environments and temperatures (up to ca. 200 °C). Potential applications include carbon dioxide removal from natural gas/biogas streams, hydrogen recovery from streams with light hydrocarbons, light olefins/paraffins separation, xenon recovery and air separation, among others [220]. Mendes' group pioneered in Portugal the development of CMS membrane technology through the work of Lagorsse et al. [62,63], who studied the transport mechanism and the ageing effect of the CMS membranes modules produced by Carbon Membranes Ltd., an Israeli company in this field. Teixeira et al. [65] produced CMS membranes from low-cost phenolic resin, with a record C_3H_6 permeability of 420 Barrer and a C_3H_6/C_3H_8 selectivity of 18.1, making these membranes a promising alternative for olefins purification. Later on, Rodrigues et al. [221] prepared renewable and low-cost cellophane-based CMS membranes with the highest reported O_2/N_2 selectivity (>800), CO_2/CH_4 and CO_2/N_2 (>2600), and H_2/CH_4 and H_2/N_2 greater (>25,000). Tailor-made CMS membranes with a well-developed microporous structure were also prepared by Rodrigues et al. [222] from an optimised, low-cost and renewable ionic liquid-regenerated cellulose. The CMS membranes surpassed the Robeson upper bound for O_2/N_2, He/N_2 and H_2/N_2 separations [222]. Very recently, Araújo et al. [67] fabricated stable cellulose-based CMS membranes, also from an ionic liquid regenerated cellulose precursor, with extremely high selectivities to H_2/CH_4 and CO_2/CH_4 (>206,000 and 14,600, respectively), due to the precise pore size. The produced CMS membranes upon a propylene post-treatment for 10 days present an ageing resistant property for oxygen chemisorption for more than 1 year without impairing the CMS membrane performance [67]. These results demonstrate the industrial attractiveness of this type of membrane for the separation of gases with high performance and stability.

3.4.2. Membranes for Pervaporation

Pervaporation is a membrane separation process with unique and adequate features to remove volatile compounds present in liquid mixtures at trace levels [223]. Pervaporation has been used by Portuguese researchers for the dehydration of organic solvents (such as ethanol [224–227], isopropanol [50] or n-butanol [228]), aroma recovery [49,87,229–234], and process intensification of ethyl lactate [235] or butyl acrylate [213,236].

Solvent dehydration processes have a high economic and environmental significance in the pharmaceutical, fine chemistry, and chemical industries. Hydrophilic pervaporation has gained progressive recognition among the available techniques, particularly for ethanol dehydration [226]. In this process, hydrophilic membranes are preferred due to their high selectivity for water, but they also may present poor mechanical stability. As an alternative, Meireles et al. [224] proposed a new microbial exopolysaccharide (EPS) membrane obtained from a low-cost and abundant carbon source. The composite membranes of the crosslinked EPS supported in polyethersulfone (EPS-PES) showed high potential for ethanol dehydration, with a water/ethanol selectivity of 3000 at 5.0 wt.% water concentration in the feed (Figure 8). The effect of biopolymer purification was also studied. The EPS-PES membrane prepared with polysaccharides purified by dia-ultrafiltration was revealed to be the best choice, with a water/ethanol selectivity of 142 for 10 wt.% water in ethanol [225].

Meireles et al. [227] also developed hybrid polysaccharide (FucoPol) membranes by incorporating a SiO_2 network. Although the hybrid FucoPol membranes showed high water selectivity values (similar to PERVAP 4101® commercial membrane), their long-term operation was compromised by an irreversible swelling in the presence of water.

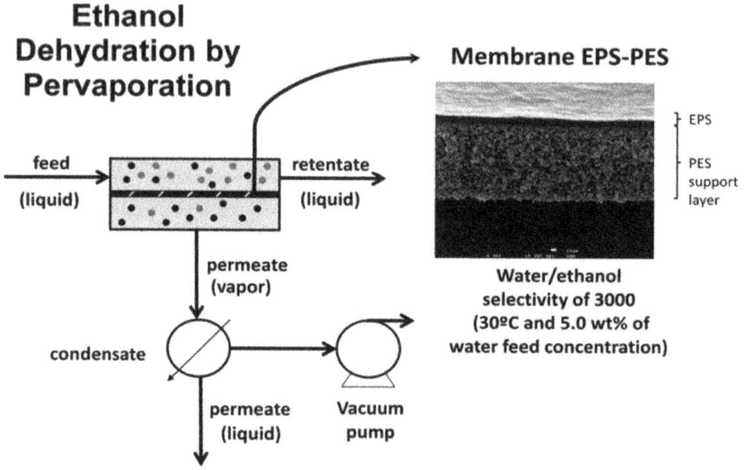

Figure 8. Scheme of ethanol dehydration by pervaporation (left) and the composite membrane of the crosslinked EPS supported in polyethersulfone (EPS-PES) (right). Reprinted with permission from Meireles et al. [224]. Copyright 2013, Elsevier.

Organophilic pervaporation is a particularly suitable process for aroma recovery since it does not require the use of any additional extracting agent and operates under mild conditions. Just to cite some examples, Brazinha et al. [230] proposed an integrated process for recovery and capture of aroma compounds from dilute natural matrices combining the advantages of sweeping gas pervaporation with gas–liquid absorption in membrane contactors. Several potential food/cosmetic compatible additives were tested, and high partitions of the aroma compounds to poly(ethylene glycol) (PEG) 300 were observed [230]. Kulkarni et al. [231] demonstrated that organophilic pervaporation enabled an efficient concentration of limonene with fewer contaminants and lower water content when compared to traditional vacuum distillation. The direct recovery of natural vanillin from post-fermentation media was also studied and optimised by Brazinha et al. [232], assuring the quantitative recovery of vanillin free of other media constituents in a single pervaporation step.

Another important contribution concerns the use of pervaporation extracting equilibrated aromas. In particular, Catarino et al. [234] used asymmetric composite membranes of polyoctylmethylsiloxane/polyetherimide (POMS/PEI) for aroma extraction from beer by pervaporation. This work allowed Super Bock Group, S.A. (formerly Unicer Bebidas S.A.) to apply this technology to develop a new industrial process for producing non-alcoholic beer with a corrected natural flavour profile [88]. This new process removes the aromas by pervaporation, and the beer without aroma is fed to an industrial distillation unit in a rotating cone column for dealcoholisation. Finally, the non-alcohol beer is again mixed with the aromas previously collected by the pervaporation unit. This process allowed the company to produce a non-alcoholic beer with an ethanol content of less than 0.5% by volume without harming the initial flavour of this product [88]. Furthermore, Catarino et al. [87] proposed an identical process for dealcoholising wine. The authors tested reverse osmosis and nanofiltration membranes to remove alcohol content from wine and pervaporation membranes to recover the frames. The use of this combined process has proven to be effective in removing alcohol from wine without impairing its taste [87].

In addition, pervaporation processes are strongly recommended for heat-sensitive products and azeotropic mixtures, as in the case of the butyl acrylate system. Constantino et al. [236] presented pervaporation data for multicomponent mixtures in the absence of reaction for the compounds involved in the esterification of acrylic acid with n-butanol at different temperatures. This experimental data is very useful for designing and optimising pervaporation-based hybrid processes for butyl acrylate synthesis [213].

4. Future Perspectives

Research is universal and, naturally, membrane research in Portugal is aligned with global concerns and expectations for sustainable social development. The United Nations Development Goals translate in a clear way the principles that are guiding us towards inclusive and sustainable development.

Considering these principles, it is natural that membrane research in Portugal is progressively focusing on the use of natural resources and renewable energy. Several examples can be given, such as:

- The development of membranes from natural, renewable resources, namely plant and bacterial cellulose and other natural (bio)polymers, which can be modified and functionalised for a diversity of applications;
- The recovery of chemical elements that are becoming rare and are required for many applications associated with a sustainable lifestyle: (bio)refining of wastewaters aiming the recovery of essential elements such as phosphorus; sea mining, seeking the recovery of a large diversity of elements, including lithium and rare metals;
- The use of renewable sources of energy, namely the integration of solar and membrane technology, but also the exploitation of salinity gradients for the harvesting of energy, supported on membrane-assisted processes.

The development of new membranes, seeking for new or improved functionalities, assuring their sustainable production, extending their service time and life cycle is one major trend on membrane research in Portugal:

- Membranes with new functionalities, including research for membranes with controlled pore size (i.e., isoporous) and pore size distribution, with extremely high hydrophilicity or extremely high hydrophobicity namely for membrane contactors applications, highly resistant to organic solvents, namely using natural polymers as starting raw materials, and responsive to external stimuli (photo and magnetic stimuli are among approaches under study) aiming different applications, notably in the biomedical area;
- Development and fabrication of membranes using less toxic, sustainable solvents, such as supercritical fluids, room temperature ionic liquids and deep eutectic solvents, among others;
- Recycling/reusing membranes after use, namely by modifying their properties extending their service time.

Additionally, research towards intensified processes, although assuring their sustainability, are critical issues under development:

- Intelligent use of energy, avoiding processes that involve phase transition and applying energy where required: at the interface where mass and energy transfer takes place. For this purpose, membrane research aiming the design of membranes with a designed topography that promotes turbulence near their surface, assuring improved mass and heat transfer conditions, are currently under development;
- Module design, as well as their components, using computational fluid dynamics is also a hot subject, aiming higher throughput by assuring better mass and heat transfer conditions, and minimising fouling with a minimum expenditure of energy. Spacer design and new (revisited) fouling control strategies are under consideration, such as the use of rotating and vibrating membranes.

- Efficient monitoring of membrane processes, seeking on-line, real-time monitoring, making possible the establishment of advanced automation and better process control strategies. This approach may require the use of different sources of information/monitoring signals, some of them rather complex that, using data-driven modelling and machine learning approaches, may lead to advanced control with an impact on process efficiency;
- Process integration in close interaction with corporate partners aiming the implementation and validation of membrane processes in all relevant domains with impact in social development: safe food and water production, valorisation and management of waste (regarded as by-products), environmental protection, energy production and storage, production of goods and health protection.

We would like conclude by emphasising the commitment of the membrane community in Portugal in education and awareness programmes at all levels, which translates our concern to develop knowledge in membrane science and engineering, transposing it to welfare and social development.

Author Contributions: Conceptualization, I.M.C., L.C.T., D.M.F.S. and S.V.; methodology, I.M.C., L.C.T., D.M.F.S. and S.V.; investigation, I.M.C., L.C.T., D.M.F.S. and S.V.; writing—original draft preparation I.M.C., L.C.T., D.M.F.S., S.V., A.M., J.G.C. and M.N.d.P.; writing—review and editing, preparation I.M.C., L.C.T., D.M.F.S., S.V., A.M., J.G.C. and M.N.d.P. All authors have read and agreed to the published version of the manuscript.

Funding: This work was supported by Associate Laboratory for Green Chemistry—LAQV, which is financed by national funds from FCT/MCTES (UIDB/50006/2020 and UIDP/50006/2020), Research project PTDC/EQU-EPQ/29579/2017 funded by FCT/MCTES "*Programa Operacional Regional de Lisboa*, FEDER", project Nanoart PTDC/CTM-BIO/6178/2014 and CeFEMA with grant number 325UID/CTM/04540/2013 funded by FCT/MCTES.

Acknowledgments: The authors acknowledge Tiago Araújo for his valuable contribution in writing—original draft preparation—the carbon molecular sieve membranes content. LCT is grateful to *Fundação para a Ciência e a Tecnologia* (FCT/MCTES) for her assistant researcher contract under Scientific Employment Stimulus (2020.01555.CEECIND). DMFS thanks FCT/MCTES for a research contract in the scope of programmatic funding UIDP/04540/2020.

Conflicts of Interest: The authors declare no conflict of interest.

Abbreviations

2D	two-dimensional
AEM	anion-exchange membrane
ASI	Advanced Study Institute
BioPEM	biopolymer electrolyte membrane
BNC	bacterial nanocellulose
BPM	bipolar membrane
CA	cellulose acetate
CAn	carbonic anhydrase
CFD	computational fluid dynamics
CMS	carbon molecular sieve
CNC	cellulose nanocrystal
CNF	cellulose nanofibril
DBFC	direct borohydride fuel cell
DCMD	direct contact membrane distillation
DD	Donnan dialysis
DES	deep eutectic solvent
DF	dia-ultrafiltration
DLFC	direct liquid fuel cell
DMFC	direct methanol fuel cell
EEM	excitation-emission matrix

EIS	electrochemical impedance spectroscopy
EPS	exopolysaccharide
FCT NOVA	Nova School of Science and Technology
FEUP	Faculdade de Engenharia da Universidade do Porto
HER	hydrogen evolution reaction
IEMB	ion-exchange membrane bioreactor
IL	ionic liquid
IST	Instituto Superior Técnico
LIB	lithium-ion battery
LNEC	Laboratório Nacional de Engenharia Civil
MBO	membrane blood oxygenator
MEA	membrane-electrode assembly
MF	microfiltration
MFC	microbial fuel cell
MOF	metal-organic framework
NAqRFB	non-aqueous redox flow battery
NF	nanofiltration
NOM	natural organic matter
OER	oxygen evolution reaction
OMW	olive mill wastewater
OTC	oxytetracycline
PAA	poly(acrylic) acid
PAC	powdered activated carbon
PCA	principal components analysis
PCL	polycaprolactone
PDMS	polydimethylsiloxane
PEG	poly(ethylene glycol)
PEI	polyetherimide
PEM	proton exchange membrane
PES	polyethersulfone
PIL	poly (ionic liquid)
POMS/PEI	polyoctylmethylsiloxane/polyetherimide
PU	polyurethane
PVA	poly(vinyl alcohol)
PVDF-TrFE	poly(vinylidene fluoride-co-trifluoroethylene)
RED	reverse electrodialysis
RO	reverse osmosis
SILM	supported ionic liquid membrane
SGO	sulphonated graphene oxide
sPEEK	sulfonated poly (ether ether ketone)
SS	silica spheres
UF	ultrafiltration

References

1. Bungay, P.M.; Londsdale, H.K.; de Pinho, M.N. (Eds.) *Synthetic Membranes: Science, Engineering and Applications*; Springer: Dordrecht, The Netherlands, 1986. [CrossRef]
2. Zhao, C.-T.; Norberta de Pinho, M. Design of polypropylene oxide/polybutadiene bi-soft segment urethane/urea polymer for pervaporation membranes. *Polymer* **1999**, *40*, 6089–6097. [CrossRef]
3. Queiroz, D.P.; Pinto, I.M.; Besteiro, M.C.F.; Silva, A.F.M.; Gil, M.H.; Guiomar, A.J.; De Pinho, M.N. Surface and Hemocompatibility Studies of Bi-Soft Segment Polyurethane Membranes. *Int. J. Artif. Organs* **2006**, *29*, 866–872. [CrossRef] [PubMed]
4. Mendes, G.; Faria, M.; Carvalho, A.; Gonçalves, M.C.; de Pinho, M.N. Structure of water in hybrid cellulose acetate-silica ultrafiltration membranes and permeation properties. *Carbohydr. Polym.* **2018**, *189*, 342–351. [CrossRef]
5. Beisl, S.; Monteiro, S.; Santos, R.; Figueiredo, A.S.; Sánchez-Loredo, M.G.; Lemos, M.A.; Lemos, F.; Minhalma, M.; de Pinho, M.N. Synthesis and bactericide activity of nanofiltration composite membranes—Cellulose acetate/silver nanoparticles and cellulose acetate/silver ion exchanged zeolites. *Water Res.* **2019**, *149*, 225–231. [CrossRef]
6. De Pinho, M.N.; Geraldes, V.M.; Rosa, M.J.; Afonso, M.D.; Figueira, H.; Taborda, F.; Almeida, G.; Ganho, R.; Creusen, R.; Hanemaaijer, J.; et al. Water recovery from bleached pulp effluents. *Tappi J.* **1996**, *79*, 117–124.

7. Geraldes, V.; de Pinho, M.N. Process water recovery from pulp bleaching effluents by an NF/ED hybrid process. *J. Membr. Sci.* **1995**, *102*, 209–221. [CrossRef]
8. Minhalma, M.; de Pinho, M.N. Development of nanofiltration/steam stripping sequence for coke plant wastewater treatment. *Desalination* **2002**, *149*, 95–100. [CrossRef]
9. Korzenowski, C.; Minhalma, M.; Bernardes, A.M.; Ferreira, J.Z.; de Pinho, M.N. Nanofiltration for the treatment of coke plant ammoniacal wastewaters. *Sep. Purif. Technol.* **2011**, *76*, 303–307. [CrossRef]
10. Minhalma, M.; de Pinho, M.N. Flocculation/Flotation/Ultrafiltration Integrated Process for the Treatment of Cork Processing Wastewaters. *Environ. Sci. Technol.* **2001**, *35*, 4916–4921. [CrossRef] [PubMed]
11. Rosa Santos, F.; Catarino, I.; Geraldes, V.; Pinho, M.N. Concentration and Rectification of Grape Must by Nanofiltration. *Am. J. Enol. Vitic.* **2008**, *59*, 446–450.
12. Minhalma, M.; de Pinho, M.N. Tannic-membrane interactions on ultrafiltration of cork processing wastewaters. *Sep. Purif. Technol.* **2001**, *22–23*, 479–488. [CrossRef]
13. Minhalma, M.; Domínguez, J.R.; de Pinho, M.N. Cork processing wastewaters treatment by an ozonization/ultrafiltration integrated process. *Desalination* **2006**, *191*, 148–152. [CrossRef]
14. Geraldes, V.; Minhalma, M.; Pinho, M.N.; Anil, A.; Ozgunay, H.; Bitlisli, B.O.; Sari, O. Nanofiltration of Cork Wastewaters and Their Possible Use in Leather Industry as Tanning Agents. *Pol. J. Environ. Stud.* **2009**, *18*, 353–357.
15. Streit, K.F.; Ferreira, J.Z.; Bernardes, A.M.; Norberta De Pinho, M. Ultrafiltration/Nanofiltration for the Tertiary Treatment of Leather Industry Effluents. *Environ. Sci. Technol.* **2009**, *43*, 9130–9135. [CrossRef]
16. Catarino, J.; Mendonça, E.; Picado, A.; Lança, A.; Silva, L.; Pinho, M.D.P.N.D. Membrane-based treatment for tanning wastewa tersA paper submitted to the Journal of Environmental Engineering and Science. *Can. J. Civ. Eng.* **2009**, *36*, 356–362. [CrossRef]
17. Brites Alves, A.M.; Norberta de Pinho, M. Ultrafiltration for colour removal of tannery dyeing wastewaters. *Desalination* **2000**, *130*, 147–154. [CrossRef]
18. Afonso, M.D.; De Pinho, M.N. Nanofiltration of Bleaching Pulp and Paper Effluents in Tubular Polymeric Membranes. *Sep. Sci. Technol.* **1997**, *32*, 2641–2658. [CrossRef]
19. Rosa, M.J.; de Pinho, M.N. The role of ultrafiltration and nanofiltration on the minimisation of the environmental impact of bleached pulp effluents. *J. Membr. Sci.* **1995**, *102*, 155–161. [CrossRef]
20. Afonso, M.D.; Geraldes, V.; Rosa, M.J.; de Pinho, M.N. Nanofiltration removal of chlorinated organic compounds from alkaline bleaching effluents in a pulp and paper plant. *Water Res.* **1992**, *26*, 1639–1643. [CrossRef]
21. Afonso, M.D.; Pinho, M.N. Membrane separation processes in pulp and paper production. *Filtr. Sep.* **1991**, *28*, 45–48. [CrossRef]
22. Afonso, M.D.; De Pinho, M.N. Ultrafiltration of bleach effluents in cellulose production. *Desalination* **1990**, *79*, 115–124. [CrossRef]
23. Magueijo, V.; de Pinho, M.N.; Geraldes, V. Numerical and experimental study of mass transfer in lysozyme ultrafiltration. *Desalination* **2002**, *145*, 193–199. [CrossRef]
24. De Pinho, M.N.; Rautenbach, R.; Herion, C. Mass transfer in radiation-grafted pervaporation membranes. *J. Membr. Sci.* **1990**, *54*, 131–143. [CrossRef]
25. Neto, J.M.; Pinho, M.N. Mass transfer modelling for solvent dehydration by pervaporation. *Sep. Purif. Technol.* **2000**, *18*, 151–161. [CrossRef]
26. Geraldes, V.; Semiao, V.; Norberta de Pinho, M. The effect on mass transfer of momentum and concentration boundary layers at the entrance region of a slit with a nanofiltration membrane wall. *Chem. Eng. Sci.* **2002**, *57*, 735–748. [CrossRef]
27. Murphy, D.; de Pinho, M.N. An ATR-FTIR study of water in cellulose acetate membranes prepared by phase inversion. *J. Membr. Sci.* **1995**, *106*, 245–257. [CrossRef]
28. Stamatialis, D.F.; Dias, C.R.; Norberta de Pinho, M. Atomic force microscopy of dense and asymmetric cellulose-based membranes. *J. Membr. Sci.* **1999**, *160*, 235–242. [CrossRef]
29. Crespo, J.; Moura, M.; Carrondo, M. Some engineering parameters for propionic acid fermentation coupled with ultrafiltration. *Appl. Biochem. Biotechnol.* **1990**, *24*, 613–625. [CrossRef]
30. Barreiros, A.; Rodrigues, C.; Crespo, J.P.; Reis, M. Membrane bioreactor for drinking water denitrification. *Bioprocess. Eng.* **1998**, *18*, 297–302. [CrossRef]
31. Velizarov, S.; Rodrigues, C.M.; Reis, M.A.; Crespo, J.G. Mechanism of charged pollutants removal in an ion exchange membrane bioreactor: Drinking water denitrification. *Biotechnol. Bioeng.* **2000**, *71*, 245–254. [CrossRef]
32. Velizarov, S.; Crespo, J.G.; Reis, M.A. Ion exchange membrane bioreactor for selective removal of nitrate from drinking water: Control of ion fluxes and process performance. *Biotechnol. Progr.* **2002**, *18*, 296–302. [CrossRef] [PubMed]
33. Velizarov, S.; Reis, M.A.; Crespo, J.G. Removal of trace mono-valent inorganic pollutants in an ion exchange membrane bioreactor: Analysis of transport rate in a denitrification process. *J. Membr. Sci.* **2003**, *217*, 269–284. [CrossRef]
34. Sousa, H.A.; Crespo, J.G.; Afonso, C.A. Asymmetric hydrolysis of a meso-diester using pig liver esterase immobilised in hollow fibre ultrafiltration membrane. *Tetrahedro Asymmetry* **2000**, *11*, 929–934. [CrossRef]
35. Sousa, H.A.; Rodrigues, C.; Klein, E.; Afonso, C.A.M.; Crespo, J.G. Immobilisation of pig liver esterase in hollow fibre membranes. *Enzym. Microb. Technol.* **2001**, *29*, 625–634. [CrossRef]
36. Schäffer, T.; Crespo, J.G. Recovery of aroma compounds from fermentation by pervaporation. *Environ. Prot. Eng.* **1999**, *25*, 73–85.
37. Schäfer, T.; Bengtson, G.; Pingel, H.; Böddeker, K.; Crespo, J. Recovery of aroma compounds from a wine-must fermentation by organophilic pervaporation. *Biotechnol. Bioeng.* **1999**, *62*, 412–421. [CrossRef]

Crespo, J.; Trotin, M.; Hough, D.; Howell, J. Use of fluorescence labelling to monitor protein fractionation by ultrafiltration under controlled permeate flux. *J. Membr. Sci.* **1999**, *155*, 209–230. [CrossRef]

Coelhoso, I.; Moura, T.; Crespo, J.; Carrondo, M. Transport mechanisms in liquid membranes with ion exchange carriers. *J. Membr. Sci.* **1995**, *108*, 231–244. [CrossRef]

Coelhoso, I.; Crespo, J.; Carrondo, M. Modeling of ion-pairing extraction with quaternary amines. *Sep. Sci. Technol.* **1996**, *31*, 491–511. [CrossRef]

Coelhoso, I.; Silvestre, P.; Viegas, R.; Crespo, J.; Carrondo, M. Membrane-based solvent extraction and stripping of lactate in hollow-fibre contactors. *J. Membr. Sci.* **1997**, *1*, 19–32. [CrossRef]

Fortunato, R.; Afonso, C.A.; Reis, M.; Crespo, J.G. Supported liquid membranes using ionic liquids: Study of stability and transport mechanisms. *J. Membr. Sci.* **2004**, *242*, 197–209. [CrossRef]

Fortunato, R.; González-Muñoz, M.J.; Kubasiewicz, M.; Luque, S.; Alvarez, J.; Afonso, C.A.; Coelhoso, I.M.; Crespo, J.G. Liquid membranes using ionic liquids: The influence of water on solute transport. *J. Membr. Sci.* **2005**, *249*, 153–162. [CrossRef]

Neves, L.A.; Crespo, J.G.; Coelhoso, I.M. Gas permeation studies in supported ionic liquid membranes. *J. Membr. Sci.* **2010**, *357*, 160–170. [CrossRef]

Neves, L.A.; Afonso, C.; Coelhoso, I.M.; Crespo, J.G. Integrated CO_2 capture and enzymatic bioconversion in supported ionic liquid membranes. *Sep. Purif. Technol.* **2012**, *97*, 34–41. [CrossRef]

Polino, M.; Portugal, C.A.M.; Le The, H.; Tiggelaar, R.; Eijkel, J.; Crespo, J.G.; Coelhoso, I.M.; Pina, M.P.; Mallada, R. Enhanced Protein Crystallization on Nafion Membranes Modified by Low-Cost Surface Patterning Techniques. *Cryst. Growth Des.* **2020**, *20*, 2174–2186. [CrossRef]

Syed, U.T.; Leonardo, I.; Lahoz, R.; Gaspar, F.B.; Huertas, R.; Crespo, M.T.B.; Arruebo, M.; Crespo, J.G.; Sebastian, V.; Brazinha, C. Microengineered Membranes for Sustainable Production of Hydrophobic Deep Eutectic Solvent-Based Nanoemulsions by Membrane Emulsification for Enhanced Antimicrobial Activity. *ACS Sustain. Chem. Eng.* **2020**, *8*, 16526–16536. [CrossRef]

Schäfer, T.; Vital, J.; Crespo, J.G. Coupled pervaporation/mass spectrometry for investigating membrane mass transport phenomena. *J. Membr. Sci.* **2004**, *241*, 197–205. [CrossRef]

Brazinha, C.; Fonseca, A.P.; Teodoro, O.M.N.D.; Crespo, J.G. On-line and real-time monitoring of organophilic pervaporation by mass spectrometry. *J. Membr. Sci.* **2010**, *347*, 83–92. [CrossRef]

Fraga, S.C.; Trabucho, L.; Brazinha, C.; Crespo, J.G. Characterisation and modelling of transient transport through dense membranes using on-line mass spectrometry. *J. Membr. Sci.* **2015**, *479*, 213–222. [CrossRef]

Wolf, G.; Crespo, J.G.; Reis, M.A. Optical and spectroscopic methods for biofilm examination and monitoring. *Rev. Environ. Sci. Biotechnol.* **2002**, *1*, 227–251. [CrossRef]

Wolf, G.; Almeida, J.S.; Pinheiro, C.; Correia, V.; Rodrigues, C.; Reis, M.A.; Crespo, J.G. Two-dimensional fluorometry coupled with artificial neural networks: A novel method for on-line monitoring of complex biological processes. *Biotechnol. Bioeng.* **2001**, *72*, 297–306. [CrossRef]

Portugal, C.A.; Lima, J.; Crespo, J.G. Probing the change of enzymatic activity of horseradish peroxidase induced by membrane permeation using tryptophan fluorescence. *J. Membr. Sci.* **2006**, *284*, 180–192. [CrossRef]

Portugal, C.A.M.; Lima, J.C.; Crespo, J.G. Fluorescence probing of structural and functional changes of proteins induced by ultrafiltration. *Desalination* **2006**, *199*, 547–549. [CrossRef]

Galinha, C.F.; Carvalho, G.; Portugal, C.A.; Guglielmi, G.; Reis, M.A.; Crespo, J.G. Two-dimensional fluorescence as a fingerprinting tool for monitoring wastewater treatment systems. *J. Chem. Technol. Biotechnol.* **2011**, *86*, 985–992. [CrossRef]

Galinha, C.; Carvalho, G.; Portugal, C.; Guglielmi, G.; Oliveira, R.; Crespo, J.; Reis, M. Real-time monitoring of membrane bioreactors with 2D-fluorescence data and statistically based models. *Water Sci. Technol.* **2011**, *63*, 1381–1388. [CrossRef]

Galinha, C.F.; Carvalho, G.; Portugal, C.A.M.; Guglielmi, G.; Reis, M.A.M.; Crespo, J.G. Multivariate statistically-based modelling of a membrane bioreactor for wastewater treatment using 2D fluorescence monitoring data. *Water Res.* **2012**, *46*, 3623–3636. [CrossRef]

Izák, P.; Godinho, M.; Brogueira, P.; Figueirinhas, J.; Crespo, J. 3D topography design of membranes for enhanced mass transport. *J. Membr. Sci.* **2008**, *321*, 337–343. [CrossRef]

Pawlowski, S.; Geraldes, V.; Crespo, J.G.; Velizarov, S. Computational fluid dynamics (CFD) assisted analysis of profiled membranes performance in reverse electrodialysis. *J. Membr. Sci.* **2016**, *502*, 179–190. [CrossRef]

Pereira, V.J.; Galinha, J.; Barreto Crespo, M.T.; Matos, C.T.; Crespo, J.G. Integration of nanofiltration, UV photolysis, and advanced oxidation processes for the removal of hormones from surface water sources. *Sep. Purif. Technol.* **2012**, *95*, 89–96. [CrossRef]

Sanches, S.; Penetra, A.; Rodrigues, A.; Cardoso, V.; Ferreira, E.; Benoliel, M.; Crespo, M.B.; Crespo, J.; Pereira, V. Removal of pesticides from water combining low pressure UV photolysis with nanofiltration. *Sep. Purif. Technol.* **2013**, *115*, 73–82. [CrossRef]

Lagorsse, S.; Magalhães, F.D.; Mendes, A. Carbon molecular sieve membranes: Sorption, kinetic and structural characterization. *J. Membr. Sci.* **2004**, *241*, 275–287. [CrossRef]

Lagorsse, S.; Magalhães, F.D.; Mendes, A. Aging study of carbon molecular sieve membranes. *J. Membr. Sci.* **2008**, *310*, 494–502. [CrossRef]

Campo, M.C.; Magalhães, F.D.; Mendes, A. Carbon molecular sieve membranes from cellophane paper. *J. Membr. Sci.* **2010**, *350*, 180–188. [CrossRef]

65. Teixeira, M.; Campo, M.C.; Tanaka, D.A.P.; Tanco, M.A.L.; Magen, C.; Mendes, A. Composite phenolic resin-based carbo molecular sieve membranes for gas separation. *Carbon* **2011**, *49*, 4348–4358. [CrossRef]
66. Teixeira, M.; Campo, M.; Tanaka, D.A.; Tanco, M.A.; Magen, C.; Mendes, A. Carbon–Al2O3–Ag composite molecular siev membranes for gas separation. *Chem. Eng. Res. Des.* **2012**, *90*, 2338–2345. [CrossRef]
67. Araújo, T.; Andrade, M.; Bernardo, G.; Mendes, A. Stable cellulose-based carbon molecular sieve membranes with very hig selectivities. *J. Membr. Sci.* **2022**, *641*, 119852. [CrossRef]
68. Pereira, A.I.; Pérez, P.; Rodrigues, S.C.; Mendes, A.; Madeira, L.M.; Tavares, C.J. Deposition of Pd–Ag thin film membranes o ceramic supports for hydrogen purification/separation. *Mater. Res. Bull.* **2015**, *61*, 528–533. [CrossRef]
69. Brandão, L.; Madeira, L.M.; Mendes, A.M. Propyne hydrogenation in a continuous polymeric catalytic membrane reactor. *Chen Eng. Sci.* **2007**, *62*, 6768–6776. [CrossRef]
70. Ribeirinha, P.; Abdollahzadeh, M.; Boaventura, M.; Mendes, A. H2 production with low carbon content via MSR in packed be membrane reactors for high-temperature polymeric electrolyte membrane fuel cell. *Appl. Energy* **2017**, *188*, 409–419. [CrossRef
71. Sá, S.; Silva, H.; Sousa, J.M.; Mendes, A. Hydrogen production by methanol steam reforming in a membrane reactor: Palladiun vs. carbon molecular sieve membranes. *J. Membr. Sci.* **2009**, *339*, 160–170. [CrossRef]
72. Sá, S.; Sousa, J.M.; Mendes, A. Methanol steam reforming in a dual-bed membrane reactor for producing PEMFC grade hydroger *Catal. Today* **2010**, *156*, 254–260. [CrossRef]
73. Silva, V.S.; Ruffmann, B.; Vetter, S.; Mendes, A.; Madeira, L.M.; Nunes, S.P. Characterization and application of composit membranes in DMFC. *Catal. Today* **2005**, *104*, 205–212. [CrossRef]
74. Silva, V.; Schirmer, J.; Reissner, R.; Ruffmann, B.; Silva, H.; Mendes, A.; Madeira, L.; Nunes, S. Proton electrolyte membran properties and direct methanol fuel cell performance: II. Fuel cell performance and membrane properties effects. *J. Power Source* **2005**, *140*, 41–49. [CrossRef]
75. Boaventura, M.; Ponce, M.; Brandao, L.; Mendes, A.; Nunes, S. Proton conductive membranes based on doped sulfonatec polytriazole. *Int. J. Hydrogen Energy* **2010**, *35*, 12054–12064. [CrossRef]
76. Campo, M.C.; Magalhães, F.D.; Mendes, A. Separation of nitrogen from air by carbon molecular sieve membranes. *J. Membr. Sc* **2010**, *350*, 139–147. [CrossRef]
77. Cruz, P.; Santos, J.C.; Magalhães, F.D.; Mendes, A. Simulation of separation processes using finite volume method. *Comput. Chem Eng.* **2005**, *30*, 83–98. [CrossRef]
78. Sousa, J.M.; Cruz, P.; Magalhães, F.D.; Mendes, A. Modeling catalytic membrane reactors using an adaptive wavelet-basec collocation method. *J. Membr. Sci.* **2002**, *208*, 57–68. [CrossRef]
79. Mendes, D.; Sá, S.; Tosti, S.; Sousa, J.M.; Madeira, L.M.; Mendes, A. Experimental and modeling studies on the low-temperatur water-gas shift reaction in a dense Pd–Ag packed-bed membrane reactor. *Chem. Eng. Sci.* **2011**, *66*, 2356–2367. [CrossRef]
80. Sousa, J.M.; Cruz, P.; Mendes, A. Modeling a catalytic polymeric non-porous membrane reactor. *J. Membr. Sci.* **2001**, *181*, 241–252 [CrossRef]
81. Ribeirinha, P.; Abdollahzadeh, M.; Pereira, A.; Relvas, F.; Boaventura, M.; Mendes, A. High temperature PEM fuel cell integratec with a cellular membrane methanol steam reformer: Experimental and modelling. *Appl. Energy* **2018**, *215*, 659–669. [CrossRef]
82. Ribeirinha, P.; Abdollahzadeh, M.; Sousa, J.; Boaventura, M.; Mendes, A. Modelling of a high-temperature polymer electrolyt membrane fuel cell integrated with a methanol steam reformer cell. *Appl. Energy* **2017**, *202*, 6–19. [CrossRef]
83. Delgado, N.M.; Monteiro, R.; Abdollahzadeh, M.; Ribeirinha, P.; Bentien, A.; Mendes, A. 2D-dynamic phenomenologica modelling of vanadium redox flow batteries—Analysis of the mass transport related overpotentials. *J. Power Sources* **2020** *480*, 229142. [CrossRef]
84. Delgado, N.M.; Monteiro, R.; Mendes, A. The first approach to dynamic modeling of a solar vanadium redox flow cell. *Nanc Energy* **2021**, *89*, 106372. [CrossRef]
85. Janknecht, P.; Lopes, A.D.; Mendes, A.M. Removal of Industrial Cutting Oil from Oil Emulsions by Polymeric Ultra- and Microfiltration Membranes. *Environ. Sci. Technol.* **2004**, *38*, 4878–4883. [CrossRef] [PubMed]
86. Catarino, M.; Mendes, A.; Madeira, L.M.; Ferreira, A. Alcohol Removal From Beer by Reverse Osmosis. *Sep. Sci. Technol.* **2007**, *42* 3011–3027. [CrossRef]
87. Catarino, M.; Mendes, A. Dealcoholizing wine by membrane separation processes. *Innov. Food Sci. Emerg. Technol.* **2011**, *12* 330–337. [CrossRef]
88. Catarino, M.; Mendes, A. Non-alcoholic beer—A new industrial process. *Sep. Purif. Technol.* **2011**, *79*, 342–351. [CrossRef]
89. Gales, L.; Mendes, A.; Costa, C. Removal of acetone, ethyl acetate and ethanol vapors from air using a hollow fiber PDMS membrane module. *J. Membr. Sci.* **2002**, *197*, 211–222. [CrossRef]
90. Brandão, L.; Madeira, L.M.; Mendes, A.M. Mass transport on composite dense PDMS membranes with palladium nanoclusters. *J Membr. Sci.* **2007**, *288*, 112–122. [CrossRef]
91. Pantaleão, I.; Portugal, A.F.; Mendes, A.; Gabriel, J. Carbon Dioxide Absorption in a Membrane Contactor with Color Change. *J Chem. Educ.* **2010**, *87*, 1377–1379. [CrossRef]
92. Portugal, A.F.; Magalhães, F.D.; Mendes, A. Carbon dioxide removal from anaesthetic gas circuits using hollow fiber membrane contactors with amino acid salt solutions. *J. Membr. Sci.* **2009**, *339*, 275–286. [CrossRef]

3. Cabral, J.; Pantateão, I.; Rodrigues, S.; Catarino, M.; Magalhães, F.; Mendes, A. Kinetics of the Carbon Dioxide Absorption and Desorption with Amino Acid Salt Solutions using Hollow Fiber Membrane Contactors. *Procedia Eng.* **2012**, *44*, 1223–1224. [CrossRef]
4. Silva, V.S.; Mendes, A.; Madeira, L.M.; Nunes, S.P. Proton exchange membranes for direct methanol fuel cells: Properties critical study concerning methanol crossover and proton conductivity. *J. Membr. Sci.* **2006**, *276*, 126–134. [CrossRef]
5. Silva, V.S.; Ruffmann, B.; Silva, H.; Silva, V.B.; Mendes, A.; Madeira, L.M.; Nunes, S. Zirconium oxide hybrid membranes for direct methanol fuel cells—Evaluation of transport properties. *J. Membr. Sci.* **2006**, *284*, 137–144. [CrossRef]
6. Brandão, L.; Rodrigues, J.; Madeira, L.M.; Mendes, A. Methanol crossover reduction by Nafion modification with palladium composite nanoparticles: Application to direct methanol fuel cells. *Int. J. Hydrogen Energy* **2010**, *35*, 11561–11567. [CrossRef]
7. Boaventura, M.; Mendes, A. Activation procedures characterization of MEA based on phosphoric acid doped PBI membranes. *Int. J. Hydrogen Energy* **2010**, *35*, 11649–11660. [CrossRef]
8. Boaventura, M.; Sander, H.; Friedrich, K.A.; Mendes, A. The influence of CO on the current density distribution of high temperature polymer electrolyte membrane fuel cells. *Electrochim. Acta* **2011**, *56*, 9467–9475. [CrossRef]
9. Boaventura, M.; Alves, I.; Ribeirinha, P.; Mendes, A. The influence of impurities in high temperature polymer electrolyte membrane fuel cells performance. *Int. J. Hydrogen Energy* **2016**, *41*, 19771–19780. [CrossRef]
0. Ponce, M.; Boaventura, M.; Gomes, D.; Mendes, A.; Madeira, L.; Nunes, S. Proton conducting membranes based on benzimidazole sulfonic acid doped sulfonated poly (oxadiazole–triazole) copolymer for low humidity operation. *Fuel Cells* **2008**, *8*, 209–216. [CrossRef]
1. Chen, X.; Zhang, Y.; Ribeiorinha, P.; Li, H.; Kong, X.; Boaventura, M. A proton conductor electrolyte based on molten CsH5(PO4)2 for intermediate-temperature fuel cells. *RSC Adv.* **2018**, *8*, 5225–5232. [CrossRef]
02. Delgado, S.; Lagarteira, T.; Mendes, A. Air bleeding strategies to increase the efficiency of proton exchange membrane fuel cell stationary applications fuelled with CO ppm-levels. *Int. J. Electrochem. Sci.* **2020**, *15*, 613–627. [CrossRef]
03. Pais, J.A.G.C.R.; Ferreira, L.M.G.A. Performance study of an industrial RO plant for seawater desalination. *Desalination* **2007**, *208*, 269–276. [CrossRef]
04. Geraldes, V.; Pereira, N.E.; Norberta de Pinho, M. Simulation and Optimization of Medium-Sized Seawater Reverse Osmosis Processes with Spiral-Wound Modules. *Ind. Eng. Chem. Res.* **2005**, *44*, 1897–1905. [CrossRef]
05. Afonso, M.D.; Jaber, J.O.; Mohsen, M.S. Brackish groundwater treatment by reverse osmosis in Jordan. *Desalination* **2004**, *164*, 157–171. [CrossRef]
06. Sanches, S.; Penetra, A.; Rodrigues, A.; Ferreira, E.; Cardoso, V.V.; Benoliel, M.J.; Barreto Crespo, M.T.; Pereira, V.J.; Crespo, J.G. Nanofiltration of hormones and pesticides in different real drinking water sources. *Sep. Purif. Technol.* **2012**, *94*, 44–53. [CrossRef]
07. Velizarov, S.; Crespo, J.G.; Reis, M.A. Removal of inorganic anions from drinking water supplies by membrane bio/processes. *Rev. Environ. Sci. Bio/Technol.* **2004**, *3*, 361–380. [CrossRef]
08. Matos, C.T.; Sequeira, A.M.; Velizarov, S.; Crespo, J.G.; Reis, M.A.M. Nitrate removal in a closed marine system through the ion exchange membrane bioreactor. *J. Hazard. Mater.* **2009**, *166*, 428–434. [CrossRef]
09. Gando-Ferreira, L.M.; Gaspar, C.S.S.; Monteiro, M.; Moreira, M.J.A. Studies on integration of ion exchange and nanofiltration for water desalination. *Sep. Sci. Technol.* **2017**, *52*, 2600–2610. [CrossRef]
10. Costa, A.R.; de Pinho, M.N. Comparison of the performance of ultrafiltration and nanofiltration in surface water treatment. *Desalination* **2006**, *199*, 73–75. [CrossRef]
11. Costa, A.R.; de Pinho, M.N. Performance and cost estimation of nanofiltration for surface water treatment in drinking water production. *Desalination* **2006**, *196*, 55–65. [CrossRef]
12. Costa, A.R.; de Pinho, M.N. Coagulation/flocculation/ultrafiltration for natural organic matter removal in drinking water production. *Water Supply* **2004**, *4*, 215–222. [CrossRef]
13. Campinas, M.; Viegas, R.M.C.; Coelho, R.; Lucas, H.; Rosa, M.J. Adsorption/Coagulation/Ceramic Microfiltration for Treating Challenging Waters for Drinking Water Production. *Membranes* **2021**, *11*, 91. [CrossRef]
14. Velizarov, S.; Reis, M.A.; Crespo, J.G. The Ion Exchange Membrane Bioreactor Developments and Perspectoves in Drinking Water Treatment Dordrecht. In *Water Purification and Management*; Springer: Dordrecht, The Netherlands, 2011; pp. 119–145.
15. Pessoa Lopes, M.; Galinha, C.F.; Crespo, J.G.; Velizarov, S. Optimisation of arsenate removal from water by an integrated ion-exchange membrane process coupled with Fe co-precipitation. *Sep. Purif. Technol.* **2020**, *246*, 116894. [CrossRef]
16. Seabra Pinto, A.; Cabral Rolo, J. Chapter 22. Trade, logistics and agro-food strategies in Portugal. In *MediTERRA 2014 (English)*; Presses de Sciences Po: Paris, France, 2014; pp. 377–386.
17. Costa, J.M.; Oliveira, M.; Egipto, R.J.; Cid, J.F.; Fragoso, R.A.; Lopes, C.M.; Duarte, E.N. Water and wastewater management for sustainable viticulture and oenology in South Portugal—A review. *Ciênc. Téc. Vitiv.* **2020**, *35*, 1–15. [CrossRef]
18. Velizarov, S.; Crespo, J.G. Membrane Processing for the Recovery of Bioactive Compounds in Agro-Industries. In *Innovation in Food Engineering: New Techniques and Products*, 1st ed.; CRC Press: Boca Raton, FL, USA, 2009; pp. 137–160. [CrossRef]
19. Fraga, M.C.; Sanches, S.; Crespo, J.G.; Pereira, V.J. Assessment of a New Silicon Carbide Tubular Honeycomb Membrane for Treatment of Olive Mill Wastewaters. *Membranes* **2017**, *7*, 12. [CrossRef] [PubMed]
20. Fraga, M.C.; Huertas, R.M.; Crespo, J.G.; Pereira, V.J. Novel Submerged Photocatalytic Membrane Reactor for Treatment of Olive Mill Wastewaters. *Catalysts* **2019**, *9*, 769. [CrossRef]

121. Kamali, M.; Suhas, D.P.; Costa, M.E.; Capela, I.; Aminabhavi, T.M. Sustainability considerations in membrane-based technologies for industrial effluents treatment. *Chem. Eng. J.* **2019**, *368*, 474–494. [CrossRef]
122. Salazar, H.; Martins, P.M.; Santos, B.; Fernandes, M.M.; Reizabal, A.; Sebastián, V.; Botelho, G.; Tavares, C.J.; Vilas-Vilela, J.L. Lanceros-Mendez, S. Photocatalytic and antimicrobial multifunctional nanocomposite membranes for emerging pollutants water treatment applications. *Chemosphere* **2020**, *250*, 126299. [CrossRef]
123. De Almeida, M.S.; Martins, R.C.; Quinta-Ferreira, R.M.; Gando-Ferreira, L.M. Optimization of operating conditions for the valorization of olive mill wastewater using membrane processes. *Environ. Sci. Pollut. Res.* **2018**, *25*, 21968–21981. [CrossRef]
124. Ribeiro, B.; Torrado, I.; Di Berardino, S.; Paixão, S.M.; Rusan, M.J.; Bani Amer, A.; Zuraiqi, S.; Eusébio, A. Jet-loop reactor with cross-flow ultrafiltration membrane system for treatment of olive mill wastewater. *Water Pract. Technol.* **2018**, *13*, 247–256. [CrossRef]
125. Antónia Nunes, M.; Pawlowski, S.; Costa, A.S.G.; Alves, R.C.; Oliveira, M.B.P.P.; Velizarov, S. Valorization of olive pomace by a green integrated approach applying sustainable extraction and membrane-assisted concentration. *Sci. Total Environ.* **2019**, *652*, 40–47. [CrossRef]
126. Gomes, L.; Borges, C.; Serralheiro, M.L.M.; Minhalma, M.; Pacheco, R. Cork processing wastewaters components fractioned by ultrafiltration membranes—Studies of antioxidant and antitumoral activity. *J. Chem. Technol. Biotechnol.* **2018**, *93*, 861–870. [CrossRef]
127. Giacobbo, A.; Moura Bernardes, A.; Filipe Rosa, M.J.; De Pinho, M.N. Concentration Polarization in Ultrafiltration/Nanofiltration for the Recovery of Polyphenols from Winery Wastewaters. *Membranes* **2018**, *8*, 46. [CrossRef] [PubMed]
128. Leandro, M.J.; Marques, S.; Ribeiro, B.; Santos, H.; Fonseca, C. Integrated Process for Bioenergy Production and Water Recycling in the Dairy Industry: Selection of Kluyveromyces Strains for Direct Conversion of Concentrated Lactose-Rich Streams into Bioethanol. *Microorganisms* **2019**, *7*, 545. [CrossRef]
129. Bastos, P.D.A.; Santos, M.A.; Carvalho, P.J.; Velizarov, S.; Crespo, J.G. Pilot scale reverse osmosis refinery wastewater treatment—A techno-economical and sustainability assessment. *Environ. Sci. Water Res. Technol.* **2021**, *7*, 549–561. [CrossRef]
130. Zakmout, A.; Sadi, F.; Portugal, C.A.M.; Crespo, J.G.; Velizarov, S. Tannery Effluent Treatment by Nanofiltration, Reverse Osmosis and Chitosan Modified Membranes. *Membranes* **2020**, *10*, 378. [CrossRef] [PubMed]
131. Viegas, R.M.C.; Mesquita, E.; Campinas, M.; Rosa, M.J. Pilot Studies and Cost Analysis of Hybrid Powdered Activated Carbon/Ceramic Microfiltration for Controlling Pharmaceutical Compounds and Organic Matter in Water Reclamation. *Water* **2020**, *12*, 33. [CrossRef]
132. Espíndola, J.C.; Cristóvão, R.O.; Mendes, A.; Boaventura, R.A.R.; Vilar, V.J.P. Photocatalytic membrane reactor performance towards oxytetracycline removal from synthetic and real matrices: Suspended vs. immobilized TiO_2-P25. *Chem. Eng. J.* **2019**, *378*, 122114. [CrossRef]
133. Cristóvão, M.B.; Tela, S.; Silva, A.F.; Oliveira, M.; Bento-Silva, A.; Bronze, M.R.; Crespo, M.T.B.; Crespo, J.G.; Nunes, M.; Pereira, V.J. Occurrence of Antibiotics, Antibiotic Resistance Genes and Viral Genomes in Wastewater Effluents and Their Treatment by a Pilot Scale Nanofiltration Unit. *Membranes* **2021**, *11*, 9. [CrossRef]
134. Cristóvão, M.B.; Torrejais, J.; Janssens, R.; Luis, P.; Van der Bruggen, B.; Dubey, K.K.; Mandal, M.K.; Bronze, M.R.; Crespo, J.G. Pereira, V.J. Treatment of anticancer drugs in hospital and wastewater effluents using nanofiltration. *Sep. Purif. Technol.* **2019**, *224*, 273–280. [CrossRef]
135. Vinoth Kumar, R.; Barbosa, M.O.; Ribeiro, A.R.; Morales-Torres, S.; Pereira, M.F.R.; Silva, A.M.T. Advanced oxidation technologies combined with direct contact membrane distillation for treatment of secondary municipal wastewater. *Process. Saf. Environ. Prot.* **2020**, *140*, 111–123. [CrossRef]
136. Arboleda Mejia, J.A.; Ricci, A.; Figueiredo, A.S.; Versari, A.; Cassano, A.; Parpinello, G.P.; De Pinho, M.N. Recovery of Phenolic Compounds from Red Grape Pomace Extract through Nanofiltration Membranes. *Foods* **2020**, *9*, 1649. [CrossRef]
137. Syed, U.T.; Brazinha, C.; Crespo, J.G.; Ricardo-da-Silva, J.M. Valorisation of grape pomace: Fractionation of bioactive flavan-3-ols by membrane processing. *Sep. Purif. Technol.* **2017**, *172*, 404–414. [CrossRef]
138. Valério, R.; Crespo, J.G.; Galinha, C.F.; Brazinha, C. Effect of Ultrafiltration Operating Conditions for Separation of Ferulic Acid from Arabinoxylans in Corn Fibre Alkaline Extract. *Sustainability* **2021**, *13*, 4682. [CrossRef]
139. Pinto, P.R.; Mota, I.F.; Pereira, C.M.; Ribeiro, A.M.; Loureiro, J.M.; Rodrigues, A.E. Separation and recovery of polyphenols and carbohydrates from Eucalyptus bark extract by ultrafiltration/diafiltration and adsorption processes. *Sep. Purif. Technol.* **2017**, *183*, 96–105. [CrossRef]
140. Giacobbo, A.; Dias, B.B.; Onorevoli, B.; Bernardes, A.M.; de Pinho, M.N.; Caramão, E.B.; Rodrigues, E.; Jacques, R.A. Wine lees from the 1st and 2nd rackings: Valuable by-products. *J. Food Sci. Technol.* **2019**, *56*, 1559–1566. [CrossRef] [PubMed]
141. Giacobbo, A.; Bernardes, A.M.; de Pinho, M.N. Sequential pressure-driven membrane operations to recover and fractionate polyphenols and polysaccharides from second racking wine lees. *Sep. Purif. Technol.* **2017**, *173*, 49–54. [CrossRef]
142. Restolho, J.A.; Prates, A.; de Pinho, M.N.; Afonso, M.D. Sugars and lignosulphonates recovery from eucalyptus spent sulphite liquor by membrane processes. *Biomass Bioenergy* **2009**, *33*, 1558–1566. [CrossRef]
143. Ghalamara, S.; Silva, S.; Brazinha, C.; Pintado, M. Valorization of Fish by-Products: Purification of Bioactive Peptides from Codfish Blood and Sardine Cooking Wastewaters by Membrane Processing. *Membranes* **2020**, *10*, 44. [CrossRef]
144. Monte, J.; Ribeiro, C.; Parreira, C.; Costa, L.; Brive, L.; Casal, S.; Brazinha, C.; Crespo, J.G. Biorefinery of Dunaliella salina: Sustainable recovery of carotenoids, polar lipids and glycerol. *Bioresour. Technol.* **2020**, *297*, 122509. [CrossRef] [PubMed]

5. Polino, M.; Portugal, C.A.M.; Di Profio, G.; Coelhoso, I.M.; Crespo, J.G. Protein Crystallization by Membrane-Assisted Technology. *Cryst. Growth Des.* **2019**, *19*, 4871–4883. [CrossRef]
6. Polino, M.; Rho, H.S.; Pina, M.P.; Mallada, R.; Carvalho, A.L.; Romão, M.J.; Coelhoso, I.; Gardeniers, J.G.E.; Crespo, J.G.; Portugal, C.A.M. Protein Crystallization in a Microfluidic Contactor with Nafion®117 Membranes. *Membranes* **2021**, *11*, 549. [CrossRef] [PubMed]
7. Polino, M.; Carvalho, A.L.s.; Juknaitė, L.; Portugal, C.A.M.; Coelhoso, I.M.; Romão, M.J.; Crespo, J.G. Ion-Exchange Membranes for Stable Derivatization of Protein Crystals. *Cryst. Growth Des.* **2017**, *17*, 4563–4572. [CrossRef]
8. Martins, C.F.; Neves, L.A.; Chagas, R.; Ferreira, L.M.; Coelhoso, I.M.; Crespo, J.G. Removing CO_2 from Xenon anaesthesia circuits using an amino-acid ionic liquid solution in a membrane contactor. *Sep. Purif. Technol.* **2021**, *275*, 119190. [CrossRef]
9. Martins, C.F.; Neves, L.A.; Chagas, R.; Ferreira, L.M.; Afonso, C.A.M.; Crespo, J.G.; Coelhoso, I.M. CO_2 removal from anaesthesia circuits using gas-ionic liquid membrane contactors. *Sep. Purif. Technol.* **2020**, *250*, 116983. [CrossRef]
0. Malankowska, M.; Martins, C.F.; Rho, H.S.; Neves, L.A.; Tiggelaar, R.M.; Crespo, J.G.; Pina, M.P.; Mallada, R.; Gardeniers, H.; Coelhoso, I.M. Microfluidic devices as gas—Ionic liquid membrane contactors for CO_2 removal from anaesthesia gases. *J. Membr. Sci.* **2018**, *545*, 107–115. [CrossRef]
1. Eusébio, T.M.; Martins, A.R.; Pon, G.; Faria, M.; Morgado, P.; Pinto, M.L.; Filipe, E.J.M.; de Pinho, M.N. Sorption/Diffusion Contributions to the Gas Permeation Properties of Bi-Soft Segment Polyurethane/Polycaprolactone Membranes for Membrane Blood Oxygenators. *Membranes* **2020**, *10*, 8. [CrossRef] [PubMed]
2. Faria, M.; Moreira, C.; Mendonça Eusébio, T.; de Pinho, M.N.; Brogueira, P.; Semião, V. Oxygen mass transfer in a gas/membrane/liquid system surrogate of membrane blood oxygenators. *AIChE J.* **2018**, *64*, 3756–3763. [CrossRef]
3. Faria, M.; Moreira, C.; Eusébio, T.; Brogueira, P.; de Pinho, M.N. Hybrid flat sheet cellulose acetate/silicon dioxide ultrafiltration membranes for uremic blood purification. *Cellulose* **2020**, *27*, 3847–3869. [CrossRef]
4. De Pascale, M.; Faria, M.; Boi, C.; Semiao, V.; de Pinho, M.N. The effect of ultrafiltration transmembrane permeation on the flow field in a surrogate system of an artificial kidney. *Exp. Results* **2021**, *2*, e16. [CrossRef]
5. Teixeira, F.C.; de Sá, A.I.; Teixeira, A.P.S.; Rangel, C.M. Nafion phosphonic acid composite membranes for proton exchange membranes fuel cells. *Appl. Surf. Sci.* **2019**, *487*, 889–897. [CrossRef]
6. Teixeira, F.C.; de Sá, A.I.; Teixeira, A.P.S.; Rangel, C.M. Enhanced proton conductivity of Nafion-azolebisphosphonate membranes for PEM fuel cells. *New J. Chem.* **2019**, *43*, 15249–15257. [CrossRef]
7. Teixeira, F.C.; de Sá, A.I.; Teixeira, A.P.S.; Ortiz-Martínez, V.M.; Ortiz, A.; Ortiz, I.; Rangel, C.M. New modified Nafion-bisphosphonic acid composite membranes for enhanced proton conductivity and PEMFC performance. *Int. J. Hydrogen Energy* **2021**, *46*, 17562–17571. [CrossRef]
8. Vilela, C.; Silvestre, A.J.D.; Figueiredo, F.M.L.; Freire, C.S.R. Nanocellulose-based materials as components of polymer electrolyte fuel cells. *J. Mater. Chem. A* **2019**, *7*, 20045–20074. [CrossRef]
9. Vilela, C.; Martins, A.P.C.; Sousa, N.; Silvestre, A.J.D.; Figueiredo, F.M.L.; Freire, C.S.R. Poly(bis[2-(methacryloyloxy)ethyl] phosphate)/Bacterial Cellulose Nanocomposites: Preparation, Characterization and Application as Polymer Electrolyte Membranes. *Appl. Sci.* **2018**, *8*, 1145. [CrossRef]
0. Vilela, C.; Gadim, T.D.O.; Silvestre, A.J.D.; Freire, C.S.R.; Figueiredo, F.M.L. Nanocellulose/poly(methacryloyloxyethyl phosphate) composites as proton separator materials. *Cellulose* **2016**, *23*, 3677–3689. [CrossRef]
1. Vilela, C.; Morais, J.D.; Silva, A.C.Q.; Muñoz-Gil, D.; Figueiredo, F.M.L.; Silvestre, A.J.D.; Freire, C.S.R. Flexible Nanocellulose/Lignosulfonates Ion-Conducting Separators for Polymer Electrolyte Fuel Cells. *Nanomaterials* **2020**, *10*, 1713. [CrossRef]
2. Vilela, C.; Silva, A.C.Q.; Domingues, E.M.; Gonçalves, G.; Martins, M.A.; Figueiredo, F.M.L.; Santos, S.A.O.; Freire, C.S.R. Conductive polysaccharides-based proton-exchange membranes for fuel cell applications: The case of bacterial cellulose and fucoidan. *Carbohydr. Polym.* **2020**, *230*, 115604. [CrossRef] [PubMed]
3. Vilela, C.; Cordeiro, D.M.; Boas, J.V.; Barbosa, P.; Nolasco, M.; Vaz, P.D.; Rudić, S.; Ribeiro-Claro, P.; Silvestre, A.J.D.; Oliveira, V.B.; et al. Poly(4-styrene sulfonic acid)/bacterial cellulose membranes: Electrochemical performance in a single-chamber microbial fuel cell. *Bioresour. Technol. Rep.* **2020**, *9*, 100376. [CrossRef]
4. Vilela, C.; Sousa, N.; Pinto, R.J.B.; Silvestre, A.J.D.; Figueiredo, F.M.L.; Freire, C.S.R. Exploiting poly(ionic liquids) and nanocellulose for the development of bio-based anion-exchange membranes. *Biomass Bioenergy* **2017**, *100*, 116–125. [CrossRef]
5. Ferreira, R.B.; Falcão, D.S.; Pinto, A.M.F.R. Simulation of membrane chemical degradation in a proton exchange membrane fuel cell by computational fluid dynamics. *Int. J. Hydrogen Energy* **2021**, *46*, 1106–1120. [CrossRef]
6. Rambabu, G.; Bhat, S.D.; Figueiredo, F.M.L. Carbon Nanocomposite Membrane Electrolytes for Direct Methanol Fuel Cells—A Concise Review. *Nanomaterials* **2019**, *9*, 1292. [CrossRef]
7. Ben Jadi, S.; El Guerraf, A.; Bazzaoui, E.A.; Wang, R.; Martins, J.I.; Bazzaoui, M. Synthesis, characterization, and transport properties of Nafion-polypyrrole membrane for direct methanol fuel cell (DMFC) application. *J. Solid State Electrochem.* **2019**, *23*, 2423–2433. [CrossRef]
8. Ben Jadi, S.; El Guerraf, A.; Kiss, A.; El Azrak, A.; Bazzaoui, E.A.; Wang, R.; Martins, J.I.; Bazzaoui, M. Analyses of scanning electrochemical microscopy and electrochemical impedance spectroscopy in direct methanol fuel cells: Permeability resistance and proton conductivity of polyaniline modified membrane. *J. Solid State Electrochem.* **2020**, *24*, 1551–1565. [CrossRef]

169. De Oliveira, P.N.; Mendes, A.M.M. Preparation and characterization of an eco-friendly polymer electrolyte membrane (PEM) Based in a Blend of Sulphonated Poly(Vinyl Alcohol)/Chitosan Mechanically Stabilised by Nylon 6,6. *Mater. Res. Bull.* **2016**, *1*5 954–962. [CrossRef]
170. Šljukić, B.; Santos, D.M.F. Chapter 10—Direct Borohydride Fuel Cells (DBFCs). In *Direct Liquid Fuel Cells*; Akay, R.G., Yurtcan, A.B Eds.; Academic Press: New York, NY, USA, 2021; pp. 203–232. [CrossRef]
171. Gouda, M.H.; Gouveia, W.; Afonso, M.L.; Šljukić, B.; El Essawy, N.A.; Nassr, A.A.A.; Santos, D.M.F. Poly(vinyl alcohol)-base crosslinked ternary polymer blend doped with sulfonated graphene oxide as a sustainable composite membrane for direc borohydride fuel cells. *J. Power Sources* **2019**, *432*, 92–101. [CrossRef]
172. Gouda, M.H.; Gouveia, W.; Elessawy, N.A.; Šljukić, B.; Nassr, A.A.A.; Santos, D.M.F. Simple design of PVA-based blend dope with SO_4(PO4)-functionalised TiO_2 as an effective membrane for direct borohydride fuel cells. *Int. J. Hydrogen Energy* **2020**, *4*5 15226–15238. [CrossRef]
173. Nunes-Pereira, J.; Kundu, M.; Gören, A.; Silva, M.M.; Costa, C.M.; Liu, L.; Lanceros-Méndez, S. Optimization of filler type withi poly(vinylidene fluoride-co-trifluoroethylene) composite separator membranes for improved lithium-ion battery performance *Compos. Part B Eng.* **2016**, *96*, 94–102. [CrossRef]
174. Costa, C.M.; Kundu, M.; Cardoso, V.F.; Machado, A.V.; Silva, M.M.; Lanceros-Méndez, S. Silica/poly(vinylidene fluoride) porou composite membranes for lithium-ion battery separators. *J. Membr. Sci.* **2018**, *564*, 842–851. [CrossRef]
175. Gonçalves, R.; Lizundia, E.; Silva, M.M.; Costa, C.M.; Lanceros-Méndez, S. Mesoporous Cellulose Nanocrystal Membranes a Battery Separators for Environmentally Safer Lithium-Ion Batteries. *ACS Appl. Energy Mater.* **2019**, *2*, 3749–3761. [CrossRef]
176. Barbosa, J.C.; Correia, D.M.; Gonçalves, R.; de Zea Bermudez, V.; Silva, M.M.; Lanceros-Mendez, S.; Costa, C.M. Enhanced ioni conductivity in poly(vinylidene fluoride) electrospun separator membranes blended with different ionic liquids for lithium io batteries. *J. Colloid Interface Sci.* **2021**, *582*, 376–386. [CrossRef] [PubMed]
177. Tsehaye, M.T.; Alloin, F.; Iojoiu, C.; Tufa, R.A.; Aili, D.; Fischer, P.; Velizarov, S. Membranes for zinc-air batteries: Recent progress challenges and perspectives. *J. Power Sources* **2020**, *475*, 228689. [CrossRef]
178. Alday, P.P.; Barros, S.C.; Alves, R.; Esperança, J.M.S.S.; Navarro-Segarra, M.; Sabaté, N.; Silva, M.M.; Esquivel, J.P. Biopolyme Electrolyte Membranes (BioPEMs) for Sustainable Primary Redox Batteries. *Adv. Sustain. Syst.* **2020**, *4*, 1900110. [CrossRef]
179. Mushtaq, K.; Lagarteira, T.; Zaidi, A.A.; Mendes, A. In-situ crossover diagnostics to assess membrane efficacy for non-aqueou redox flow battery. *J. Energy Storage* **2021**, *40*, 102713. [CrossRef]
180. Pawlowski, S.; Galinha, C.F.; Crespo, J.G.; Velizarov, S. 2D fluorescence spectroscopy for monitoring ion-exchange membrane based technologies—Reverse electrodialysis (RED). *Water Res.* **2016**, *88*, 184–198. [CrossRef] [PubMed]
181. Merino-Garcia, I.; Kotoka, F.; Portugal, C.A.M.; Crespo, J.G.; Velizarov, S. Characterization of Poly(Acrylic) Acid-Modified Heterogenous Anion Exchange Membranes with Improved Monovalent Permselectivity for RED. *Membranes* **2020**, *10*, 134 [CrossRef] [PubMed]
182. Xu, J.; Amorim, I.; Li, Y.; Li, J.; Yu, Z.; Zhang, B.; Araujo, A.; Zhang, N.; Liu, L. Stable overall water splitting in an asymmetric acid/alkaline electrolyzer comprising a bipolar membrane sandwiched by bifunctional cobalt-nickel phosphide nanowire electrodes. *Carbon Energy* **2020**, *2*, 646–655. [CrossRef]
183. Dias, A.C. Chlor-Alkali Membrane Cell Process: Study and Characterization. Ph.D. Thesis, University of Porto, Porto Portugal, 2010.
184. Dias, A.C.; Pereira, M.J.; Brandão, L.; Araújo, P.; Mendes, A. Characterization of the Chlor-Alkali Membrane Process by EIS. *J Electrochem. Soc.* **2010**, *157*, E75. [CrossRef]
185. Franco, F.; Prior, J.; Velizarov, S.; Mendes, A. A Systematic Performance History Analysis of a Chlor-Alkali Membrane Electrolyser under Industrial Operating Conditions. *Appl. Sci.* **2019**, *9*, 284. [CrossRef]
186. Tomé, L.C.; Marrucho, I.M. Ionic liquid-based materials: A platform to design engineered CO_2 separation membranes. *Chem. Soc Rev.* **2016**, *45*, 2785–2824. [CrossRef]
187. Albo, J.; Santos, E.; Neves, L.A.; Simeonov, S.P.; Afonso, C.A.M.; Crespo, J.G.; Irabien, A. Separation performance of CO_2 through Supported Magnetic Ionic Liquid Membranes (SMILMs). *Sep. Purif. Technol.* **2012**, *97*, 26–33. [CrossRef]
188. Santos, E.; Albo, J.; Daniel, C.I.; Portugal, C.A.M.; Crespo, J.G.; Irabien, A. Permeability modulation of Supported Magnetic Ionic Liquid Membranes (SMILMs) by an external magnetic field. *J. Membr. Sci.* **2013**, *430*, 56–61. [CrossRef]
189. Tomé, L.C.; Patinha, D.J.S.; Freire, C.S.R.; Rebelo, L.P.N.; Marrucho, I.M. CO_2 separation applying ionic liquid mixtures: The effect of mixing different anions on gas permeation through supported ionic liquid membranes. *RSC Adv.* **2013**, *3*, 12220–12229 [CrossRef]
190. Tomé, L.C.; Patinha, D.J.S.; Ferreira, R.; Garcia, H.; Silva Pereira, C.; Freire, C.S.R.; Rebelo, L.P.N.; Marrucho, I.M. Cholinium-based Supported Ionic Liquid Membranes: A Sustainable Route for Carbon Dioxide Separation. *ChemSusChem* **2014**, *7*, 110–113 [CrossRef]
191. Tomé, L.C.; Florindo, C.; Freire, C.S.R.; Rebelo, L.P.N.; Marrucho, I.M. Playing with ionic liquid mixtures to design engineered CO_2 separation membranes. *Phys. Chem. Chem. Phys.* **2014**, *16*, 17172–17182. [CrossRef]
192. Couto, R.; Neves, L.; Simões, P.; Coelhoso, I. Supported Ionic Liquid Membranes and Ion-Jelly® Membranes with [BMIM][DCA] Comparison of Its Performance for CO_2 Separation. *Membranes* **2015**, *5*, 13–21. [CrossRef]
193. Gouveia, A.S.L.; Tomé, L.C.; Marrucho, I.M. Towards the potential of cyano and amino acid-based ionic liquid mixtures for facilitated CO_2 transport membranes. *J. Membr. Sci.* **2016**, *510*, 174–181. [CrossRef]

4. Gouveia, A.S.L.; Tomé, L.C.; Lozinskaya, E.I.; Shaplov, A.S.; Vygodskii, Y.S.; Marrucho, I.M. Exploring the effect of fluorinated anions on the CO_2/N_2 separation of supported ionic liquid membranes. *Phys. Chem. Chem. Phys.* **2017**, *19*, 28876–28884. [CrossRef] [PubMed]
5. Abdelrahim, M.Y.M.; Martins, C.F.; Neves, L.A.; Capasso, C.; Supuran, C.T.; Coelhoso, I.M.; Crespo, J.G.; Barboiu, M. Supported ionic liquid membranes immobilized with carbonic anhydrases for CO_2 transport at high temperatures. *J. Membr. Sci.* **2017**, *528*, 225–230. [CrossRef]
6. Tomé, L.C.; Gouveia, A.S.L.; Ab Ranii, M.A.; Lickiss, P.D.; Welton, T.; Marrucho, I.M. Study on Gas Permeation and CO_2 Separation through Ionic Liquid-Based Membranes with Siloxane-Functionalized Cations. *Ind. Eng. Chem. Res.* **2017**, *56*, 2229–2239. [CrossRef]
7. Craveiro, R.; Neves, L.A.; Duarte, A.R.C.; Paiva, A. Supported liquid membranes based on deep eutectic solvents for gas separation processes. *Sep. Purif. Technol.* **2021**, *254*, 117593. [CrossRef]
8. Castro, A.M.d.; Prasavath, D.; Bevilaqua, J.V.; Portugal, C.A.M.; Neves, L.A.; Crespo, J.G. Role of water on deep eutectic solvents (DES) properties and gas transport performance in biocatalytic supported DES membranes. *Sep. Purif. Technol.* **2021**, *255*, 117763. [CrossRef]
9. Tomé, L.C.; Gouveia, A.S.L.; Freire, C.S.R.; Mecerreyes, D.; Marrucho, I.M. Polymeric ionic liquid-based membranes: Influence of polycation variation on gas transport and CO_2 selectivity properties. *J. Membr. Sci.* **2015**, *486*, 40–48. [CrossRef]
0. Tomé, L.C.; Mecerreyes, D.; Freire, C.S.R.; Rebelo, L.P.N.; Marrucho, I.M. Pyrrolidinium-based polymeric ionic liquid materials: New perspectives for CO_2 separation membranes. *J. Membr. Sci.* **2013**, *428*, 260–266. [CrossRef]
1. Tomé, L.C.; Aboudzadeh, M.A.; Rebelo, L.P.N.; Freire, C.S.R.; Mecerreyes, D.; Marrucho, I.M. Polymeric ionic liquids with mixtures of counter-anions: A new straightforward strategy for designing pyrrolidinium-based CO_2 separation membranes. *J. Mater. Chem. A* **2013**, *1*, 10403–10411. [CrossRef]
2. Teodoro, R.M.; Tomé, L.C.; Mantione, D.; Mecerreyes, D.; Marrucho, I.M. Mixing poly(ionic liquid)s and ionic liquids with different cyano anions: Membrane forming ability and CO_2/N_2 separation properties. *J. Membr. Sci.* **2018**, *552*, 341–348. [CrossRef]
3. Tomé, L.C.; Guerreiro, D.C.; Teodoro, R.M.; Alves, V.D.; Marrucho, I.M. Effect of polymer molecular weight on the physical properties and CO_2/N_2 separation of pyrrolidinium-based poly(ionic liquid) membranes. *J. Membr. Sci.* **2018**, *549*, 267–274. [CrossRef]
4. Tomé, L.C.; Isik, M.; Freire, C.S.R.; Mecerreyes, D.; Marrucho, I.M. Novel pyrrolidinium-based polymeric ionic liquids with cyano counter-anions: High performance membrane materials for post-combustion CO_2 separation. *J. Membr. Sci.* **2015**, *483*, 155–165. [CrossRef]
5. Gouveia, A.S.L.; Ventaja, L.; Tomé, L.C.; Marrucho, I.M. Towards Biohydrogen Separation Using Poly(Ionic Liquid)/Ionic Liquid Composite Membranes. *Membranes* **2018**, *8*, 124. [CrossRef]
6. Gouveia, A.S.L.; Malcaitè, E.; Lozinskaya, E.I.; Shaplov, A.S.; Tomé, L.C.; Marrucho, I.M. Poly(ionic liquid)–Ionic Liquid Membranes with Fluorosulfonyl-Derived Anions: Characterization and Biohydrogen Separation. *ACS Sustain. Chem. Eng.* **2020**, *8*, 7087–7096. [CrossRef]
7. Gouveia, A.S.L.; Yáñez, M.; Alves, V.D.; Palomar, J.; Moya, C.; Gorri, D.; Tomé, L.C.; Marrucho, I.M. CO_2/H_2 separation through poly(ionic liquid)–ionic liquid membranes: The effect of multicomponent gas mixtures, temperature and gas feed pressure. *Sep. Purif. Technol.* **2021**, *259*, 118113. [CrossRef]
8. Gouveia, A.S.L.; Oliveira, V.; Ferraria, A.M.; Do Rego, A.M.B.; Ferreira, M.J.; Tomé, L.C.; Almeida, A.; Marrucho, I.M. Processing of poly(ionic liquid)–ionic liquid membranes using femtosecond (fs) laser radiation: Effect on CO_2 separation performance. *J. Membr. Sci.* **2022**, *642*, 119903. [CrossRef]
9. Nabais, A.R.; Martins, A.P.S.; Alves, V.D.; Crespo, J.G.; Marrucho, I.M.; Tomé, L.C.; Neves, L.A. Poly(ionic liquid)-based engineered mixed matrix membranes for CO_2/H_2 separation. *Sep. Purif. Technol.* **2019**, *222*, 168–176. [CrossRef]
10. Sampaio, A.M.; Nabais, A.R.; Tomé, L.C.; Neves, L.A. Impact of MOF-5 on Pyrrolidinium-Based Poly(ionic liquid)/Ionic Liquid Membranes for Biogas Upgrading. *Ind. Eng. Chem. Res.* **2020**, *59*, 308–317. [CrossRef]
11. Monteiro, B.; Nabais, A.R.; Casimiro, M.H.; Martins, A.P.S.; Francisco, R.O.; Neves, L.A.; Pereira, C.C.L. Impact on CO_2/N_2 and CO_2/CH_4 Separation Performance Using Cu-BTC with Supported Ionic Liquids-Based Mixed Matrix Membranes. *Membranes* **2018**, *8*, 93. [CrossRef] [PubMed]
12. Ferreira, I.C.; Ferreira, T.J.; Barbosa, A.D.S.; de Castro, B.; Ribeiro, R.P.P.L.; Mota, J.P.B.; Alves, V.D.; Cunha-Silva, L.; Esteves, I.A.A.C.; Neves, L.A. Cr-based MOF/IL composites as fillers in mixed matrix membranes for CO_2 separation. *Sep. Purif. Technol.* **2021**, *276*, 119303. [CrossRef]
13. Pardo, F.; Gutiérrez-Hernández, S.V.; Hermida-Merino, C.; Araújo, J.M.M.; Piñeiro, M.M.; Pereiro, A.B.; Zarca, G.; Urtiaga, A. Integration of Stable Ionic Liquid-Based Nanofluids into Polymer Membranes. Part II: Gas Separation Properties toward Fluorinated Greenhouse Gases. *Nanomaterials* **2021**, *11*, 582. [CrossRef]
14. Hermida-Merino, C.; Pardo, F.; Zarca, G.; Araújo, J.M.M.; Urtiaga, A.; Piñeiro, M.M.; Pereiro, A.B. Integration of Stable Ionic Liquid-Based Nanofluids into Polymer Membranes. Part I: Membrane Synthesis and Characterization. *Nanomaterials* **2021**, *11*, 607. [CrossRef]
15. Lito, P.F.; Cardoso, S.P.; Rodrigues, A.E.; Silva, C.M. Kinetic Modeling of Pure and Multicomponent Gas Permeation Through Microporous Membranes: Diffusion Mechanisms and Influence of Isotherm Type. *Sep. Purif. Rev.* **2015**, *44*, 283–307. [CrossRef]

216. Cardoso, S.P.; Azenha, I.S.; Lin, Z.; Portugal, I.; Rodrigues, A.E.; Silva, C.M. Inorganic Membranes for Hydrogen Separation. *Sep. Purif. Rev.* **2018**, *47*, 229–266. [CrossRef]
217. Bernardo, G.; Araújo, T.; da Silva Lopes, T.; Sousa, J.; Mendes, A. Recent advances in membrane technologies for hydrogen purification. *Int. J. Hydrogen Energy* **2020**, *45*, 7313–7338. [CrossRef]
218. Cardoso, S.P.; Azenha, I.S.; Portugal, I.; Lin, Z.; Rodrigues, A.E.; Silva, C.M. Single and binary surface diffusion permeation through zeolite membranes using new Maxwell-Stefan factors for Dubinin-type isotherms and occupancy-dependent kinetics. *Sep. Purif. Technol.* **2017**, *182*, 207–218. [CrossRef]
219. Cardoso, S.P.; Lin, Z.; Portugal, I.; Rodrigues, A.E.; Silva, C.M. Synthesis, dynamic characterization, and modeling studies of an AM-3 membrane for light gases separation. *Microporous Mesoporous Mater.* **2018**, *261*, 170–180. [CrossRef]
220. Araújo, T.; Bernardo, G.; Mendes, A. Cellulose-Based Carbon Molecular Sieve Membranes for Gas Separation: A Review. *Molecules* **2020**, *25*, 3532. [CrossRef] [PubMed]
221. Rodrigues, S.C.; Andrade, M.; Moffat, J.; Magalhães, F.D.; Mendes, A. Carbon membranes with extremely high separation factors and stability. *Energy Technol.* **2019**, *7*, 1801089. [CrossRef]
222. Rodrigues, S.C.; Andrade, M.; Moffat, J.; Magalhães, F.D.; Mendes, A. Preparation of carbon molecular sieve membranes from an optimized ionic liquid-regenerated cellulose precursor. *J. Membr. Sci.* **2019**, *572*, 390–400. [CrossRef]
223. Crespo, J.G.; Brazinha, C. 1—Fundamentals of pervaporation. In *Pervaporation, Vapour Permeation and Membrane Distillation*; Basile, A., Figoli, A., Khayet, M., Eds.; Woodhead Publishing: Oxford, UK, 2015; pp. 3–17. [CrossRef]
224. Meireles, I.T.; Brazinha, C.; Crespo, J.G.; Coelhoso, I.M. A new microbial polysaccharide membrane for ethanol dehydration by pervaporation. *J. Membr. Sci.* **2013**, *425–426*, 227–234. [CrossRef]
225. Meireles, I.T.; Portugal, C.; Alves, V.D.; Crespo, J.G.; Coelhoso, I.M. Impact of biopolymer purification on the structural characteristics and transport performance of composite polysaccharide membranes for pervaporation. *J. Membr. Sci.* **2015**, *493*, 179–187. [CrossRef]
226. Meireles, I.T.; Brazinha, C.; Coelhoso, I.M.; Crespo, J.G. 10—Membranes for ethanol dehydration. In *Membrane Technologies for Biorefining*; Figoli, A., Cassano, A., Basile, A., Eds.; Woodhead Publishing: Oxford, UK, 2016; pp. 241–262. [CrossRef]
227. Meireles, I.T.; Huertas, R.M.; Torres, C.A.V.; Coelhoso, I.M.; Crespo, J.G. Development and characterisation of hybrid polysaccharide membranes for dehydration processes. *Carbohydr. Polym.* **2018**, *191*, 216–224. [CrossRef]
228. Boutikos, P.; Pereira, C.S.M.; Silva, V.M.T.M.; Rodrigues, A.E. Performance evaluation of silica membrane for water–n-butanol binary mixture. *Sep. Purif. Technol.* **2014**, *127*, 18–28. [CrossRef]
229. Brazinha, C.; Crespo, J.G. Aroma recovery from hydro alcoholic solutions by organophilic pervaporation: Modelling of fractionation by condensation. *J. Membr. Sci.* **2009**, *341*, 109–121. [CrossRef]
230. Brazinha, C.; Alves, V.D.; Viegas, R.M.C.; Crespo, J.G. Aroma recovery by integration of sweeping gas pervaporation and liquid absorption in membrane contactors. *Sep. Purif. Technol.* **2009**, *70*, 103–111. [CrossRef]
231. Kulkarni, P.S.; Brazinha, C.; Afonso, C.A.M.; Crespo, J.G. Selective extraction of natural products with benign solvents and recovery by organophilic pervaporation: Fractionation of d-limonene from orange peels. *Green Chem.* **2010**, *12*, 1990–1994. [CrossRef]
232. Brazinha, C.; Barbosa, D.S.; Crespo, J.G. Sustainable recovery of pure natural vanillin from fermentation media in a single pervaporation step. *Green Chem.* **2011**, *13*, 2197–2203. [CrossRef]
233. Fraga, S.C.; Kujawska, A.; Kujawski, W.; Brazinha, C.; Crespo, J.G. Transport of dilute organics through dense membranes: Assessing impact on membrane-solute interactions. *J. Membr. Sci.* **2017**, *523*, 346–354. [CrossRef]
234. Catarino, M.; Ferreira, A.; Mendes, A. Study and optimization of aroma recovery from beer by pervaporation. *J. Membr. Sci.* **2009**, *341*, 51–59. [CrossRef]
235. Pereira, C.S.M.; Silva, V.M.T.M.; Pinho, S.P.; Rodrigues, A.E. Batch and continuous studies for ethyl lactate synthesis in a pervaporation membrane reactor. *J. Membr. Sci.* **2010**, *361*, 43–55. [CrossRef]
236. Constantino, D.S.M.; Faria, R.P.V.; Ribeiro, A.M.; Loureiro, J.M.; Rodrigues, A.E. Performance Evaluation of Pervaporation Technology for Process Intensification of Butyl Acrylate Synthesis. *Ind. Eng. Chem. Res.* **2017**, *56*, 13064–13074. [CrossRef]

membranes

MDPI

·view

·ast, Present and Future of Membrane Technology in Spain

sé I. Calvo [1,*], Clara Casado-Coterillo [2,*] and Antonio Hernández [1]

1 Surfaces and Porous Materials (SMAP) Group, Associated Research Unit to CSIC, UVa-innova Bldg, P. Belén, 11 and Institute of Sustainable Processes (ISP), Dr. Mergelina, s/n, University of Valladolid, 47071 Valladolid, Spain; antonio.hernandez@uva.es

2 Department of Chemical and Biomolecular Engineering, Universidad de Cantabria, Av. Los Castros s/n, 39005 Santander, Spain

* Correspondence: joseignacio.calvo@uva.es (J.I.C.); clara.casado@unican.es (C.C.-C.)

Abstract: The following review aims at analyzing the contribution of Spanish researchers to membrane science and technology, with a historical compilation of the main milestones. We used a bibliometric analysis based on the Scopus database (1960–2020) dealing with 8707 documents covering the different disciplines and subject areas where membranes are involved. Furthermore, the information has been updated to the present moment of writing this manuscript in order to include the latest research lines and the different research groups currently active in Spain, which may lead the way to the development of the field in the coming years.

Keywords: membrane technology; membrane processes; Spanish research groups; bibliometric analysis

1. Introduction

1.1. Early Historic Background

The filtration process can indeed be considered the oldest existing separation technique. It was already known and exploited by civilizations as ancient as the Egyptians and the Chinese. However, limiting ourselves to Europe, we can consider the work of Lucretius (95–55 BC), "De Rerum Natura", as the beginning of the science of separation in our continent [1].

However, this beginning was quite slow, since more than 15 centuries had to pass in order to find new works related to the subject. Among these works, that of the French Abbé Nollet is often considered a pioneer in membranology. In 1748, Nollet made the first experiment with semi-permeable membranes (porcine bladders, in fact) [2]. The separation studies carried out throughout the nineteenth century, already more numerous and systematic, led to a better understanding of the separation processes and the laws that govern them. Very important in these aspects are the works of Fick on diffusion [3], Traube [4], Pfeffer [5], and Van't Hoff [6,7] with pressure osmotic, and many other interesting works from Raoult, Graham, etc.

From Traube's pioneering work, Zsigmondy was the first to develop a cellulose nitrate membrane suitable for the sterilization of small amounts of fluid, [8], later marketed by Sartorius [9].

A really entertaining and interesting insight into these early years of membranology can be found in [10]. Figures such as Abbé Nollet, Dutrochet, Doebereiner, Graham, and many other pioneers of membrane science parade through this pleasant historical "review". A work more related to membrane technology, although already classic is that of Lonsdale, the first Editor in Chief of the *Journal of Membrane Science* [11].

However, from a practical point of view, the true starting point of what we can without fear call the membrane technology revolution can be found in the asymmetric cellulose acetate membrane developed by Loeb and Sourirajan in 1960 [12], which allowed solving some of the necessary challenges to obtain a material (the membrane) with the selectivity and stability features necessary to be applied on a large scale in desalination of seawater by reverse osmosis.

Before the 1980s different processes with membranes are used industrially and proposed scientifically as promising alternatives for other applications. In the late 70s and 80s, membrane technology was starting to be presented as a strategic science and technology with definitions, history, and agenda [13].

In all these pioneering studies, the contribution of Spanish researchers in membrane technology is practically absent [14,15].

By the end of the 1970s, when membrane technology began to be mature, research in Spain was also ready to make its input. As it occurred in the rest of the world, the problems addressed by membrane processes in southern countries focused on the poor quality of drinking water and water supply [16], as well as on wastewater management from dairy and paper industries [17], and membrane researchers started to formulate this in terms of membrane technology for example by using reverse osmosis. The first Spanish contributions to membrane processes were thus found in general chemical engineering journals [18–20] or more specific journals dealing with actual industrial or environmental challenges in food, paper, and metallurgical industries. Thus, it is not surprising to find that the first Spanish contributions to membrane research will be related to the then recently discovered cellulose acetate membranes [21–24] as well as others dealing with reverse osmosis water desalination plants [25,26]. Two previous articles on ion-exchange membranes could be remarked [27,28] although, apparently without subsequent continuity. Regarding the work of Muro et al. [26] on seawater desalination plants, this is mostly a national desalination policies review and does not present actual research. Likewise, Guzmán's article (followed by a later one in 1984, [29], are descriptive works on the underlying engineering in two water desalination plants in the Canary Islands. Thus, we can consider the group born at the Complutense University of Madrid, by the hand of Juan Ignacio Mengual, as the first active research group on synthetic membranes in Spain. On his part, Cristóbal Fernández Pineda, upon obtaining his teaching position at the University of Malaga, would create his own group at that university, also very active in these early years of membranology in Spain.

These first names already allow us to note a curious characteristic of membrane research in Spain. Surprisingly most of the research groups on the subject come from departments of Applied Physics of Spanish Universities. So are Mengual and Fernández Pineda, but also Fernando Tejerina and Javier Garrido and Julio Pellicer in the University of Valencia. José Antonio Ibáñez and Antonio Hernández were disciples of Fernando Tejerina in The Autonomous University of Barcelona and afterwards in Murcia and Valladolid, respectively. All of these published works are related to the application of the Thermodynamics of Irreversible Processes on the explanation of the different electrokinetic phenomena observed in charged membranes.

Thus, these first works dealt with salt diffusion [30], transport models [31], membrane current, membrane potential, charge densities, or ionic permeabilities. In parallel, the study of the properties of permeability [32] and solute separation [33] in membranes would continue its course. In the next years, Fernández-Pineda, very active in publishing, would produce several papers on hydrodynamics and separation properties of cellophane membranes.

A colleague of Fernández-Pineda at the same University (Málaga) as—with whom she also publishes several articles—professor Juana Benavente specialized in electrokinetic properties, firstly using cellophane membranes [34], but later with many other membrane materials [35].

Meanwhile, Antonio Hernández from the University of Valladolid published works on the permeation of ionic salts species through various polymeric membranes [36,37]. Another group of papers, such as Garrido et al., 1989, Díaz et al., 1989, or Alegret et al., 1989 [38–40], focused on membranes not in industrial separation but as an important part of electrodes and sensors.

In 1987, new names (not linked to Applied Physics) began to appear dedicated to membrane research (apart from new articles associated with desalination companies [41]) especially chemical engineers [42], which would give rise to a thriving group of researchers at the Polytechnic University of Valencia, for example.

The end of the 80s registers the appearance of one of the largest membrane groups in Spain, that led by Inmaculada Ortiz, initially at the University of the Basque Country, and since the early 90s, at the University of Cantabria. Being part of a purely Chemical Engineering department, Prof. Ortiz published her first paper on liquid membranes in 1988 in *Industrial and Engineering Chemistry Research* [43]. This group of chemical engineers introduced a strong focus on the design and development of membrane processes using process simulation tools and explored the scale-up of membrane processes with numerous pilot plant case studies in collaboration with the chemical industry.

The young but increasing membrane technology field attracts many researchers due to the big range of applications where synthetic membranes can offer a solution for separation problems in the industry. A paper in 1988, by researchers of the Instituto de la Grasa from the Spanish Superior Research Council, CSIC [44], presented the application of Ultrafiltration to the regeneration of olive brines.

Alonso et al. used membrane filters to control the population of *Pseudomonas aeruginosa* in water [45].

At the University of Granada, there was from the beginning a group active in electrokinetic studies on porous plugs [46,47], which later evolved into a group using network thermodynamics to model membrane transport [48]. More studies continue to evaluate the performance of RO in water desalination [49,50].

The group of Organic Chemistry from the Instituto de Ciencia y Tecnología de Polímeros (CSIC) started to publish on the properties of UF membranes cast from polyamide, being one of the first groups in Spain researching on membrane preparation [51,52].

In 1992, the group of Jose Coca, from the University of Oviedo, started to publish regarding the application of reverse osmosis membrane processes to dairy whey treatment [53], a research line which later has been revealed as very productive [54,55].

The mid-90s saw other research groups with different backgrounds starting to contribute to membrane science as an emerging technology. One of those was the group of University of Zaragoza, led by Jesús Santamaría, who published in 1994 their first paper on ceramic membrane reactors [56] and soon became an internationally recognized group on zeolite and other ceramic membranes [57].

Step by step, the 80s and early 90s would see a progressive increase in the applications of membrane processes, and with them, of the publications of Spanish researchers, increasingly associated with groups of chemical engineers or chemists, mostly interested in the different processes in which membranes were finding increasing potential but also in the way to develop new membranes with selected features.

In fact, from 1990 onwards, there is a clear increase in the number of publications from Spanish researchers related to membrane technology as can be seen in Figure 1. This trend has been maintained and even increased in the following decades, also linked to an important general rise of scientific research in Spain. Several factors contributed to this increase:

(1) University Reform Law, issued in 1983, was the first law in democracy that sought to modernize the obsolete and undemocratic structure of the Spanish university. On the other hand, the opening and democratization of the access to the university in these years led to an almost exponential increase in the number of students, rising from 170,000 in the 1959–60 academic year to more than 1,440,000 in the year 1994–95 [58].
(2) The increase in the number of students led to the creation of new universities and the proportional increase in the number of professors who, in addition, were now required to have a research curriculum to stabilize their positions.

(3) Considering that in Spain, the research work was always closely linked to the University, with little contribution until this century from the industrial sector, such a notable increase in university size and quality, led to a consequent expansion of research in general and research in applied fields such as membrane technology in particular.
(4) In parallel, the 90s would see an important effort from the various democratic governments to provide Spanish researchers with funding. Thus, the consolidated groups and new groups created to cover new emerging topics found more financing possibilities that allowed them to acquire increasingly complex instrumental equipment, take part regularly in internationally recognized congresses and send their best young researchers to foreign research centers establishing collaborations that enriched the research background of the national groups.

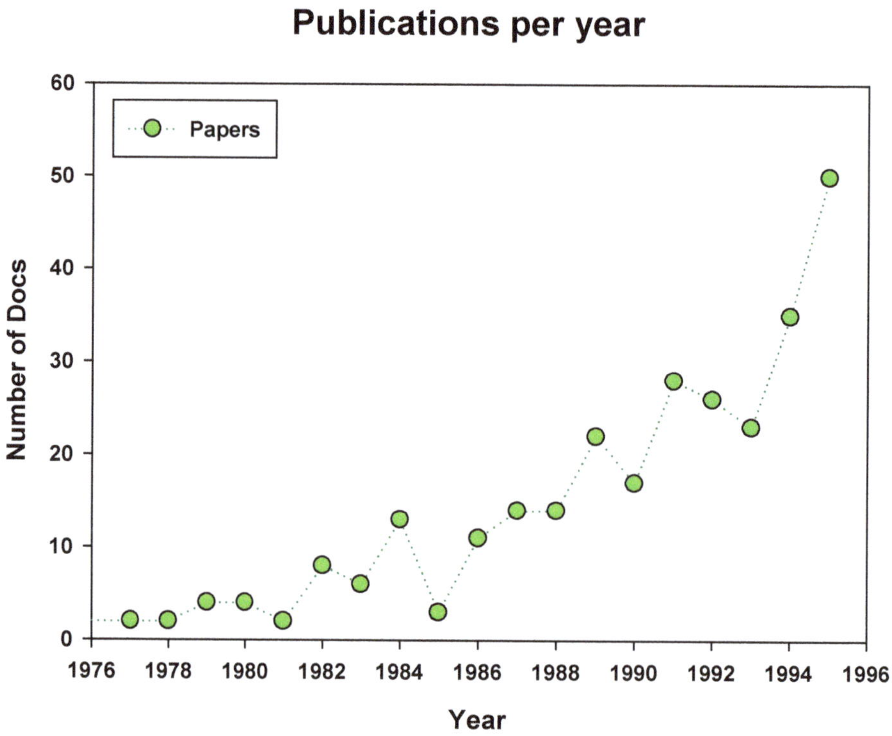

Figure 1. Spanish based membrane publications in international journals from 1970 to 1995.

This clearly somewhat superficial analysis of the first decades of membrane research in Spain can be completed with a view to the main source of publication of Spanish research reports dealing with membranes. Between 1970 and1995, 38 out of the 286 articles reported were published in the *Journal of Non-equilibrium Thermodynamics*, a journal clearly focused on the thermodynamic analysis of irreversible processes, including electrokinetic processes. The outcome of this journal is closely followed (36 documents) by the main reference on membrane science, the *Journal of Membrane Science*, founded in 1976 "to draw together a new field, tentatively called 'membranology', which turned to 'membrane technology'", and has included contributions from Spanish researchers from almost the very beginning [23]. The relative importance of these journals can be seen in Figure 2, where the number of papers in the main international journals during these pioneering years is depicted.

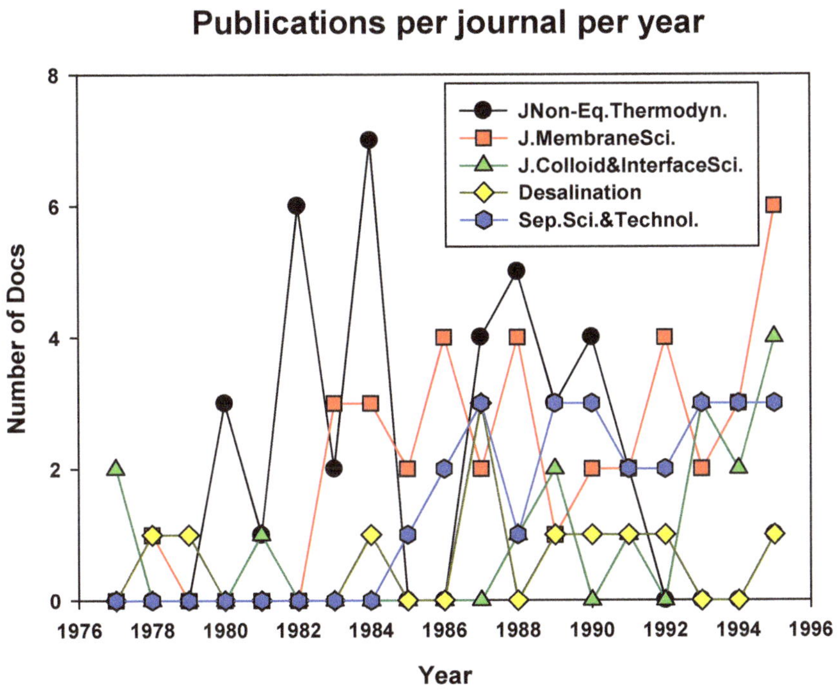

Figure 2. Spanish based membrane-related publications in selected international journals from 1970 to 1995.

Journals mostly devoted to physical properties of separation of charged species as *Journal of Non-Equilibrium Thermodynamics* or *Journal Colloid and Interface Science*, will be preferred in the initial part of the period while, as more and more researchers move to study the process itself and its applications, the journals mainly devoted to separation as *Desalination*, *Separation Science and Technology* or, the most important one, *Journal of Membrane Science* would start to attract maximum attention to the Spanish membrane researchers. From 1988 onwards, Spanish researchers started contributing to the development of membrane processes in journals focused on chemical engineering and chemical technology, that already included sections devoted to separation such as *Industrial & Engineering Chemistry Research* (7), *AIChE Journal* (4), and *Chemical Engineering Science* (4). However, these latter are not included in Figure 2 for clarification.

The fact that a broader list of journals was available for authors to publish their contributions shows us the rise of applications and approaches related to membranes that Spanish researchers were beginning to explore. Thus, in 1977 or 1978, while the total number of publications in the journals selected in Figure 2 almost equaled 100% of the existing ones, in 1995, only 30% of the total articles were published in those 5 journals in Figure 2, the rest being published elsewhere. They were thus distributed among 27 different journals and magazines, covering a wide variety of fields, from catalysis, materials science to chemical or food engineering.

Another important milestone was the creation of a Spanish Membrane Group (Grupo Español de Membranas, GEM), a consequence of the interest of Spanish researchers who had begun to meet in several reputed international congresses on the topic (Euromembrane or the International Congress on Membranes and Membrane Processes, ICOM, as flag representation, and their specific sub-sections on Inorganic Membranes, ICIM, or Membrane Reactors, ICCMR, among others) and wanted to be aware of mutual research efforts and synergies. Certainly, the birth of the GEM was influenced by the European Society of Membrane Science and Technology (now European Membrane Society, EMS) founded in 1982, and in some way imitated it. The group, founded in the beginning of the 90s, was initially chaired by Juan Ignacio Mengual, with Antonio Hernández as secretary, but the unfortunate loss of its president and the growing drive of chemical engineers would result in the slow languishment of GEM until its effective disappearance. The spirit of association could be then maintained for the duration of the thematic network on selective membranes promoted by Ane Urtiaga and funded by the Spanish Ministry of Science and Innovation in 2002–2004, which fostered the exchange of ideas and the transfer of knowledge on membrane science and technology from the academy to the industry and society.

An important impulse pushed by the research groups actively involved in national associations would translate into the necessity of the creation of some type of international scientific congress that allows Iberian researchers to meet and discuss their common work. This resulted in the series of Ibero-American Congresses on Membrane Science and Technology (CITEM) of which X editions have been held to date. The first one was organized by the Universidad de Murcia in Murcia in 1992. Further editions were held at: Rio de Janeiro, 1994; Aveiro, 2001; Florianópolis, 2003; Valencia, 2005; Campina Grande, 2007; Sintra, 2010; Salta, 2012; Santander 2014 and Mexico DF, 2016. In all cases the CITEM congresses have proved a genuine ability to promote the mutual collaboration between Ibero-American research groups, obviously with more Iberian researchers was major in those celebrated in Spain or Portugal while Ibero-American researchers were the majority in the meetings held at the other side of the Atlantic Sea.

1.2. Data Resource and Methodology

The aim of this work is the bibliometric analysis of the research performed in the last century in Spain in the light of how they can address the challenges of the present. Therefore, this paper will be useful to focus the global panorama of the research in membranes with Spain as one of the authors' affiliation countries.

As commented before, after this slow introduction of membrane technology in the Spanish research, the subject was tackled by many other researchers of various working fields. In order to grasp a preliminary idea of the relevance of Spanish contributions in the scientific literature on membrane technology, we have carried out a non-exhaustive bibliometric analysis search to find the number of papers published in international journals authored by Spanish researchers and technologists. A bibliometric analysis is a systematic evaluation of the records found under specific keywords is used to quantitatively determine the characteristics of the research and provide an overview of the trends in this topic regarding most emerging applications, which has not been the object of similar studies before, as far as we know. The tool used is a bibliometric analysis of the literature indexed in the Scopus database [59] up to 2021. In this search we have used following the marks:

- Keyword: Membrane
- Affiliation country: Spain

In an initial search, this led to a significantly large amount of references, including many devoted to biological membranes or other fields not related with the use of synthetic membranes for separations. We also tried more specific keywords as "Membrane Technology" or Membrane Processes". However, these searches left out a number of very important publications. For example, using "Membrane Technology" most papers published in the Journal of Membrane Science were not selected. Therefore, we decided to continue using this general keyword but eliminate all references coming from journals specializing in Medicine, Immunology, Therapies or similar subjects. This finally led to a total amount of 13,492 papers, of which the analysis was then limited up to 8707 documents after eliminating those not related to membrane science and technology. For this further reduction, the keywords included by authors were used and also the name of the journals where the papers were published. Therefore, papers containing keywords such as: apoptosis, gene expression, cytology, rats, mice, or similar words, were excluded.

Similarly, those papers published in journals as: *Physiologia Plantarum, Journal of Experimental Botany, Theoretical Computer Science* or *Neuroscience Letters*, among others, were also discarded.

Obviously, in this selection, there still remain many papers not really related to what we usually consider membrane technology while surely a large number of interesting papers is also missing. Yet, this selection serves to grasp a quite approximative idea about the membrane scientific production and evolution in Spain (and certainly not only in Spain, as many of the papers included are international collaborations).

The quality of the approach for this search can be observed by the leading journals and citations. In Figure 3, those 8707 papers have been classified by the name of the publication journal, and the 20 journals more represented are plotted in Figure 3.

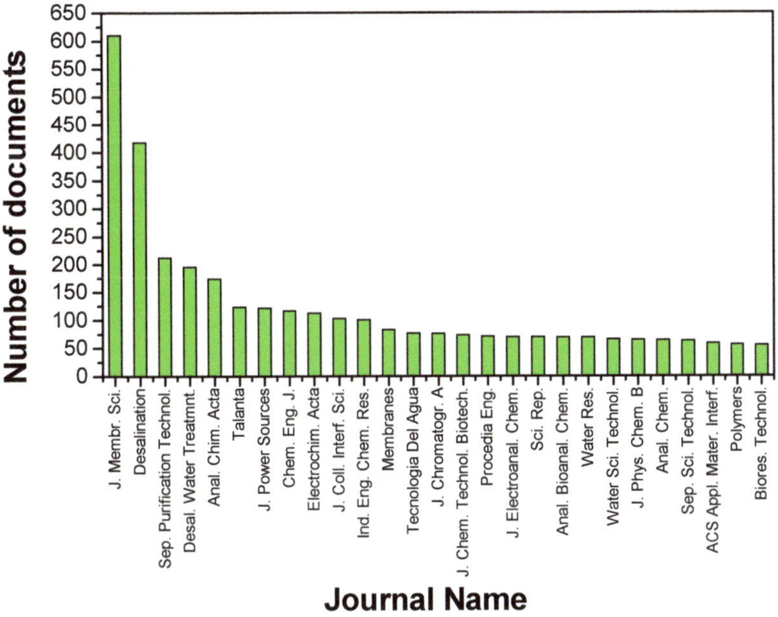

Figure 3. Spanish-based publications by the most productive early journals from 1960. (Source: Scopus, Keyword "Membrane" and Affiliation "Spain").

2. Results and Discussion

2.1. Scientific Production of Spanish Researchers on Membrane Technology

As commented above, Figure 3 includes the most frequently encountered journals from 1960 to the moment of writing this review. Only those journals with more than 50 documents found in the sample analyzed in this work are represented in the figure. Most importantly, almost half of them are clearly recognized as reputed journal publications about membranes. In fact, *Journal of Membrane Science, Desalination* and *Separation and Purification Technology*, all Q1 journals and highly considered among the most reputed journals in the field, are the three most frequently encountered. It is worth noting that a very common journal during the early stages of Spanish membrane research, as *Journal of Non-equilibrium Thermodynamics*, is not represented in Figure 3 as in Figure 2, which shows that as years passed and more researchers started to work in membrane technology the focus of Spanish research moved from Irreversible Thermodynamics and charged membranes to the wide range of membrane processes appearing year by year, represented by chemical engineering and water treatment and separation technologies in general, or membrane materials and characterization in particular.

On the temporal issue, Figure 4 presents the evolution in the total number of papers published per year during those last 50 years. A slow increase is observed in the first years (1975–1996), as was depicted above in Figure 1, but along the 90s the increase is continuous and uninterrupted (the number of papers rose from around 200 in 1995 to more than 600 in 2010). This increase corresponds to the already commented growth of Spanish research in general. It is obvious that a continuous increase in the funding devoted to research allowed more researchers to work in membrane science and technology. Moreover, more importantly, this allowed young researchers to have the opportunity to carry out significant parts of their work in leading research centers around the world. Thus, the obvious improvement in their training led to an increase in knowledge and expertise that they led back to their research groups after the return from these stays. Likewise, the increase in funds allowed greater participation of Spanish researchers in international congresses, favoring scientific exchange with top-level groups. The result is that a linear increase in funding translated into an almost exponential increase in scientific production, which also improved quality and visibility. Similarly, the figure also reflects the sudden slow-down of the pace of increasing the number of publications occurring during the economic crisis in 2010. The severe impact of the crisis and the subsequent cuts in research funding of the successive Spanish governments has not recovered and continues moving in almost a plateau shape (Figure 4).

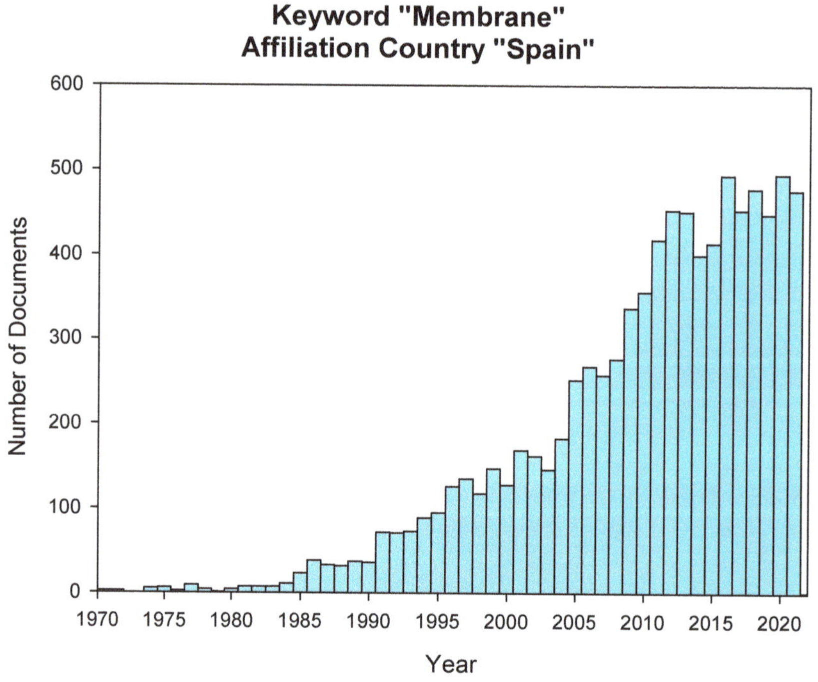

Figure 4. Spanish based publications in international journals from 1960. (Source: Scopus, Keyword "Membrane").

A common indicator of the quality of a scientific article is the number of citations that the paper has received. Table 1 lists, in descending order, the 20 most cited articles from among those present in the non-exhaustive sample analyzed in this work. The papers included in this table are all very reputed works, which have attracted more than

400 citations each. Also noteworthy in this table is the fact that 12 out 20 of the most cited papers are reviews covering a wide range of membrane-related applications from traditional gas separation or water filtration to most recently, PEM fuel cells and other electrochemical applications using membranes. In the general sample analyzed, 82% of the documents retrieved are original scientific articles and only 6% were reviews.

Table 1. The articles cited more than 400 times.

Ranking	Most Cited Article	Type	Times Cited
1	Rodenas, T., Luz, I., Prieto, G., Seoane, B., Miro, H., Corma, A., Kapteijn, F., Llabrés i Xamena, F.X., Gascon, J. Metal-organic framework nanosheets in polymer composite materials for gas separation, Nature Materials, 14(1) 48–55, 2015	Article	1143
2	El-Bourawi, M.S., Ding, Z., Ma, R., Khayet, M. A framework for better understanding membrane distillation separation process, Journal of Membrane Science, 285, 4–29, 2006	Review	921
3	Navarro, R.M., Peña, M.A., Fierro, J.L.G. Hydrogen production reactions from carbon feedstocks: Fossil fuels and biomass, Chemical Reviews, 107(10) 3952–3991, 2007	Review	871
4	Moreira, F.C., Boaventura, R.A.R., Brillas, E., Vilar, V.J.P. Electrochemical advanced oxidation processes: A review on their application to synthetic and real wastewaters, Applied Catalysis B: Environmental, 202, 217–261, 2017	Review	868
5	Sunarso, J., Baumann, S., Serra, J.M., Meulenberg, W.A., Liu, S., Lin, Y.S., Diniz da Costa, J.C. Mixed ionic-electronic conducting (MIEC) ceramic-based membranes for oxygen separation, Journal of Membrane Science, 320, 13–41, 2008	Review	865
6	Mecerreyes, D. Polymeric ionic liquids: Broadening the properties and applications of polyelectrolytes, Progress in Polymer Science, 36(12) 1629–1648, 2011	Review	843
7	Garcia-Ochoa, F., Gomez, E. Bioreactor scale-up and oxygen transfer rate in microbial processes: An overview, Biotechnology Advances, 27(2) 153–176, 2009	Review	809
8	Radjenović, J., Petrović, M., Barceló, D. Fate and distribution of pharmaceuticals in wastewater and sewage sludge of the conventional activated sludge (CAS) and advanced membrane bioreactor (MBR) treatment, Water Research, 43(3) 831–841, 2009	Review	762
9	Khayet, M. Membranes and theoretical modeling of membrane distillation: A review, Advances in Colloid and Interface Science, 164, 56–88, 2011	Article	746
10	Heinzel, A., Barragán, V.M. Review of the state-of-the-art of the methanol crossover in direct methanol fuel cells, Journal of Power Sources, 84(1) 70–74, 1999	Review	742
11	Sirés, I., Brillas, E. Remediation of water pollution caused by pharmaceutical residues based on electrochemical separation and degradation technologies: A review, Environment International, 40(1) 212–229, 2012	Article	659
12	Klahr, B., Gimenez, S., Fabregat-Santiago, F., Hamann, T., Bisquert, J. Water oxidation at hematite photoelectrodes: The role of surface states, Journal of the American Chemical Society, 134(9) 4294–4302, 2012	Review	614
13	Ursúa, A., Gandía, L.M., Sanchis, P. Hydrogen production from water electrolysis: Current status and future trends, Proceedings of the IEEE, 100(2) 5898382, 410–426, 2012	Article	589
14	Petrović, M., Gonzalez, S., Barceló, D. Analysis and removal of emerging contaminants in wastewater and drinking water, TrAC - Trends in Analytical Chemistry, 22(10) 685–696, 2003	Conference Paper	523

Table 1. *Cont.*

Ranking	Most Cited Article	Type	Times Cited
15	Seoane, B., Coronas, J., Gascon, I., Benavides, M.E., Karvan, O., Caro, J., Kapteijn, F., Gascon, J. Metal-organic framework based mixed matrix membranes: A solution for highly efficient CO_2 capture?, Chemical Society Reviews, 44(8) 2421–2454, 2015	Article	509
16	Asensio, J.A., Sánchez, E.M., Romero, P.-G. Proton-conducting membranes based on benzimidazole polymers for high-temperature PEM fuel cells. A chemical quest, Chemical Society Reviews, 39(8) 3210, 2010	Review	505
17	Pérez-González, A., Urtiaga, A.M., Ibáñez, R., Ortiz, I. State of the art and review on the treatment technologies of water reverse osmosis concentrates, Water Research, 46(2) 267–283, 2012	Review	489
18	Sorribas, S., Gorgojo, P., Téllez, C., Coronas, J., Livingston, A.G. High flux thin film nanocomposite membranes based on metal-organic frameworks for organic solvent nanofiltration, Journal of the American Chemical Society, 135(40) 15201–15208, 2013	Article	448
19	Falguera, V., Quintero, J.P., Jiménez, A., Muñoz, J.A., Ibarz, A. Title: Edible films and coatings: Structures, active functions and trends in their use, Trends in Food Science and Technology, 22(6) 292–303, 2011	Article	416
20	Quintana, J.B., Weiss, S., Reemtsma, T. Pathways and metabolites of microbial degradation of selected acidic pharmaceutical and their occurrence in municipal wastewater treated by a membrane bioreactor, Water Research, 39(12) 2654–2664, 2005	Review	411

The 12 most productive journals (among those depicted in Figure 3) are further analyzed in Table 2, in decreasing order of the number of publications among JCR indexed journals. The most cited paper within the current search for each of these journals is also given in Table 2. The citations vary largely from one journal to another and largely depend on the date of publication and the age of the journal, encumbering fair comparison between the *Journal of Membrane Science* or *Desalination* with novel journals as *Membranes* or *Desalination and Water Treatment*, which were created in 2011. It reflects the scientific relevance of the journal, as most researchers try (as the first option) to publish their results in the top journals. Table 2, along with Figure 3, indicates that most of the research produced in membrane technology in Spain is published in the most recognized journals. However, it also reflects the historical background of the first established research groups dealing with membrane applications in industrial challenges such as wastewater treatment, as *Tecnología del agua* and *Water Research*, providing reviews highly consulted by their peers that have already been included in Table 1.

This significant amount of research publications in membrane technology in Spain covers almost all the fields of membranes, both fundamental and applied. Table 3 shows in fact that the major subject categories involved are Chemistry and Chemical Engineering, with transversal contributions from other categories as Materials and Environmental Science, which aim at the appointed potential of membrane technology as an alternative solution to existing processes in industry as will be discussed later. Furthermore, there is an important contribution of Biochemistry, Genetics and Molecular Biology that gives an idea of the close relation between membranes and Biological Sciences, although in this analysis, out of about 2700 documents, some could be surely more related with biological than synthetic membranes.

Table 2. The top most productive journals in Figure 3.

Ranking	Journal	IF 2020 (WoS)	SJR 2020 (Scopus)	No. Documents (%)	Most Cited Article	Times Cited
1	Journal of Membrane Science	7.183	1.929	610 (7.0%)	El-Bourawi, M.S., Ding, Z., Ma, R., Khayet, M. A framework for better understanding membrane distillation separation process, J. Membr. Scie., 285, 4–29, 2006	921
2	Desalination	9.550	1.794	418 (4.8%)	Qtaishat, M., Matsuura, T., Kruczek, B., Khayet, M. Heat and mass transfer analysis in direct contact membrane distillation, Desal., 219, 272–292, 2008	318
3	Separation and Purification Technology	7.132	1.279	212 (2.4%)	Toledano, A., García, A., Mondragon, I., Labidi, J. Lignin separation and fractionation by ultrafiltration, Sep. Purif. Technol., 71(1) 38–43, 2010	318
4	Desalination and Water Treatment	0.854	0.251	195 (2.2%)	Albo, J., Luis, P., Irabien, A. Absorption of coal combustion flue gases in ionic liquids using different membrane contactors, Desal. Water Treatmnt., 27, 54–59, 2011	48
5	Analytica Chimica Acta	6.558	1.403	174 (2.0%)	Herrera-Herrera, A.V., González-Curbelo, M.T., Hernández-Borges, J., Rodríguez-Delgado, M.T. Carbon nanotubes applications in separation science: a review. Anal. Chim. A., 734, 1–30, 2012	205
6	Talanta	6.057	1.181	124 (1.4%)	Payán, M.R., López, M.A.B., Torres, R.F., Navarro, M.V., Mochón, M.C. Electromembrane extraction (EME) and HPLC determination of non-steroidal anti-inflammatory drugs (NSAIDs) in wastewater samples, Talanta, 85(1):394–399, 2011	103
7	Journal of Power Sources	8.247	2.139	122 (1.4%)	Heinzel, A., Barragán, V.M. Review of the state-of-the-art of the methanol crossover in direct methanol fuel cells, J. Power Sources, 84(1) 70–74, 1999	742
8	Chemical Engineering Journal	13.273	2.528	117 (1.3%)	Skouteris, G., Hermosilla, D., López, P., Negro, C., Blanco, Á. Anaerobic membrane bioreactors for wastewater treatment: A review, Chem. Eng. J., 198-199, 138–148, 2012	210
9	Electrochimica Acta	6.901	1.534	113 (1.3%)	Lobato, J., Cañizares, P., Rodrigo, M.A., Linares, J.J. PBI-based polymer electrolyte membranes fuel cells. Temperature effects on cell performance and catalyst stability, Electrochim. A., 52(12) 3910–3920, 2007	137
10	Journal of Colloid And Interface Science	8.128	1.538	103 (1.2%)	García-Payo, M.C., Izquierdo-Gil, M.A., Fernández-Pineda, C. Wetting study of hydrophobic membranes via liquid entry pressure measurements with aqueous alcohol solutions, J. Coll. Interf.ce Sci., 230(2) 420–431, 2000	163
11	Industrial and Engineering Chemistry Research	3.720	0.878	101 (1.2%)	Gadipelly, C., Pérez-González, A., Yadav, G.D., Ortiz, I., Ibáñez, R., Rathod, V.K., Marathe, K.V. Pharmaceutical industry wastewater: Review of the technologies for water treatment and reuse, Ind. Eng. Chem. Res., 53(29) 11571–11592, 2014	297

Table 2. *Cont.*

Ranking	Journal	IF 2020 (WoS)	SJR 2020 (Scopus)	No. Documents (%)	Most Cited Article	Times Cited
12	Membranes	4.106	0.609	83 (0.95%)	Blandin, G., Verliefde, A.R.D., Comas, J., Rodriguez-Roda, I., Le-Clech, P. Efficiently combining water reuse and desalination through forward osmosis-reverse osmosis (FO-RO) hybrids: A critical review, Membranes, 6(3) 10.3390/membranes6030037, 2016	61
13	Tecnología del Agua	-	0.209	77 (0.88%)	Artiga Acuña, P., Garcia-Toriello Romero, G., Garrido Fernández, J.M., Méndez Pampin, R. Membrane bioreactor: An advanced technology for the treatment and reuse of waste waters, Tecnologia del Agua, 26(269) 54–60, 2006	6
14	Journal of Chromatography A	4.390	1.011	76 (0.87%)	Rodil, R., Schrader, S., Moeder, M. Non-porous membrane-assisted liquid-liquid extraction of UV filter compounds from water samples, J. Chromatogr. A, 1216(24) 4887–4894, 2009	96
15	Journal of Chemical Technology and Biotechnology	2.750	0.650	73 (0.84%)	Rivera-Utrilla, J., Bautista-Toledo, I., Ferro-Garca, M.A., Moreno-Castilla, C. Activated carbon surface modifications by adsorption of bacteria and their effect on aqueous lead adsorption, J. Chem. Technol. Biotech., 76(12) 1209–1215, 2001	344
16	Procedia Engineering	1.880	0.320	71 (0.82%)	Stoller, M., Ochando-Pulido, J.M. Going from a critical flux concept to a threshold flux concept on membrane processes treating olive mill wastewater streams, Proc. Eng., 44, 607–608, 2012	41
17	Journal of Electroanalytical Chemistry	4.280	0.845	70 (0.80%)	Bisquert, J., Garcia-Belmonte, G., Fabregat-Santiago, F., Bueno, P.R. Theoretical models for ac impedance of finite diffusion layers exhibiting low frequency dispersion, J. Eletroanal. Chem., 475(2) 152–163, 1999	201
18	Scientific Reports	0.38 (*)	1.240	70 (0.80%)	Ivanova, M.E., Escolástico, S., Balaguer, M., Palisaitis, J., Sohn, Y.J., Meulenberg, W.A., Guillon, O., Mayer, J., Serra, J.M. Hydrogen separation through tailored dual phase membranes with nominal composition $BaCe_{0.8}Eu_{0.2}O_{3-\delta}$:$Ce_{0.8}Y_{0.2}O_{2-\delta}$ at intermediate temperatures, Sci. Rep., 6, 34773, 2016	32
19	Water Research	11.236	3.099	69 (0.79%)	Radjenović, J., Petrović, M., Barceló, D. Fate and distribution of pharmaceuticals in wastewater and sewage sludge of the conventional activated sludge (CAS) and advanced membrane bioreactor (MBR) treatment, Water Res., 43(3) 831–884, 2009	762
20	Analytical and Bioanalytical Chemistry	3.286	0.321	69 (0.79%)	Radjenovic, J., Petrovic, M., Barceló, D. Analysis of pharmaceuticals in wastewater and removal using a membrane bioreactor, Anal. Bioanal. Chem., 387(4) 1365–1377, 2007.	361

(*) 2-year impact factor.

Table 3. The top most popular subject categories.

Ranking	Subject Categories	Documents	Percentage (%) *
1	Chemistry	4366	50
2	Chemical Engineering	3425	39
3	Biochemistry, Genetics and Molecular Biology	2690	30
4	Materials Science	2455	28
5	Engineering	2088	24
6	Environmental Science	2043	23
7	Energy	799	9.2

* This percentage is calculated regarding the final 8707 records after scrutiny of the initial non-exhaustive search as explained in Section 1.2.

This can be complemented by a search on the feature words present in the title of the papers included in this database selection. We have counted the papers containing up to 200 keywords covering all aspects of membrane research. The entry "membrane materials" is highlighted because it involves inorganic and polymeric membranes, which are also represented separately. For the sake of simplicity, the results have been grouped in a shorter classification and they are presented, finally, in Figure 5.

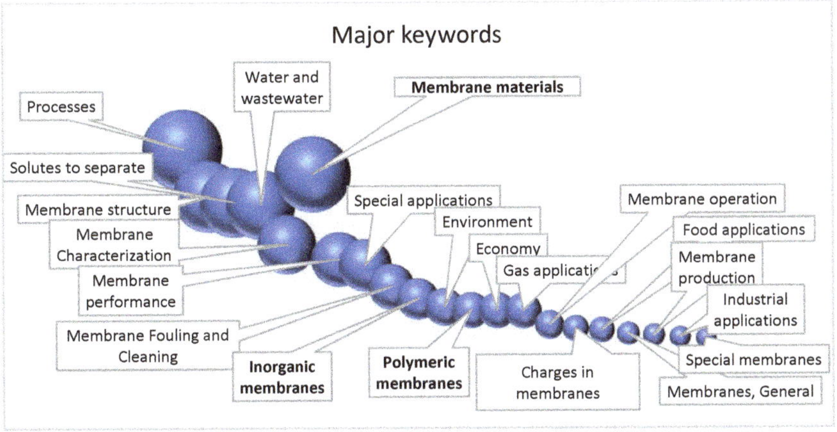

Figure 5. Comparative plot of relative number of papers according to the title keywords.

Most of the papers selected referred in their title to one of the many membrane processes (1842, 21.1% of the total number of papers), being UF, RO, and NF the most frequently studied. Thus, 1412 papers (16.2%) are related to membrane structure and the different configurations a membrane can be made. There is also a significant number of papers devoted to fundamental research in membrane characterization (969, 11.1%) covering all analytical techniques available. The important contribution to the study of membrane fouling and cleaning (580, 6.7%), a key aspect of membrane applicability, cannot be missed from this review. There is also a significant number of papers devoted to membrane materials, which we have classified in this review as devoted to organic polymeric membranes (366, 4.2%) and inorganic or composite membranes (469, 5.4%). The fact that our selected sample identified more papers on inorganic than organic membranes, contrary to the practical reality, may be somehow surprising. This can be explained by the fact of the existence of some leading groups in ceramic membranes (in particular, the group from the University of Zaragoza, an international benchmark in zeolites and ceramics in general) that also attract a greater number of researchers and associated works. On the other hand, the classification of inorganic membranes contains also composite and mixed matrix membranes that have attracted growing interest in recent years.

Fundamental research can be completed with the 834 papers (9.6%) devoted to membrane performance, including studies on flux, rejection, permeability, or theoretical modelling of transport mechanisms.

Regarding the membrane applications, most of the studies dealt with several aspects of water and wastewater treatment (1338, 15.4%), including desalination or brackish water treatment. But many other membrane applications have been covered in these more than 40 years, such as: 185 papers on food applications, 119 on industrial applications, or 361 in gas applications.

Last but not least, environmental studies are not forgotten (396 papers) covering all types of pollutants and contaminants. Finally, economic studies (365 papers), membrane production (135), charged membranes (197), or membrane operation (232) cover all aspects of membrane research. A great number of publications (1357, 15.6%) can be grouped under the term of different solutes to separate (remember effective separation is the key factor of membrane success).

Since research is a non-stand-alone process almost 60% of these works were performed in collaboration with researchers working at institutions in other countries, of which 6% were with the United States of America, 5% with Germany, France, or Italy, 4%, the United Kingdom and 3% Portugal. Figure 6 depicts the distribution of the most productive collaborations obtained from the present analysis search, above 2.4% of the total of the documents analyzed in this work. Collaborative works with Belgian institutions amount to 1.6% of the works delivered by the analyzed document samples analyzed in this work, same as the percentage of collaborations noted with Mexico or Argentina, which was to be expected in view of the historical and cultural connections between South American countries and Spain. There is in fact no general rule for international collaboration, but the common interest in a specific target and membrane application, as revealed by the polymer-inclusion membranes in environmental applications of the University of Girona with several institutions in France and Tunisia [60] and the University of Melbourne (Australia) [61]. To conclude with Figure 6, as expected, collaboration with European countries (taken together) is the most important (accounting for up to 60% of the total number of collaborative works) due surely to proximity reasons but, more importantly, to the impulse that European funding has given to international collaboration projects, through the various Framework Programs.

To conclude with this analysis of the Spanish publications on membranes and membrane processes, Table 4 shows the 10 most frequent authors in our list. For all of them, their affiliation is included, which represents the institution where the most productive research groups up to the moment of carrying out the present analysis belong. Several authors are currently affiliated with the same academic institution, which gives an idea of the size and continuity of the historical membrane science and technology research groups. It also reflects the fact that was mentioned earlier in this text, that Spanish research is mostly carried out in public universities and the Consejo Superior de Investigaciones Científicas (CSIC), the public council for scientific research. The creation of multiple mixed research institutes between the CSIC and a public university has contributed to the apparent preeminence of the CSIC in the membrane research field, in terms of the number of documents provided by the Scopus database in the current search (1207, 13.8%), which are shared with the universities also participating in the mixed research institutes, as the ITQ in Table 4. This collaborative combination allows concentrating and maximizing the research effort of the Universities.

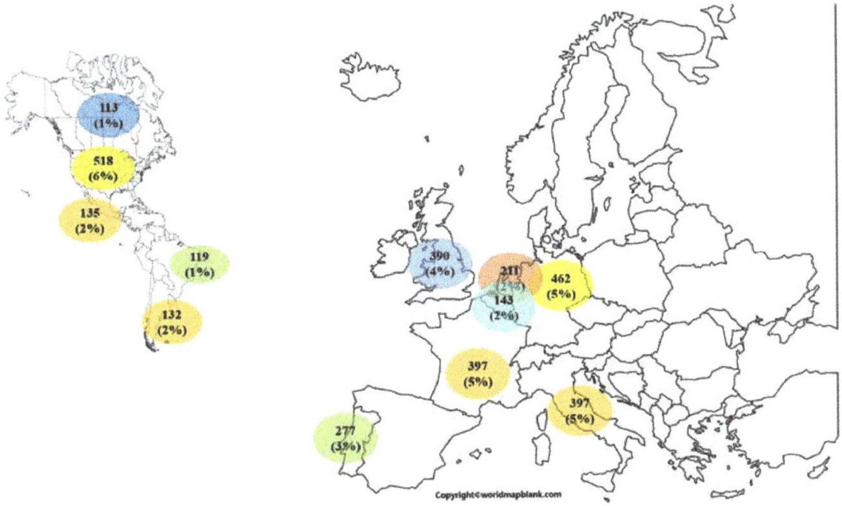

Figure 6. Top most collaborative countries with Spanish institutions worldwide.

Table 4. List of 10 most prolific authors with their affiliation and relevance in the period 1960–2021 analyzed the present non-exhaustive overview performed in this work.

Ranking	Author	Current Affiliation	Documents	H-Index	Citations
1	Coronas, J.	Universidad de Zaragoza (UZ)	141	58	10,802
2	Ortiz, I.	Universidad de Cantabria (UC)	141	55	11,312
3	Khayet, M.	Universidad Complutense de Madrid (UCM)	135	60	11,320
4	Hernández, A.	Universidad de Valladolid (UVA)	115	37	4685
5	Benavente, J.	Universidad de Málaga (UMA)	92	28	2341
6	Prádanos, P.	Universidad de Valladolid (UVA)	92	35	3754
7	Urtiaga, A.	Universidad de Cantabria (UC)	90	43	5729
8	Irabien, A.	Universidad de Cantabria (UC)	88	52	9435
9	Téllez, C.	Universidad de Zaragoza (UZ)	88	48	7480
10	Serra, J.M.	Universidad Politécnica de Valencia, UPV—Consejo Superior de Investigaciones Científicas, CSIC (Instituto de Tecnología Química, ITQ)	77	39	5479

Furthermore, the number of documents authored by them as reported in the present sample studied, as well as two parameters conventionally used to calibrate the scientific relevance of a research author. These parameters are the H-index and the total number of citations that this particular author has generated throughout his scientific production. The values of both parameters included in Table 4 are those provided by the Scopus search and are referred to the total number of publications of these authors in the database in the period considered in this work (1969–2021). Therefore, it should be kept in mind that, for some of these productive authors, research solely on membranes may only be a part of their total research on separation processes, while others with extensive publication on liquid membranes such as professor M. Valiente, from the Autonomous University of Barcelona, do not give more than 38 documents in the non-exhaustive search performed for this work, in contrast with the 189 documents on membranes a single search provides in Scopus. The same occurs with Dr. C. Fontàs (from the University of Girona), presently the editor-in-chief of the section devoted to membrane analysis and characterization of the "Membranes" journal. This highlights the deficiency of the Scopus database for bibliometric analysis with a few selected keywords, being more focused on algorithms such as the H-index than the real research of the authors that must be evaluated manually and for 8707 documents delivered by the present study almost inapprehensible.

2.2. Outlook on Present and Future Development

In order to observe the continuity of the active authors and groups in the membrane science and technology field, the Supporting Information includes a directory of the research groups operational in membrane research and innovation. This directory has been formed thanks to the contribution of the interested actors and actresses themselves responding to a brief survey prepared on purpose for this work. The survey only asked for the institution, name, and acronym (when available) of the group, PhD-holding member components, and a brief description of their research lines on membranes and membrane processes. Some correspondents added some relevant results as the preferred papers published in the subject area, and these were completed on the writing of the manuscript including the most recent publications of each group for homogenization and update purposes.

From there, it is clear that as in other scientific fields, membrane research in Spain is still mainly conducted at national universities (19 institutions out of 24, see directory in the Supporting Information), specifically in departments of chemical engineering and applied chemistry. These departments have grown in the number of recipients over the years, mostly due to the lack of positions for trained researchers in other economic entities in Spain, and the atomization of membrane research into the specialty applications solving specific problems facing changing challenges of industry, health, and environment, as well as the advancements of materials characterization techniques. At least 15 groups report research lines or keywords devoted to technologies to improve environmental remediation or wastewater treatment using membranes and membrane processes. Of the groups from CSIC's research institutes that responded to the survey, it should be considered the absence of ITQ (included in Table 4 above), and the usual collaboration with researchers from national universities, usually close in geographical proximity such as the University of Valladolid, along their historical development, thus strengthening the research efforts of both, as mentioned also in the previous section.

The size growth of the research groups on membrane science and development has led to the subdivision of the former departments into subgroups specialized in a certain approach of membrane research and the type of publications and journals their members divulgate their results on, belonging to a wide range of applications and membrane materials and module fabrication methods. See, for instance in the directory, the IEC and PROMETEO groups of the Polytechnic University of Valencia, distinguish themselves between electrochemical applications and membrane processes for water treatment [62–64], the subgroups of the GIQA group at Rey Juan Carlos University, one devoted to polymeric membranes for wastewater treatment, the other on inorganic membranes for gas separation [65–67], or the division of the Chemical and Biomolecular Engineering Department at the University of Cantabria into four subgroups covering almost all the membrane topics [68–76]. Likewise, the CREG group at the University of Zaragoza has been divided by the results obtained in the later years in two subgroups, one devoted to membrane reactors and inorganic membranes, and the other on the synthesis and characterization of mixed matrix membranes for molecular separations [77–79], while the consolidation of membranes in health issues such as bone regeneration, grows in groups similar to that from the University of Seville here [80–82].

The most commonly encountered are the membrane processes devoted to wastewater treatment, as Spain belongs to the Southern countries facing increasing scarcity of water supplies that will be aggravated by the actual climate emergency. In fact, desalination and water treatment industrial plants using membranes have already been implemented at a large scale for decades in Spain, with particular focus in the Mediterranean and Canary Islands. This follows the initial trend highlighted at the early period of the membrane technology field in international and Spanish research, where research was devoted to solving challenges faced in the industry and not as a technology in itself, which has not fully taken off in Spain yet.

3. Conclusions and Final Thought

The main challenge to membrane development is the persistent gap between high standard research performed at the scientific laboratories and the large-scale implementation, with no membrane company so far opened in the national territory This can be attributed to the general idiosyncrasy of Spanish economic, political, and historical development of the last century and the outcome of the economic crisis of 2010 whose effects we are still facing. Nevertheless, it is remarkable that most research groups active in membrane research—most of them present in the directory on the Supporting Information—are involved in European consortia or other internationally funded projects. The the recent announcement this year 2021 of a novel company producing metallic membranes for hydrogen purification, launched by the research technological center Tecnalia, may lead the way to a more active role of Spanish membrane research in the industrial development contribution to the national economic growth.

Supplementary Materials: The following are available online at https://www.mdpi.com/2077-0375/11//808/s1.

Author Contributions: Conceptualization, J.I.C. and C.C.-C.; methodology, C.C.-C.; investigation, J.I.C. and C.C.-C.; writing—original draft preparation, J.I.C. and C.C.-C.; writing—review and editing, A.H. All authors have read and agreed to the published version of the manuscript.

Funding: This work did not receive any external funding.

Acknowledgments: Ana Maria Urtiaga (Universidad de Cantabria) is gratefully acknowledged for her insightful comments upon article revision.

Conflicts of Interest: The authors declare no conflict of interest.

References

1. Gelman, C. Microporous membrane technology. I. Historical development and applications. *Anal. Chem.* **1965**, *37*, 29–37.
2. Nollet, A. Leçons de Physique Experimentale. Gueriny, H.L., Delatour, L.F., Eds.; Guérin: París, France, 1748.
3. Fick, A. Uber difussion. *Ann. Phys. Chem.* **1855**, *94*, 59–86.
4. Traube, M. Physiologie und wisserchlaftliche medizin. In *Archiv für Anatomie*; Reischert, C.B., Bois-Reymond, D., Eds.; Verlag: Leipzig, Germany, 1867.
5. Pfeffer, W. *Osmotic Untersuchungen*; Wilhelm Engelmann: Leipzig, Germany, 1877.
6. Hoff, J.H.V. Die Rolle des osmotischen Druckes in der Analogie zwischen Lösungen und Gasen. *Z. Für Phys. Chem.* **1887**, *1U*, 481–508. [CrossRef]
7. Van´t Hoff, J.H. The function of osmotic pressure in the analogy between solutions and gases. *Lond. Edinb. Dublin Phil. Mag.* **1888**, *26*, 81–105. [CrossRef]
8. Zsigmondy, R. Filter for Ultramicroscopic Particles. U.S. Patent 1,421,341, 27 June 1922.
9. Peinador, R.; Kaabouch, M.; Ben Aim, R.; Calvo, J. Non-Destructive Characterization of Industrial Membrane Cartridges by Using Liquid–Liquid Displacement Porosimetry (LLDP). *Membranes* **2020**, *10*, 369. [CrossRef]
10. Mason, E. From pig bladders and cracked jars to polysulfones: An historical perspective on membrane transport. *J. Membr. Sci.* **1991**, *60*, 125–145.
11. Lonsdale, H. The growth of membrane technology. *J. Membr. Sci.* **1982**, *10*, 81–181. [CrossRef]
12. Loeb, S.; Sourirajan, S. Sea Water Demineralization by Means of a Semipermeable Membrane. *UCLA Dept. Eng.* **1960**.
13. Baker, R.W. *Membrane Technology and Applications*; John Wiley & Sons, Ltd.: London, UK, 2004; ISBN 0470854456.
14. Lente, H.v.; Rip, A. The Rise of Membrane Technology: From Rhetorics to Social Reality. *Soc. Stud. Sci.* **1998**, *28*, 221–254.
15. Mulder, M. *Basic Principles of Membrane Technology*, 2nd ed.; Kluwer Academic Publishers: Dordrecht, The Netherlands, 1996.
16. Fernandez-Polanco, F. Aplicaciones industriales de la ósmosis inversa. *Ing. Química* **1973**, *56*, 85–90.
17. Galtes-Ros, J.M. Osmosis inversa. *Ing. Química* **1981**, *147*, 151–155.

18. Reimers, G. Ultrafiltración y ósmosis inversa aplicada a la industria quesera: Un caso de aplicación práctica. *Rev. Mens. De Las Ind. Lácteas Españolas* **1982**, *44*, 73–76.
19. Alvarez, R.; Coca, J.; Rodríguez, M.B. La tecnología de membranas en la industria papelera. *Ing. Química* **1990**, *253*, 137–143.
20. Urtiaga, A.; Ortiz, M.I.; Irabien, A. Phenol Recovery with Supported Liquid Membranes: Experimental Study. *Chem. Prot. Environ.* **1991**, *119*, 647–651. [CrossRef]
21. Fernández-Pineda, C.; Mengual, J. Permeation of water at low pressure in cellulose acetate membranes. I. Experimental results and their variation with temperature. *J. Colloid Interface Sci.* **1977**, *61*, 95–101.
22. Fernández-Pineda, C.; Mengual, J. Permeation of water at low pressure in cellulose acetate membranes. II. Nature of flow. *J. Colloid Interface Sci.* **1977**, *61*, 102–108.
23. Mengual, J.; Aguilar, J.; Fernandez-Pineda, C. Thermoosmosis of water through cellulose acetate membranes. *J. Membr. Sci.* **1978**, *4*, 209–219.
24. Mengual, J.; Fernández-Pineda, C. Frictional coefficients and activation energies for permeation through cellulose acetate membranes. *Eur. Polym. J.* **1979**, *15*, 197–202.
25. Guzman, J. Las Palmas phase II desalination plant in the Canary Islands (Spain). *Desalination* **1978**, *27*, 175–187. [CrossRef]
26. Muro, S.; Sterner, R.; Ugarte, J. Investigation on sea water desalination plants in Spain. *Desalination* **1979**, *28*, 109–116.
27. Riande, E.; Mateos, A.M.; Guzmán, G.M. Ion-exchange membranes prepared by grafting glycidyl acrylate on paper. *Eur. Polym. J.* **1970**, *6*, 437–442. [CrossRef]
28. Riande, E.; Mateos, A.M.; Guzmán, G.M. Ion-exchange membranes prepared by grafting glycidyl acrylate on paper-II. Electropositive membranes. *Eur. Polym. J.* **1970**, *6*, 1247–1252. [CrossRef]
29. Guzman, J. Rehabilitation of an uneconomic MSF plant in the Canary Islands, Spain. *Desalination* **1984**, *52*, 75–85.
30. Ibáñez, J.A.; Tejerina, A.F.; Garrido, J.; Pellicer, J. Diffusion Salt Flow Through Membranes and Permeability Determination from Cell Potential Measurements. *J. Non-Equilib. Thermodyn.* **1980**, *5*, 213–220.
31. Ibáñtez-Mengual, J.A.; Tejerina-García, A.F. Quantum model for transport through membranes. *Int. J. Quantum Chem.* **1980**, *17*, 1191–1200.
32. Fernández-Pineda, C.; Serrano, F. A study of membrane hydraulic permeability obtained by different procedures. *J. Membr. Sci.* **1984**, *19*, 325–332.
33. Fernández-Pineda, C.; Serrano, F. Hydraulic and osmotic permeabilities of water and water/sucrose solutions through cellophane membranes. *J. Membr. Sci.* **1984**, *19*, 309–323. [CrossRef]
34. Benavente, J. A Study of Membrane Potentials Across a Cellophane Membrane for Different Electrolytes. *J. Non-Equilib. Thermodyn.* **1984**, *9*, 245–254.
35. Benavente, J.; Fernandez-Pineda, C. Electrokinetic phenomena in porous membranes: Determination of phenomenological coefficients and transport numbers. *J. Membr. Sci.* **1985**, *23*, 121–136.
36. Martínez, L.; Hernández, A.; Tejerina, F. Diffusion coefficients of Polycarbonate microporous membranes. *Il Nuovo Cim.* **1987**, *9*, 174. [CrossRef]
37. Martinez, L.; Gigosos, M.; Hernandez, A.; Tejerina, F. Study of some electrokinetic phenomena in charged microcapillary porous membranes. *J. Membr. Sci.* **1987**, *35*, 1–20. [CrossRef]
38. Garrido, J.; Mafé, S.; Aguilella, V. Transport processes through membranes with gaseous electrodes. *Electrochim. Acta* **1989**, *34*, 1385–1386.
39. Diaz, C.; Vidal, J.C.; Galban, J.; Urarte, M.; Lanaja, J. A double-membrane ion-selective electrode for the potentiometric determination of potassium. *Microchem. J.* **1989**, *39*, 289–297. [CrossRef]
40. Alegret, S.; Florido, A.; Lima, J.; Machado, A. Flow-through tubular iodide and bromide selective electrodes based on epoxy resin heterogeneous membranes. *Talanta* **1989**, *36*, 825–829. [CrossRef]
41. Torres, M.; Vera, J. Reverse osmosis seawater desalting plants built by the Ministerio de Obras Publicas Y Urbanismo (Spain). *Desalination* **1987**, *64*, 443–458. [CrossRef]
42. Lora, J.; Soriano, E. Composite membranes of aromatic-polyamide for desalination: Membrane preparation and characterization. *Desalination* **1987**, *64*, 375–386. [CrossRef]
43. Uribe, I.O.; Wongswan, S.; de Ortiz, E.S.P. A systematic method for the study of the rate-controlling mechanisms in liquid membrane permeation processes. Extraction of zinc by bis(2-ethylhexyl)phosphoric acid. *Ind. Eng. Chem. Res.* **1988**, *27*, 1696–1701. [CrossRef]
44. Balbuena, M.B.; García, P.G.; Fernández, A.G. Regeneration of Spanish Style Green Table Olive Brines by Ultrafiltration. *J. Food Sci.* **1988**, *53*, 1733–1736. [CrossRef]
45. Alonso, J.; Garay, E.; Hernandez, E. Membrane filter procedure for enumeration of Pseudomonas aeruginosa in water. *Water Res.* **1989**, *23*, 1499–1502. [CrossRef]
46. Hidalgo-Álvarez, R.; Caballero, F.G.; Bruque, J.; Pardo, G. Irreversible thermodynamics of transport processes through porous media composed of particles of different size. *J. Colloid Interface Sci.* **1981**, *82*, 45–52. [CrossRef]
47. Caballero, F.G.; Nieves, F.D.L. On the relation between the electrokinetic behavior of heterogeneous ion-exchange membranes and their structural characteristics. *J. Membr. Sci.* **1983**, *16*, 225–235.
48. Horno, J.; González-Fernández, C.; Hayas, A.; Caballero, F.G. Application of network thermodynamics to the computer modelling of nonstationary diffusion through heterogeneous membranes. *J. Membr. Sci.* **1989**, *42*, 1–12.

49. Saavedra, E. Graphic evolution of the 24.000 hours (3 years) operating data of a RO brackish water desalination plant, in Las Palmas, Canary Islands, Spain. *Desalination* **1989**, *76*, 15–26.
50. Lora, J.; Arnal, J.M. The effect of agitation on reverse osmosis desalination. *Desalination* **1990**, *79*, 720–724. [CrossRef]
51. Torres, M.R.; Soriano, E.; Abajo, J.d.; Campa, J.G.d. Comparative study of the behaviour of experimental polyamide UF membranes. The effect of polyvinylpyrrolidone used as an additive. *J. Membr. Sci.* **1993**, *81*, 31–42.
52. Prádanos, P.; Abajo, J.d.; Campa, J.G.d.l.; Hernández, A. A comparative analysis of flux limit models for ultrafiltration membranes. *J. Membr. Sci.* **1995**, *108*, 129–142.
53. Suárez, E.; San Martín, F.; Alvarez, R.; Coca, J. Reverse osmosis of whey. Determination of mass transfer coefficients. *J. Membr. Sci.* **1992**, *68*, 301–305.
54. Mehaia, M.A.; Alvarez, J.; Cheryan, M. Hydrolysis of whey permeate lactose in a continuous stirred tank membrane reactor. *Int. Dairy J.* **1993**, *3*, 179–192.
55. Pérez, A.; Andrés, L.J.; Álvarez, R.; Coca, J.; Hill, C.G., Jr. Electrodialysis of whey permeates and retentates obtained by Ultrafiltration. *J. Food Process. Eng.* **1994**, *17*, 177–190.
56. Coronas, J.; Menendez, M.; Santamaria, J. Development of ceramic membrane reactors with a non-uniform permeation pattern. Application to methane oxidative coupling. *Chem. Eng. Sci.* **1994**, *49*, 4749–4757.
57. Piera, E.; Giroir-Fendler, A.; Alain Dalmon, J.; Moueddeb, H.; Coronas, J.; Menéndez, M.; Santamaría, J. Separation of alcohols and alcohols/O2 mixtures using zeolite MFI membranes. *J. Membr. Sci.* **1998**, *142*, 97–109.
58. Bricall, J.M. Informe Universidad 2000, Barcelona, CRUE. 2000. Available online: http://www.oei.es/oeivirt/bricall.htm (accessed on 1 August 2021).
59. Scopus Scopus (Article Title, Abstract, Keywords). Available online: https://www.scopus.com/search/form.uri?display=basic (accessed on 16 October 2020).
60. Fontàs, C.; Tayeb, R.; Dhahbi, M.; Gaudichet, E.; Thominette, F.; Roy, P.; Steenkeste, K.; Fontaine-Aupart, M.-P.; Tingry, S.; Tronel-Peyroz, E. Polymer inclusion membranes: The concept of fixed sites membrane revised. *J. Membr. Sci.* **2007**, *290*, 62–72. [CrossRef]
61. Garcia-Rodríguez, A.; Matamoros, V.; Kolev, S.; Fontàs, C. Development of a polymer inclusion membrane (PIM) for the preconcentration of antibiotics in environmental water samples. *J. Membr. Sci.* **2015**, *492*, 32–39.
62. Pavón, S.; Fortuny, A.; Coll, M.; Bertau, M.; Sastre, A. Permeability dependencies on the carrier concentration and membrane viscosity for Y(III) and Eu(III) transport by using liquid membranes. *Sep. Purif. Technol.* **2020**, *239*, 116573.
63. Rotta, E.H.; Marder, L.; Perez-Herranz, V.; Bernardes, A.M. Characterization of an anion-exchange membrane subjected to phosphate and sulfate separation by electrodialysis at overlimiting current density condition. *J. Membr. Sci.* **2021**, *635*, 119510. [CrossRef]
64. Cifuentes-Cabezas, M.; Carbonell-Alcaina, C.; Vincent-Vela, M.C.; Mendoza-Roca, J.A.; Alvarez-Blanco, S. Comparison of different ultrafiltration membranes as first step for the recovery of phenolic compounds from olive-oil washing wastewater. *Process. Saf. Environ. Prot.* **2021**, *149*, 724–734.
65. Alguacil, F.J.; Lopez, F.A. Dispersion-free extraction of In(III) from HCl solutions using a supported liquid membrane containing the HA324H+Cl− ionic liquid as the carrier. *Sci. Rep.* **2020**, *10*, 13868. [CrossRef]
66. Qiu, Y.; Lv, Y.; Tang, C.; Liao, J.; Ruan, H.; Sotto, A. Sustainable recovery of high-saline papermaking wastewater: Optimized separation for salts and organics via membrane-hybrid process. *Shen. J. Desalination* **2021**, *507*, 114938.
67. Diez, B.; Sotto, A.; Martin, A.; Arsuaga, J.; Rosal, R. Poly(vinyl chloride)-hyperbranched polyamidoamine ultrafiltration membranes with antifouling and antibiofouling properties. *React. Funct. Polym.* **2020**, *154*, 104669. [CrossRef]
68. Basauri, A.; González-Fernández, C.; Fallana, M.; Bringas, E.; Fernández-López, R.; Giner, L.; Moncalián, G.; Cruz, F.d.l.; Ortiz, I. Biochemical interactions between LPS and LPS-binding molecules. *Crit. Rev. Biotechnol.* **2020**, *40*, 292–305.
69. Yañez, M.; Ortiz, A.; Gorri, D.; Ortiz, I. Comparative performance of commercial polymeric membranes in the recovery of industrial hydrogen waste gas streams. *Int. J. Hydrog. Energy* **2021**, *46*, 17507–17521.
70. Corredor, J.; Perez-Peña, E.; Rivero, M.J. Ortiz performance of rgo-tio2 photocatalytiuc membranes for H2 separation. *Membranes* **2020**, *10*, 218. [CrossRef]
71. Pardo, F.; Gutierrez-Hernandez, S.V.; Zarca, G.; Urtiaga, A. Toward the Recycling of Low-GWP Hydrofluorocarbon/Hydrofluoroolefin Refrigerant Mixtures Using Composite Ionic Liquid-Polymer Membranes. *ACS Sustain. Chem. Eng.* **2021**, *9*, 7012–7021.
72. Arguillarena, A.; Margallo, M.; ARrruti-Fernandez, A.; Pinedo, J.; Gomez, P.; Urtiaga, A. Scale-up of membrane-based zinc recovery from spent pickling acids of hot-dip galnizing. *Membranes* **2020**, *10*, 444. [PubMed]
73. Abejón, R.; Fernández-Ríos, A.; Domínguez-Ramos, A.; Laso, J.; Ruiz-Salmón, I.; Yáñez, M.; Ortiz, A.; Gorri, D.; Donzel, N.; Jones, D.; et al. Hydrogen Recovery from Waste Gas Streams to Feed (High-Temperature PEM) Fuel Cells: Environmental Performance under a Life-Cycle Thinking Approach. *Appl. Sci.* **2020**, *10*, 7461. [CrossRef]
74. Vadillo, J.M.; Hospital-Benito, D.; Moya, C.; Gomez-Coma, L.; Palomar, J.; Garea, A.; Irabien, A. Modelling and simulation of hollow fiber membrane vacuum regeneration for CO2 desorption processes using ionic liquids. *Sep. Purif. Technol.* **2021**, *277*, 119465. [CrossRef]
75. García-Cruz, L.; Casado-Coterillo, C.; Iniesta, J.; Montiel, V.; Irabien, Á. hitosan:poly (vinyl) alcohol composite alkaline membrane incorporating organic ionomers and layered silicate materials into a PEM electrochemical reactor. *J. Membr. Sci.* **2016**, *498*, 395–407. [CrossRef]

76. Fernandez-Gonzalez, C.; Zhang, B.; Dominguez-Ramos, A.; Ibanez, R.; Irabien, A.; Chen, Y. Enhancing fouling resistance of polyethylene anion exchange membranes using carbon nanotubes and iron oxide nanoparticles. *Desalination* **2017**, *411*, 19–27 [CrossRef]
77. Julian, I.; Herguido, J.; Menendez, M. Gas permeation effect on the Two-Section Two-Zone Fluidized Bed Membrane Reactor (TS-TZFBMR) fluid dynamics: A CFD simulation study. *Chem. Eng. J.* **2016**, *305*, 201–211. [CrossRef]
78. Esteras-Saz, J.; de la Iglesia, O.; Pena, C.; Escudero, A.; Tellez, C.; Coronas, J. Theoretical and practical approach to the dealcoholization of water-ethanol mixtures and red wine by osmotic distillation. *Sep. Purif. Technol.* **2021**, *270*, 118793.
79. Martinez-Izquierdo, L.; Malankowska, M.; Tellez, C.; Coronas, J. Phase inversion method for the preparation of Pebax®3533 thin film membranes for CO_2/N_2 separation. *J. Environ. Chem. Eng.* **2021**, *9*, 105624.
80. Tanis-Kanbur, M.B.; Peinador, R.I.; Calvo, J.I.; Hernandez, A.; Chew, J.W. Porosimetric membrane characterization techniques: A review. *J. Membr. Sci.* **2021**, *619*, 118750.
81. Hernandez-Suarez, A.; Rizo-Gorrita, M.; Suarez-Vega, D.; Velazco, G.; Gelfenstein, I.R.; Vazquez-Pachon, C.; Serrera-Figallo, M.-A.; Torres-Lagares, D. Effectiveness of silicon platelet-rich fibrin and autologous bone on bone regeneration in rabbit calvarian defects: A radiological and histological study. *Appl. Sci.* **2021**, *11*, 4074.
82. Toledano, M.; Gutierrez-Perez, J.L.; Gutierrez-Corrales, A.; Serrera-Figallo, M.A.; Toledano-Osorio, M.; Rosales-Leal, J.I.; Aguilar, M.; Osorio, R.; Torres-Lagares, D. Novel non-resorbable polymeric-nanostructured scaffolds for guided bone regeneration. *Clin. Oral Investig.* **2020**, *24*, 2037–2049. [PubMed]

MDPI AG
Grosspeteranlage 5
4052 Basel
Switzerland
Tel.: +41 61 683 77 34

Membranes Editorial Office
E-mail: membranes@mdpi.com
www.mdpi.com/journal/membranes

www.ingramcontent.com/pod-product-compliance
Lightning Source LLC
LaVergne TN
LVHW070459061126
851858LV00011B/260